Cardiovascular Molecular Imaging

Cardiovascular Molecular Imaging

Edited by

Robert J. Gropler
Washington University School of Medicine
St. Louis, Missouri, USA

David K. Glover
University of Virginia School of Medicine
Charlottesville, Virginia, USA

Albert J. Sinusas
Yale University School of Medicine
New Haven, Connecticut, USA

Heinrich Taegtmeyer
The University of Texas Houston Medical School
Houston, Texas, USA

CRC Press
Taylor & Francis Group
Boca Raton London New York

CRC Press is an imprint of the
Taylor & Francis Group, an **informa** business

CRC Press
Taylor & Francis Group
6000 Broken Sound Parkway NW, Suite 300
Boca Raton, FL 33487-2742

© 2007 by Taylor & Francis Group, LLC
CRC Press is an imprint of Taylor & Francis Group, an Informa business

First issued in paperback 2019

No claim to original U.S. Government works

ISBN 13: 978-0-367-45300-8 (pbk)
ISBN 13: 978-0-8493-3377-4 (hbk)

Visit the Taylor & Francis Web site at
http://www.taylorandfrancis.com

and the CRC Press Web site at
http://www.crcpress.com

Foreword

The past two decades have witnessed a remarkable expansion in the utilization of a variety of techniques for imaging the cardiovascular system. These imaging techniques have provided major diagnostic and prognostic insights into patients with known or suspected cardiovascular disease. These approaches are based upon classic anatomic and physiologic principles.

We are now firmly embedded in the era of molecular medicine. Consequently, the next generation of imaging technology should harness this new knowledge to forge a new imaging discipline. The wedding of new molecular concepts with the classic principles of imaging defines the new field of cardiovascular molecular imaging. Still in its early stages, molecular imaging derives its scientific basis from the principles of molecular and cell biology rather than the classic systems physiology of the previous era. This new field is quite demanding, requiring the integration of molecular biology, molecular genetics, chemistry, and imaging. Rarely can all expertise be found in one individual, or even one investigative group. This is truly a multidisciplinary endeavor requiring diverse talent and diverse input. The stakes are high, but the goals are substantial. In the near future, cardiovascular molecular imaging should provide new definitions of vascular and myocardial disease, gene products, cell death, and cellular metabolism, to name but a few fundamental areas with highly relevant clinical implications. Molecular imaging techniques can also partner with pharmacogenomics in defining therapeutic optimization in individual patients.

As one moves into the new molecular imaging era, certain concepts are clear. Molecular imaging techniques require utilization and integration of multiple imaging modalities. Individual imaging techniques can no longer remain parochially in their respective silos. Rather, as noted by the contents of this book, multiple approaches, maximizing the strength of each, should be utilized to answer relevant questions. The targets have also changed and will continue to change. Evaluating blood flow, perfusion, and ventricular function will remain important imaging parameters, but these analyses will be augmented by an understanding of more fundamental biological processes such as angiogenesis, inflammation, vessel wall matrix biology, basic metabolism, and gene transfer. To answer the important questions posed by this new discipline, new small animal models must be utilized. There are distinct technical and physiologic challenges posed by imaging small animals such as mice, but these challenges are now being addressed. Imaging the blood vessel wall noninvasively and at multiple sites, particularly the coronary arteries, also will require major advances in imaging technology.

We are currently at a very early stage of the field. Future advances are eminently and imminently feasible. There is no movement backward. Molecular imaging, a classic example of so-called "translational research," in order to remain relevant will need to keep pace with the current molecular understanding of basic cardiovascular pathophysiology. The basic science is there to exploit. The imaging technology is also improving. Proof of principle will be followed by more detailed studies in larger animal models and then in man. Finally, by collaborating with the discipline of outcomes research, we should understand the clinical role this new biology-based imaging technology will play in the management of patients with cardiovascular disease. It is indeed an exciting time. This volume embodies the early seminal thinking in the field of cardiovascular molecular imaging and is a harbinger of a bright future.

Barry L. Zaret, MD
Robert W. Berliner Professor of Medicine
Section of Cardiovascular Medicine
Yale University School of Medicine
New Haven, Connecticut, U.S.A.

Preface

Interest in clinical cardiovascular imaging has skyrocketed over the past quarter century. Dozens of new imaging modalities have arisen from advances in the physical sciences, electrical engineering, information technology, chemical synthesis, and in pharmacology. Mostly driven by newly developed technology, these advances have been of great benefit to clinical cardiology. This is especially true in the field of structural imaging of the heart and great vessels. This book will educate the reader about a new frontier: the molecular basis of functional imaging of the cardiovascular system.

The reader will appreciate the enormous progress made in functional imaging of a complex biological system. Indeed, we are now able to visualize such dynamic processes as changes in gene expression, changes in immune response, changes in intermediary metabolism of energy-providing substrates, and changes in programmed cell death and programmed cell survival—to name just a few examples.

Will the knowledge collected and presented in this book make the reader a better, more effective clinical decision maker? The editors believe the answer is a clear "yes." For example, the use of positron emitting tracers combined with computed tomography has already significantly improved the diagnosis and management of patients with ischemic heart disease. The use of ultrasound contrast agents for left ventricular opacification and myocardial perfusion has revolutionized echocardiography. The use of magnetic resonance imaging defines not only scar tissue, but also fibrosis. Add to this the impressive array of techniques, including three-dimensional echocardiography, multi-slice computed tomography, and multi-modality imaging. Also consider the new ways to detect stem cells in dysfunctional or infarcted myocardium and inflammatory processes in the arterial wall. The opportunities for the precise diagnosis and management of cardiovascular diseases seem unlimited.

With all this enthusiasm, a word of caution is in order: in spite of enormous progress, there is still a paucity of research on the reproducibility of imaging, quality assessment, and the incremental benefits of imaging in medical decision making. A reasonable starting point is therefore a critical review of the molecular basis for contemporary cardiovascular imaging. Understanding the molecular principles of specific imaging techniques also provides an appreciation of the strengths and limitations of each new technique. *Cardiovascular Molecular Imaging* meets the needs of clinicians and translational scientists working at the very forefront of an exciting, rapidly growing field of biomedical research. Right now there is still a striking disconnect between a high demand and a low level of evidence for cardiovascular imaging. The book is a step in the right direction to eliminate this disconnect.

The editors are especially grateful to Vanessa Sanchez and Alyssa Fried as well as the entire publishing staff at Informa Healthcare for their forbearance.

Robert J. Gropler
David K. Glover
Albert J. Sinusas
Heinrich Taegtmeyer

Contents

Contributors

Juan Gilberto S. Aguinaldo   Department of Radiology, Mount Sinai School of Medicine, New York, New York, U.S.A.

Smbat Amirbekian   Department of Radiology, Mount Sinai School of Medicine, New York, New York, and Emory University School of Medicine, Atlanta, Georgia, U.S.A.

Vardan Amirbekian   Department of Radiology, Mount Sinai School of Medicine, New York, New York, Johns Hopkins University School of Medicine, Baltimore, Maryland, and The Sarnoff Endowment for Cardiovascular Science, Great Falls, Virginia, U.S.A.

Carolyn J. Anderson   Mallinckrodt Institute of Radiology, Washington University School of Medicine, St. Louis, Missouri, U.S.A.

Harrison H. Barrett   Department of Radiology and Optical Sciences Center, Tucson, Arizona, U.S.A.

Carolyn Z. Behm   University of Virginia, Charlottesville, Virginia, U.S.A.

Jeffrey R. Bender   Divisions of Cardiovascular Medicine and Immunobiology, Raymond and Beverly Sackler Foundation Cardiovascular Laboratory, Yale University School of Medicine, New Haven, Connecticut, U.S.A.

Frank M. Bengel   Division of Nuclear Medicine, Russell H. Morgan Department of Radiology and Radiological Science, Johns Hopkins University School of Medicine, Baltimore, Maryland, U.S.A.

Rene M. Botnar   Cardiovascular Division, Department of Medicine, Harvard Medical School, Boston, Massachusetts, U.S.A., and Department of Nuclear Medicine, Technical University Munich, Munich, Germany

Brett E. Bouma   Wellman Center for Photomedicine, Harvard Medical School Massachusetts General Hospital, Boston, Massachusetts, U.S.A.

James H. Caldwell   Divisions of Cardiology and Nuclear Medicine, University of Washington, Seattle, Washington, U.S.A.

Ignasi Carrió   Nuclear Medicine Department, Autonomous University of Barcelona, Hospital Sant Pau, Barcelona, Spain

Shelton D. Caruthers   Cardiovascular Division, Department of Medicine, Washington University School of Medicine, St. Louis, Missouri, U.S.A.

Michael Courtois   Center for Cardiovascular Research, Washington University School of Medicine, St. Louis, Missouri, U.S.A.

Kathy C. Crowder   Cardiovascular Division, Department of Medicine, Washington University School of Medicine, St. Louis, Missouri, U.S.A.

Ebo D. de Muinck   Department of Medicine and Department of Physiology, Angiogenesis Research Center and Section of Cardiology Dartmouth-Hitchcock Medical Center, Dartmouth Medical School, Lebanon, New Hampshire, U.S.A.

Lawrence W. Dobrucki   Yale University School of Medicine, New Haven, Connecticut, U.S.A.

William C. Eckelman   Molecular Tracer, LLC, Bethesda, Maryland, U.S.A.

Igor R. Efimov   Department of Biomedical Engineering, Washington University School of Medicine, St. Louis, Missouri, U.S.A.

Arye Elfenbein   Departments of Medicine and Pharmacology and Toxicology, Angiogenesis Research Center and Section of Cardiology, Dartmouth-Hitchcock Medical Center, Dartmouth Medical School, Lebanon, New Hampshire, U.S.A.

Zahi A. Fayad   Department of Radiology, Imaging Science Laboratories, and Department of Cardiology, The Zena and Michael A. Wiener Cardiovascular Institute, Mount Sinai School of Medicine, New York, New York, U.S.A.

John R. Forder   Advanced Magnetic Resonance Imaging and Spectroscopy Facility, The McKnight Brain Institute, and Department of Radiology, The University of Florida, Gainesville, Florida, U.S.A.

Lars R. Furenlid   Department of Radiology and Optical Sciences Center, Tucson, Arizona, U.S.A.

Valentin Fuster   Department of Cardiology, The Zena and Michael A. Wiener Cardiovascular Institute, Mount Sinai School of Medicine, New York, New York, U.S.A.

Sanjiv S. Gambhir   Molecular Imaging Program at Stanford (MIPS), Division of Nuclear Medicine, Department of Radiology, Bioengineering, and Bio-X Program, Stanford University School of Medicine, Stanford, California, U.S.A.

David K. Glover   Cardiovascular Division, Department of Medicine, University of Virginia School of Medicine, Charlottesville, Virginia, U.S.A.

David S. Goldstein   Clinical Neurocardiology Section, National Institute of Neurological Disorders and Stroke, National Institutes of Health, Bethesda, Maryland, U.S.A.

Ik-Kyung Jang   Cardiology Division, Harvard Medical School, Massachusetts General Hospital, Boston, Massachusetts, U.S.A.

Lynne L. Johnson   Division of Cardiology, Department of Medicine, Columbia University, New York, New York, U.S.A.

Juhani Knuuti   Turku PET Centre, Turku University Hospital, Turku, Finland

Attila Kovacs   Center for Cardiovascular Research, Washington University School of Medicine, St. Louis, Missouri, U.S.A.

Gregory M. Lanza   Cardiovascular Division, Department of Medicine, Washington University School of Medicine, St. Louis, Missouri, U.S.A.

Wayne C. Levy   Division of Cardiology, University of Washington, Seattle, Washington, U.S.A.

Jonathan R. Lindner   Cardiovascular Division, Oregon Health and Science University, Portland, Oregon, U.S.A.

Jeanne M. Link   Division of Nuclear Medicine, University of Washington, Seattle, Washington, U.S.A.

Zhonglin Liu   Department of Radiology and Optical Sciences Center, Tucson, Arizona, U.S.A.

Warren J. Manning   Cardiovascular Division, Department of Medicine, Harvard Medical School, Boston, Massachusetts, U.S.A.

Carolyn Mansfield   Center for Cardiovascular Research, Washington University School of Medicine, St. Louis, Missouri, U.S.A.

Koichi Morita   Department of Nuclear Medicine, Hokkaido University Graduate School of Medicine, Sapporo, Japan

Anne M. Neubauer   Cardiovascular Division, Department of Medicine, Washington University School of Medicine, St. Louis, Missouri, U.S.A.

Matthew O'Donnell   Department of Biomedical Engineering, University of Michigan, Ann Arbor, Michigan, U.S.A.

Bradley E. Patt   Gamma Medica Inc., Northridge, California, U.S.A.

Justin D. Pearlman   Department of Medicine and Department of Radiology, Angiogenesis Research Center and Section of Cardiology, Dartmouth-Hitchcock Medical Center, Dartmouth Medical School, Lebanon, New Hampshire, U.S.A.

Andrea Pichler   Molecular Imaging Center, Mallinckrodt Institute of Radiology, Washington University School of Medicine, St. Louis, Missouri, U.S.A.

David Piwnica-Worms   Molecular Imaging Center, Mallinckrodt Institute of Radiology, and Department of Molecular Biology and Pharmacology, Washington University School of Medicine, St. Louis, Missouri, U.S.A.

Jeanne Poole   Division of Cardiology, University of Washington, Seattle, Washington, U.S.A.

Daniel Pryma   Department of Radiology, Memorial Sloan Kettering Cancer Center, New York, New York, U.S.A.

Crystal M. Ripplinger   Department of Biomedical Engineering, Washington University School of Medicine, St. Louis, Missouri, U.S.A.

Erik L. Ritman   Department of Physiology and Biomedical Engineering, Mayo Clinic College of Medicine, Rochester, Minnesota, U.S.A.

Mehran M. Sadeghi   Divisions of Cardiovascular Medicine and Immunobiology, Raymond and Beverly Sackler Foundation Cardiovascular Laboratory, Yale University School of Medicine, New Haven, Connecticut, U.S.A.

Guy Salama   Department of Cell Biology and Physiology, University of Pittsburgh, Pittsburgh, Pennsylvania, U.S.A.

Michael Simons   Departments of Medicine and Pharmacology and Toxicology, Angiogenesis Research Center and Section of Cardiology, Dartmouth-Hitchcock Medical Center, Dartmouth Medical School, Lebanon, New Hampshire, U.S.A.

Albert J. Sinusas   Yale University School of Medicine, New Haven, Connecticut, U.S.A.

David N. Smith   Divisions of Cardiovascular Medicine and Immunobiology, Raymond and Beverly Sackler Foundation Cardiovascular Laboratory, Yale University School of Medicine, New Haven, Connecticut, U.S.A.

Elmar Spuentrup   Department of Diagnostic Radiology, Technical University of Aachen, Aachen, Germany

David Stout   UCLA Department of Molecular and Medical Pharmacology, Crump Institute for Molecular Imaging, Los Angeles, California, U.S.A.

John R. Stratton   Division of Cardiology, Veterans Administration Medical Center, Seattle, Washington, U.S.A.

H. William Strauss   Department of Radiology, Memorial Sloan Kettering Cancer Center, New York, New York, U.S.A.

Heinrich Taegtmeyer   Division of Cardiology, Department of Internal Medicine, The University of Texas Houston Medical School, Houston, Texas, U.S.A.

Yuan-Chuan Tai   Mallinckrodt Institute of Radiology, Washington University School of Medicine, St. Louis, Missouri, U.S.A.

Nagara Tamaki   Department of Nuclear Medicine, Hokkaido University Graduate School of Medicine, Sapporo, Japan

Guillermo J. Tearney   Wellman Center for Photomedicine, Department of Pathology, Harvard Medical School, Massachusetts General Hospital, Boston, Massachusetts, U.S.A.

Flordeliza S. Villanueva   Cardiovascular Institute, University of Pittsburgh, Pittsburgh, Pennsylvania, U.S.A.

Carla J. Weinheimer   Center for Cardiovascular Research, Washington University School of Medicine, St. Louis, Missouri, U.S.A.

Michael J. Welch   Division of Radiological Sciences, Mallinckrodt Institute of Radiology, Washington University School of Medicine, St. Louis, Missouri, U.S.A.

Samuel A. Wickline   Cardiovascular Division, Department of Medicine, Washington University School of Medicine, St. Louis, Missouri, U.S.A.

Andrea J. Wiethoff   EPIX Pharmaceuticals, Cambridge, Massachusetts, U.S.A.

Patrick M. Winter   Cardiovascular Division, Department of Medicine, Washington University School of Medicine, St. Louis, Missouri, U.S.A.

Joseph C. Wu   Division of Cardiology, Department of Medicine, and Division of Nuclear Medicine, Department of Radiology, Stanford University School of Medicine, Stanford, California, U.S.A.

Susan B. Yeon   Cardiovascular Division, Department of Medicine, Harvard Medical School, Boston, Massachusetts, U.S.A.

Overview of Cardiovascular Molecular Imaging

Andrea Pichler

Molecular Imaging Center, Mallinckrodt Institute of Radiology, Washington University School of Medicine, St. Louis, Missouri, U.S.A.

David Piwnica-Worms

Molecular Imaging Center, Mallinckrodt Institute of Radiology, and Department of Molecular Biology and Pharmacology, Washington University School of Medicine, St. Louis, Missouri, U.S.A.

DEFINITION

The 21st century has witnessed an explosion of molecular biology techniques, amazing advances in imaging, and the design of unique imaging probes. With these accomplishments, the era of molecular imaging has started. Molecular imaging is broadly defined as the characterization and measurement of biological processes in living animals, model systems, and humans at the cellular and molecular level (1–8). It differs from functional diagnostic imaging that examines the effects of anatomic or bulk physiologic alterations rather than the underlying molecular abnormalities that are the basis of disease. Development of noninvasive, high-resolution, in vivo molecular imaging technologies requires collaboration of basic research scientists, who discover new genes and their function, and imaging scientists, who exploit new imaging techniques based on these findings. Both groups have a common interest in developing and using state-of-the-art imaging technology and developing molecular imaging assays for studying intact biological systems. The goal is to advance the understanding of biology and medicine through noninvasive in vivo investigation of the cellular and molecular events mediating normal physiology and pathologic processes. Imaging of specific molecular targets will lead in the future to earlier detection and characterization of disease, earlier and direct molecular assessment of treatment effects, and a better understanding of the disease process.

Molecular imaging also can be useful for drug discovery and development (9). By labeling a drug with an imagable emittor (e.g., radiolabel, fluorophor) or by constructing an imaging agent that monitors the molecular target of the drug, one can evaluate the presence, efficacy, and biodistribution of the drug in vivo over time without the need for tissue samples. The development of versatile and sensitive assays that do not require destructive tissue samples will be of considerable value for monitoring molecular and cellular processes in animal models of human disease and in humans in the future.

IMAGING STRATEGIES

Development of molecular imaging strategies for a particular disease or pathway requires the presence of four key aspects: (*i*) a molecular target that is relevant to the disease or pathway, (*ii*) an affinity ligand or a reporter gene specific for that molecular target, (*iii*) the potential to synthesize an imaging agent or to use a genetically encoded reporter based on the ligand or gene, and (*iv*) insight into which imaging technology will give the best sensitivity and spatial resolution relevant to the biological signal under investigation. For development of an imaging agent, pharmacokinetics is a major challenge, including efficient in vivo specificity, in vivo stability, and high target-to-background ratio. Furthermore, efficient intracellular and organ targeting by overcoming biological delivery barriers, and biological or chemical amplification techniques aid execution of imaging studies.

Imaging technologies allow for noninvasive visualization of the body based on different forms of energy interacting with tissues. Current image outputs are moving away from an observational status, removed from the underlying mechanism, to more and more detailed analysis, reflecting the fundamental molecular event (Fig. 1). Therefore, various imaging modalities can be used to detect anatomical abnormalities in organs (mass effect or hemorrhage), to monitor changes in bulk physiological events (blood volume or perfusion), to track specific markers on the cellular level (cell markers and receptor interactions) or to give information of molecular events (genetic mutations, gene expression, signal transduction, and

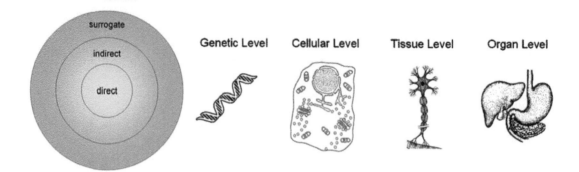

Figure 1 Different levels of molecular focus. Molecular imaging focuses in from an observational status with surrogate markers/techniques to more and more detailed molecular analysis, leading to direct imaging of the underlying fundamental biochemical event. Functional imaging of the organ (e.g., tumor perfusion) is being supplemented with direct molecular imaging at the cellular (e.g., receptor detection and drug targeting) and genetic (e.g., genetic mutations and gene expression) levels.

protein–protein interactions). Molecular imaging is driven primarily by biological questions and thus, selection of the imaging technology will be determined by various advantages of different modalities rather than by use of one technique for all investigations.

A strategic choice (as depicted in Fig. 2) is required to execute molecular imaging experiments, taking into consideration the biochemical context of the signal source, the imaging probe, and the desired end point of the study. The signal source can either arise from endogenous genes or from exogenous transgenes. One consideration relates to the number of target molecules and their impact on generating sufficient signal-to-noise ratios (6). For example, direct imaging of DNA poses a considerable challenge because of low abundance (only two molecules per cell) and sequence variabilities between individuals.

Furthermore, imaging DNA provides no information on gene expression and contributions to the physiological state of the imaged cell or tissue. Proteins, on the other hand, can be present at significantly higher levels and thus, direct imaging is achieved readily, for example, by direct molecular imaging of receptor subtypes with radiopharmaceuticals in a laboratory or clinical setting. To image exogenous transgene expression, a reporter gene must be introduced into the target cell or tissue, using a variety of methods, including transfection of plasmid DNA, transduction with viral vectors, or incorporation of reporter constructs into the DNA of genetically engineered animals. Imaging probes are either injectable agents or genetically encoded reporters.

Although one ultimate goal of molecular imaging is clinical application, a great deal of basic research

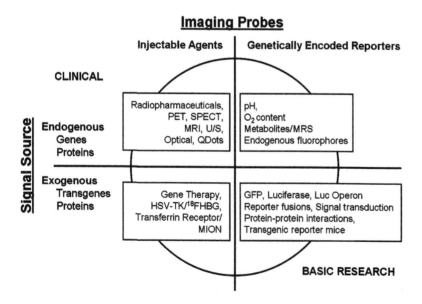

Figure 2 Molecular imaging strategies. Different imaging probes are depicted with respect to signal source and detection strategy in relation to clinical or basic research. *Abbreviations*: FHBG, ^{18}F-labeled 9-[4-fluoro-3-(hydroxymethyl)butyl] guanine; GFP, green fluorescent protein; HSV-TK, herpes simplex virus-1 thymidine kinase; MION, monocrystalline iron oxide nanoparticles; MRI, magnetic resonance imaging; MRS, magnetic resonance spectroscopy; PET, positron emission tomography; Qdots, quantum dots; SPECT, single photon emission computed tomography; U/S, ultrasound. *Source*: From Ref. 8.

is performed with animal models of disease. In practice, molecular imaging can be performed at differing levels of molecular specificity and noninvasiveness with increasing difficulties and challenges. Routinely used methodologies in the laboratory and in vitro settings are based on destructive sampling of cells or tissue samples, which yield only a static snapshot at a given experimental endpoint. New imaging technologies allow for noninvasive, repeated in vivo imaging of dynamic processes. Reporter constructs can be made to study transcriptional regulation, signal transduction, protein–protein interactions, cell differentiation, cell trafficking, and targeted drug action, and thus, genetically encoded reporters are excellent for interrogation of a broad array of molecular pathways (8). Genetically encoded reporters can produce signal intrinsically (e.g., fluorescent proteins), through enzymatic activation of an inactive substrate (luciferases), by enzymatic modification of an active (e.g., radiolabeled) substrate with selective retention in reporter cells, or by direct binding or import of an active (e.g., radiolabeled) reporter substrate or probe. A fundamental advantage of reporters is that once validated, the reporter gene can theoretically be cloned into an appropriate vector and any gene of interest can be interrogated with the same validated reporter probe. For PET, this eliminates the constraints inherent to traditional routes of synthesizing, labeling, and validating a new and different radioligand for every new receptor or protein of interest. However, except in the case of gene therapy, reporters are not likely to be used routinely in humans. On the other hand, imaging reagents comprising injectable radiopharmaceuticals, fluorescent probes and contrast agents have the potential to be used in the clinic (10).

IMAGING MODALITIES

Imaging modalities most useful in molecular imaging can be divided into three major groups: (*i*) optical imaging (fluorescence and bioluminescence imaging) (7,11–15); (*ii*) nuclear imaging (SPECT or PET) (2,4, 16–23); and (*iii*) magnetic resonance imaging (MRI) (24–27). Bioluminescence imaging is becoming more widely available and because of the lack of background activity, it is excellent for low-signal imaging. Bioluminescence imaging is relatively simple and cost effective, but is dependent on substrate pharmacokinetics. Furthermore, bioluminescence imaging is not suitable for deep tissue imaging in larger animals and the obtained signal contains limited information about the exact location of the internal source. Recent advances in the fluorescence field with red-shifted proteins or near-infrared fluorescent probes and development of highly sensitive detection devices have propagated new imaging modalities, such as fluorescence tomography, spectrally resolved whole-body imaging and intravital multiphoton imaging. However, high autofluorescence in the blue–green optical window, photo-bleaching, and limited quantification of photon output are major pitfalls for fluorescence imaging. In contrast to bioluminescence imaging, the lack of a requirement for an injectable substrate is highly advantageous. Nuclear imaging (SPECT and PET) is highly sensitive, quantitative, and tomographic. However, it demands sophisticated instrumentation and readily available in-house production of radiopharmaceuticals. MRI does not involve ionizing radiation, generates high spatial resolution and combined physiologic-anatomic information can be obtained simultaneously. However, the low sensitivity of MRI requires high concentration of contrast material or signal amplification to monitor molecular events in vivo. While targeted bubbles in principle may be used in molecular imaging, ultrasound will most likely be confined to a niche in targeted delivery and therapeutics.

In practice, the choice of imaging modality and probe is usually determined by the examined biological process. Sometimes a multimodality approach is the best way to answer biological questions, keeping in mind the advantages of research performed with cells in their intact environment in living subjects to potentially develop new ways to studies mechanisms of pathogenesis, diagnose diseases, and monitor therapies in patients.

COLLABORATIVE EFFORT

Molecular imaging is a broad based platform that unifies multiple disciplines. The tools and strategies of molecular biology and biochemistry are combined with chemistry for the generation of new imaging probes to use with sophisticated imaging technologies (Table 1). This multidisciplinary research involves a variety of fields, including target discovery and molecular biology, chemistry, cell biology, modeling of disease, imaging, physics and engineering, data processing, and clinical research (Fig. 3).

MOLECULAR CARDIOVASCULAR IMAGING

Several initiatives exist to further the development and application of novel cellular and molecular imaging probes and technologies to image the cardiovascular system in vivo (28–31). The goal is to detect and quantify at the molecular and cellular level the pathways that regulate heart and blood vessel function, and to understand abnormalities in these pathways. Current

Table 1 Integrated Strategy for Molecular Imaging

Molecular biology	Chemistry	Animal/patient imaging
Reporter genes	Organic synthesis	Optical imaging
Expression analysis	Bioconjugate chemistry	MRI
Gene delivery	Combinatorial chemistry	Nuclear imaging
Protein expression	Peptide chemistry	PET/CT
Amplification strategy	Nanoparticle chemistry	Intravital microscopy
Bioinformatics	Modeling rational design	Ultrasound
Mouse transgenic models		Pharmacokinetics

methods for imaging the cardiovascular system, such as CT, MRI, and ultrasound focus predominantly on anatomical and bulk functional measurements (e.g., flow, perfusion, and wall motion). In contradistinction to "classical" diagnostic imaging, molecular imaging probes the molecular abnormalities that are the basis of disease rather than imaging the result of these molecular alterations. The full potential for molecular and cellular imaging has not yet been realized for imaging heart and blood.

Imaging the cardiovascular system poses unique technical problems, such as cardiac motion and the circulation of blood cells through the body. These technical issues impose additional requirements such as the addition of gated image acquisition to novel imaging technologies currently being developed for static imaging. Sensitivity needs to be optimized to compensate for signal loss due to image gating or cell circulation, and improved coregistration of molecular and anatomical images is also required.

Areas of focused research include:

- Imaging of probes targeted to specific cells or molecules involved in cardiovascular and blood disease processes; for example, plaque components, integrins expressed in angiogenesis, and molecular receptors expressed in disease or stem cell therapy.
- Use of probes activated by specific enzymes involved in cardiovascular disease processes to allow regional assessment of enzyme activity, for example, metalloproteinase activation in the development of aneurysms.
- Development of new or improved software or hardware approaches to compensate for cardiac motion, facilitating quantitative measurement of tracers.
- Development of new or improved software or hardware approaches to increase sensitivity and facilitate tracking of circulating blood cells in vasculature and in tissues.
- Development and imaging of probes activated by specific environments relevant to cardiovascular disease, for example, redox and pH-sensitive probes.

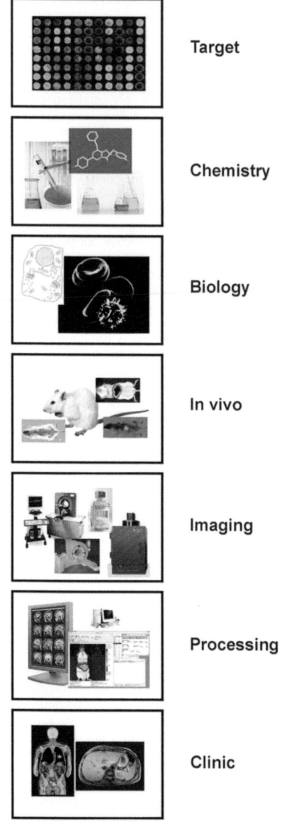

Figure 3 Multidisciplinary approach/collaborative effort. Molecular imaging combines multiple disciplines and only in a collaborative effort can the ultimate goals of early detection of disease and therapeutic response in patients be accomplished.

- Adaptation of optical techniques (e.g., fluorescence molecular tomography) to cardiovascular and blood cell imaging.
- Imaging methods aimed at specific molecular targets to visualize the anatomical distribution of specific cell types in the heart, vessels, and circulation.
- In situ generation of magnetic resonance probes by biocatalysis to monitor gene therapy and gene expression in the cardiovascular and hematopoietic systems.
- Development of improved temporal and spatial stem cell localization using novel or existing cellular tracking technologies, for example, in vivo bioluminescence, multiphoton microscopy, and MRI.

At the present time, substantial interest has developed for the application and expansion of molecular cardiovascular imaging (32). This represents a potential paradigm shift from a physiology-based (functional) imaging discipline to a cell and molecular biology-driven (molecular) imaging discipline. Current topics and areas for molecular imaging research include: novel approaches to ischemic myocardium (33), myocyte integrity, stem cell therapy and gene therapy imaging (34), imaging of cell death (35), new approaches to hibernating and stunned myocardium, angiogenesis (36,37), receptor occupancy (38), vascular integrity, imaging of unstable plaque and imaging inflammatory atherosclerosis (39). Such approaches can form the basis for the key translational research necessary for moving adeptly from bench to bedside as new drugs and therapeutic approaches, such as gene therapy, are developed. Molecular imaging may provide an ideal vehicle for assessing therapeutic efficacy of newly developed biologically-based therapies.

Development of centers of excellence, such as has been done in molecular oncology imaging is of fundamental importance. New training paradigms are necessary for the advancement of a new generation of imaging scientists and clinicians who utilize the principles of modern biology to diagnose and evaluate cardiovascular disease with imaging strategies. Such physician-scientists will need expertise in multiple imaging areas as well as nuclear cardiology and modern molecular and cell biology.

EXAMPLE OF TRANSLATING FROM FUNCTIONAL IMAGING TO MOLECULAR IMAGING: 99mTC-SESTAMIBI

99mTc-Sestamibi was originally developed as a radiopharmaceutical for clinical imaging of myocardial perfusion, an important functional assessment of cardiovascular physiology (40,41). 99mTc-Sestamibi is a lipophilic, monovalent cation and enters the cell via passive diffusion across plasma and mitochondrial membranes (42,43). Because of this property, 99mTc-Sestamibi shows high first-pass extraction in vivo and thus, favorable characteristics as a perfusion marker (41). Studying multidrug resistance (MDR) with 99mTc-Sestamibi has lead to a deeper understanding of the mechanistic basis of how this imaging agent works at the molecular level, thereby opening opportunities for novel applications in molecular imaging. MDR is a major obstacle to successful chemotherapy in cancer, mediated by overexpression of the MDR P-glycoprotein (Pgp) (44,45). Pgp is a transmembrane protein and a member of the ATP-binding cassette transporter superfamily (45,46). Pgp confers cross-resistance to many unrelated drugs, including many natural product agents, that differ widely with respect to molecular structure and target specificity (47). Pgp acts as an energy-dependent efflux transporter that efficiently enhances outward transport and/or prevents entry of these compounds, thereby resulting in decreased intracellular accumulation and decreased cytotoxicity of anticancer drugs in tumors (4,48). Furthermore, Pgp is normally present in various tissues including the intestinal epithelium, choroid plexus epithelium, kidney, liver, adrenal, placenta, and capillary endothelial cells of the blood–brain and blood–testis barriers (49–52).

During the search for imaging agents that allowed noninvasive interrogation of Pgp function, 99mTc-Sestamibi was one of the early gamma-emitting compounds shown to be a Pgp substrate (53). In the absence of Pgp expression, 99mTc-Sestamibi accumulates within cells in response to the physiologically negative mitochondrial and plasma membrane potentials (43,54). At equilibrium, the intracellular target for 99mTc-Sestamibi in living tissues is the mitochondrial inner matrix. Thus, heart, a Pgp-negative tissue, retains a high cellular content of 99mTc-Sestamibi, and this mechanism likely accounts for the initial distribution of the agent into heart as well as other mitochondrial-rich tissues such as kidney, liver, skeletal muscle, and tumors in vivo. However, in Pgp-expressing cells, net cellular accumulation levels of 99mTc-Sestamibi are reduced or the complex is rapidly excreted (53,55–58). Thus, expression of Pgp in the liver is responsible for the rapid hepatobiliary excretion of 99mTc-Sestamibi (59,60), and forms the basis of clinical MDR imaging in cancer patients (61–65). This mechanism also renders the imaging agent highly valuable when evaluating heart perfusion because liver clearance of the tracer enables visualization of the inferior myocardium in patients. Therefore, this radiopharmaceutical is currently used successfully in the clinics to assess perfusion patterns, myocardium at risk, infarct size, and treatment efficacy in acute myocardial infarction (66–72). Understanding these underlying molecular events in vivo also provide the basis for use of 99mTc-Sestamibi in assessing the role of Pgp in chronic ischemia and dedifferentiation of myocardium.

This example should encourage the quest to understand the molecular basis of all imaging agents, leading to further translation of molecular imaging into the clinics.

ACKNOWLEDGMENTS

National Institutes of Health grant P50 CA94056 and the U.S. Department of Energy grant 94ER61885 (D. Piwnica-Worms).

REFERENCES

1. Weissleder R, Mahmood U. Molecular imaging. Radiology 2001; 219:316–333.
2. Luker GD, Piwnica-Worms D. Beyond the genome: molecular imaging in vivo with PET and SPECT. Acad Radiol 2001; 8:4–14.
3. Wagenaar DJ, Weissleder R, Hengerer A. Glossary of molecular imaging terminology. Acad Radiol 2001; 8:409–420.
4. Sharma V, Luker G, Piwnica-Worms D. Molecular imaging of gene expression and protein function in vivo with PET and SPECT. J Mag Reson Imaging 2002; 16:336–351.
5. Massoud T, Gambhir S. Molecular imaging in living subjects: seeing fundamental biological processes in a new light. Gene Dev 2003; 17:545–580.
6. Blasberg RG, Tjuvajev JG. Molecular-genetic imaging: current and future perspectives. J Clin Invest 2003; 111:1620–1629.
7. Weissleder R, Ntziachristos V. Shedding light on live molecular targets. Nat Med 2003; 9:123–128.
8. Gross S, Piwnica-Worms D. Spying on cancer: molecular imaging in vivo with genetically encoded reporters. Cancer Cell 2005; 7:5–15.
9. Rudin M, Weissleder R. Molecular imaging in drug discovery and development. Nat Rev Drug Discov 2003; 2:123–131.
10. Jaffer FA, Weissleder R. Molecular imaging in the clinical arena. JAMA 2005; 293:855–862.
11. Contag C, Spilman S, Contag P, et al. Visualizing gene expression in living mammals using a bioluminescent reporter. Photochem Photobiol 1997; 66:523–531.
12. Bhaumik S, Gambhir S. Optical imaging of Renilla luciferase reporter gene expression in living mice. Proc Natl Acad Sci U S A 2001; 99:377–382.
13. Contag C, Ross B. It's not just about anatomy: in vivo bioluminescence imaging as an eyepiece into biology. J Mag Reson Imaging 2002; 16:378–387.
14. Contag CH, Bachmann MH. Advances in vivo bioluminescence imaging of gene expression. Annu Rev Biomed Eng 2002; 4:235–260.
15. Shaner NC, Campbell RE, Steinbach PA, Giepmans BN, Palmer AE, Tsien RY. Improved monomeric red, orange and yellow fluorescent proteins derived from Discosoma sp. red fluorescent protein. Nat Biotechnol 2004; 22:1567–1572.
16. Rogers BE, Zinn KR, Buchsbaum DJ. Gene transfer strategies for improving radiolabeled peptide imaging and therapy. Q J Nucl Med 2000; 44:208–223.
17. Doubrovin M, Ponomarev V, Beresten T, et al. Imaging transcriptional regulation of p53-dependent genes with positron emission tomography in vivo. Proc Natl Acad Sci U S A 2001; 98:9300–9305.
18. Jacobs A, Tjuvajev J, Dubrovin M, et al. Positron emission tomography-based imaging of transgene expression mediated by replication-conditional, oncolytic herpes simplex virus type 1 mutant vectors in vivo. Cancer Res 2001; 61:2983–2995.
19. Luker G, Sharma V, Pica C, et al. Noninvasive imaging of protein-protein interactions in living animals. Proc Natl Acad Sci U S A 2002; 99:6961–6966.
20. Liang Q, Gotts J, Satyamurthy N, et al. Noninvasive, repetitive, quantitative measurement of gene expression from a bicistronic message by positron emission tomography, following gene transfer with adenovirus. Mol Ther 2002; 6:73–82.
21. Groot-Wassink T, Aboagye EO, Glaser M, Lemoine NR, Vassaux G. Adenovirus biodistribution and noninvasive imaging of gene expression in vivo by positron emission tomography using human sodium/iodide symporter as reporter gene. Human Gene Therapy 2002; 13:1723–1735.
22. Doubrovin M, Ponomarev V, Serganova I, et al. Development of a new reporter gene system--dsRed/xanthine phosphoribosyltransferase-xanthine for molecular imaging of processes behind the intact blood-brain barrier. Mol Imaging 2003; 2:93–112.
23. Morin K, Duan W, Xu L, et al. Cytotoxicity and cellular uptake of pyrimidine nucleosides for imaging herpes simplex type-1 thymidine kinase (HSV-1 TK) expression in mammalian cells. Nucl Med Biol 2004; 31:623–630.
24. Louie A, Huber M, Ahrens E, et al. In vivo visualization of gene expression using magnetic resonance imaging. Nat Biotechnol 2000; 18:321–325.
25. Weissleder R, Moore A, Mahmood U, et al. In vivo magnetic resonance imaging of transgene expression. Nat Med 2000; 6:351–355.
26. Moore A, Josephson L, Bhorade R, Basilion J, Weissleder R. Human transferrin receptor gene as a marker gene for MR imaging. Radiology 2001; 221:244–250.
27. Zhao M, Beauregard D, Loizou L, Davletov B, Brindle K. Non-invasive detection of apoptosis using magnetic resonance imaging and a targeted contrast agent. Nat Med 2001; 7:1241–1244.
28. Wu JC, Tseng JR, Gambhir SS. Molecular imaging of cardiovascular gene products. J Nucl Cardiol 2004; 11:491–505.
29. Sinusas AJ. Imaging of angiogenesis. J Nucl Cardiol 2004; 11:617–633.
30. Jaffer FA, Weissleder R. Seeing within: molecular imaging of the cardiovascular system. Circ Res 2004; 94:433–445.
31. Dobrucki LW, Sinusas AJ. Cardiovascular molecular imaging. Semin Nucl Med 2005; 35:73–81.

32. Dobrucki LW, Sinusas AJ. Molecular imaging. A new approach to nuclear cardiology. Q J Nucl Med Mol Imaging 2005; 49:106–115.

33. Wu JC, Chen IY, Wang Y, et al. Molecular imaging of the kinetics of vascular endothelial growth factor gene expression in ischemic myocardium. Circulation 2004; 110:685–691.

34. Kastrup J, Jorgensen E, Ruck A, et al. Direct intramyocardial plasmid vascular endothelial growth factor-A165 gene therapy in patients with stable severe angina pectoris A randomized double-blind placebo-controlled study: the Euroinject One trial. J Am Coll Cardiol 2005; 45:982–988.

35. Narula J, Kietselaer B, Hofstra L. Role of molecular imaging in defining and denying death. J Nucl Cardiol 2004; 11:349–357.

36. Cristofanilli M, Charnsangavej C, Hortobagyi GN. Angiogenesis modulation in cancer research: novel clinical approaches. Nat Rev Drug Discov 2002; 1:415–426.

37. Meoli DF, Sadeghi MM, Krassilnikova S, et al. Noninvasive imaging of myocardial angiogenesis following experimental myocardial infarction. J Clin Invest 2004; 113:1684–1691.

38. Elsinga PH, van Waarde A, Vaalburg W. Receptor imaging in the thorax with PET. Eur J Pharmacol 2004; 499:1–13.

39. Choudhury RP, Fuster V, Fayad ZA. Molecular, cellular and functional imaging of atherothrombosis. Nat Rev Drug Discov 2004; 3:913–925.

40. Abrams MA, Davison A, JonesAG, Costello CE, Pang H. Synthesis and characterization hexakis(alkyl isocyanide) and hexakis(arylisocyanide) complexes of technetium(I). Inorg Chem 1983; 22:2798–2800.

41. Wackers FJ, Berman D, Maddahi J, et al. Tc-99m-hexakis 2-methoxy isobutylisonitrile: human biodistribution, dosimetry, safety and preliminary comparison to thallium-201 for myocardial perfusion imaging. J Nucl Med 1989; 30:301–309.

42. Chiu ML, Kronauge JF, Piwnica-Worms D. Effect of mitochondrial and plasma membrane potentials on accumulation of hexakis (2-methoxyisobutyl isonitrile) technetium(I) in cultured mouse fibroblasts. J Nucl Med 1990; 31:1646–1653.

43. Piwnica-Worms D, Kronauge J, Chiu M. Uptake and retention of hexakis (2-methoxy isobutyl isonitrile) technetium(I) in cultured chick myocardial cells: mitochondrial and plasma membrane potential dependence. Circulation 1990; 82:1826–1838.

44. Gottesman M, Fojo T, Bates S. Multidrug resistance in cancer: role of ATP-dependent transporters. Nat Rev Cancer 2002; 2:48–58.

45. Ambudkar S, Dey S, Hrycyna C, Ramachandra M, Pastan I, Gottesman M. Biochemical, cellular, and pharmacological aspects of the multidrug transporter. Annu Rev Pharmacol Toxicol 1999; 39:361–398.

46. Riordan JR, Ling V. Genetic and biochemical characterization of multidrug resistance. Pharmacol Ther 1985; 28:51–75.

47. Ford JM, Hait WN. Pharmacology of drugs that alter multidrug resistance in cancer. Pharmacol Rev 1990; 42:155–199.

48. Sauna ZE, Smith MM, Muller M, Kerr KM, Ambudkar SV. The mechanism of action of multidrug-resistance-linked P-glycoprotein. J Bioenerg Biomembr 2001; 33:481–491.

49. Thiebaut F, Tsuruo T, Hamada H, Gottesman MM, Pastan I, Willingham MC. Cellular localization of the multidrug-resistance gene product P-glycoprotein in normal human tissues. Proc Natl Acad Sci U S A 1987; 84:7735–7738.

50. Hitchins RN, Harman DH, Davey RA, Bell DR. Identification of a multidrug resistance associated antigen (P-glycoprotein) in normal human tissues. Eur J Cancer Clin Oncol 1988; 24:449–454.

51. Cordon-Cardo C, OBrien J, Casals D, et al. Multidrug-resistance gene (P-glycoprotein) is expressed by endothelial cells at blood-brain barrier sites. Proc Natl Acad Sci U S A 1989; 86:695–698.

52. Rao V, Dahlheimer J, Bardgett M, et al. Choroid plexus epithelial expression of MDR1 P-glycoprotein and multidrug resistance-associated protein contribute to the blood-cerebrospinal fluid drug-permeability barrier. Proc Natl Acad Sci U S A 1999; 96:3900–3905.

53. Piwnica-Worms D, Chiu M, Budding M, Kronauge J, Kramer R, Croop J. Functional imaging of multidrug-resistant P-glycoprotein with an organotechnetium complex. Cancer Res 1993; 53:977–984.

54. Backus M, Piwnica-Worms D, Hockett D, et al. Microprobe analysis of Tc-MIBI in heart cells: calculation of mitochondrial potential. Am J Physiol (Cell) 1993; 265:C178–C187.

55. Piwnica-Worms D, Rao V, Kronauge J, Croop J. Characterization of multidrug-resistance P-glycoprotein transport function with an organotechnetium cation. Biochemistry 1995; 34:12210–12220.

56. Ballinger J, Hua H, Berry B, Firby P, Boxen I. ^{99m}Tc-Sestamibi as an agent for imaging P-glycoprotein-mediated multi-drug resistance: in vitro and in vivo studies in a rat breast tumour cell line and its doxorubicin-resistant variant. Nucl Med Comm 1995; 16:253–257.

57. Ballinger JR, Sheldon KM, Boxen I, Erlichman C, Ling V. Differences between accumulation of Tc-99m-MIBI and Tl-201-thallous chloride in tumor cells: role of P-glycoprotein. Q J Nucl Med 1995; 39:122–128.

58. Cordobes M, Starzec A, Delmon-Moingeon L, et al. Technetium-99m-Sestamibi uptake by human benign and malignant breast tumor cells: correlation with mdr gene expression. J Nucl Med 1996; 37:286–289.

59. Luker GD, Fracasso PM, Dobkin J, Piwnica-Worms D. Modulation of the multidrug resistance P-glycoprotein: detection with Tc-99m-Sestamibi in vivo. J Nucl Med 1997; 38:369–372.

60. Dyszlewski M, Blake H, Dalheimer J, Pica C, Piwnica-Worms D. Characterization of a novel Tc-99m-carbonyl complex as a functional probe of MDR1 P-glycoprotein transport activity. Molec Imaging 2002; 1:24–35.

61. Del Vecchio S, Ciarmiello A, Potena MI, et al. In vivo detection of multidrug resistance (MDR1) phenotype by technetium-99m-sestamibi scan in untreated breast cancer patients. Eur J Nucl Med 1997; 24:150–159.

62. Zhou J, Higashi K, Ueda Y, et al. Expression of multidrug resistance protein and messenger RNA correlate

with ⁹⁹ᵐTc-MIBI imaging in patients with lung cancer. J Nucl Med 2001; 42:1476–1483.

63. Bates SF, Chen C, Robey R, Kang M, Figg WD, Fojo T. Reversal of multidrug resistance: lessons from clinical oncology. Novartis Found Symp 2002; 243:83–96; discussion 96–102, 180–105.

64. Del Vecchio S, Salvatore M. Tc-99m-MIBI in the evaluation of breast cancer biology. Eur J Nucl Med Mol Imaging 2004; 31:S88–S96.

65. Bates SE, Bakke S, Kang M, et al. A phase I/II study of infusional vinblastine with the P-glycoprotein antagonist valspodar (PSC 833) in renal cell carcinoma. Clin Cancer Res 2004; 10:4724–4733.

66. Maddahi J, Kiat H, Berman DS. Myocardial perfusion imaging with technetium-99m-labeled agents. Am J Cardiol 1991; 67:27D–34D.

67. Gibbons RJ, Miller TD, Christian TF. Infarct size measured by single photon emission computed tomographic imaging with (99m)Tc-sestamibi: a measure of the efficacy of therapy in acute myocardial infarction. Circulation 2000; 101:101–108.

68. Roe MT. Treatment strategies for microvascular dysfunction following acute myocardial infarction. Curr Cardiol Rep 2000; 2:405–410.

69. Persson E, Palmer J, Pettersson J, et al. Quantification of myocardial hypoperfusion with 99m Tc-sestamibi in patients undergoing prolonged coronary artery balloon occlusion. Nucl Med Commun 2002; 23:219–228.

70. Gibbons RJ, Valeti US, Araoz PA, Jaffe AS. The quantification of infarct size. J Am Coll Cardiol 2004; 44:1533–1542.

71. Kontos MC, Wackers FJ. Acute rest myocardial perfusion imaging for chest pain. J Nucl Cardiol 2004; 11:470–481.

72. Travin MI, Bergmann SR. Assessment of myocardial viability. Semin Nucl Med 2005; 35:2–16.

2

PET/microPET Imaging

Yuan-Chuan Tai

Mallinckrodt Institute of Radiology, Washington University School of Medicine, St. Louis, Missouri, U.S.A.

INTRODUCTION

Positron emission tomography (PET) is a nuclear medicine imaging technique that provides three-dimensional (3D) tomographic images of radiotracer distribution within a living subject (1, 2). Proton-rich radionuclides such as ^{11}C, ^{13}N, ^{15}O, ^{18}F, ^{64}Cu, ^{68}Ga, and [82]Rb can be used to label biological molecules of interest. Typically, trace quantities (pico-molar) of bio-molecules are administered into subjects to be imaged. Radiotracer concentration can be measured quantitatively over a period of time. These quantitative measurements, combined with tracer kinetic modeling technique, permit different biological processes to be modeled without intervention (3–6). PET has been widely accepted for clinical diagnosis of various types of diseases (neurological, cardiac, and oncological). Equally important is PET's role in basic sciences, in which carefully engineered and labeled bio-molecules are employed to study various biological processes in human subjects and laboratory animals (7). The development of bio-molecules and their applications in cardiovascular research will be discussed in more detail in other chapters of this book. The underlying physics and detector technologies of PET and high-resolution animal PET, as well as the techniques involved in quantitative PET imaging, will be introduced in this chapter.

FUNDAMENTALS OF PET

Physics

The physical property to be measured by PET is the activity concentration of radiolabeled tracer in a subject. Figure 1 illustrates the source of signal and the theory of detection. Bio-molecules [e.g., ^{18}F-fluorodeoxyglucose (FDG) in Fig. 1] are labeled with a positron-emitting radionuclide and administered to a subject. The proton-rich nucleus spontaneously converts one of its protons to a neutron, resulting in the emission of a positron, which is the antimatter of an electron, and a neutrino. Electrons and positrons have the same mass but carry opposite charges. The emitted positron collides with the surrounding matter and slows down to finally combine with an electron before they annihilate. Upon annihilation, the mass of positron and electron is converted into two 511 keV photons that travel in approximately opposite directions. Coincidence detection of these annihilation photons followed by mathematical reconstruction of the images form the basis of PET.

A positron travels away from its mother nucleus with a range determined by its initial kinetic energy. Because the original location of the bio-molecule and the location of annihilation are not the same, there is an uncertainty that limits the image resolution achievable by PET. The average range of a positron depends on the emission spectrum of the radionuclide. Table 1 lists the half-life, average emission energy and the corresponding positron range of commonly used radionuclides for PET imaging. The average range of the positron is significantly larger than the resolution loss due to positron range effect (8–10).

At annihilation, the positron is not at complete rest. Therefore, conservation of energy and momentum requires the two 511 keV photons to be emitted at an angle slightly deviated from 180°. The angular distribution of the deviation has been determined to have a full-width-at-half-maximum (FWHM) of approximately 0.5° that leads to another uncertainty in identifying the source of origin (11). The degradation of image resolution as the result of acolinearity effect is proportional to the distance between the object and the detectors, and can be expressed as $FWHM_{acolinearity} = 0.0022D$ for a pair of detectors separated by D (12). When this equation is applied to a typical human PET scanner with a diameter greater than 80 cm, the acolinearity effect contributes to approximately 2 mm FWHM degradation in image resolution.

The annihilation photons, though generated within the body of a subject, have a finite probability of escaping the subject without any interaction with tissues because the 511 keV photons are highly penetrative. At this energy, photons interact with matter primarily through Compton scattering and photoelectric absorption. With Compton scattering, a photon is deflected from its original path by an electron. Part of the original photon energy is transferred to the recoil

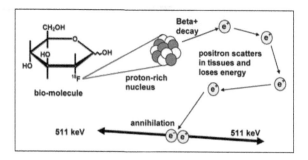

Figure 1 Positron is emitted from radiolabeled molecules. It travels away from its origin and annihilates with an electron to provide two 511 keV photons.

electron while the remaining energy is carried away by the scattered photon. With photoelectric absorption, a photon interacts with an electron in the inner shell of an atom, causing the electron's ejection. The ejected photoelectron carries the initial photon energy minus the binding energy of its original shell. Therefore, photons that interact with tissues are either absorbed or deflected from its original path, resulting in attenuation of the signal or mispositioning of the event, as will be described later. The probability of photon attenuation and scatter increases exponentially as the dimension of the subject increases. For clinical PET applications, this translates to reduced counts and degraded image contrast for large patients compared to small patients. It also means that the amount of attenuation and scatter in animal PET studies is significantly less than in human PET studies.

Detector and Scanner

The detection of annihilation photons relies on gamma ray detectors that can effectively determine the energy, time, and position (direction) of incoming photons of 511 keV energy. Several physical properties affect photons' interaction with material and efficiency of signal detection. While detector materials differ greatly from tissues, 511 keV photons still interact with the detectors through photoelectric absorption and Compton scatter. Gamma ray detectors typically convert the energy of the photoelectron or recoil electron into light or charge signals that can be measured and recorded. Photoelectric absorption is the preferred detection mechanism because the measured signal corresponds

to the energy of the incoming photon, which differentiates 511 keV annihilation photons from scattered photons that undergo Compton interactions with tissues. This information can be used to reject scattered coincidences, a process described in more detail later. Materials of higher density generally have more efficient interactions with photons. Materials with a higher atomic number (Z) have a higher probability that the interaction will be photoelectric absorption (photofraction) than materials of lower Z. Most PET scanners use high-density, high-Z inorganic scintillating material to convert 511 keV photons to light, then use photomultiplier tubes or avalanche photodiode to convert light to an electrical signal.

In addition to measuring the energy of an incoming photon, PET detectors need to accurately measure the arrival time of individual photons, a basic requirement of coincidence detection to identify annihilation photon pairs. Materials with fast scintillation processes provide good timing performance and are usually favored for PET applications. Light yield of a scintillation material affects both the energy and timing resolutions of the detector. Therefore, high light yield scintillators are preferred. Furthermore, the emission wavelength of the scintillation material needs to match the spectral response of the light detector in order to achieve optimal performance. Accommodating all these requirements leaves only a handful of scintillation materials suitable for PET applications. The potential candidates are listed in Table 2. NaI(Tl), though not ideal for PET applications, is widely used in gamma cameras and is listed as a reference to compare with other materials.

Even with these carefully chosen, high-density and high-Z scintillators, the primary interaction of 511 keV photons with the detector is still Compton scatter. A photon scattered by a detector may escape the scanner and never be detected. In such a case, the measured energy of the photon is less than 511 keV, and the system can not determine whether the loss of energy is due to scatter occurring in the detector system or in the body of the subject. Alternatively, the scattered photon may be detected by additional interactions in the same detector such that the total energy deposited in the series of interactions corresponds to the energy of the incoming photon. However, the position of the

Table 1   Half-life, Average Energy, and Positron Range in Soft Tissue of Common Positron Emitters

Radionuclide	^{11}C	^{13}N	^{15}O	^{18}F	^{64}Cu	^{76}Br	^{82}Rb	^{86}Y	^{124}I
Half-life	20.38 m	9.96 m	2.04 m	109.77 m	12.70 h	16.20 h	1.27 m	14.74 h	4.176 d
Energy$_{avg}$[a] (MeV)	0.386	0.492	0.735	0.250	0.278	1.184	1.475	0.666	0.818
Range$_{avg}$[b] (mm)	1.52	2.05	3.28	0.83	0.97	5.55	7.02	2.93	3.70

[a]Data from http://www.nndc.bnl.gov/mird/
[b]Data interpolated for average beta particle energy in CSDA (Continuous Slowing Down Approximation) tables from http://www.physics.nist.gov/PhysRefData/Star/Text/contents.html for soft tissue with a density of 1.05 g/cm3.

Table 2 Physical and Optical Properties of Scintillation Materials Commonly Used in PET

Scintillation material	Density (g/cm3)	Effective atomic number (Z)	Primary decay constant (ns)	Light yield (% relative to NaI)	Peak emission wavelength (nm)	Attenuation coefficient at 511 keV (cm⁻¹)
NaI(Tl)	3.67	51	230	100	410	0.35
BGO	7.13	75	300	15	480	0.95
LSO	7.40	65	40	75	420	0.86
GSO	6.71	59	60	30	430	0.70
YAP	5.55	32	27	40	350	0.37
LuAP	8.34	64	17	30	365	0.87
BaF2	4.88	53	2	12	220,310	0.45
LGSO	7.23	65	60	40	420	0.84

annihilation event may be misidentified if the detector system can not detect multiple interactions at different locations simultaneously. This type of event (inter-detector scatter) leads to degradation of detector intrinsic spatial resolution and also limits the image resolution of PET.

Most PET scanners use block detector technology consisting of four single-channel PMTs that decode a block of scintillation crystals (13). The crystal block is cut multiple times to different depths and is partitioned into many small segments, shown in Figure 2A. When a gamma ray interacts with the block detector, scintillation light generated in one of the segments will propagate and spread over the surface of the four PMTs. The signal intensities detected by these four PMTs depend on the origin of the light signal. By taking the ratio of the four PMT signals, one can estimate where the light originated, thus identifying which segment of the block interacted with the incoming photon. The sum of the four PMT signals is proportional to the total amount of light generated, which corresponds to the energy deposited by the photon. This design has the advantages of reasonable performance and low cost, and has been widely used in clinical PET scanners.

The primary disadvantage of the block detector is its significant level of light sharing among neighboring segments, resulting in the so-called "block effect" that limits its intrinsic spatial resolution (12). However, techniques developed over the years have improved the performance of block detectors without significantly increasing the cost. Figure 2B shows one variation of block detector design that uses discrete crystal

elements assembled into an array, coupled to PMTs through a light guide that still permits some degree of light sharing to decode multiple crystals. This design reduces the block effect and is used in many modern clinical PET scanners.

Figure 2C shows a discrete detector design that uses multi-channel PMT or an array of APD to decode the discrete scintillation crystal array (14). When each crystal is read out individually by a PMT channel or an APD, inter-detector scatter can be identified. Therefore, the discrete detector has no block effect and the highest intrinsic spatial resolution. The primary limitation of this design is its high cost, due to the large number of light detectors and associated electronics. The resolution of this type of detector is subject to the availability of light detector, not the scintillation detector. Hybrid detectors that combine features of different designs can overcome various limitations and have been developed for both human and animal PET scanners (15).

Typical PET scanners consist of large number of scintillation detector arrays arranged in multiple rings. An object is placed in the scanner and radiotracer is administered. As illustrated in Figure 3, a coincidence event is recorded when two 511 keV photons are registered by a pair of detectors within a predefined coincidence time window (typically between 2 and 12 ns).

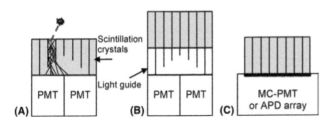

Figure 2 Side view of a (**A**) block detector, (**B**) discrete block coupled to PMT through light guide, and (**C**) discrete detector.

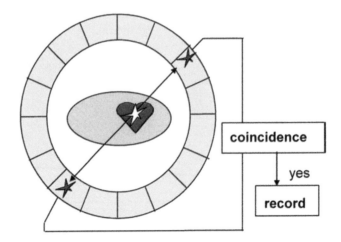

Figure 3 Coincidence detection of annihilation photons.

Each time a valid coincidence event is detected, the system assumes that a β+ decay has occurred somewhere along the line defined by the two detectors, the coincidence line of response (LOR). A set of LORs can be combined to form a projection. The simplest projection, known as a parallel projection, groups a set of parallel LORs to sample activity distribution along a particular angle (Fig. 4). A collection of parallel projections from different angles forms a matrix known as a parallel-beam sinogram (Fig. 4), because a point source produces a sinusoidal curve in the matrix.

As a PET scanner acquires data, each valid coincidence event is sorted based on the radial and angular offsets (s and θ) of the LOR defined by the two detectors in coincidence. The number of accumulated events stored in a sinogram element is proportional to the line integral of the activity concentration along the corresponding LOR. Recall the definition of Radon transform of a 2-dimensional function $f(x,y)$:

$$g(s,\theta) = \int_{-\infty}^{\infty} \int_{-\infty}^{\infty} f(x,y)\delta\left(x\cos\theta + y\sin\theta - s\right) \cdot dx \cdot dy,$$
$$-\infty < s < \infty, 0 \leq \theta < \pi \tag{1}$$

where $g(s,\theta)$ is the Radon transform of $f(x,y)$ and $\delta(\ )$ is a delta function. If we consider the radioactivity distribution measured by a PET scanner to be the 2D function $f(x,y)$, the collection of line integrals of the radioactivity distribution along different LORs (i.e., a sinogram) is precisely the Radon transform of the $f(x,y)$ in a discrete representation. Mathematical reconstruction of the original 2D function $f(x,y)$ to give the distribution of radioactivity in an image plane defined by a particular ring of detectors is done by inverse Radon transform of the measured sinogram $g(s,\theta)$.

Image Reconstruction

The rigorous derivation and implementation of inverse Radon transform is beyond the scope of this chapter and can be found in several references (16,17). However, a simplified description is provided to aid understanding

of the requirements and limitations of PET image reconstruction. The basic idea starts with the Fourier Slice Theorem, which is:

> The 1D Fourier transform of a projection $g(s,\theta_0)$ of an image $f(x,y)$ taken at an angle θ_0 is equivalent to a line profile through the center of the 2D Fourier transform of the image, $F(u,v)$, along the same angle θ_0.

This idea is illustrated in Figure 5, where a 2D image $f(x,y)$ is projected along an angle θ_0 to obtain a 1D projection function $g(s,\theta_0)$. If we apply 1D Fourier transform to this projection function, we obtain a 1D function $G(t) = FT\{g(s,\theta_0)\}$. Alternatively, if we apply 2D Fourier transform to the original image, we obtain a 2D function $F(u,v) = FT\{f(x,y)\}$. Fourier Slice Theorem says that if we extract a line profile from $F(u,v)$ through its origin along the same angle , this 1D function will be identical to $G(t)$.

Using Fourier Slice Theorem, one can approximate the 2D Fourier representation of the image, $F(u,v)$, by the collection of 1D Fourier transform of the projection data at different angles. If we could collect an infinite number of projections with infinite resolution, we would obtain the exact representation of $F(u,v)$, and so the original image $f(x,y)$ is simply the inverse Fourier transform of the function $F(u,v)$. In reality, every detector system has finite resolution and the number of projections collected is also finite. Each pair of detectors in a PET scanner measures the radioactivity distribution in an object along the corresponding LOR, which is the discrete representation of the analytical form in Equation (1). Fourier transform of a parallel projection measured by a PET scanner (i.e., each row of the sinogram in Fig. 4) corresponds to a set of discrete samples in the frequency domain, shown as solid dots in Figure 6A. Collection of projection data from all angles fills the frequency domain with a nonuniform sampling pattern, shown in Figure 6B. The low-frequency region (near the center) is over-sampled (over-represented) while the high-frequency region (far from the center) is under-sampled. To reconstruct this frequency representation into the original image $f(x,y)$, individual frequency components need to be properly weighted. That is, the over-represented low-frequency data need to be weighted down while the under-represented high-frequency data need to be weighted up, illustrated by the pie-shaped weighting function in Figure 6C. This is commonly accomplished in the frequency domain by multiplying the 1D Fourier transform of a projection by a weighting function that grows linearly with the frequency, commonly known as a Ramp filter (Fig. 6D). Once done, the filtered projection data can be back-projected to approximate the original image (radioactivity distribution). This is, indeed, the basis of the most common image reconstruction algorithm, filtered backprojection (FBP), used in tomographic imaging systems.

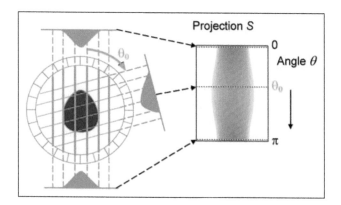

Figure 4 Parallel LORs are grouped together to form a parallel projection. The collection of parallel projections along different angles forms a matrix called "sinogram," which can be reconstructed to form a tomographic image.

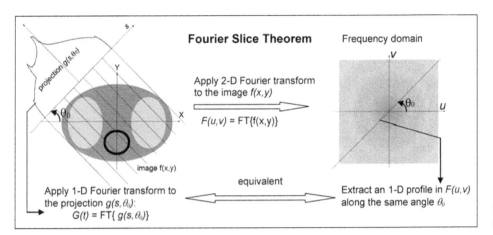

Figure 5 Graphical representation of Fourier Slice Theorem.

FBP is relatively fast and its operation is linear, both favorable features that have made FBP widely used in PET and computed tomography for image reconstruction. However, FBP has an inherent disadvantage because its filtering operation is subject to noise amplification. Noise, by its nature, is a high-frequency signal. Because FBP assigns the biggest weight to the highest frequency signal, it produces poor image quality when the data contains a lot of noise. This is a particular problem for PET because the measurement is based on photon counting, which has inherent statistical noise. Depending on the counting statistics, different filters may need to be used in place of the theoretical Ramp filter to limit the amplification of high-frequency signal and noise. Iterative reconstruction methods, such as the maximum likelihood expectation maximization (ML-EM) algorithm, that take into account the statistical nature of the signal generally produce images of better quality. Variations of the ML-EM algorithm that include different types of regularization and penalty functions have been developed to further improve conversion speed and image quality. More details on reconstruction techniques can be found in several references (16–19).

QUANTITATION OF PET IMAGES

The accuracy of biological models derived from PET imaging experiments relies on quantitation of PET images. Data measured by a PET scanner is subject to many physical perturbations (such as attenuation and scatter of the annihilation photons) and instrumental limitations (such as nonuniform sensitivity and system dead time). These factors need to be taken into account and compensated for in order to provide quantitative measurement of the true radioactivity distribution within the subject. Common correction techniques are discussed below.

Random and Scatter Corrections

The coincidence events detected by a PET scanner can be divided into three categories (Fig. 7). If the annihilation photons escape the object without any interaction and subsequently become detected by the scanner, this is defined as a true coincidence, illustrated as event A. If at least one of the two 511 keV photons is scattered by the object, such as event B in Figure 7, this is a scattered coincidence. If two uncorrelated 511 keV photons are detected within the predefined timing window, shown as event C, it is a random coincidence. True coincidences are the desired signal and are proportional to the amount of activity within the field of view (FOV). Scatter coincidences are proportional to the activity within and near the FOV, but also depend on the amount of scattering medium that annihilation photons encounter. For human imaging, particularly the whole-body imaging protocol, this can be a significant portion of the total coincidences measured by a

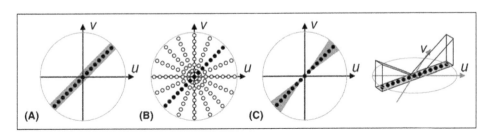

Figure 6 (**A**) One-dimensional Fourier transform of a parallel projection measured by a PET scanner. (**B**) The collection of the 1D Fourier transform of all projection data from a PET scanner fills the frequency domain with a nonuniform sampling pattern. (**C**) Different frequency components need to be weighted to properly represent their contributions to the frequency representation of the original image. (**D**) Weighting is commonly accomplished by a filtering step for each projection data in the frequency domain.

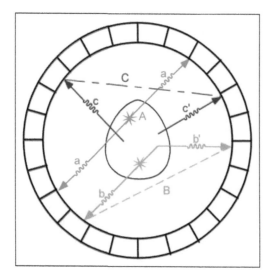

Figure 7 (**A**) True coincidence. (**B**) Scatter coincidence. (**C**) Random coincidence.

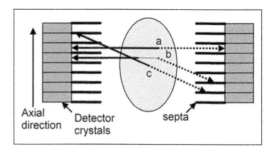

Figure 8 Inter-crystal septa are used in PET scanners running in 2D acquisition mode. It allows true coincidence (event a) to be detected and most of the scatter coincidence (event b) to be shielded. However, some true coincidences (event c) will be sacrificed, thus reducing system sensitivity.

scanner. Random coincidences are proportional to the width of the coincidence timing window and the square of the singles event rate of detectors. That is, when the amount of activity within and near the FOV doubles, the random coincidences increase by nearly a factor of 4. Both scatter and random coincidences need to be subtracted from the measured coincidences before the images are reconstructed.

To reduce or eliminate the error from scatter coincidences, two actions are commonly taken. The first is to prevent the scatter coincidences from being accepted. Ideally, one would use detectors of good energy resolution and only accept events whose energies fall within a narrow window around 511 keV. This approach, however, has not been very effective because a detector can not differentiate whether a photon undergoes Compton scatter in the subject or a Compton interaction in the detector material when the output signal shows energy less than 511 keV. In fact, many true coincidences will be rejected if a tight energy window is used, resulting in poor system sensitivity. Conventional PET systems use inter-crystal septa made of tungsten or lead to limit the acceptance angle of coincidences. Figure 8 shows that septa effectively shield detectors from inter-plane scatter coincidences (event b) and random coincidences, but prohibit detection of some true coincidences (event c). When septa are removed, coincidence detection is permitted between crystals of any detector planes. This wider acceptance angle is referred to as "3D acquisition" that has much higher sensitivity than the "2D acquisition" using septa (20). Most modern PET scanners either do not include septa or make them retractable because 3D mode is becoming a standard operation. In the event that excessive scatter coincidences are present, such as when imaging a very large patient, scanners equipped

with septa can be operated in 2D mode to reduce the impact of scatter.

For scatter coincidences detected by a PET scanner, scatter correction techniques need to be applied to minimize their effects. Two types of scatter correction techniques exist. One calculates the scatter contamination analytically by deconvolution and tail fitting the data in the sinogram. The other is based on Monte Carlo techniques that use the reconstructed activity distribution of uncorrected data to estimate the scatter distribution based on known object density distribution (21–24).

To reduce the impact of random coincidences, two similar approaches can be taken. If fast scintillation detectors are employed to allow a smaller coincidence timing window, the number of random events is reduced proportionally. In addition, reducing the number of singles events that reach the detectors can reduce the number of random coincidences quadratically. This is why inter-crystal septa also reduce the random coincidences. Correction for random coincidences is relatively easy compared to scatter correction. Random coincidences can be estimated if the singles event rates of individual detectors are measured. Alternatively, delay-window technique (25) can be used to directly measure the random coincidences' contribution in the total events.

Another item of note is the "noise equivalent count" (NEC) (26) that relates the noise level of data to the number of true, random, and scatter coincidences. Imaging techniques based on radionuclide decay, including PET, rely on photon counting for signal detection. The detected signal follows Poisson statistics in which the variance equals the mean. Therefore, the signal-to-noise ratio of a photon counting measurement is proportional to the square root of the number of counts (27). When the random and scatter events are subtracted from the total coincidences, the statistical noise inherent in individual measurements propagates to the final result. NEC is defined as:

$$NEC = \frac{T^2}{T + S + kR} \qquad (2)$$

where T, S, and R are the number of true, scatter, and random coincidences, respectively. k is the fraction of the transverse FOV occupied by the object. The factor 2 is used when random coincidences are measured by delay-window technique, which has a higher statistical error from the measurement. A factor of 1 is used if, instead, the random coincidences are estimated from the detector singles rate and the coincidence timing window, which is assumed to have very low statistical error. Although NEC does not directly correlate to the image quality at different counting rates, it has become a commonly accepted measure of the counting rate performance of a PET scanner.

Attenuation Correction

When either or both annihilation photons are absorbed by the object or scattered out of the FOV, the original decay event is undetectable. The probability that a photon survives (is not attenuated) as it travels through matter is inversely proportional to the exponential of the line integral of the attenuation coefficient along its path. Therefore, the attenuation correction for detecting a single photon depends on the depth of the source. However, with coincidence detection in PET, the probability that both photons survive and become detected depends on the line integral of the attenuation coefficient along the entire LOR, and not on the depth of the source. This is illustrated in Figure 9 where the detection probabilities of event A and B are the same regardless of their depths. Similarly, an external source C is subject to the same amount of attenuation if detected by the same pair of detectors.

This observation means that attenuation correction factors can be measured using an external positron-emitting source. The initial probability of detection, p_0, can be measured by a blank scan using the source C without the subject in place. The probability of detection is then reduced in a transmission scan, in which the subject is placed between the source C and detector. The ratio of the blank to the transmission

scan provides the attenuation correction factor for any given LOR. Alternatively, if the 3D distribution of the attenuation coefficient of the object is known, the attenuation correction factor can be calculated for each LOR by integrating the μ(s) along the LOR (28,29). This approach is now widely adopted by clinical PET/CT scanners in which the distribution of the attenuation coefficients of the object is given by the CT images (30,31).

Normalization

Normalization is a procedure that compensates for systemic variation in measurements due to non-uniform detector response and system geometry (32,33). This type of systemic variation is measured and stored in the system normalization file. When a PET study is performed, the measured data is "normalized" before it can be reconstructed into images. Because the gain of detectors and electronics may drift over time, detector response should be checked on a routine basis. If detector response drifts or detectors have been replaced or serviced, the systemic variation of the scanner should be re-measured to generate new normalization files. Failure to normalize measured data or keep the normalization file up to date may result in systemic errors in images. Therefore, the importance of this simple correction should not be overlooked.

Component-based normalization (34) has been adopted by most PET systems to replace the inversion-based approach. The inversion-based technique simply acquires coincidence events from a uniform positron-emitting source and inverts the sinograms to get the correction factors. The statistical error of individual measurements limits the accuracy of this method. Component-based normalization assumes that systemic variation consists of: (i) geometric efficiency of detector, (ii) crystal interference pattern within a detector module, and (iii) detector efficiency. Each of these three components can be measured with high accuracy, resulting in a normalization procedure

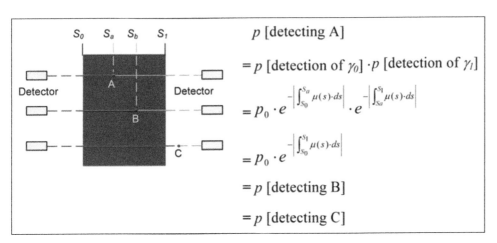

Figure 9 Detection probability of the annihilation photons is independent of source depth.

more consistent and more accurate than the inversion-based approach.

Dead Time Correction

A photon counting system takes a certain amount of time to process an event. The period of time that a system can not accept an additional event is called dead time. For simple systems, the system dead time may be characterized as paralyzable or non-paralyzable (27). A paralyzable system takes time to recover after the latest event, even if that event is not detected. For example, if a second event occurs before the system finishes processing the first event, the system will be dead for an extended period of time trying to restore its function. If the counting rate is extremely high and the system never has a chance to fully recover before the next event occurs again, the system becomes completely paralyzed after the initial trigger. In contrast, a non-paralyzable system can not accept new events for a fixed amount of time after it starts processing an event, but events occurring during this period are simply ignored. The system resumes its counting capability once this fixed period of time passes. For systems that follow either of these two simple models, the true counting rate can be accurately determined from the measured counting rate, as long as the system's dead time model and its parameters are known.

Complex systems such as PET scanners usually exhibit a combination of paralyzable and non-paralyzable characteristics. The dead time of a PET scanner may vary with activity level, spatial distribution of the activity, and imaging conditions. Its behavior can be characterized by models that range from relatively simple (34) to extremely complicated (35,36). It is important to understand the counting rate capability of a PET system and operate well under its limit to ensure that the system response is fairly linear.

Decay Correction

In order to collect a sufficient number of counts, a PET study may take several minutes or even an hour to complete. The accumulated coincidences represent the activity concentration averaged over the period of time the data is acquired. To find out the activity concentration at a particular time, the measured number needs to be decay-corrected to the desired time point. This is important for imaging studies that are designed to follow the radiotracer distribution over time to establish the kinetics of the tracer. This type of study is typically broken down into multiple time frames, each representing a different time point of the kinetics. The measured activity concentration from each frame is then decay-corrected to the same reference time point, typically the beginning of the study, to establish the kinetics of the biological system rather than the radiotracer decay.

HIGH-RESOLUTION PET IMAGING OF SMALL ANIMALS

Challenges

Laboratory animals have long been used for biomedical research. Recent advances in molecular biology and genomics have made it possible to study human disease and develop treatment interventions using genetically modified mice. These transgenic mice can be extremely delicate and expensive. Non-invasive imaging techniques that allow longitudinal study of the same subject repeatedly have the potential to expedite research and lower costs. Also, when individual animals serve as their own control, inter-subject variation is eliminated and the research outcome may be more accurate. All these factors lead to a strong demand for high-resolution animal imaging systems, including animal PET.

A rodent is two to three orders of magnitude smaller than a human (25 g for a mouse, 250 g for a rat, and 70 kg for a human). In order to image a rodent with sufficient detail, image resolution of an animal PET scanner needs to be a factor of ten higher than for a human PET scanner for each of the three dimensions. Current state-of-the-art human PET scanners have a resolution ranging from 3 to 6 mm FWHM. Limited by the counting statistics of a whole-body imaging protocol, most clinical PET images are reconstructed with smoothing to achieve an image resolution ranging from 6 to 10 mm FWHM. Ideally, animal PET scanners should have a resolution of 1.0 mm FWHM or less. In reality, most animal PET scanners, except one (37), have not achieved this goal. Despite the physics such as positron range and acolinearity effect that limit image resolution of PET, detector intrinsic spatial resolution remains the primary limiting factor for all PET scanners when short-range radionuclides such as F-18 or Cu-64 are employed (9).

When detectors of smaller dimensions are used in high-resolution PET systems, the number of counts detected by individual detectors is reduced. Since the signal-to-noise ratio of PET images depends on counting statistics, high-resolution images may not provide the desired advantages in image quality unless sufficient events are collected. To improve the counting statistics of an animal study, one may extend the scan duration, increase the radiotracer concentration administered or improve the sensitivity of the PET scanner.

Extending the scan duration works for static imaging protocols in which the kinetics of the radiotracer is assumed to have reached a steady state that does not change during the scan. However, many imaging studies are designed to obtain the kinetics of the radiotracer, and so this assumption can not be made. In such cases, the counting statistics can not be improved by increasing the duration without sacrificing the temporal resolution of the dynamic images. In other occasions, such as when short-lived radionuclides such as O-15 or C-11 are used, the physical half-life may limit the scan duration and counting statistics.

Increasing the radiotracer concentration is acceptable when the radiation dose to animals does not affect the outcome of the study and the quantity of the radiotracer still satisfies the tracer principle (i.e., trace quantity that does not disturb the biological system). A dose of 9.25 MBq (250 μCi) FDG injected into a 25 g mouse will result in a radiotracer concentration that is 46 to 70 times higher than that of a typical human whole-body PET study in which 370–555 MBq (10–15 mCi) of FDG is injected (assuming 70 kg for body weight). This may cause a significant radiation dose to the mouse and alter the biology, for example affecting tumor development or progression. For receptor studies, care must be taken to ensure that the amount of radioligand injected does not saturate the binding sites. Even when both conditions are satisfied, the total activity that can be injected into an animal may still be limited by the maximum counting rate that a system can handle without causing severe dead time. In short, more activity does not always guarantee a better imaging study.

Current state-of-the-art human PET scanners have a sensitivity of approximately 5%. It is, in theory, possible to develop an animal PET scanner that is an order of magnitude more sensitive than a human PET system. In reality, it is extremely difficult to improve both image resolution and system sensitivity at the same time. Many PET scanners dedicated to animal imaging have been developed since the 1990s (37–50). While the image resolution of animal PET scanners has been improved significantly in the last decade, sensitivity of all existing animal PET scanners remains less than 10% and needs more improvement (15).

microPET Technology

The prototype microPET scanner was developed by Cherry et al. (42) at UCLA in 1996. Its technology was made commercially available by Concorde Microsystems Inc., Knoxville, Tennessee, U.S.A. (now part of Siemens Molecular Imaging) in 2000. The second prototype system, microPET II, was completed in 2002 at UC Davis by the same group (49), and the third generation microPET developed at Concorde was revealed in 2003 (Focus 220) (51) and 2004 (Focus 120) (52). Figure 10 shows photographs of the micro PET families. With more than 80 systems installed world-wide, microPET is currently the most widely used animal PET systems in the biomedical research community.

The design of the microPET detector is similar to the discrete detector in Figure 2 that minimizes the optical cross-talk between crystals and light detectors. However, microPET multiplexes the output signals of the light detectors, while a standard discrete detector would read out individual light detector signals separately. In either case, the detector intrinsic spatial resolution is limited by the width of the crystal element. Therefore, the scintillation crystals need to be cut

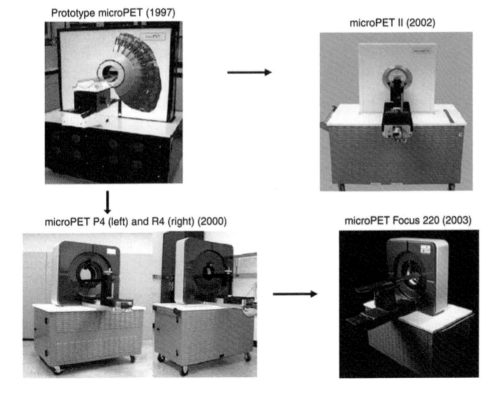

Figure 10 The microPET family. (*Top*) Research prototypes—original microPET (*left*) and microPET II (*right*). (*Bottom*) Commercial microPET from CTI-Concorde Microsystems Inc. (Knoxville, Tennessee, U.S.A.)—first generation (P4 and R4, left) and third generation (Focus 220). *Source*: Courtesy of Dr. Simon R. Cherry at UC Davis and Mr. Robert E. Nutt at CTI-Concorde Microsystems Inc.

to a small width in order to achieve high spatial resolution. The crystal length, on the other hand, has to remain long in order to maintain good detection efficiency. This geometry results in poor light collection efficiency if the scintillation light needs to travel along the finger-like crystals to be readout from one of the small ends (53). As a result, BGO ($Bi_4Ge_3O_{12}$) that is widely used in conventional human PET scanners, but has low light yield, is not ideal for high-resolution animal PET applications.

The prototype microPET was the first PET scanner that utilized a new type of scintillation material, LSO (Lu_2SiO_5:Ce) (54). LSO has higher light yield and faster scintillation process than BGO. The high light yield of LSO allows microPET detectors to collect a sufficient number of scintillation photons from each 511 keV photon interaction to maintain good energy resolution and timing resolution. A fast scintillation process permits shorter integration time for pulse processing, which improves the counting rate performance and reduces the dead time of the scanner.

The prototype microPET detector uses 64-channel PMT to decode a LSO array of 8-by-8 elements each measuring $2 \times 2 \times 10$ mm^3. The 64 outputs of the multi-channel (MC) PMT are multiplexed by a 2D charge-division resistor network to generate four position-encoded outputs (55). Commercial microPET scanners use position-sensitive (PS) PMT in place of the MC-PMT to reduce the cost of the system. The outputs of PS-PMT are also multiplexed using two 1D charge-division resistor networks to generate four position-encoded outputs. Compared to the discrete detector design in which the individual outputs of PMT channels are processed and digitized separately, this discrete-crystal multiplexed-readout design reduces the number of electronic channels from 64 to 4. Therefore, it significantly lowers the cost and complexity of a scanner. The drawback is its inability to accurately identify an inter-crystal scatter event that produces two or more interactions in a detector block. The positions of individual interactions are weighted by the corresponding deposited energy in the multiplexing process to calculate the averaged position. The net result is a slight degradation in spatial resolution compared to discrete detectors of the same dimension. On the other hand, the resolution of discrete detector design is often limited by the physical dimension of available light detectors rather than by the scintillation crystals. Most systems that use discrete detector design still use crystals of 2 mm width, while the microPET II and some other systems are already using crystals of sub-millimeter width (49,56,57). It is therefore debatable which design provides better image resolution, but the cost saving of multiplexed readout is apparent.

Optical fiber coupling between the LSO array and MC-PMT or PS-PMT is a unique feature of microPET technologies. It permits detectors to be packaged into multiple rings without inter-detector gaps. This provides two advantages: (*i*) more solid-angle coverage that yields higher system sensitivity and (*ii*) complete sampling of the imaging FOV that simplifies the image reconstruction and minimizes the potential of artifacts due to insufficient sampling. However, the additional interfaces between individual components, limited numerical aperture and transmission efficiency of the optic fibers reduce the overall collection efficiency of scintillation light from the detector crystals. This further emphasizes the need for high-light-yield scintillators such as LSO. The prototype microPET used one-to-one coupling between individual components. That is, each LSO crystal was coupled to one anode of the MC-PMT through a single fiber. The early commercial microPET models (P4 and R4) used individual optical fibers to couple individual LSO crystals to the surface of PS-PMT. The microPET II used a fused optical fiber bundle to couple an LSO array of 14×14 elements to an MC-PMT of 8×8 elements. The latest commercial microPETs (F220 and F120) use an 8×8 element fiber bundle to couple a 12×12 element LSO array to a PS-PMT. Figure 11 shows photographs of various detectors in microPET systems. Despite differences in implementation, the underlying design remains the same for all microPET scanners. Systems' primary differences are the intrinsic spatial resolution of the detectors, transverse FOV, and axial extend of the systems. Table 3 summarizes the specifications of all existing microPET scanners.

The performances of various models of microPET have been evaluated by different groups (49,51,52,58–61) and are summarized in Table 4. Because there are no commonly accepted standards for performance evaluation of animal PET scanners, precluding direct comparison of the systems. The experimental details are particularly important for system sensitivity and NEC measurements. Nevertheless, the table provides a quick reference for the imaging capability of all microPET systems. Also, even though the NEC rate does not peak until a large amount of activity is used, microPET scanners should not be operated at this activity level. Because of dead time, a PET system should be operated well below its limit to ensure that the system response is fairly linear and the results are quantitative.

Implementation of Correction Techniques in microPET
Random Correction

Random coincidences were initially measured by all microPET systems using the delay-window technique.

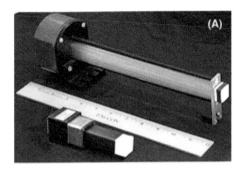
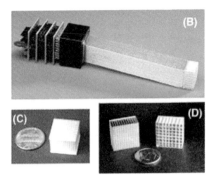

Figure 11 Photographs of various microPET detectors. (**A**) Detectors of the prototype microPET (*top*) and microPET II (*bottom*). The MC-PMT in the prototype detector was much larger and required long optical fibers. microPET II detector uses fused optical fiber bundle, compact MC-PMT and 14 × 14 elements LSO array with sub-millimeter crystal width (as shown in **C** next to a penny). (**B**) Detector in the first generation commercial microPET systems. (**D**) LSO arrays used in commercial microPET systems. 12 × 12-element array (*left*) is used in F120 and F220 models. 8 × 8-element array (*right*) is used in R4 and P4 models.

The coincidence processor of a scanner measures prompt and delayed coincidences independently. Both types of events are transmitted to the host computer where the listmode data is recorded. The delayed coincidences correspond to randoms and are subtracted from the prompt events for random correction.

While the delay-window technique adequately corrects random coincidences in most imaging conditions, it is not the most efficient way of utilizing the system bandwidth. Current microPET scanners use Firewire (IEEE 1394) to link the coincidence processors and host computers. The bandwidth of the system is approximately 400 Mbps. When the total data rate combining the prompt and delayed coincidences exceeds this limit, a fraction of both types of events are lost. Under extremely high counting rate conditions, there are more randoms than trues. As the prompt and delay coincidences compete for the limited system bandwidth, approximately half of the bandwidth is used to transmit the delay coincidences. Correction techniques that estimate randoms based on singles rates do not need to transmit individual randoms events and provide more efficient use of system resources. If the random coincidences are calculated from the width of

Table 3 Specifications of microPET Systems

Category	Parameter	Prototype	microPET II	P4	R4	Focus 220	Focus 120
Detector	Crystal material	LSO	LSO	LSO	LSO	LSO	LSO
	Crystal size (mm)	2 × 2 × 10	0.975 × 0.975 × 12.500	2.2 × 2.2 × 10.0	2.2 × 2.2 × 10.0	1.51 × 1.51 × 10.00	1.51 × 1.51 × 10.00
	Crystal pitch (mm)	2.25	1.15	2.45	2.45	1.59	1.59
	Crystal array	64 (8 × 8)	196 (14 × 14)	64 (8 × 8)	64 (8 × 8)	144 (12 × 12)	144 (12 × 12)
	Photomultiplier tube	Philips XP1722	Hamamatsu H7546-M64	Hamamatsu R5900-C12	Hamamatsu R5900-C12	Hamamatsu R5900-C12	Hamamatsu R5900-C12
System	Number of detectors	30	90	168	96	168	96
	Number of crystals	1,920	17,640	10,752	6,144	24,192	13,824
	Number of rings	8	42	32	32	48	48
	Number of crystals/ring	240	420	336	192	504	288
	Ring diameter (cm)	17.2	16.0	26.1	14.8	25.8	14.7
	Gantry aperture (cm)	16.0	15.3	22.0	13.0	22.0	13.0
	Axial FOV (cm)	1.8	4.9	7.8	7.8	7.6	7.6
	Transaxial FOV (cm)	11.25	8.00	19.00	9.40	19.00	9.4
Dataset	Number of sinogram (3D)	64	1,764	1,024	1,024	2,304	2,304
	Number of sinogram (2D)	not used	83	63	63	95	95
	Sinogram size	100 × 120	140 × 210	192 × 168	84 × 96	288 × 252	128 × 144
	Data size (3D) (MB)	1.54	207	126	33	638	162
	Data size (2D) (MB)	not used	9.7	7.8	2.0	26.3	6.7
	Sampling pitch (mm)	1.125	0.575	1.225	1.225	0.795	0.795

Table 4   Performance of microPET Systems

Characteristics	Prototype (58)	microPET II (49,61)	P4 (59)	Focus 220 (51)	R4 (60)	Focus 120 (52)
Energy resolution at 511 keV	19%	42%	26%	18%	23%	Same as Focus 220
Timing resolution (FWHM)	2.4 ns	3.0 ns	3.2 ns	N/A	Same as P4	N/A
Intrinsic spatial resolution (FWHM)	1.58 mm	1.05 mm	1.75 mm	1.36 mm	Same as P4	Same as Focus 220
Image resolution near CFOV	~1.8 mm	~1.1 mm	~2.2 mm	~1.75 mm	~2.2 mm	~1.75 mm
Tangential resolution at 2 cm from CFOV	~1.8 mm	~1.25 mm	~2.4 mm	~1.75 mm	~2.3 mm	~1.8 mm
Radial resolution at 2 cm from CFOV	~2.3 mm	~2.0 mm	~2.5 mm	~1.85 mm	~2.9 mm	~2.4 mm
Axial resolution at 2 cm from CFOV	~2.0 mm	~1.5 mm	~2.4 mm	~1.75 mm	~2.8 mm	~2.0 mm
Sensitivity at CFOV w/ 250 keV lower energy threshold[a]	0.56%	2.25%	2.25%	3.4%	4.4%	7.1%
Peak NEC rate for mouse phantom[a]	N/A	235 kcps at 80 MBq	290 kcps at 100 MBq	645 kcps at 147 MBq	174 kcps at 77 MBq	809 kcps at 88 MBq[b]
Peak NEC rate for rat phantom[a]	4.1 kcps at 65 MBq	25 kcps at 80 MBq	100 kcps at 100 MBq	177 kcps at 143 MBq	94 kcps at 60 MBq	300 kcps at 120 MBq[b]
Peak NEC rate for monkey phantom[a]	4.2 kcps at 74 MBq	N/A	120 kcps at 110 MBq	44 kcps at 242 MBq	N/A	N/A

[a]Refer to individual references for methods of measurement and phantom specifications.
[b]Based on NEC-1R.

the coincidence timing window and the singles rates of individual detectors, the amount of data that needs to be transmitted across the data link will be significantly reduced. Therefore, the same system bandwidth will be able to handle imaging conditions with higher counting rate. Another benefit of this technique is that the statistical noise associated with delayed-window measurement does not contribute to the final image. Therefore, a unity factor replaces the factor 2 in Equation 2 for NEC calculation. These advantages have led to the implementation of calculated random correction in commercial microPET scanners to improve their noise property and counting rate capability.

Attenuation Correction

Prototype microPET and microPET II systems do not support measured attenuation correction because most studies are performed on mice which have relatively low attenuation. Commercial microPET systems do support measured attenuation correction. In these systems, a point source of a positron-emitter (^{68}Ge) or a single-photon emitter (^{57}Co) is orbited along the edge of gantry opening in a helical trajectory. When the system operates in coincidence detection mode, the counting rate of detectors near the ^{68}Ge point source limits the amount of activity one can use for transmission scan. When the system operates in singles mode, it has higher efficiency and detects more counts. In addition, half of the detectors near the point source are electronically disabled to avoid the dead

time problem. Therefore, more activity can be used for the point source in singles transmission scan. Furthermore, the difference in attenuation coefficients of different types of tissues is larger at 122 keV (energy of ^{57}Co source) than that at 511 keV. This results in better contrast between different types of tissue. All these factors contribute to better transmission image quality when the system operates in a singles mode using ^{57}Co as the transmission source. Figure 12 shows a transmission image of a typical rat using a ^{57}Co source. The acquisition time was 17 minutes for the transmission scan. The lungs of the rat can be clearly delineated from other types of tissues. Re-mapping the attenuation coefficients of tissues to 511 keV or segmentation of different tissue types, followed by forward projection of the transmission images, allows

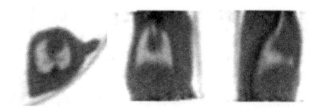

Figure 12   (*Left to right*) Transverse, coronal, and sagittal views of transmission images of a 340 g rat, acquired with a 57Co point source in singles mode. The images were rotated around the Z-axis (the central axis of the cylindrical gantry) so that the images follow the commonly used orientation.

one to calculate the attenuation correction factors for the animal (62,63).

Scatter Correction

All microPET systems operate in 3D mode only. Although small animals have less scatter than humans do, scatter coincidences can still account for 12 to 40% of total coincidences. This ratio, the scatter fraction, depends on the size of animal and the energy window used for the study. For small animals that have little scatter, it may be beneficial to widen the energy window so that maximum sensitivity can be achieved.

Scatter correction currently implemented in microPET is an image-based direct-calculation technique (64). It requires the emission images to be pre-reconstructed with attenuation correction. It also requires the attenuation coefficient map of the object to be known. The algorithm assumes that all scatter events only undergo a single scatter in the object and divides the image into small sub-regions. The scatter contribution to sinograms is analytically calculated for each region based on the region's average activity concentration and overall tissue attenuation coefficient distribution. Once the scatter component in the sinogram is calculated, it is subtracted from the original sinogram before the corrected images are reconstructed. Since the original emission images are contaminated by scatter events, this process may need to be repeated to get an accurate estimate of scatter contamination and correction.

Dead Time Correction

Current microPET systems use a non-paralyzable dead time model based on average detector singles count rate to estimate the average dead time fraction for the entire system. This approach does not account for local variation in count density or spatially variant dead time. Fortunately, small animals are usually positioned at the CFOV so that the highest image resolution can be obtained. Under this condition, the singles count rate shall not vary significantly among detectors in the same image plane. For experiments in which large amounts of activity are located outside the scanner's axial FOV or within only one half of the FOV, the current dead time model may not accurately address the spatial varying of dead time. For most other imaging conditions, study has shown that the current dead time model provides less than 5% error for injected activity up to ~74 MBq (2 mCi) in a small animal (52).

Other Corrections

Normalization of microPET scanners are based on component-based normalization as described earlier. Decay correction is built-in as part of the reconstruction process and is transparent to the users.

Imaging Capability of microPET

Resolution Capability

High-resolution imaging capability of microPET is demonstrated in Figure 13. A micro-Derenzo phantom with hot rods of different sizes was imaged by P4, Focus 220 and microPET II. The diameter of rods is 0.8, 1.0, 1.25, 1.5, 2.0, and 2.5 mm respectively. The center-to-center distance between adjacent rods is 2 times the rod diameter. It is evident that Focus 220 has a higher resolution than P4, while the microPET II has the highest resolution. With the maximum a posteriori (MAP) (65,66) reconstruction algorithm, microPET II can clearly resolve the 1 mm rods.

Dynamic Imaging Capability

All microPET scanners are capable of dynamic imaging, which is critical for acquiring time activity curves (TAC) of organs or sub-organ structures to establish kinetic models of biological functions and tracer distribution. List mode data can be sorted into frames of arbitrary durations. Typically, users choose short frame duration for early time points in order to catch the rapidly changing dynamics when radiotracer is first injected. Longer frame duration is used for later time points to obtain better counting statistics when the change in radiotracer distribution slows down.

Figure 14A shows dynamic images of a 26 g mouse injected with 55.5 MBq (1.5 mCi) of ^{18}F-FDG

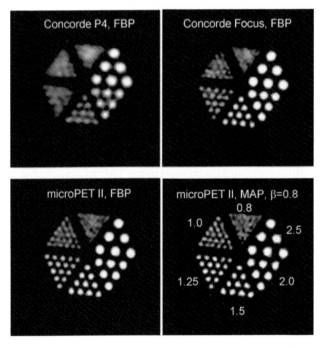

Figure 13 Micro-Derenzo phantom imaged by three different microPET systems and reconstructed with FBP and MAP (*lower right*). *Source*: From Ref. 61.

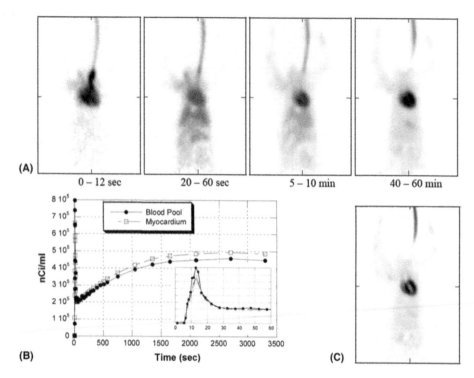

(A)

0 – 12 sec   20 – 60 sec   5 – 10 min   40 – 60 min

(B)   Time (sec)   **(C)**

Figure 14 A 26 g mouse was injected with 55.5 MBq (1.5 mCi) of ¹⁸FDG and imaged for 1 hr. (**A**) Summed frames that correspond to the first 12 sec, 20–60 sec, 5–10 min, and last 20 min of the study. (**B**) TAC of the LV (blood pool) and myocardium. Small inset corresponds to the first 60 sec of the curves. (**C**) MAP reconstructed image corresponding to 40–60 min of the study.

and scanned in a microPET F-120 system. Data from a 60-min imaging session was sorted into dynamic frames of 10×1 sec, 2×2 sec, 2×3 sec, 8×5 sec, 8×30 sec, 5×1 min, 4×5 min, and 3×10 min. Regions of interest (ROI) were drawn in the left ventricle (LV) of the heart and the myocardium. TAC of LV and myocardium were extracted and plotted in Figure 14B. These curves show that activity concentration of organ or sub-organ structures can be measured with good temporal resolution. However, the quantitative accuracy of this particular experiment suffered from partial volume effects, that is, activity within a small structure appears to spread out over a wider area when the structure is smaller than the resolving capability of the imaging system. In this case, activity in the LV appeared to spill into the myocardium in images at early time points. The sharp peak of the TAC of the LV due to bolus injection of highly concentrated radiotracer can also be seen in the TAC of myocardium because of the spill-over. Similarly, the gradual uptake of ¹⁸F-FDG in myocardium also appears in the TAC of the LV at late time points. Partial volume effects and spill-over can be reduced if higher resolution PET systems (such as microPET II) or better image reconstruction algorithms are employed. Figure 14C shows an image reconstructed from the same data that corresponds to 40–60 min of the study using MAP reconstruction algorithm. It is evident that better reconstruction algorithms can significantly improve the image resolution and better resolve small organs and sub-organ structures, thereby improving the quantitative accuracy of the study.

Gating Capability

All microPET systems (except the prototype) are equipped with two input connectors for gating signals. When a trigger signal is sent to the coincidence processor, the arrival time of the trigger is inserted in the data stream and recorded by the host computer. Post-acquisition sorting programs can extract the timing of gate signals and divide the listmode data into an arbitrary number of gates. Cardiac studies are typically gated by the ECG signals. Figure 15 shows a rat imaged by microPET Focus 220 with and without cardiac gating. It is evident that the left and right ventricles are better defined in the gated images, despite greater noise due to reduced counts in each gate. Simultaneous cardiac and respiratory gating has also been demonstrated by Yang et al. (67) using microPET II.

Blood Input Function

The activity concentration in blood as a function of time, the so-called blood input function, is often required for kinetic modeling using PET images.

Figure 15 ¹⁸FDG cardiac imaging of a rat in microPET Focus 220. (*Left*) Gated at end-systole. (*Middle*) Gated at end-diastole. (*Right*) Non-gated. *Source*: From Ref. 51.

For human PET studies, it can be estimated from the dynamic images or measured from blood samples fairly easily. For a mouse study, the thickness of myocardium and the size of the left and right ventricles are small relative to the image resolution of microPET. Therefore, the image-based calculation suffers from severe partial volume effects, as illustrated in Figure 14. A mouse also has an extremely small blood volume which limits the total number of blood samples one may withdraw to acquire the blood input function. The surgical procedure for taking arterial blood samples from a mouse is also highly invasive and reduces the survival rate in longitudinal studies. Several approaches have been proposed for acquiring blood input function in animal studies without surgical procedures, or at least with reduced invasiveness (68). These techniques appear promising in larger animals such as rats, but have limited success for mice. A convenient and accurate method for measuring blood input function of mice remains a challenge for microPET imaging experiments.

CONCLUSIONS

Human PET imaging has been widely adopted in clinical diagnosis of various types of diseases. microPET systems provide high-resolution, high-sensitivity, and dynamic whole-body imaging (with the option of gating) of small laboratory animals such as mice and rats. Scientists have used this in-vivo imaging technique to study normal development, disease models, pharmacokinetics, and treatment interventions using transgenic mice. The quantitative nature of PET imaging is extremely powerful for biological research. However, users need to understand the underlying physics and mathematics in order to fully comprehend the capabilities and limitations of this technology and to improve the design of imaging experiments to achieve the highest image quality and quantitation.

REFERENCES

1. Phelps ME, Hoffman EJ, Mullani NA, Ter-Pogossian MM. Application of annihilation coincidence detection to transaxial reconstruction tomography. J Nucl Med 1975; 16:210–224.
2. Hoffman EJ, Phelps ME. An analysis of some of the physical aspects of positron transaxial tomography. Comput Biol Med 1976; 6:345–360.
3. Phelps ME, Hoffman EJ, Coleman RE, et al. Tomographic images of blood pool and perfusion in brain and heart. J Nucl Med 1976; 17:603-612.
4. Hoffman EJ, Phelps ME, Weiss ES, et al. Transaxial tomographic imaging of canine myocardium with 11C-palmitic acid. J Nucl Med 1977; 18:57–61.
5. Phelps ME, Hoffman EJ, Huang SC, Kuhl DE. Positron tomography: "in vivo" autoradiographic approach to measurement of cerebral hemodynamics and metabolism. Acta Neurol Scand Suppl 1977; 64:446–447.
6. Phelps ME, Huang SC, Hoffman EJ, Selin C, Sokoloff L, Kuhl DE. Tomographic measurement of local cerebral glucose metabolic rate in humans with (F-18)2-fluoro-2-deoxy-D-glucose: Validation of method. Ann Neurol 1979; 6:371–388.
7. Phelps ME. PET: The merging of biology and imaging into molecular imaging. J Nucl Med 2000; 41:661–681.
8. Phelps ME, Hoffman EJ, Huang S-C, Ter-Pogossian MM. Effect of positron range on spatial resolution (in radioisotope scanning). J Nucl Med 1975; 16:649–652.
9. Levin CS, Hoffman EJ. Calculation of positron range and its effect on the fundamental limit of positron emission tomography system spatial resolution. Phys Med Biol 1999; 44:781–799.
10. Laforest R, Rowland DJ, Welch MJ. MicroPET imaging with nonconventional isotopes. IEEE Trans Nucl Sci 2002; 49:2119–2126.
11. Hoffman EJ, Phelps ME. An analysis of some of the physical aspects of positron transaxial tomography. Comput Biol Med 1976; 6:345–360.
12. Moses WW, Derenzo SE. Empirical observation of resolution degradation in positron emission tomographs utilizing block detectors. J Nucl Med 1993; 34:101P.
13. Casey ME, Nutt R. A multicrystal two dimensional BGO detector system for positron emission tomography. IEEE Trans Nucl Sci 1986; NS-33:460–463.
14. Lecomte R, Cadorette J, Richard P, Rodrigue S, Rouleau D. Design and engineering aspects of a high-resolution positron tomograph for small animal imaging. IEEE Trans Nucl Sci 1994; 41:1446–1452.
15. Tai Y-C, Laforest R. Instrumentation aspects of animal PET. Ann Rev Biomed Engg 2005; 7:255–285.
16. Kak AC, Slaney M. IEEE/Engineering in Medicine and Biology Society. Principles of Computerized Tomographic Imaging. New York: IEEE Press, 1988; 329pp.
17. Natterer F, Wübbeling F. Mathematical Methods in Image Reconstruction. Philadelphia: Society for Industrial and Applied Mathematics, 2001.
18. Ollinger JM, Fessler JA. Positron-emission tomography. IEEE Signal Proces Mag 1997; 14:43–55.
19. Leahy RM, Qi JY. Statistical approaches in quantitative positron emission tomography. Stat Comput 2000; 10:147–165.
20. Cherry SR, Meikle SR, Hoffman EJ. Correction and characterization of scattered events in three-dimensional PET. J Nucl Med 1993; 34:671–678.
21. Bergstrom M, Eriksson L, Bohm C, Blomqvist G, Litton J. Correction for scattered radiation in a ring detector positron camera by integral transformation of the projections. J Comput Assist Tomogr 1983; 7:42–50.
22. Levin CS, Dahlbom M, Hoffman EJ. A Monte Carlo correction for the effect of Compton scattering in 3-D PET brain imaging. IEEE Trans Nucl Sci 1995; 42:1181–1185.
23. Bentourkia M, Msaki P, Cadorette J, Lecomte R. Nonstationary scatter subtraction-restoration in high-resolution PET. J Nucl Med 1996; 37:2040–2046.

24. Ollinger JM. Model-based scatter correction for fully 3D PET. Phys Med Biol 1996; 41: 153–176.

25. Hoffman EJ, Huang S-C, Phelps ME, Kuhl DE. Quantitation in positron emission computed tomography. IV. Effect of accidental coincidences. J Comput Assist Tomogr 1981; 5: 391–400.

26. Strother SC, Casey ME, Hoffman EJ. Measuring PET scanner sensitivity: Relating countrates to image signal-to-noise ratios using noise equivalent counts. IEEE Trans Nucl Sci 1990; 37:783–788.

27. Knoll GF. Radiation Detection and Measurement. New York: Wiley, 1989; 754pp.

28. Huang SC, Hoffman EJ, Phelps ME, Kuhl DE. Quantitation in positron emission computed tomography: 2. Effects of inaccurate attenuation correction. J Comput Assist Tomogr 1979; 3:804–814.

29. Huang SC, Carson RE, Phelps ME, Hoffman EJ, Schelbert HR, Kuhl DE. A boundary method for attenuation correction in positron computed tomography. J Nucl Med 1981; 22:627–637.

30. Beyer T, Kinahan PE, Townsend DW, Sashin D. The use of X-ray CT for attenuation correction of PET data. Nuclear Science Symposium and Medical Imaging Conference, 1994. IEEE Conference Record (Cat. No.94CH35762) 1995; 4:1573–1577.

31. Kinahan PE, Townsend DW, Beyer T, Sashin D. Attenuation correction for a combined 3D PET/CT scanner. Med Phys 1998; 25:2046–2053.

32. Casey ME, Hoffman EJ. Quantitation in positron emission computed tomography: 7. A technique to reduce noise in accidental coincidence measurements and coincidence efficiency calibration. J Comput Assist Tomogr 1986; 10:845–850.

33. Hoffman EJ, Guerrero TM, Germano G, Digby WM, Dahlbom M. PET system calibrations and corrections for quantitative and spatially accurate images. IEEE Trans Nucl Sci 1989; 36:1108–1112.

34. Casey ME, Gadagkar H, Newport D. A component based method for normalization in volume PET. Proceedings of the 1995 International Meeting on Fully Three-Dimensional Image Reconstruction in Radiology and Nuclear Medicine, 1995; 67–71.

35. Badawi RD, Marsden PK. Developments in component-based normalization for 3D PET. Phys Med Biol 1999; 44:571–594.

36. Bai B, Li Q, Holdsworth CH, et al. Model-based normalization for iterative 3D PET image reconstruction. Phys Med Biol 2002; 47:2773–2784.

37. Jeavons AP, Chandler RA, Dettmar CAR. A 3D HIDAC-PET camera with sub-millimetre resolution for imaging small animals. IEEE Trans Nucl Sci 1999; 46:468–473.

38. Cutler PD, Cherry SR, Hoffman EJ, Digby WM, Phelps ME. Design features and performance of a PET system for animal research. J Nucl Med 1992; 33:595–604.

39. Watanabe M, Uchida H, Okada H, et al. A high resolution PET for animal studies. IEEE Trans Med Imag 1992; 11:577–580.

40. Marriott CJ, Cadorette JE, Lecomte R, Scasnar V, Rousseau J, van Lier JE. High-resolution PET imaging and quantitation of pharmaceutical biodistributions in a small animal using avalanche photodiode detectors. J Nucl Med 1994; 35:1390–1396.

41. Lecomte R, Cadorette J, Rodrigue S, et al. Initial results from the Sherbrooke avalanche photodiode positron tomograph. IEEE Trans Nucl Sci 1996; 43:1952–1957.

42. Cherry SR, Shao Y, Silverman RW, et al. MicroPET: A high resolution PET scanner for imaging small animals. IEEE Trans Nucl Sci 1997; 44:1161–1166.

43. Watanabe M, Okada H, Shimizu K, et al. A high resolution animal PET scanner using compact PS-PMT detectors. IEEE Trans Nucl Sci 1997; 44:1277–1282.

44. Bloomfield PM, Myers R, Hume SP, Spinks TJ, Lammertsma AA, Jones T. Three-dimensional performance of a small-diameter positron emission tomograph. Phys Med Biol 1997; 42:389–400.

45. Weber S, Terstegge A, Herzog H, et al. The design of an animal PET: Flexible geometry for achieving optimal spatial resolution or high sensitivity. IEEE Trans Med Imaging 1997; 16: 684–689.

46. Del Guerra A, Di Domenico G, Scandola M, Zavattini G. YAP-PET: First results of a small animal positron emission tomograph based on YAP:Ce finger crystals. IEEE Trans Nucl Sci 1998; 45:3105–3108.

47. Ziegler SI, Pichler BJ, Boening G, et al. A prototype high-resolution animal positron tomograph with avalanche photodiode arrays and LSO crystals. Eur J Nucl Med 2001; 28:136–143.

48. Seidel J, Vaquero JJ, Green MV. Resolution uniformity and sensitivity of the NIH ATLAS small animal PET scanner: Comparison to simulated LSO scanners without depth-of-interaction capability. IEEE Trans Nucl Sci 2003; 50:1347–1350.

49. Tai Y, Chatziioannou A, Yang Y, et al. MicroPET II: design, development and initial performance of an improved microPET scanner for small-animal imaging. Phys Med Biol 2003; 48:1519–1537.

50. Miyaoka RS, Janes ML, Lee K, Park B, Kinahan PE, Lewellen TK. Development of a Prototype Micro Crystal Element Scanner (MiCES): QuickPET II. Mol Imaging 2005; 4:117–127.

51. Tai Y-C, Ruangma A, Rowland D, et al. Performance Evaluation of the microPET-Focus: A third generation microPET scanner dedicated to animal imaging. J Nucl Med 2005; 46:455–463.

52. Laforest R, Longford D, Siegel S, Newport D, Yap J. Performance Evaluation of the microPET-Focus - F120. Presented at 2004 IEEE Nuclear Science Symposium Conference Record, Rome, Italy, 2005.

53. Cherry SR, Shao Y, Tornai MP, Siegel S, Ricci AR, Phelps ME. Collection of scintillation light from small BGO crystals. IEEE Trans Nucl Sci 1995; 42:1058–1063.

54. Melcher CL, Schweitzer JS. Cerium-doped lutetium oxyorthosilicate: A fast, efficient new scintillator. IEEE Trans Nucl Sci 1992; 39:502–505.

55. Siegel S, Silverman RW, Yiping S, Cherry SR. Simple charge division readouts for imaging scintillator arrays using a multi-channel PMT. IEEE Trans Nucl Sci 1996; 43:1634–1641.

56. Miyaoka RS, Kohlmyer SG, Lewellen TK. Performance characteristics of micro crystal element (MiCE) detectors. IEEE Trans Nucl Sci 2001; 48:1403–1407.

57. Rouze NC, Schmand M, Siegel S, Hutchins GD. Design of a small animal PET imaging system with 1 microliter volume resolution. IEEE Trans Nucl Sci 2004; 51: 757-63

58. Chatziioannou AF, Cherry SR, Shao Y, et al. Performance evaluation of microPET: a high-resolution lutetium oxy-orthosilicate PET scanner for animal imaging. J Nucl Med 1999; 40:1164–1175.

59. Tai YC, Chatziioannou A, Siegel S, et al. Performance evaluation of the microPET P4: A PET system dedicated to animal imaging. Phys Med Biol 2001; 46:1845–1862.

60. Knoess C, Siegel S, Smith A, et al. Performance evaluation of the microPET R4 PET scanner for rodents. Eur J Nucl Med Mol Imaging 2003; 30:737–747.

61. Yang Y, Tai Y-C, Siegel S, et al. Optimization and performance evaluation of the microPET II scanner for in vivo small-animal imaging. Phys Med Biol 2004; 49:2527–2545.

62. Xu EZ, Mullani NA, Gould KL, Anderson WL. A segmented attenuation correction for PET. J Nucl Med 1991; 32:161–165.

63. deKemp RA, Nahmias C. Attenuation correction in PET using single photon transmission measurement. Med Phys 1994; 21:771–778.

64. Watson CC. New, faster, image-based scatter correction for 3D PET. IEEE Trans Nucl Sci 2000; 47:1587–1594.

65. Qi J, Leahy RM, Hsu C, Farquhar TH, Cherry SR. Fully 3D Bayesian image reconstruction for the ECAT EXACT HR+. IEEE Trans Nucl Sci 1998; 45:1096–1103.

66. Chatziioannou A, Qi J, Moore A, et al. Comparison of 3-D maximum a posteriori and filtered backprojection algorithms for high-resolution animal imaging with microPET. IEEE Trans Med Imaging 2000; 19:507–512.

67. Yang Y, Rendig S, Siegel S, Newport DN, Cherry SR. Cardiac PET imaging in mice with simultaneous cardiac and respiratory gating. Phys Med Biol 2005; 50:2979–2989.

68. Laforest R, Sharp TL, Engelbach JA, et al. Measurement of input functions in rodents: Challenges and solutions. Nucl Med Biol 2005; 32:679–685.

3

SPECT/microSPECT Imaging

Lars R. Furenlid, Zhonglin Liu, and Harrison H. Barrett
Department of Radiology and Optical Sciences Center, Tucson, Arizona, U.S.A.

INTRODUCTION

Single photon emission computed tomography (SPECT) and microSPECT are valuable tools for molecular imaging in cardiovascular applications. They have certain advantages over positron emission tomography (PET) such as more readily available isotopes that have longer half-lives, and there are potential advantages in spatial resolution, which in PET is limited by positron range and photon non-colinearity. The corresponding disadvantage, however, is the need to use an image-forming structure such as a collimator, pinhole, or coded aperture. The fundamental concepts of emission tomography are nicely reviewed in the recent volume by Wernick and Aarsvold (1).

Full realization of the potential of SPECT/micro-SPECT for molecular imaging requires advances in image formation, gamma-ray detectors, electronics for data acquisition, and image-reconstruction algorithms. Research in our own laboratory, the Center for Gamma-ray Imaging (CGRI) at the University of Arizona, and other academic institutions is dedicated to advancing the state of the art in all of these respects. The purpose of this chapter is to summarize the approaches used at CGRI and other laboratories, describe several representative small-animal SPECT systems, and demonstrate their applicability to cardiovascular molecular imaging.

TECHNOLOGICAL REQUIREMENTS

Objective Image Quality

It is the premise of the research at CGRI that any meaningful metric of image quality must specify the information that one wants to extract from the image (the task), how that information will be extracted (the observer) and the statistical properties that limit the extraction. Tasks of interest can be either classification (including defect detection) or estimation (also called quantitation). The distinction is that for a classification task the output is one or more labels (e.g., normal vs. abnormal tissue, defect present or absent), while in an estimation task the output is one or more numbers.

Classification tasks in nuclear cardiovascular imaging include detection of perfusion defects, detection of wall-motion abnormalities, classification of myocardium as viable vs. nonviable, distinguishing transmural from subendocardial infarcts, detecting the presence and type of atherosclerotic plaque, and distinguishing apoptotic from oncotic myocardial necrosis. Estimation tasks in nuclear cardiology include measurement of ejection fraction, absolute ventricular volumes or blood flow, quantitation of wall motion, or estimation of the volume of an infarct.

Observers for classification tasks are usually humans, and their performance can be measured by psychophysical studies and receiver operating characteristic (ROC) analysis. Considerable progress has also been made in the development of mathematical model observers that either predict human performance or estimate the maximum possible performance (the ideal observer).

Statistical limitations that influence observer performance include Poisson noise in the data, excess electronic noise in some kinds of detectors, and random variability in the objects themselves. For a review of all of these aspects of image quality and guidance on implementing objective image quality assessment methodology, see chapter 14 in Ref. (2).

Spatial Resolution

Small animals are, of course, small, so the need for high spatial resolution is intuitive. More rigorously, performance on almost any clinical or biomedical task is improved by higher spatial resolution. The detectability of small lesions or defects is a strong function of the spatial resolution of the image-forming aperture and the image detector, but the contribution of the image-reconstruction algorithm is much less important. The performance of the ideal observer is invariant to the algorithm, virtually by definition, unless the algorithm removes information. Performance of a human observer is influenced by the algorithm, which serves to match the raw data to the characteristics of the human visual and cognitive system, but the effect is usually small.

Similarly, estimation of small-scale parameters, such as activity in a small region of interest or small wall motions, is significantly enhanced by improved spatial resolution. Indeed, it becomes impossible in principle to find an unbiased estimator of parameters associated with details smaller than the resolution, and in those situations the usual metrics of estimation performance, such as mean-square error, are not even defined (see chapter 15 in Ref. 2).

For both classification and estimation tasks, the relevant measure of spatial resolution is usually volumetric. One wants to know what volume is encompassed under the 3D point spread function of the system. Current small-animal SPECT systems have linear resolutions on the order of 1 mm, which corresponds to a volume resolution of about 1 μL if the resolution is isotropic; such systems are rightfully referred to as microSPECT. A major goal at CGRI is to improve the linear resolution of small-animal SPECT to 100 μm, which would correspond to a volume resolution of 1 nL and justify the term nanoSPECT (3).

Temporal Resolution

Temporal resolution is needed for many clinical imaging applications, but none more so than cardiac imaging. The ideal heart imager would have a frame time short compared to the duration of a cardiac cycle and provide a snapshot 3D image for each frame. A temporal sequence of 3D images would then constitute a 4D data set.

The usual alternative to true 4D imaging is gated imaging, in which it is implicitly assumed that each heartbeat is identical in motion (4) and that a modest number of time slices (<24) adequately describe the cardiac cycle. Though the resulting image display appears to be 4D, deviations from the assumed periodic motion cannot be discerned, and the passage of the initial tracer bolus through the heart cannot be seen. Current clinical first-pass studies of ventricular function are not tomographic (5).

The need for high temporal resolution is greater for small animals than for humans because of their higher heart rate. Moreover, molecular imaging studies often require investigation of washout rates and other kinetic parameters. The temporal resolution of the system needs to be high in order to extract accurate estimates of these parameters.

Sensitivity

Photon-collection efficiency is always important in nuclear medicine, but again it is especially critical in nuclear cardiovascular imaging. If one wants to acquire enough photons for good 3D reconstruction within a small fraction of a heartbeat, or even within a tracer washout time, then the imaging aperture must subtend a reasonable total solid angle from each point in the object.

Gated acquisition alleviates the sensitivity problem but does not eliminate it; if the total number of detected photons is divided over, say, 20 time slices for each cardiac cycle, then 20 times the sensitivity is needed to get the same statistical accuracy as one would get in a time-integrated (3D) image with the same total acquisition time.

OPPORTUNITIES FOR IMPROVING PERFORMANCE

Image Formation

Image formation in SPECT is usually based on multi-bore collimators or pinholes. The latter are particularly advantageous for small-animal studies since they can be placed close to the subject and therefore subtend a relatively large solid angle for a given spatial resolution. Moreover, many pinholes can be used, increasing the total solid angle still further (6). Another feature of small-animal imaging is that low-energy isotopes can be used with negligible self-absorption within the body. This opens up the possibility of using glancing angle reflective or diffractive focusing optics which do not work at higher energies.

Detectors

For any task, continuing improvement in performance ultimately requires improvements in the image detector. The single detector parameter that has the most influence on task performance, for a properly optimized system, is the detector space-bandwidth product, defined basically as the number of resolvable elements in the detector area. Increased space-bandwidth product can be obtained by increasing the detector area, improving its intrinsic resolution, or both.

One attractive way of increasing the detector space-bandwidth product is to use a large number of small, modular detectors. A module, like an individual head on a multi-head clinical camera, is electrically, optically and mechanically independent of the other modules. All else being equal, a system of N modules has N times the sensitivity, N times the count-rate capability and N times the space-bandwidth product of a system that uses a single module.

Pinholes and High-Resolution Detectors

A little-recognized aspect of multiple-pinhole imaging is that it allows us to gain sensitivity without sacrificing final image resolution, provided we can improve the detector performance (7). If we develop a detector with improved resolution, we can use it with a smaller pinhole or finer collimator and improve the final image resolution at the expense of sensitivity, but we can also use it to improve sensitivity. To do so, we move the detector closer to the pinhole, thereby reducing the magnification and leaving room for more pinholes without running into problems from overlapping (multiplexed) images. Even though smaller pinholes are then needed for the same final resolution, the number of pinholes increases faster than the area of each decreases, so the overall sensitivity is actually increased.

SYSTEMS FOR microSPECT

An excellent survey of the current state of the art in microSPECT systems is provided by Schelbert and Glover (8). Systems can be divided into three basic categories: adaptations of commercial clinical Anger cameras, dedicated small-animal imagers that use a small number of high-resolution cameras, and stationary systems with many cameras that require no (or only limited) motions for acquisition of tomographic data sets. All of these approaches can give satisfactory images of mouse- and rat-scale cardiac tissues, especially when there is plenty of activity and long acquisition times are possible. The advantages of stationary systems are the increase in sensitivity from acquiring all views at once, the ability to collect and reconstruct 4D data from non-periodic processes, and that direct system calibration is easily accomplished.

Systems Developed at CGRI
Detectors
The workhorse gamma-ray detectors used at CGRI are photomultipliers (PMT)-based scintillation cameras (modcams) (9,10) and cadmium zinc telluride (CZT) detector arrays (11,12). Both are modular, and both use true list-mode readout in which all collected information about each event is stored as an entry in a list. In both cases, the list-mode data are used to estimate the energy and position of the interaction event, including possibly the depth of interaction. Maximum-likelihood (ML) estimation methods are used based on carefully measured calibration data.

The original modcam used a 10 cm × 10 cm × 5 mm NaI(Tl) crystal and four square 5 cm × 5 cm PMTs (9). A more recent version uses a 12 cm × 12 cm × 5 mm NaI(T1) crystal and nine round PMTs, each 3.8 cm in diameter, in a 3 × 3 array (13). Event detection is carried out by algorithms implemented in Field Programmable Gate Arrays (FPGAs) on list-mode event processor boards. A list-mode entry consists of the 9 PMT signals for each event, a time stamp to indicate to 30 nsec precision when the event occurred, and a tag to identify the module in which it occurred (13).

The CZT detector consists of a 26.9 mm × 26.9 mm × 2 mm slab of CZT metallized with a continuous electrode on one side and a 64 × 64 array of pixel electrodes on the other side. The pixel electrodes are 330 μm × 330 μm, and the pixel pitch is 380 μm. The pixelated side is bonded with indium cold welds to an application-specific integrated circuit (ASIC) with

input pads on the same pitch as the pixels. Each ASIC input is connected to a gated integrator that integrates the leakage current and any current resulting from a gamma ray on a small (150 fF) capacitor. The array integrates for 1 msec and is then read out, whether or not a gamma-ray event has occurred (14).

During readout, each frame is parsed by an FPGA to detect pixel neighborhoods with signals above some threshold, which are tentatively assumed to be associated with gamma-ray events. The list-mode entry then consists of the address of the pixel that had the largest signal in its neighborhood, the value of the signal on that pixel and its eight neighbors, a frame counter that specifies the event time to 1 msec precision, and a module tag. Thus, the data-acquisition architecture utilized for the CZT detectors has the same structure as used with the modcam, and all subsequent processing can employ common techniques (15).

Another detector under study at CGRI is a modular scintillation camera based on the Hamamatsu H8500 multi-anode PMT (MAPMT) (16), which is essentially an 8 × 8 array of PMTs in a single glass envelope. This MAPMT and others of similar design are used in a number of gamma-camera configurations by other laboratories, typically in conjunction with segmented crystal arrays. We are investigating its potential as a sensor for continuous scintillator crystals when combined with ML position-estimation techniques. List-mode data storage in this case requires some pre-processing to reduce the 64 PMT outputs to a smaller number of signals (called sufficient statistics) that convey the essential information about each event.

Finally, we are investigating the possibility of using new low-noise CCD cameras for reading out scintillator crystals simply by imaging the light onto the CCD. If the gamma-ray flux is sufficiently small, video frames can be parsed for events, just as we do with the CZT detectors, and store the signals from some neighborhood in a list-mode entry for subsequent ML processing. Simulation studies show that quite respectable energy resolution and depth-of-interaction resolution should be obtainable this way, and the lateral resolution can be excellent since it is based on direct optical imaging. It may also prove advantageous to work in an integration mode.

Systems Based on Semiconductor Arrays
Several imaging systems based on CZT arrays have been constructed at CGRI. The simplest, called the Spot Imager, uses a parallel-bore collimator with 260 μm × 260 μm square bores on a 380-μm pitch, matching that of the CZT pixel array. The Spot Imager serves as a handheld device for small-area clinical imaging, but it has also been built into a dual-modality CT/SPECT mouse imager as shown in Figure 1 (17). The mouse is rotated around a vertical axis to acquire projection data

Figure 1 Dual-modality CT/SPECT system based on a CZT detector array.

for both modalities, and the SPECT component delivers sub-millimeter spatial resolution (18). This makes it possible to visualize detailed functional features, such as the renal calices in mouse kidneys shown in Figure 2, that have not been accessible via gamma-ray imaging techniques before. However, since the system has only

a single camera, the sensitivity of our CT/SPECT mouse imager is limited and tomographic acquisitions typically require 30 minutes.

A more sensitive mouse imager, called SemiSPECT, uses a ring of eight of the CZT arrays and 0.5 mm diameter pinholes. The system, shown in Figure 3, acquires eight projections at once and requires only a set of minor motions to acquire enough angular samples for tomographic reconstruction. The performance of the imager is enhanced by the direct measurement of the point spread function as a function of voxel location. A very small calibration source is scanned throughout the imager's field of view in a regular Cartesian three-dimensional grid. The acquired calibration data are then processed to form a system imaging matrix that can be used for iterative 3D reconstruction. This procedure compensates for any imperfections in the system's fabrication and results in a partial elimination of pinhole blur from the final tomographic images.

Systems Based on Modular Scintillation Cameras

FastSPECT I and II are two complete systems based on the modcams. FastSPECT I was originally designed as a clinical brain imager. It uses 24 of the older 4-PMT cameras arranged in two rings of 11 and 13. The system implements ML position estimation via fast precomputed lookup tables that make the system capable

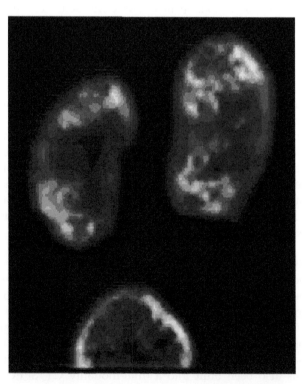

Figure 2 Volumetric rendering of ^{99m}Tc glucarate image of mouse kidney taken with dual-modality CT/SPECT system.

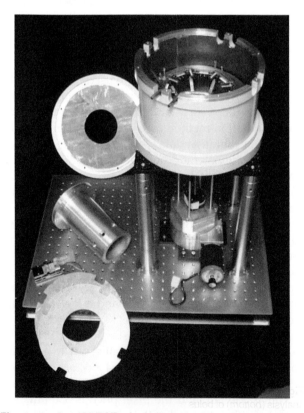

Figure 3 SemiSPECT: A dedicated mouse imager with 8 CdZnTe detectors.

of forming image frames in 50-ms intervals. In work with a torso phantom built around a cardio-west artificial heart, we were able to demonstrate first-pass tomographic imaging of the passage of a bolus of activity through the right ventricle as shown in Figure 4.

FastSPECT I was converted to a microSPECT scanner via construction of a new aperture, demonstrating that the difference between SPECT and micro-SPECT is a matter of optical design if the detector technology employed in the system has sufficient space-bandwidth product. The tradeoff is a smaller field of view as a result of the higher magnification.

FastSPECT II, shown in Figure 5, uses 16 of the newer 9-PMT cameras. Both FS I and FS II are capable of acquiring projection data without any motion of the object or detection system, making them true 4D imaging systems. FastSPECT II allows for slight rotation of the animal ($\pm 7.5°$) to increase the angular sampling. Various pinhole apertures can be used, depending on the imaging task, that yield magnifications as high as 30:1 for image resolution in the 100–200 µm range.

Systems Based on CCD Cameras

As an initial test of the feasibility of using CCD cameras to read out scintillation light, we have used a Roper

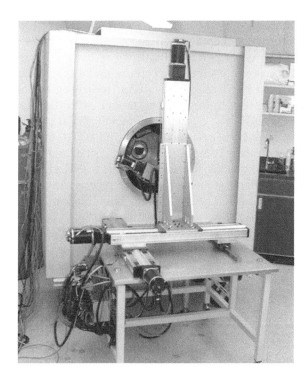

Figure 5 FastSPECT II, shown with the mechanical stage used for calibration and animal handling in place.

Scientific camera cooled to $-100°C$ and a pair of F/2 lenses operated at 1:1 magnification to image the scintillation light. To date, no attempt has been made to count individual gamma-ray photons or to do energy resolution, but excellent images have been obtained in a simple integrating mode. Since the same system is capable of detecting fluorescent or bioluminescent light, we refer to it as LumiSPECT.

CARDIOVASCULAR IMAGING APPLICATIONS

The successful development of high-resolution small-animal imaging systems provides a powerful new means for the basic research in cardiovascular science. Many experimental procedures that could previously only be employed in large-animal cardiac models now can be used in rat or mouse models. Large animals, such as canines, have often been used for in vivo experimental imaging to study cardiovascular disease. The disadvantages of using large animals are obvious: the cost is high, and the researcher must master complex surgical techniques. On the contrary, the cost of a small-animal heart model is much lower, and the surgery is simpler, yet with a higher success rate. More importantly, many gene-therapy strategies for the treatment of cardiac failure and long-term myocardial protection can be studied in rodent models with noninvasive imaging.

Numerous cardiac studies on mice and rats are underway at CGRI. Rat heart models have been well

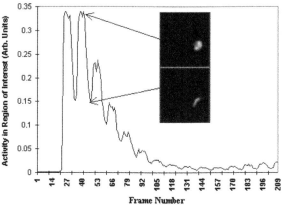

Figure 4 First pass FastSPECT tomographic images and ROI analysis (*bottom*) of bolus of activity through right ventricle of human torso phantom (*top*) incorporating artificial heart. *Source*: Irene Pang, University of Arizona.

established and applied in our laboratory to investigate myocardial ischemia-reperfusion injury in vivo with our high-resolution SPECT imaging systems. The acquisition procedures, imaging process, and reconstruction in the rat heart model have been well programmed. Meanwhile, a mouse heart model with ischemia-reperfusion is also under development. To date, most of them have been performed on FastSPECT I, but FastSPECT II offers some advantages in spatial and temporal resolution. We are in the process of designing a protocol for non-tomographic imaging of a Langendorff isolated perfused heart using the Spot Imager.

High-quality cardiac images can be achieved on the FastSPECT systems with only 5–10 minute acquisitions, 30–120 minutes following intravenous injection of 99mTc-labeled sestamibi or tetrofosmin. As shown in Figure 6, in the short-axis and vertical and horizontal

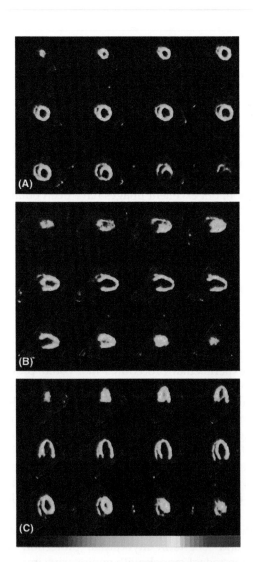

Figure 6   Representative FastSPECT 99mTc-sestamibi images in normal rat heart 2 hours after injection. (**A**) Short-axis slices. (**B**) Vertical-axis slices. (**C**) Horizontal-axis slices.

long-axis images, the left and right ventricular myocardium can be imaged fully and clearly with uniform distribution of 99mTc-sestamibi in all images. The extent of the left ventricular perfusion defect on 99mTc-sestamibi images reflects the size of the myocardial infarction in patients with heart attack. Similar to what is reported in clinical studies, 99mTc-sestamibi FastSPECT images of infarcts in the ischemic-reperfused rat myocardium have been demonstrated to have a significant correlation with biochemical staining measurements (see Fig. 7) (19).

Ideally, an acute ischemic-reperfused rat/mouse heart model should be closely related to acute clinical heart attack and revascularization cases. In the small-animal acute myocardial infarct models, the duration of anesthesia and the volume of imaging agent injected intravenously are limited. Thus, fast, repeated imaging combining high sensitivity and spatial resolution is required so that the ongoing myocardial injury produced by ischemia-reperfusion can be detected effectively. With the advantages of our stationary SPECT systems, the dynamic cardiac images can be acquired using similar procedures to those required by clinical physicians. The myocardial radioactivity can be quantified accurately and the washout kinetics of radiolabeled agents can be determined effectively.

As an example of cardiac molecular imaging with FastSPECT, Figure 8 shows dynamic images of 99mTc-glucarate (GLA), an agent developed by Molecular Targeting, Inc., in a rat heart subjected to 30 min ischemia followed by 30 min reperfusion. Serial images of the same short-axis tomographic slice from 1 min to 120 min post-injection are displayed. The blood pool was well visualized in the first minute. In the second minute after injection, the radioactivity began to localize in the stenotic zone in the lateral wall of the left ventricle. Good infarct definition was achieved 10 min after injection, and persisted for at least 2 hrs (20). The location and size of the hot spot was consistent with the unstained area on TTC (see Fig. 9).

99mTc-glucarate can mark nonviable regions by hot spot imaging in myocardium with acute necroses (21–24). The imaging results obtained from the rat heart models indicate that the severity of myocardial injury induced by different durations of ischemia following reperfusion can be assessed using 99mTc-glucarate noninvasively and quantitatively (20). 99mTc-glucarate imaging may not only provide an imaging tool to diagnose equivocal myocardial infarct in patients with heart attacks and allow differentiation of acute from recent infarcts, but also direct the use of thrombolytic therapy. Quantitative analyses on dynamic images with 99mTc-glucarate would make it possible to identify myocardial acute necrosis earlier and more accurately, and provide a unique, noninvasive tool for evaluation of patient prognosis. The results

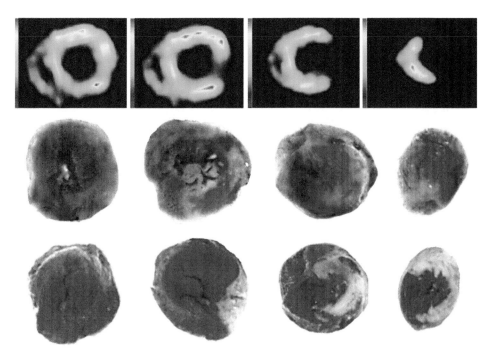

Figure 7 Perfusion defects were exhibited on ^{99m}Tc-sestamibi images in an ischemic-reperfused heart (*top row*), which were consistent with myocardial ischemic area at risk evaluated by Evans blue (unstained by blue dye) (*middle row*) and infarct myocardium determined by TTC staining (unstained by TTC) (*bottom row*).

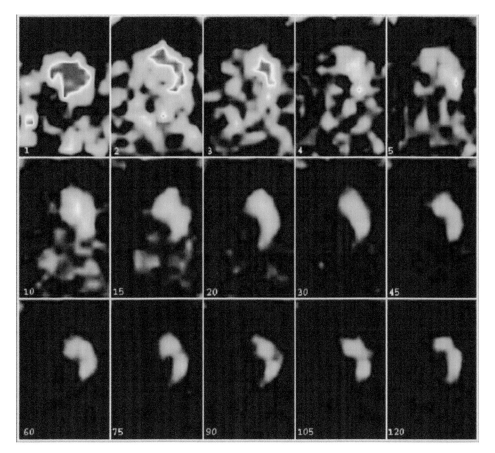

Figure 8 Serial glucarate images of a single tomographic slice in a rat with ischemia-reperfusion injury. (*Upper left*) 1 min p.i. (*Lower right*) 2 hrs p.i. *Abbreviation*: p.i., post injection.

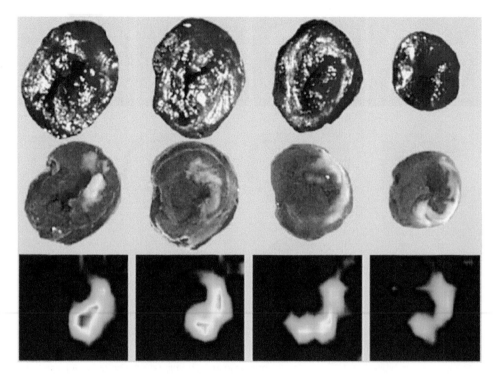

Figure 9 Comparison of SPECT images with postmortem staining for a rat with 30-minute ischemia. (*Top*) Evans blue stain showing perfused area as blue. (*Middle*) TTC stain showing damaged myocardium as light color. (*Bottom*) ^{99m}Tc-glucarate images.

of the small-animal imaging with ^{99m}Tc-glucarate provide new evidence of the suitability of microSPECT for performing preclinical investigations of cardiovascular radiopharmaceuticals.

Small-animal SPECT may also play a significant role in exploring strategies for cardioprotection. Using selected myocardial imaging agents, we have adopted rat/mouse heart models to assess cardioprotective effects of myocardial ischemic preconditioning (IPC) treatment. Our recent results show that significant tolerance to myocardial ischemia-reperfusion injury, as determined by biochemical assay, can be induced with an IPC protocol in the rat heart model. IPC significantly decreased the infarct size as measured by ^{99m}Tc-labeled glucarate imaging with high-resolution dynamic SPECT. Furthermore, we applied ^{99m}Tc-glucarate in vivo imaging to determine whether the adenosine A1 receptor is involved in the cardioprotection induced by the IPC protocol.

Adenosine has been recognized to be one of mediators involved in IPC cardioprotection by stimulating the adenosine A1 receptor. The vasodilatation of adenosine-mediating IPC has been extensively described in a variety of animal models (25,26). However, for rat hearts such a role is controversial. We hypothesized that the adenosine A1 receptor is involved in protection by IPC, in which case a receptor agonist, 2-chloro-N6-cyclopentyladenosine (CCPA), should give a protection similar to IPC, while

the receptor antagonist, 8-(p-sulfophenyl)-theophylline (SPT), should block protection by IPC. We quantified the kinetic washout of ^{99m}Tc-glucarate from the ischemic-reperfused zone and normal myocardium in rat hearts subjected to varied treatments. The hot spot of ^{99m}Tc-glucarate showed significantly higher fractional washout in the heart with IPC and CCPA compared to the control heart without IPC (No-IPC) and SPT. As a result, the hot spots in IPC and CCPA exhibited significantly lower radioactive retention (% peak) than in No-IPC and SPT, respectively (see Fig. 10). The hot spots in CCPA demonstrated similar radioactive retention as in IPC. The increased hot-spot retention in the SPT hearts was as high as in the No-IPC heart. Our data indicate that cardioprotection by IPC could be simulated by the adenosine receptor agonist CCPA, or blocked by the antagonist SPT.

As evident from their major role in above studies using ischemic-reperfused rat heart models, the small-animal SPECT systems developed in our laboratory provide flexible molecular imaging approaches to important problems due to their high spatial resolution and fast dynamic acquisition. Each system can be applied to the investigation of the kinetics of cardiovascular radiopharmaceuticals, the success of cardioprotective and repair strategies, and other detection and estimation tasks. In order to expand the cardiovascular applications of the imaging systems that have been developed and pave the way for new collaborative

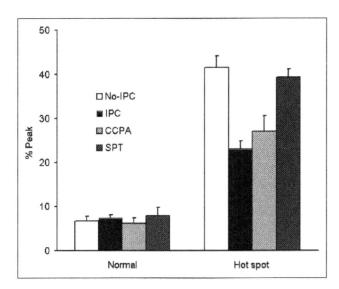

Figure 10 Fractional retention of GLA in the infarcted (hot spot) and normal region for 30-minute ischemia alone (No-IPC), with IPC, with the adenosine A1 receptor agonist (CCPA) and antagonist (SPT).

research in cardiology, we shall continue to develop specialized small-animal models for basic investigations, acquire and evaluate new radiopharmaceuticals, and develop new methods of data analysis.

REFERENCES

1. Wernick MN, Aarsvold JN, eds. Emission Tomography: The Fundamentals of PET and SPECT. Philadelphia: Elsevier Mosby, 2004.

2. Barrett HH, Myers KJ. Foundations of Image Science. Hoboken: John Wiley and Sons, 2004.

3. Beekman FJ, Vastenhouw J. Design and simulation of a high-resolution stationary SPECT system for small animals. Phys Med Bio 2004; 49:4579–4592.

4. Germano G, Berman S. Regional and global ventricular function and volumes from single-photon emission computed tomography perfusion imaging. In: Zaret BL, Beller GA, eds. Clinical Nuclear Cardiology, chap. 12. Philadelphia: Elsevier Mosby, 2004.

5. Nichols K, DePuey EG, Gooneratne N, Salensky H, Fredman M, Cochoff S. First pass ventricular ejection fraction using a single-crystal nuclear camera. J Nucl Med 1994; 35(8):1292–1300.

6. Barrett HH, Swindell W. Radiological Imaging. New York: Academic Press, 1981.

7. Kupinski MA, Barrett HH. Detectors for small animal SPECT I. Small-Animal SPECT Imaging. New York: Springer Science 2005.

8. Schelbert HR, Glover DK. State-of-the-art instrumentation for positron emission tomography and single-photon emission computed tomography imaging in small animals. In: Zaret BL, Beller GA, eds. Clinical Nuclear Cardiology, chap. 10. Philadelphia: Elsevier Mosby, 2004.

9. Milster TD, Aarsvold JN, Barrett HH, et al. A full field modular gamma camera. J Nucl Med 1984; 31(5): 632–639.

10. Milster TD, Aarsvold JN, Barrett HH, Landesman AL, Mar LS, Patton DD, Roney TJ, Rowe RK and Seacat 3rd RH. A full field modular gamma camera. J Nucl Med 1990; 31(5):632–639.

11. Marks DG, Barber HB, Barrett HH, Dereniak EL, Eskin JD, Matherson KJ, Woolfenden JM, Young ET, Augustine FL, Hamilton WL, Venzon JE, Apotovsky EA, Doty FP. A 48 × 48 CdznTe array with multiplexer readout. IEEE Trans Nucl Sci 1996; 43(3):1253–1259.

12. Matherson KJ, Barber HB, Barrett HH, Eskin JD, Dereniak EL, Marks DG, Woolfenden JM, Young ET, Augustine FL. Progress in the development of large area modular 64 × 64 CdznTe arrays for nuclear medicine. IEEE Trans on Nucl Sci 1998; 45(3):354–358.

13. Furenlid LR, Wilson DW, Chen Y, et al. FastSPECT II: A second-generation high-resolution dynamic SPECT imager. IEEE Trans Nucl Sci 2004; 51(3):631–635.

14. Barber HB, Apotovsky BA, Augustine FL, Barrett HH, Dereniak EL, Doty FP, Eskin JD, Marks DG, Matherson KJ, Venzon JE, Woolfenden JM, Young ET. Semiconductor pixel detectors for gamma ray imaging in nuclear medicine. Nucl Inst Meth 1997; A395:421–428.

15. Kim H, Furenlid LR, Crawford MJ, Wilson DW, Barber HB, Peterson TE, Hunter WCJ, Liu Z, Woolfenden JM, Barrett HH. SemiSPECT: a small-animal SPECT imager based on eight CZT detector arrays. Med Phys 2006; 33(2):465–474.

16. Pani R, Cintib MN, Pellegrinia R, Trottaa C, Trottaa G, Montania L, Ridolfia S, Garibaldic F, Scafèd R, Belcarie N, Del Guerra A. Evaluation of flat panel PMT for gamma ray imaging. Nucl Meth Inst A 2003; 504(1–3):262–268.

17. Kastis GA, Furenlid LR, Wilson DW, Peterson TE, Barber HB, Barrett HH. Compact CT/SPECT small-animal imaging system. IEEE Trans Nucl Sci 2004; 51(1):63–67.

18. Kastis GA, Wu MC, Wilson DW, Furenlid LR, Stevenson G, Barrett HH, Barber HB, Woolfenden JM, Kelly P, Appleby M. Tomographic small animal imaging using a high-resolution semiconductor camera. IEEE Trans Nucl Sci 2002; 49(1):172–175.

19. Liu Z, Kastis GA, Stevenson GD, et al. Quantitative analysis of acute myocardial infarction in rat hearts with ischemia-reperfusion using a high-resolution stationary SPECT system. J Nucl Med 2002; 43: 933–939.

20. Liu Z, Stevenson GD, Kastis GA, et al. High-resolution imaging with 99mTc-glucarate for assessing myocardial injury in rat models exposed to different durations of ischemia followed by reperfusion. J Nucl Med 2004; 45:1251–1259.

21. Khaw BA, Nakazawa A, O'Donnell SM, Pak KY, Narula J. Avidity of technetium 99m glucarate for the necrotic myocardium: In vivo and in vitro assessment. J Nucl Cardiol 1997; 4:283–290.

22. Narula J, Petrov A, Pak KY, Lister BC, Khaw BA. Very early noninvasive detection of acute experimental

nonreperfused myocardial infarction with 99mTc-labeled glucarate. Circulation 1997; 95:1577–1584.

23. Mariani G, Villa G, Rossettin PF, et al. Detection of acute myocardial infarction by 99mTc-labeled D-glucaric acid imaging in patients with acute chest pain. J Nucl Med 1999; 40:1832–1839.

24. Khaw BA, Silva JD, Petrov A, Hartner W. Indium 111 antimyosin and Tc-99m glucaric acid for noninvasive identification of oncotic and apoptotic myocardial necrosis. J Nucl Cardiol 2002; 9:471–481.

25. David AL, Mirella A, van den Doel, et al. Role of adenosine in ischemic preconditioning in rats depends critically on the duration of the stimulus and involves both A1 and A3 receptors. Cardiovas Res 2001; 51: 701–708.

26. Stadler B, Phillips J, Toyoda Y, et al. Adenosine-enhanced ischemic preconditioning modulates necrosis and apoptosis: Effects of stunning and ischemia-reperfusion. Ann Thorac Surg 2001; 72:555–563.

4

Imaging of Microvascular Systems with Micro-CT

Erik L. Ritman

Department of Physiology and Biomedical Engineering, Mayo Clinic College of Medicine, Rochester, Minnesota, U.S.A.

INTRODUCTION

Micro-CT is a complement, collaborator and competitor of radionuclide-based micro-tomographic imaging methods. It complements in that it provides high spatial resolution micro-anatomic information which cannot be provided by radionuclide tomographic imaging methods, it collaborates in that it provides anatomic location of the radionuclide accumulation provided by the radionuclide image as well as enabling attenuation correction of the radiation emitted by the radionuclide and it is a competitor in that vascular transport (especially perfusion) can be imaged by both modalities. For these multi-modality strengths to be maximally synergistic the different images must be accurately registered in both space and time. While this chapter focuses on the strengths of micro-CT, its intent is to highlight their relationship to radionuclide imaging as applied to the cardiovascular system.

The rapid increase in interest in tomographic imaging of small animals and organ biopsies (1) is conveyed by recent publications, the appearance of dedicated specialty meetings, National Institutes of Health (NIH) call for a number of grant proposals for developing small-animal imaging capabilities, and the growing number of newly emerging companies which build such specialty imaging scanners. The following is a brief overview of micro-CT technology and its applications, especially to blood vessels.

DEVELOPMENT OF MICRO-CT

State-of-the-art clinical CT scanners involves 3D imaging with voxels of the order of 1 mm^3. Although these scanners are also of some use for imaging small animals, special purpose, high resolution, CT scanners are necessary for many studies of intact small-animals and reasonably large (>1 cm^3) tissue specimens (2). In the early 1980s, the first micro-CT scanners were developed (3–7) using "bench-top" X-ray CT sources. With the introduction of synchrotron radiation sources (8,9) computed tomography at micrometer resolutions became possible. An important landmark was the publication of a cone-beam reconstruction algorithm by Feldkamp (6), who was developing a CT scanner, which advanced from fan-beam to cone-beam geometry. This algorithm was central in enabling use of large-angle cone-beam so that bench-top micro-CT scanners could now complete scans within reasonable time-periods. This is because a number of contiguous slices can be scanned simultaneously as compared to scanning one slice at the time (10,11). The cone-beam method, however, limits the axial extent of the specimen that can be scanned in one scan. The most recently developed method, pioneered by Kalendar and others involves "helical" scanning (12–14), an approach that has the advantage that very long specimens (such as mouse or arterial segments) can be scanned intact, independent of the scanner's imaging array's axial height.

Current bench-top scanners, which can scan intact small rodents were developed in response to the rapidly developing need to screen small animals for drug discovery, cancer detection and monitoring, and phenotype identification (15–19). These needs are seen to be well met by use of special-purpose micro-imaging methods because they have the potential for accurate measurement of organ anatomy and the spatio-temporal distribution of various functions within the organs in a minimally invasive manner. The minimally invasive aspect is required when longitudinal studies, which are statistically preferable to cross-sectional studies, are desired. Moreover, in genomics research this capability is particularly important because once the genetically expressed modification of interest is identified, the (now) valuable animal is left intact for more detailed study, for use as a disease model and/or for reproduction.

RESOLUTION REQUIREMENTS FOR IMAGING SMALL ANIMALS

The volume of mammals' internal organs scale with their body weight (20). Hence, if we wish to merely scale the current clinical images [adult human approximately 75 kg and CT image voxel dimensions of the order of (1 mm)^3] to a mouse, we need to have a scanner that has voxels of the order of (100 μm)³ for a 25 gm adult mouse. In addition to dimensional scaling, which broadly speaking is proportional to body mass $M^{0.3}$, the functional aspects such as heart and respiratory cycle durations scale as $M^{-0.25}$ so that scan aperture durations should be reduced from the <300 msec in humans to <75 msec in mice if, for instance, the heart is to be imaged in the diastolic phase.

RESOLUTION CAPABILITIES OF MICRO-CT

Although bench-top micro-CT scanners can achieve meaningful cubic voxel sizes as small as 5 μm, the synchrotron radiation-based scanners provide the highest resolution, with voxels down to the submicrometer range. This is because of their brilliance and narrow bandwidth, energy-selective X-ray. Moreover, the synchrotrons generate X-ray intensity sufficiently high to permit rapid scanning (21). An important contribution that micro-CT imaging can make to micro 3D imaging is that it can image organs' Basic Functional Units (BFU—the smallest aggregation of diverse cells within an organ that functions like the organ, such as a hepatic lobule or a nephron). The BFUs are on the order of (100–200 μm)³ and are approximately the same size in humans and mice. Hence, with a voxel size of less than (50 μm)³ it is possible to obtain information not achievable in clinical or mini-CT scanners, such as the number, size, and packing of the BFUs in an entire mouse organ. With voxel sizes less than (10 μm)³ even cellular dimensions, such as membranes of few cells thick (22,23), and with voxels less than (1 μm)³, subcellular dimensions can be imaged (24–26). Figure 1 illustrates the detail with which the microvasculature within organs can be imaged with bench-top micro-CT.

Although scaling down the voxel size involves machining the scanner components down to a smaller size (which has technological implications for precision tolerances and component packing) it also means that different aspects of the well understood physics underlying CT scanning must be accommodated (27,28). Thus, if the X-ray photon-energy used is less than 25 keV, the X-ray interaction with matter is predominantly via the photo-electric effect, whereas in clinical and heart mini-CT scanners (in which the photon energy generally exceeds 50 keV) it is predominantly Compton scatter (29). A desirable feature of the low photon energy is that the X-ray attenuation is up to two decades higher than it is in clinical CT images, thereby allowing better discrimination of soft tissue types. Another consequence is, however, that if maximum signal-to-noise ratio is to be achieved the X-ray photon energy must be adjusted to match the object diameter and the radiation should be essentially monochromatic if excessive imaging artifacts such as X-ray beam hardening (30) are to be minimized.

In an attempt to generate quasi monochromatic radiation in bench-top micro-CT scanners some X-ray source anode materials can be selected for their characteristic K_α emission radiation (e.g., Molybdenum 17.5 keV). This radiation output can then be filtered through a thin metal foil selected to have its K absorbtion edge at an energy just above the K_α of the anode material (e.g., Zirconium 18.0 keV with a Molybdenum anode), which therefore predominantly attenuates the X-ray photons with energies above and below the K_α characteristic radiation energy. Some promising approaches to developing bench-top monochromatic X-ray sources suitable for micro-CT are in development, an example is the use of a powerful laser to elicit characteristic K_α emission from a suitable target material, so that combined with a suitable foil filter, little bremsstrahlung contamination and <400 eV bandwidth at FWHM of the K_α peak is possible (31).

For mini-CT scanners (i.e., 50–100 μm on-a-side cubic voxels) applied to living animals, the speed of scanning can be critically important. This can be achieved by several mechanisms (21,32,33). One is by use of higher X-ray photon energies (e.g., >50 keV), usually generated with a tungsten anode with considerable filtering through an aluminum or a copper foil, so that adequate X-ray transmission signal is achieved in a shorter time. This increased speed can also be enhanced by bringing the X-ray source closer to the object so as to take advantage of the increase in X-ray

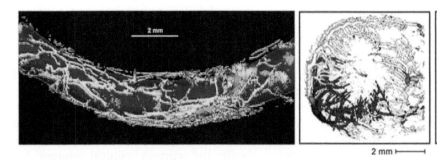

Figure 1 Computer-generated display of 3D micro-CT images. The 3D images consist of 21 μm on-a-side cubic voxels. The *left panel* is of a porcine coronary artery's vasa vasorum after contrast injected into the vascular tree. The *right panel* is of a mouse LV wall and coronary arterial micro-circulation. *Source*: Courtesy of Kassab GS.

intensity with $1/(\text{source-to-specimen distance})^2$. However, this also increase the cone-beam angle—with all the difficulties that result (e.g., penumbral blurring due to focal spot diameter and the reconstruction algorithm needed). Scan speed can be further increased by using multi X-ray source/detector array scanners so that the scan duration is proportionately decreased, such as the micro-CT system built by Bioimaging Research, Inc. (BIR, Inc., Lincolnshire, IL, U.S.A.) (34). Another approach to increasing the X-ray intensity at the specimen, and to some extent reducing the photon energy bandwidth, is the use of polycapillary focused X-ray optics (35).

BASIC COMPONENTS OF MICRO-CT

All micro-CT scanners follow the same general plan. An X-ray source, a specimen that is to be imaged, an X-ray-to-electronic signal converting imaging array and a device that generates an X-ray image at multiple angles of view about the specimen. This is achieved either by rotating the specimen within the stationary scanner or by rotating the scanner about the stationary specimen. The bench-top X-ray sources are generally microfocus (i.e., focal spot diameter <100 μm). Line focus spectroscopy X-ray tubes (focal "spot" several mm long) can be used to produce a stack of parallel fan beam exposures (36) or, when held at an acute angle for use as a "contact" illuminating system (37). The power output (P) of the microfocus tubes is limited (38), hence, increased spatial resolution is achieved at the expense of increased scan duration.

A thin (<0.5 mm) fluorescent crystal plate [e.g., CsI(Tl)] which is cut from a large crystal (or it is built up via vapor deposition) is commonly used for converting X-ray to light. However, there are several tradeoffs. A thicker crystal plate will convert more X-ray photons to light, but the thicker crystal plate provides more opportunity for the light to stray from the X-ray-to-light interaction site before it leaves the crystal, hence poorer spatial resolution. Another concern of any fluorescent material is that of "lag," the duration the light continues to be emitted from the crystal after the X-ray exposure has ceased. This can result in superposition of different angles of view or in blurring due to cardiogenic or respiratory motion during the X-ray exposure. A fluorescent screen made up of a compressed layer of small granules of fluorescing material can be made very efficient as far as X-ray-to-light conversion is concerned. Although this has spatial resolution limited by the granule size, this medium is useful for cone-beam-magnified images as the granule size has diminished impact the greater the magnification. Fiber optics doped with fluorescent material and/or tipped with fluorescent material has great potential for increased efficiency due to depth of fluorescent material and the fibers result

in constrained light transmission—that is, maintain spatial resolution (39). Image intensifiers are still used, but most micro-CT scanners use a fluorescent image which is coupled to a CCD imaging array [light to electrons (40)] via an optical lens or a fiber optic (41). The lens is a relatively inefficient optical coupling but has the advantage of variable magnification (42). The fiber optic coupling is much more efficient but magnification is fixed. The image intensifier does provide more rapid illumination of the imaging array (but not necessary more X-ray photon statistics) which allows for more rapid bench-top CT scanning (43). The number of levels of analog-to-digital conversion needed, depends on the noise in the image being digitized and the dynamic range of signal within the image (44).

For the micro-CT scanners, specimens are generally rotated about a vertical axis within a fixed X-ray/detector system. However, anesthetized animals could be physiologically challenged if held in this vertical orientation for any extended period of time. Rotation of the animal horizontally about its cephalo-caudal axis would result in gross distortions and movements of the animal during the scan. Hence, in the mini-CT scanner, the torso of intact animals requires a fixed horizontal support table for the animal and the X-ray/detector system should rotate about the animal—resulting in a true mini clinical CT scanner.

MICRO-ARCHITECTURAL INDICES OF FUNCTION

Vascular trees, cardiac geometry, and even myocardial fibers and their orientation can be imaged with micro-CT and quantitated, as illustrated in Figure 2. These data can then be used in models to compute functional consequences such as blood flow through vessels and mechanical stresses in myocardium. Similarly, the presence of blood vessels not normally in a certain location and/or the size of vessels may differ from normal, as images of certain patho-physiological processes—such as illustrated in Figure 3.

RADIOPAGUE INDICATORS OF PHYSIOLOGICAL SPACES AND PROCESSES

Essentially all current X-ray micro-CT scanners, like clinical CT scanners, use the attenuation of the X-ray by tissues as the signal for generating the X-ray images. In micro-CT scanners, this means that for a given X-ray photon energy the contrast signal is most closely related to the atomic number (Z) and the concentration of that element (45). At the X-ray exposures tolerated by living tissues, this means that the signal-to-noise in the CT image is adequate primarily for differentiating air, fatty tissue (e.g., brain white matter), non fatty tissue [e.g., muscle, brain gray matter (46)], and bone [in which

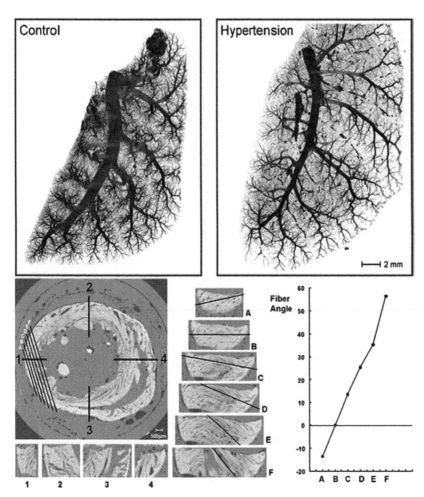

Figure 2 Illustration of function deducible from micro-CT imaged anatomic micro-schematic. The *upper two panels* are of the pulmonary arterial tree of a rat left control and right with monocrotaline-induced pulmonary hypertension. The greatly narrowed lumen diameters (*right panel*) translate to changes in arterial resistance that can be computed from the arterial tree branching geometry. The lower panels relate to measurement of myocardial fiber direction as a function of transmural location in the LV heart wall. The *left* shows a short axis CT image of a rat heart and the labeled panels around it are images of transmural and tangential sections through heart wall computed from the 3D image of the heart. The *right-most* shows the fiber direction as a function of transmural location. This information allows analysis of regional wall stresses and strains. *Source*: From Ref. 37.

several layers, at different stages of mineralization, can be distinguished (10)]. In some tissues there are normally fairly high concentrations of heavy elements (e.g., iodine in the thyroid, iron in hemoglobin and in the hemochromatotic liver) at concentrations which just reach a level at which a pathological increase, or decrease, can be detected by change in CT image contrast (47,48).

The most commonly used clinical contrast agents are based on the highly attenuating element iodine. Iodine is attractive because it is readily attached to biologically relevant molecules, such as

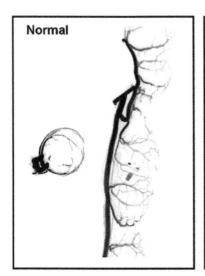

Figure 3 Computer-generated display of 3D CT images of the opacified colonic vasculature in mice. In the *left panel* the normal colonic lumen, which is clear. In the *right panel* there are several adenomas within the colonic lumen. Note that with adenomas the native vessels increase in diameter as well as showing the new vessels within the adenomas. *Source*: Courtesy of Chinery RK.

sugars, which are tolerated at relatively high concentrations in blood (49). The relatively low sensitivity (relative to radionuclide imaging) of X-ray attenuation imaging of these agents requires that relatively high concentrations are needed for quantitation of local concentrations of the contrast agent in small regions-of-interest. The discrimination of physiological spaces, other than the intravascular, can be achieved by selectively opacifying those spaces with administered contrast agents. For instance, as a fraction of soluble intravascular contrast agent passes through the vascular endothelium, especially if it is impaired by reduced oxygen levels, inflammation or because it is newly formed, as is often the case in malignancies. As the contrast media remains in the extravascular space for up to several minutes endothelial permeability can also be estimated from the images. Moreover, as these contrast agents are generally preferentially excreted through the kidney, the opacification of the nephrons (which occurs over many minutes) could be used to quantitate several aspects of renal function as well. An intra-peritoneal injection of iodinated contrast agents has two effects. One is the immediate ability to better delineate the gut and liver within the abdominal cavity and, because of fairly rapid transfer of the contrast medium to the blood stream, the cardiac chambers and large vessels as well as the filtration by the kidneys are subsequently opacified (2). Iodinated contrast agents encapsulated in micrometer-diameter liposomes, can be used as markers of the intravascular space or to be conveyed into the lymphatic space or biliary ducts (50–52). Chylomicron-like sub-micrometer diameter particles can also be iodinated and serve as obligatory intravascular contrast agents which do not enter the extravascular space (or the renal tubules) but are excreted selectively through the biliary system and hence are a liver-enhancing contrast agent (52).

COMBINATION OF MICRO IMAGING MODALITIES

None of the imaging modality provides all the information about all aspects of tissue microstructure, function, and molecular processes in one scan. Indeed scanners of each modality need special emphasis of a particular physics-aspect of that modality in order to optimize the imaging of a particular aspect of organ microstructure and/or function. Consequently, none of the micro-CT image is likely to provide even the full range of the applications that are possible for that imaging modality. However, such limitations can often be mitigated by combined use of micro-CT with another imaging modality and thereby greatly increases the sensitivity and/or specificity of the resulting image data. A necessary aspect of this synergistic use of multiple modalities is that accurate spatial and temporal registration of the 3D anatomy is achieved. This presents a challenging technological problem which can, in part, be solved by software registration algorithms which warp, rotate, and/or translate an image within the 3D image. Three examples of this combined approach follow.

Radionuclide-based imaging methods, whose great strength is the very high specificity and sensitivity of the images (53,54), allows use of pico-molar concentrations of the radionuclide-labeled biologically-active molecules and of labeling of relatively sparsely distributed specific cells. Micro-CT image data can be used to improve the accuracy of the radionuclide imaging methods by enabling correction for the gamma-ray attenuation through the tissues. Similarly, if there is a priori knowledge as to what physiological space the radionuclide is likely to accumulate in, then back projection of radionuclide activity in the tomographic image reconstruction process can be constrained to that space (which is delineated by the X-ray CT image data) so that the local concentration of that radionuclide can be more accurately estimated.

Histological techniques, with their broad spectrum of tissue element-specific stains and immuno-histochemical methods, can provide much useful information about molecular and cellular processes in cells and tissues. Micro-CT image of the microanatomy which can be used to register the histological sections obtained after the micro-CT scan is completed. This could greatly extend the information content beyond that of either method used alone.

Molecular analysis methods, such as immunohistochemistry or mRNA quantitation, often require fresh specimens and generally involve destruction of the specimen. Some tissue-analysis methods can tolerate prior fixation whereas others cannot. Respectively, these methods provide either sparsely sampled information or values averaged over an entire biopsy sample which generally consists of many tissue types in addition to the one of interest. The 3D microanatomy obtained via micro-CT can be used to determine the fraction of the tissue sample that has, and/or the likely location where, the molecule of interest accumulates so that a more meaningful comparison between animals can be made. The latter case can be dealt with by doing a micro-CT scan of the freshly frozen specimen, while frozen, so that the specimen can subsequently be processed for molecular analysis without loss of the 3D microanatomic information (55). Here, too, the average value of the molecular concentration in the tissue sample can be corrected for the physiological space's volume. The 3D micro-CT images can be used to direct where the "biopsy" within the specimen should be obtained so as to enhance the likelihood that the

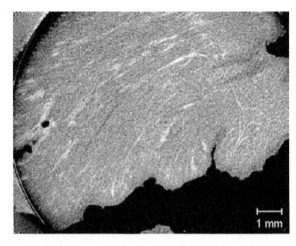

Figure 4 Cryogenic micro-CT image of a 100 μm thick slice through a pig heart wall immediately after contrast injection into the coronary artery. The white streaks are microvessels and the surrounding myocardium is opacified in proportion to the contrast concentration in the extravascular space, black being zero contrast concentration. *Source*: Courtesy of Gössl M, et al. Proc of Flucome, 2003; 7th Triennial International Symposium on Fluid Control, Measurement and Visualization, August 2003, CD ROM Proc ISBN 0-9533991-4-1.

concentration of the molecule of interest is highest in or, conversely, how the concentration distribution is related to microarchitectural features. As illustrated in Figure 4 in the case of extravascular accumulation of contrast agents around the imaged microvessels can be used to estimate the microvascular endothelial permeability.

SUMMARY

Mini and micro-CT imaging are rapidly developing fields stimulated by the recent increased demand for small-animal imaging. These imaging methods are useful because they non-destructively image a 3D volume. However, they are limited by relatively poor contrast of most biologically important molecules as well as by radiation damage during high-resolution scans (29,56). Synergistic use of micro-CT with radionuclide or histological imaging methods can greatly expand the role for micro-CT.

ACKNOWLEDGMENTS

Dr. Erik L. Ritman's micro-CT research was supported, in part, with NSF grant, BIR-9317816, and NIH grants: RR11800, EB000305, HL65342 and his use of micro-CT at the National Synchrotron Light Source, Brookhaven National Laboratory, was supported in part by DOE #DE-AC0276H00016.

REFERENCES

1. Service RF. Scanners get a fix in lab animals. *Science* 1999; 286:2261–2262.
2. Paulus MJ, Sari-Sarraf H, Gleason SS. A new X-ray computed tomography system for laboratory mouse imaging. IEEE Trans Nucl Sci 1999; 46:558–564.
3. Kujoori MA, Hillman BJ, Barrett H. High resolution computed tomography of the normal rat nephrogram. Invest Radiol 1980; 15:148–154.
4. Sato T, Ikeda O, Yamakoshi Y, Tsubouchi M. X-ray tomography for microstructural objects. Appl Opt 1981; 20: 3880–3883.
5. Elliott JC, Dover SD. X-ray tomography. J Microsc 1982; 126(2):211–213.
6. Feldkamp LA, Davis LC, Kress JW. Practical cone-beam algorithm. J Opt Soc Am 1984; A1:612–619.
7. Burstein P, Bjorkholm PJ, Chase RC, Seguin FH. The largest and the smallest X-ray computed tomography systems. Nucl Instr Methods Phys Res 1984; 221:207–212.
8. Bowen DK, Elliott JS, Stock SR, Dover SD. X-ray microtomography with synchrotron radiation. Proc SPIE 1986; 691:94–98.
9. Margaritondo G. Introduction to Synchrotron Radiation. Oxford: University Press, 1988.
10. Elliott JC, Dover SD. Three dimensional distribution of mineral in bone at resolution of 15 μm determined by X-ray microtomography. Metab Bone Rehab Res 1984; 5:219–221.
11. Rüegsegger P, Köller B. A micro CT system for the nondestructive analysis of bone samples. Bone Min 1994; 25:S4.
12. Kalendar W, Klotz W, Vock E. Spiral volumetric CT with single-breathhold technique, continuous transport, and continuous scanner rotation. Radiol 1990; 176:181–183.
13. Wang G, et al. Scanning cone-beam reconstruction algorithms for X-ray microtomography. Proc SPIE 1991; 1556:99–112.
14. Corrigan NM, Chavez AE, Wisner ER, Boone JM. A multiple detector array helical X-ray microtomography system for specimen imaging. Med Phys 1999; 26(8):1708–1713.
15. Paulus MJ, Gleason SS, Kennel SJ, Hunsicker PR, Johnson DK. High resolution X-ray computed tomography: An emerging tool for small animal cancer research. Neoplasia 2000; (1-2):62–70.
16. Holdsworth DW, Thornton MM. Micro CT in small animal and specimen imaging. Trends Biotechnol 2002; 20(8):S34–39.
17. Weber DA, Ivanovic M. Ultra-high resolution imaging of small animals: Implications for preclinical and research studies. J Nucl Med 1999; 6:332–334.
18. Wang G, Vannier MW. Micro-CT scanners for biomedical applications—An overview. Advan Imag 2001; 16:18–27.
19. Paulus MJ, Gleason SS, Easterly ME, Foltz CJ. A review of high resolution X-ray Computed tomography and other imaging modalities for small animals. Lab Animals 2001; 30:36–45.
20. Calder WA III. Size Function and Life History. Cambridge: Harvard University Press, 1984.

21. Wang Y, DeCarlo F, Foster I, et al. A quasi-real time X-ray microtomography system at the advanced photon source. Proc SPIE 1999; 3772:318–327.

22. van Spaendonck MP, Cryns K, van de Heyning PH, Scheuermann DW, van Camp G, Timmermans JP. High resolution imaging of the mouse inner ear by microtomography: A new tool in inner ear research. Anatom Rec 2000; 259:229–236.

23. Ritman EL, Bolander ME, Fitzpatrick LA, Turner RT. Micro-CT imaging of structure-to-function relationship of bone microstructure and associated vascular involvement. Technol and Health Care 1998; 6:403–412.

24. Stampanoni M, Borchert G, Abela R, Rüegsegger P. Bragg magnifier: A detector for submicrometer X-ray computer tomography. J Appl Physics 2002; 92: 7630–7635.

25. Haddad WS, McNulty GI, Trebes JE, Anderson EH, Levesque RA, Yang L. Ultrahigh resolution X-ray tomography. Science 1994; 266:1213–1215.

26. Wang Y, Jacobson C, Maser J, Osanna A. Soft X-ray microcopy with cryoscanning transmission x-ray microscope: II. Tomography. J Microscopy 1999; 197(Pt. 1): 80–93.

27. Grodzins L. Optimum energies for x ray transmission tomography of small samples. Nucl Instr Methods 1983; 206(3):541–545.

28. Spanne P. X-ray energy optimization in computed tomography. Phys Med Biol 1989; 34(6):679–690.

29. Dowseth DJ, Kenny PA, Johnston RE. The Physics of Diagnostic Imaging. London: Chapman and Hall Medical, 1998.

30. Kak AC, Slaney M. Principles of Computerized Tomographic Imaging. New York: IEEE Press, 1998.

31. Chen L, Forget P, Toth R, et al. Laser based intense hard X-ray source for medical imaging. Proc of SPIE 2003; 5031: 923–928.

32. Kohlbrenner A, Koller B, Hämmerle S, Rüegsegger P. In vivo microtomography. Adv Exp Med Biol 2001; 496: 213–224.

33. Sasov A. In vivo micro-CT for small animal imaging. Proc 2002 IEEE Intl Symp Biomed Imaging 2002; 377–380.

34. www.bio-imaging.com

35. Jorgensen SM, Reyes DA, MacDonald CA, Ritman EL. Micro-CT scanner with a focusing polycapillary X-ray optic. Proc SPIE 1999; 3772:158–166.

36. Kohlbrenner A, Hämmerle S, Laib A, Koller B, Rüegsegger P. Fast 3D multiple fanbeam CT systems. Proc SPIE 1999; 3772:44–53.

37. Jorgensen SM, Demirkaya O, Ritman EL. Three-dimensional imaging of vascular and parenchyma intact rodent organs with X-ray micro-CT. Am J Physiol 1998; 44:H1103–H1114.

38. Flynn MJ, Hames, SM, Reimann DA, Wilderman SJ. Microfocus X-ray sources for 3D microtomography. Nucl Instrum Methods Phys Res 1994; Sect A 352(1–3): 312–315.

39. Bueno C, Rairden RL, Betz RA. Hybrid scintilators for X-ray imaging. Proc SPIE 1996; 2708:1–14.

40. Reimers P, Goebbels J, Weise HP, Wilding K. Some aspects of industrial nondestructive evaluation by X-ray and gamma ray computed tomography. Nucl Inst Methods 1984; A-221:201–216.

41. Castelli CM, Allinson NM, Moon KJ, Watson DL. High spatial resolution scintillator screens coupled to CCD detectors for X-ray imaging applications. Nucl Instr Meth Phys Res 1994; A348:649–653.

42. Flannery BP, Deckman HW, Roberge WG, D'Amico KL. Three dimensional X-ray microtomography. Science 1987; 237:1439–1444.

43. Johnson RH, Karau KL, Molther RC, Dawson CA. Quantification of pulmonary arterial wall distensibility using parameters extracted form volumetric micro-CT images. Proc SPIE 1999; 3772:15–23.

44. Dunsmuir JH, Ferguson SR, D'Amico KL. Design and operation of an X-ray imaging X-ray detector for microtomography. Proc Conf Photo-electronic Image Devices Sect 1992; 3:257–264.

45. Torikoshi M, Tsumoo T, Sasaki M, et al. Electron density measurement with dual energy X-ray CT using synchrotron radiation. Phys Med Biol 2003; 7:673–685.

46. Takeda T, Momose A, Hirano K, Haraoka S, Watannabe T, Itai Y. Human carcinoma, early experience with phase contrast X-ray CT with synchrotron radiation-comparative specimen study with optical microscopy. Radiol 2000; 214:298–301.

47. Bonse U, Johnson Q, Nichols M, Nußhardt R, Krasnicki S, Kinney J. High resolution tomography with chemical specificity. Nucl Instrum Methods Phys Res 1986; A246(1-3):644–648.

48. Dilmanian FA. Computed tomography with monochromatic X-rays. Am J Physiol Imaging 1992; 3/4:175–193.

49. Paroz Z, Moncada R, Sovak M. Contrast media: Biological effects and clinical applications, vols 1–3. Boca Raton: CRC Press, 1987.

50. Ketai LH, Muggenberg BA, McIntire GL, et al. CT imaging of intrathoracic lymph nodes in dogs with bronchoscopically administered iodinated nanoparticles. Acad Radiol 1999; 6:49–54.

51. Kao C-Y, Hoffman EA, Beck KC, Billamkonda RV, Annapragada AV. Long residence time nano-scale liposomal iohexiol for X-ray-based blood pool imaging. Acad Radiol 2003; 10:475–483.

52. Weichert JP, Lee FT, Jr, Chosy SG, et al. Combined hepatocyte-selective and blood pool contrast agents for the CT detection of experimental liver tumors in rabbits. Radiology 2000; 216:865–871.

53. Williams MB, Guimin Z, More MJ, et al. Integrated CT-SPECT system for small animal imaging penetrating radiation systems and application II. Proc SPIE 2000; 4142:265–274.

54. Cherry SR, Shao Y, Silverman RW, Chatziioannou A, Meadors K. Micro PET: A high resolution PET scanner for imaging small animals. IEEE Trans Nucl Sci 1997; 44:1161–1166.

55. Maran A, Khosla S, Riggs BL, Zhang M, Ritman EL, Turner RT. Measurement of gene expression following cryogenic μ-CT scanning of human iliac crest biopsies. J Musculoskeletal Neuronal Interactions 2003; 3:83–88.

56. Chow PL, Goertzen AL, Berger F, DeMarco JJ, Chatziioannou AF. Monte Carlo model for estimation of dose delivered to small animals during 3D high resolution X-ray computed tomography. Nucl Sci Symp Conf Record IEEE 2002; 3:1678–1681.

5

Targeted MRI

Susan B. Yeon
Cardiovascular Division, Department of Medicine, Harvard Medical School, Boston, Massachusetts, U.S.A.

Andrea J. Wiethoff
EPIX Pharmaceuticals, Cambridge, Massachusetts, U.S.A.

Warren J. Manning
Cardiovascular Division, Department of Medicine, Harvard Medical School, Boston, Massachusetts, U.S.A.

Elmar Spuentrup
Department of Diagnostic Radiology, Technical University of Aachen, Aachen, Germany

Rene M. Botnar
Cardiovascular Division, Department of Medicine, Harvard Medical School, Boston, Massachusetts, U.S.A., and Department of Nuclear Medicine, Technical University Munich, Munich, Germany

MOLECULAR MRI

MRI and Molecular Imaging

The high spatial resolution and structural definition obtainable using magnetic resonance imaging (MRI) makes it a promising modality for imaging processes at the molecular and cellular level (1). Furthermore, the availability of a wide range of magnetic resonance (MR) scanners, ranging from small bore animal scanners to whole body clinical systems provide a means to bridge the gap between experimental models and clinical application. The availability of high field animal scanners is especially important as many basic molecular imaging experiments are first performed in small animals for which particularly high spatial resolution is required for meaningful detection and localization. In contrast to some of the other imaging modalities (e.g., SPECT, PET), however, MRI has inherently lower sensitivity for probe detection which complicates its use for imaging low quantities of molecular markers.

The goals of this chapter are to provide an overview of MR physics important for molecular imaging and to review the current state of molecular MRI. The first sections will focus on the basic principles of MR with an emphasis on signal and contrast manipulation and instrumentation requirements followed by a discussion of how the properties of the instrumentation impact the development of contrast agents. The later sections will focus on the basic properties of targeted contrast agents including target identification, targeting approaches, and signal amplification strategies. Finally, current and potential applications of cardiovascular molecular MRI will be described.

Principles of MR

MRI is based on the principle of nuclear magnetic resonance. Nuclei consisting of an odd number of protons and/or neutrons have a magnetic moment. When placed inside a strong magnetic field some of the nuclei align with the magnetic field establishing a net longitudinal magnetization with the proton nuclei of water (hydrogen nuclei from water are the most abundant in the body). The nuclei precess at a frequency directly proportional to the strength of the main magnetic field. Recovery of magnetization in the longitudinal direction (T1) and decay of magnetization in the transverse plane (T2, T2*) are the basis of soft tissue contrast in MRI.

MR Signal Intensity

Signal intensity in MRI primarily depends on the local values of the longitudinal (1/T1) and transverse (1/T2) relaxation rate of water protons. Depending upon the pulse sequence, signal usually tends to increase with shorter T1 (higher 1/T1) and decrease with shorter T2 (higher 1/T2) relaxation times. The environment in which the nuclei are located determine the MR signals created. Therefore, by manipulating the chemical environment around the protons, the signal can be altered. MR contrast agents have been developed as a way to modulate the chemical environment inside an organism. The relaxivities r1 and r2, which are commonly expressed in $mM^{-1}s^{-1}$ indicate the increase in $1/T1$ and $1/T2$ per concentration of contrast agent as demonstrated in Figure 1 for relaxivities of $r1 = 20$, 40, and 80 mM^{-1}s^{-1}

$$1/T1 = 1/T1_0 + r1 \text{ [contrast agent]}$$

$$1/T2 = 1/T2_0 + r2 \text{ [contrast agent]}$$

with $T1_0$ and $T2_0$ being the relaxation times of native tissue (i.e., tissue devoid of exogenous contrast agent).

Gadolinium (Gd) based contrast agents usually increase $1/T1$ and $1/T2$ in similar amounts ($r2/r1 \cong$ 1–2) (2–4) whereas iron particle based contrast agents have a much stronger effect on increasing $1/T2$ ($r2/r1 > 10$) (5). Gadolinium based contrast agents therefore lead to a positive contrast effect (detected as an increase in signal intensity or brightness) whereas iron particle based contrast agents usually cause a negative contrast effect (detected as a decrease in signal intensity or darkness). MR pulse sequences that emphasize differences in T1 and T2 are commonly referred to as T1 and T2 weighted sequences. Apart from their effect in increasing $1/T2$, iron particles also increase $1/T2^*$ due their effect on the local magnetic field B_0 thus causing local field inhomogeneities ΔB_0. This additional effect leads to even more severe signal decay.

$$1/T2^* = 1/T2 + \gamma \Delta B_0$$

Iron based contrast agents are therefore often imaged using $T2^*$ weighted imaging sequences. For signal quantification, T2 weighted multi echo spin echo sequences can be used to generate T2 maps. Typical r1 and r2 values of currently approved Gd based contrast agents are in the range of $r1 = 3$–5 mM^{-1}s^{-1} and $r2 = 5$–6 mM^{-1}s^{-1}. The relaxivities of iron based contrast agents are significantly higher $r1 = 20$–25 mM^{-1}s^{-1} and $r2 = 100$–200 mM^{-1}s^{-1}. Due to the low concentrations at which molecular imaging targets are generally found, relaxivity is important in the design of molecular contrast agents and will be discussed further in the second part of this chapter.

MR Image Acquisition

In this chapter, we mainly focus on cardiac applications, which are subject to cardiac and respiratory motion. Requirements on motion compensation for large vessels such as the aorta or the carotid arteries are less stringent.

Cardiac and Respiratory Motion Compensation

The physics of MRI require significant time to collect the data necessary for image generation. Often the time required for data acquisition is long enough to encompass several cardiac or respiratory cycles. Motion occurring during the acquisition period can dramatically diminish image quality. Therefore significant work has gone into developing "gated" imaging protocols to eliminate motion artifacts during data acquisition. When data are acquired over multiple cardiac and respiratory cycles (segmented data acquisition), synchronization with the ECG and/or the position of the diaphragm is mandatory. Since the ECG is distorted (elevated T-wave) when recorded from a patient in a high magnetic field, state of the art MR scanners use ≥4 leads R-wave detection (≥2 ECG traces) to differentiate between the R-wave and the so called T-wave artifact (6). The T-wave artifact is caused by the magneto hydrodynamic effect (MHD). Deflection of rapidly moving ions by the main magnetic field produces additional voltage that is superimposed on the ECG signal. The MHD artifact is strongest during maximal flow in systole and increases with increasing field strengths.

Small bore animal scanners are commonly equipped with less sophisticated ECG gating hardware and software. In most cases, R-wave detection is performed by simple threshold algorithms. In addition, both clinical and small animal systems are equipped with respiratory sensors (bellows) that enable respiratory motion gating. However, simple gating mechanisms may cause interruption of MR data acquisition and thus signal variations due to altered MR steady state conditions. This can lead to artifacts especially if inversion recovery sequences are used in concert with contrast agents. Gating schemes that acquire data at a near constant TR and subsequently label data as accepted or rejected based on the respiratory position of the diaphragm help overcome this limitation. A drawback of all gating schemes is the increased scanning time they entail.

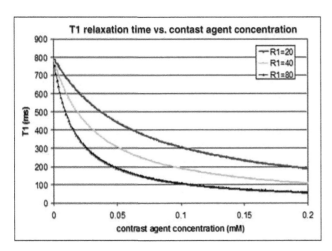

Figure 1 T1 relaxation time for various r1 values plotted for increasing contrast agent concentrations; $1/T1 = 1/T1_0 + r1 \times$ [contrast agent]; $T1_0$ of native tissue (i.e., myocardium was assumed to be 800 ms). To achieve a T1 of 200 ms, that is usually sufficient for contrast agent detection, <50 µM Gd is required for $r1 \geq 80$ mM^{-1}s^{-1}. For a r1 of ≤ 20 mM^{-1}s^{-1}, this concentration increases almost 4-fold to ≥ 200 µM Gd.

Molecular MRI Sequences

MR imaging sequences can be divided into spin echo (SE) (typically 2D) and gradient echo (GRE) (typically 3D) sequences. Several types of SE or GRE imaging sequences are used for optimal contrast depending on the type of agent, location, motion, etc. Table 1 outlines the general parameters for the most common classes of sequences.

Spin Echo Sequences

ECG triggered and non-triggered T1 and T2 weighted spin echo (SE) sequences belong to the standard sequence repertoire of every MR scanner. These sequences are used extensively for neuro, body, and musculoskeletal imaging since they provide excellent image quality and can provide variable T1 or T2 weighting by adjusting the echo time (TE) and repetition time (TR). T1 weighted SE sequences are characterized by short TEs (5–15 ms) and TRs (300–700 ms) whereas T2 weighted SE sequences have long TEs (50–150 ms) and TRs (>2000 ms). In the presence of a T1 lowering contrast agent, high resolution images with excellent soft tissue contrast with concomitant T1 weighting can be achieved. In applications for which morphologic details or hypo intense blood (black blood) appearance in concert with visualization of contrast uptake are required, spin echo approaches are often the method of choice. A disadvantage of fast spin echo sequences is their inability to demonstrate increasing contrast effect from higher contrast agent concentrations (Fig. 2) and frequently observed suboptimal contrast induced signal enhancement due to the relatively high signal from surrounding tissues.

The maximum MR signal is reached at Gd concentrations of approximately 1 mM for a typical contrast agent with a relaxivity r1 of 4 mM^{-1}s^{-1}. For higher concentrations, the T2 effect begins decreasing the maximal achievable signal due to the finite achievable TE, whereas T2 weighting increases with increasing contrast agent concentrations.

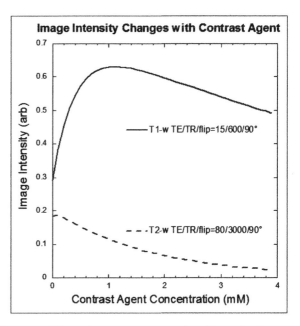

Figure 2 Effect of contrast agent on signal intensity. Impact of contrast agent on T1-weighted (TR = 600 ms) and T2-weighted spin-echo images. *Source*: From Ref. 4.

T1 Weighted 3D Gradient Echo Sequences

Non ECG triggered fast radio frequency (RF) spoiled 3D gradient echo sequences (TE < 5 ms, TR < 10 ms, flip angle = 30–50°) are heavily T1 weighted and exhibit a near linear relationship between contrast agent concentration and MR signal intensity (Fig. 3).

These sequences are therefore especially suited for higher contrast agent concentrations. Due to their short scan times (5–60 s) and excellent background suppression, these sequences are the workhorse in first pass contrast enhanced angiography of the large vessels and in molecular imaging of non-moving tissues and organs. A disadvantage of this approach is that it generally produces a hyper intense (bright) appearance of blood, which makes it a suboptimal candidate for molecular imaging of the vessel wall. The use of saturation pulses can help minimize the inflow (blood

Table 1 MR imaging sequences

	Spin echo		Steady state GRE	IR GRE	GRE
Weighting	T1w	T2w	T1w	T1w	T2*
Prepulses	(DIR)	(DIR)	(REST)	IR	—
ECG gating	Yes	Yes	No	Yes	Yes
Resp. gating	Yes	Yes	Yes	Yes	Yes
TR	TE	>5 ms	Shortest	~5–10 ms	>2000 ms
TR effective	300–700 ms	>2000 ms	—	TR$_{IR}$ ≥ 1 HB	—
TE	5–15 ms	50–150 ms	Shortest	<5 ms	5–50 ms
Flip angle	90	90	30–50	30–50	20–70

Abbreviations: DIR, double inversion prepusle; REST, regional saturation band; IR, inversion recovery prepusle; GRE, gradient echo; HB, heart beat; TR, repetition time; TE, echo time; (), optional

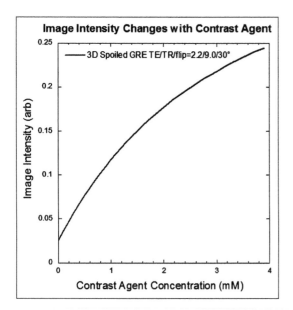

Figure 3 Effect of contrast agent on signal intensity using a short-TR fast spoiled gradient-echo sequence. *Source*: From Ref. 4.

signal enhancing) effect as demonstrated in a study of molecular MRI of fibrin (7).

T1 Weighted Inversion Recovery 3D Gradient Echo Sequences

T1 weighted inversion recovery sequences are particularly useful if ECG triggering or respiratory gating is required for suppression of cardiac or respiratory motion artifacts. Typical scan parameters include TE < 5 ms, TR = 5–10 ms, flip angle = 30–50°, 10–30 RF excitations per heart beat, and bandwidth = 100–300 Hz/ pixel. The choice of the inversion repetition time TR_{IR} (≥1 heart beat) determines the optimum inversion delay TI

$$TI = \ln 2 \times T1 - T1 \times \ln(\exp(-TR_{IR}/T1) + 1)$$

and thus the maximum achievable signal intensity of the administered contrast agent. T1 is the longitudinal relaxation time of the suppressed tissues. Longer inversion repetition times TR_{IR} (>1 cardiac cycle) lead to longer optimal inversion delays TI and thus to higher signal intensities (Fig. 4) at the site of contrast uptake. The drawback of this approach is the increased scanning time that it requires. An example of this approach was provided by early proof-of-concept work (8) in which an IR gradient echo sequence was used in the detection of coronary thrombi that had been labeled with Gd-DTPA fibrinogen in vitro (Fig. 5).

Advantages of inversion recovery sequences are excellent background suppression and flow insensitivity. Due to little signal contamination from surrounding tissues, these sequences are particularly useful for visualization of small amounts of contrast uptake at a specific target site. A limitation is the lack of morphologic information provided, though it is possible to acquire morphologic information in a separate scan and overlay the images to obtain anatomic localization. Unlike images produced by non-triggered 3D gradient echo sequences, this approach provides images in which blood signal is well suppressed and thus allows for targeted imaging of the endothelium, vessel wall or thrombus.

T2* Weighted Gradient Echo Sequences

Due to the prominent T2* effect of iron based contrast agents, imaging using these agents is done using predominantly T2* weighted gradient echo sequences. Typical parameters include TE = 5–50 ms, TR ≥ 2000 ms heart beat, and flip angle = 20–70°. Due to the overall low signal in T2* weighted images, morphologic information is limited and sometimes requires additional scans to allow for co-localization between the site of contrast uptake and the corresponding morphologic image.

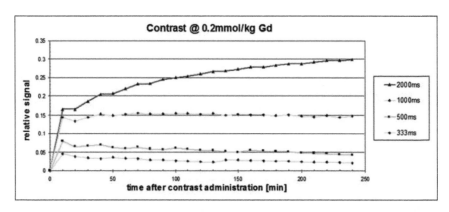

Figure 4 Effect of MR signal strength on IR images for various IR repetition times and with respect to time of contrast administration. Simulation of the Bloch equations based on expected Gd concentrations in blood (assuming kinetics similar to gadopentetate dimeglumine) demonstrating the relative signal vs. time after contrast agent injection (minutes) for four different IR repetition times (333–2000 ms). For longer IR repetition times, a higher relative signal level can be expected. Furthermore, the simulation demonstrates that the relative signal level remains constant over a relatively long period of time after contrast administration.

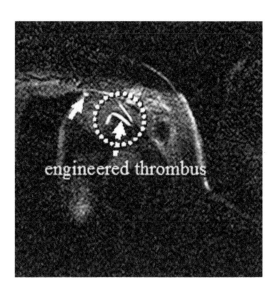

Figure 5   Example of an ECG gated inversion recovery (IR) sequence for in-vivo imaging of Gadolinium-labeled coronary thrombus. Note the high contrast-to-noise due to the excellent suppression of surrounding tissues and blood. This approach is well suited for determining whether the chosen target is present or absent. *Source*: From Ref. 8.

Spatial Resolution

Due to the small size of the molecules and cells to be investigated by molecular MRI, spatial resolution is an important limiting factor. The spatial resolution necessary to identify molecules or cells that underlie biologic processes depends heavily on the imaging sequence employed and the question to be answered. If the task is to identify the presence or absence of biological markers, low resolution images (~1–2 mm) may be sufficient and imaging is best done with T1 weighted IR sequences in conjunction with the use of Gd based contrast agent or with T2* weighted sequences in conjunction with the use of iron oxide based contrast agent. If identification and localization of a molecular marker with respect to small anatomical structures is desired, spatial resolution and partial volume effects become crucial. Because limited data are currently available, general recommendations cannot be made. However, for minimization of partial volume effects, IR sequences appear very promising, as native tissues can be quite effectively nulled over a wide T1 range (e.g., myocardium = 800; blood =1200 ms) thereby creating high contrast between the agent and the surrounding tissues (e.g., Fig. 5).

MRI Scanner
Field Strength

The field strength of most clinical scanners is 1.5T (64 MHz). Recently introduced high field systems operate at 3T (128 MHz) and have been shown to be advantageous for neuro applications because of the 2-fold gain in signal-to-noise (SNR). Body imaging at 3T has not yet been as successful due to greater technical hurdles, which are primarily related to B0 field inhomogeneities (suboptimal shimming), limited RF penetration and increased RF heating. Nevertheless, the potential increase in SNR that has higher field strength can provide, may prove useful for, imaging at the molecular and cellular level using Gd based targeted contrast agents.

For development of T1 lowering contrast agents, the field strength dependency of the longitudinal relaxivity r1 plays a critical role. This field strength dependency is also referred to as nuclear magnetic resonance dispersion (4). If the product between the Larmor frequency ω (=field strength) and the correlation time (fluctuation of local magnetic field induced by contrast agent) τ_c exceeds 1, r1 begins to decrease (4). Most contrast agents that are in clinical use today were optimized for 1.5T ($\tau_c \cong 2.5$ ns). When developing contrast agents for 3T or higher field systems, lower correlation times should be sought in order to achieve maximal longitudinal relaxivities r1. Conversely, the transverse relaxivity r2 has a different behavior and may even increase at higher field strengths. Thus, at higher field strengths, the ratio r2/r1 usually increases. Iron particle based contrast agents (=T2* agents) therefore should be well suited for use in 3T and higher field MR systems. On the other hand, Gd based agents (=T1 agents) demonstrate less T1 lowering effect at higher field strengths.

Physiology Signals and Motion Compensation

Most clinical MR scanners are equipped with ECG sensors and peripheral pulse units to allow for scan synchronization with the patient's heartbeat. In addition, respiratory bellows or MR navigators are used to monitor and correct for respiratory motion. The newest scanner generations are also equipped with blood pressure cuffs and blood oxygen saturation sensors to allow constant monitoring of vital signs in more critically ill patients.

Receiver Coils

In cardiac MRI, phased array coils (4–6 coil elements) are required to meet current standards for imaging. This is because of the need for sub-millimeter resolution, especially when coronary artery or vessel wall imaging is included. Imaging of processes at the molecular level is likely to require even higher spatial resolution. Cardiovascular molecular imaging will benefit from improvements in phased array coil technology, which has recently advanced with the advent of 16–32 channel receiver technology. Analagous to developments in multislice CT technology, such improvements can be expected to reduce imaging time, which will likely translate into reduced motion and image artifacts. Furthermore, there is evidence from recent publications (9) that SNR is likely to increase using this novel technology, which is especially beneficial for imaging of small molecular and cellular targets.

Data Analysis
T1 Measurements
Measurements of the T1 relaxation time are usually performed with inversion recovery sequences. By changing the inversion delay TI between the non-selective inversion prepulse and data acquisition, signal from tissue A (T1 = $T1_A$) will be nulled if the inversion delay TI fulfills the condition $TI = T1_A \times \ln 2$. Most T1 measurements approaches are based on the Look and Locker sequence, which acquires multiple images (6–11) along the T1 relaxation curve after an initial inversion prepulse (10). Several new approaches have been proposed to reduce imaging time (11) and to enable T1 measurements in moving organs such as the heart (12).

T2* Maps
T2* maps are acquired by sampling the signal along the free induction decay (FID) curve using multiple TE at a constant TR. The most common approaches are based on gradient echo sequences with signal sampling along Cartesian trajectories (13,14). The drawback is the relatively long scanning time with this approach. In a recent study, Schaeffter and coworkers proposed a faster approach by taking advantage of the undersampling properties of radial imaging (15). Undersampled radial sub-images with differing TE were reconstructed from a complete radial data set that was acquired with multiple TEs. An exponential pixel-by-pixel fit of the FID as derived from the under-sampled sub-images then allows generation of T2* maps (Fig. 6) (16).

Endogenous Contrast
Recently, groups have taken advantage of the intrinsic T1 and T2 differences of diseased tissues to image both thrombus (17,18) and vulnerable plaque (19–21). The presence of methemoglobin in thrombi produces T1 shortening leading to high signal intensity on T1-weighted images of thrombi such as in deep vein thrombosis (22) and pulmonary embolism (23). However, this effect declines over time as the thrombus becomes organized.

Toussaint and Yuan have characterized plaques with 4 types of contrast weightings (T1, T2, proton-density, and 3D time-of-flight) and shown good correlation with histopathologic examination in carotid endarterectomy specimens (19,24). Fayad and colleagues have performed similar studies in the aorta with T1, T2 and proton density weighted sequences (25).

Targeted Contrast Agents
While advances in MRI hardware are important to improved molecular imaging, advances in molecular imaging with MRI are increasingly dependent on the development of exogenous probes. Successful molecular probe development requires selection of appropriate biologically and clinically relevant targets and effective strategies for meeting the challenges of sensitivity, specificity, spatial localization and safety required for accurate diagnosis. Methods that provide for significant signal amplification are needed to detect molecular markers effectively.

Probe Detection by MR
The sensitivity of a molecular imaging probe depends upon the distribution of the probe, the strength of the probe signal, and the means of detection of the probe. The distribution of a probe is determined by the mode of its administration, its compartmentalization (degree of passage into various intravascular, extracellular, and intracellular spaces), and its mechanisms and speed of clearance. Probe distribution together with probe-target affinity determines the concentration of probe at the target for detection. For magnetic resonance, the strength of the achievable probe signal is determined by the concentration of contrast agent and the relaxivity of the agent. However probe signal may not increase proportionately to increases in probe concentration (see below). In addition, the distribution of probe in non-targeted regions is an important consideration for timing and methods of probe detection to optimize the target-to-background signal ratio. The plasma half-life of an agent must be sufficient to expose target receptors to the agent. On the other hand, it may be necessary to wait for plasma concentrations of agent to fall sufficiently to distinguish luminal from vessel wall contrast uptake, although methods may be employed to reduce this requirement (26).

MR Signal Amplification—General Considerations
MR contrast agents are not detected directly but by their effect on water protons. Molar concentrations of water protons are required to provide sufficient effect on signal intensity so high local concentrations of contrast agents (e.g., typically μM to mM for those agents having r1 in the 4–20 $mM^{-1}s^{-1}$ range) are needed to alter the chemical environment sufficiently for detectable signal effects. Since many molecular targets are found in the nM range in the body, new methods of signal generation are being developed to enable MRI of such molecular targets.

T1 Effects
Most MR contrast agents are based on gadolinium (Gd) complexes (3,4) (Fig. 7A) or less commonly, iron oxide particles (Fig. 7B) (5). Gd(III) is ideally suited for

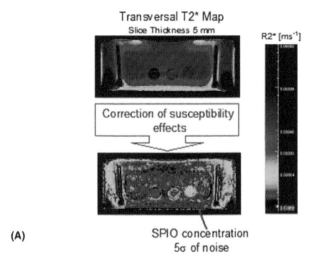

(A)

SPIO concentration
5σ of noise

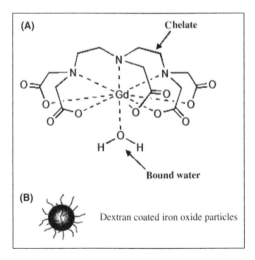

(B)

Figure 6 T2* maps of water tubes fill with SPIO contrast agents (A). Without susceptibility correction, no significant differences can be observed between the different tubes. After correction, SPIO concentrations as low as 5 standard deviation above the noise level could be detected. In-vivo T2* maps of the liver before and after susceptibility correction (16).

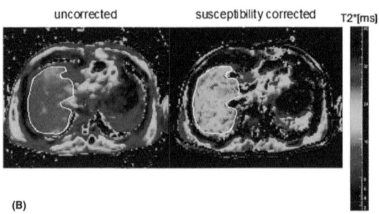

Figure 7 (**A**) Example of a Gadolinium complex demonstrating the chelation of the otherwise toxic Gd^{3+} metal. The free water interacts with the surrounding protons thereby locally changing the T1 and thus increasing the MR signal strength on T1 weighted images. (**B**) Example of a dextran coated iron oxide particle. *Source*: From Ref. 4.

use as an MRI contrast agent because it not only has seven unpaired electrons but the symmetry of its electronic states produces an electron spin relaxation time slow enough to interact significantly with neighboring water protons (4).

Relaxivity is impacted by a number of contrast agent properties including hydration number, the distance between the ion and the solvent proton, solvent exchange rate, electronic relaxation time, and rotational correlation time. The hydration number (number of water molecule coordination sites) for Gd chelates is generally >1 (4). Obtaining the minimal distance between the ion and bound protons (r) is important because the relaxation rate is inversely proportional to r^6. The ability of the bound protons to rapidly exchange with free water allows for distribution of the relaxation effects throughout the bulk water (Fig. 8). A longer electronic relaxation time also leads to higher relaxivity.

The rotational correlation time (τ_R) is ~0.1 ns for approved agents. Since increases in τ_R enhance relaxivity,

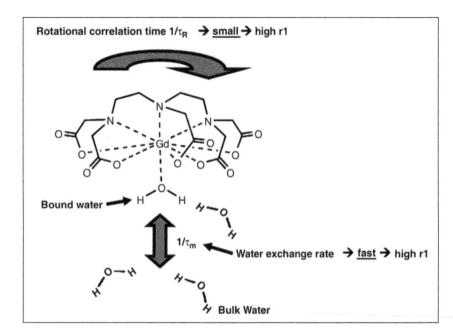

Figure 8 Schematic of water exchange between bound and bulk water. Fast water exchange leads to a more efficient energy transfer between Gadolinium and bulk water, thereby creating a higher relaxivity r1. Reduction of the tumbling motion (rotational relaxation rate) of the Gd-chelate increases the relaxivity r1 as well. Binding to a target or construction of macromolecular agents (e.g., dendrimers) are a means of slowing down the rotational motion of Gd-compounds.

various efforts in contrast agent design have focused on increasing this parameter. T_R is lengthened by formation of conjugates between the metal ion complex and slowly moving structures such as proteins, polymers or dendrimers.

Molecular MRI probes frequently involve attachment of Gd complexes to small ligands (e.g., small molecules, peptides) that in turn attach to slowly moving targets (e.g., proteins); thus, lengthening of τ_R is accomplished, providing a convenient means of amplifying the detection of contrast agents positioned at molecular targets. Since the unbound fraction of molecular probe will retain a lower r1, a good target-to-background signal ratio is achieved. This has been termed receptor-induced magnetization enhancement (27,28). An early example of this type of contrast agent is MS-325 (EPIX Pharmaceuticals, Lexington, Massachusetts U.S.A.). MS-325 is an intravascular contrast agent that reversibly binds to albumin in plasma. When bound to albumin the relaxivity increases to ~20–23 mM^{-1}s^{-1} from the 6 mM^{-1}s^{-1} observed in buffer at 65 MHz (4).

A different approach to improve T1 effects without improvement in the relaxivity per Gd atom is by increasing the number of Gd per targeting complex (=nanoparticle or other ligand) thus producing multivalency of Gd attachment to ligand. Such multivalency within the targeting complex may be sufficient to offset a flat or even diminished relaxivity effect of individual Gd atoms. This effect can be exploited for both Gd chelate-antibody (or peptide) probes as well as for nanoparticle probes, although the magnitude of the effect is much greater for nanoparticles (e.g., Gd/antibody ratio of ~20 Gd/antibody molecule vs. ~100 Gd/20nm colloid nanoparticle vs. ~300,000 Gd/300

nm nanoparticle). Thus nanoparticles offer the potential for greatly improved sensitivity over conventional chelated complexes because the much higher achievable concentration of Gd (or iron oxide).

This principle appears to apply despite the fact that only a small fraction of the Gd atoms in these complexes are located at the surface of the particle and exposed to water. Since contrast agent relaxation theory emphasizes the importance of the interaction of electrons from Gd's outer shell with surrounding water molecules, one might anticipate that the T1 signal generated for a colloid would be significantly less than the same mole concentration of chelated Gd. However, since this may not be the case, other factors that are not currently fully understood must be involved in signal generation. A drawback of this type of agent is the large size of the contrast agent loaded particles. With sizes ranging upto 10 to 500 nm in size, accessibility to the target site may be limited.

Estimates for minimal Gd concentration required for detection depend upon the relaxivity of the given Gd complex, which varies with the field strength. As noted earlier, for T1 agents, the maximum relaxivity attainable decreases with increasing field strength. Aime et al. found that for an agent with high relaxivity (~80 mM^{-1}s^{-1}), the threshold for detection was 4 + 1 × 10^7 complexes/cell or ≈15 µM (29). In an animal study of coronary thrombosis, we found that gadolinium concentrations between 100–150 µM translated into an SNR of approximately 11 ± 2, which allowed for target detection (8).

As alluded to above, the effects of MR contrast agent concentration are nonlinear. While contrast agent distribution impacts contrast concentration, contrast agent compartmentalization and local phenomena may alter local agent relaxivity. As Gd concentration increases, T1 falls. However if T1 < TR/2 (TR = repetition time for imaging), tissue will recover nearly fully before the subsequent RF pulse. Also, at high concentrations, Gd will reduce T2 to the order of the TE (echo time) so increasing concentrations will decrease MR signal intensity. As noted above and demonstrated in Figures 2 and 3, fast T1-weighted gradient-echo sequences (especially 3D ones) typically have a larger scalable range than spin-echo sequences (4).

T2*

The synthesis and use of stable, nanosized iron oxide particles for use as MR contrast agent have been extensively described (5,30). Iron oxide particles have differential effects on 1/T1 and 1/T2 depending on their size. Superparamagnetic iron oxide (SPIO) particles produce much larger increases in 1/T2 than in 1/T1 so they are best imaged with T2-weighted scans, which reveal signal decrease (30). SPIO particles produce a marked disturbance in surrounding magnetic field homogeneity, especially apparent when a nonhomogenous distribution produces a T2* susceptibility effect. On the other hand, ultra-small superparamagnetic particles of iron oxide (USPIOs) have a greater effect on 1/T1 than SPIO particles, so they can be used for T1-weighted imaging (31).

Although iron oxide based agents have greater relaxivity per metal atom than Gd based agents, Gd based agents provide positive T1 signal enhancement which is more readily distinguished from artifact than negative signal effects and has a larger potential scalable range for detection.

Safety

Although $GdCl_3$ is very toxic, chelating Gd(III) (as with DTPA) largely reduces its toxic profile (32–34). With a high thermodynamic stability constant ($K_{Gd-L} = 10^{26}$ for Gd-DTPA), the chelated metal complex is essentially inert and is cleared at a significantly faster rate than the rate of dissociation of the metal ion from the complex. However, the plasma clearance rate of chelated Gd is still sufficiently slow to provide ample contrast for MRI applications. Thus, the chelation process transforms an otherwise highly toxic salt into a useful diagnostic reagent (35).

Safety considerations for potential molecular imaging probes include speed and mode of clearance from the blood and body compartments. While conventional Gd chelates are not hydrophobic and therefore generally remain extracellular, some novel molecular contrast agents are more likely to enter cells, raising concern for greater potential for toxicity. Additional concerns include antigenicity and biological interaction. When selecting a suitable target for imaging, the selected target would preferably not be involved in biologic processes that might be inhibited by the binding of a targeting molecular imaging probe. Additionally, for probes designed for clinical use it is important to ascertain whether binding of the molecular probe affects binding of therapeutic agents.

Target Identification

Molecular probe target selection criteria include substantiality of biologic and clinical relevance and feasibility of target identification, which is dependent upon factors such as adequacy of target density and accessibility of targets to probes.

Targeting

A specific disease process or cell type may be targeted by a contrast agent using either passive or active means. While the term "molecular imaging" may be most accurately applied to probes that bind to specific molecular markers ("active" targeting), similar targeting effects may be obtained using probes having distribution characteristics that favor concentration in particular cell types or in regions of disease activity (i.e., "passive" targeting).

Passive Targeting

Examples of passive targeting agents are iron oxide particles that may be used to label components of the reticuloendothelial system (RES). SPIO nanoparticles (diameter ~200 nm) are recognized by the RES and are rapidly removed from the blood stream (36). SPIO uptake detectable by MRI and histology has been found in atherosclerotic plaques of hyperlipidemic rabbits in regions of high macrophage content (37). USPIOs (diameter ~20 nm) are not immediately scavenged by the hepatic and splenic RES, so they have a longer intravascular half-life, and are small enough to transmigrate the capillary wall via vesicular transport and interendothelial junctions (5) USPIOs are also phagocytosed by macrophages in plaque in hyperlipidemic rabbits (38) and humans (39).

Active Targeting

A particular cell type or disease process may be identified by targeting specific molecular markers. Potential targets may be selected from a variety of molecular markers associated with various diseases. Cells surface markers (such as CD4, CD8, and, Mac1) on immune cells may be targeted by labeled antibodies to image areas of immune response such as murine encephalitis (40). Monoclonal antibodies against tumor-associated antigens (e.g., Ra96, HER-2/neu receptor) may serve as ligands to identify the presence of tumor cells (41,42).

However, even if contrast enhancement with a targeted probe is observed in or around the specified target, the results must be initially interpreted with caution since non-targeted contrast agents may accumulate in areas of interest, for example, accumulation of non-targeted Gd based agents in areas of atherosclerotic plaque (26,43–45). That is, contrast may accumulate at sites of interest due to the effects of kinetics rather than molecular targeting. Therefore careful use of controls and competition experiments is necessary to help establish the basis of contrast effect.

Ligand Identification

Once an appropriate molecular target has been identified, appropriate candidate ligands must be identified and selected. Methods for creating and screening libraries of peptides, such as phage display technology, are useful for identification of optimal peptide ligands for various proteins, such as receptors and antibodies. Using such an approach, a ligand with high affinity (low K_d value) for the target can be identified.

Imaging Probe Construction

Following identification of an appropriate targeting ligand, a molecular imaging probe may be constructed by attaching the ligand to a suitable signal element (typically Gd(III) or Fe oxide). The agent carrier may be the ligand itself or a larger construct to which the ligand and contrast agent is attached (Fig. 9). For magnetic resonance imaging, the imaging probe is generally in the form of a metal chelate bound to a ligand (such as a peptide) or a carrier such as a nanometer-sized particle or liposome with associated signal elements (Gd or Fe oxide) and attached targeting ligand.

The avidity of the molecular probe to target may be optimized by conjugating multiple ligands to the contrast agent carrier (multivalency of ligands). An important consideration in such attachment is preservation of specificity and affinity of binding of the ligand to the target molecule following carrier and signal element attachment. In vitro binding studies are useful to assess whether specific ligand-target binding has been retained.

On the other hand, the signal effect of the molecular probe may be enhanced by attaching multiple signal enhancing molecules (such as Gd chelate or iron oxide) per ligand or carrier (as discussed above under MR Signal Amplification).

To verify that the Gd uptake and consequent T1 shortening effect of the molecular probe is due to paramagnetic Gd molecules bound to its specific ligand, a non-paramagnetic analog of these probe can be synthesized (i.e., gadolinium (Gd) replaced by yttrium (Y) or lanthanum (La)). Y and La have negligible MR contrast properties. Displacement of Gd-labeled ligands by non-paramagnetic labeled ligands can then be demonstrated by MR imaging in cell cultures or animal models using an excess (e.g., 10-fold dose) of the non-paramagnetic labeled agents.

SOME POTENTIAL APPLICATIONS FOR CARDIOVASCULAR MOLECULAR MR

Targeted Imaging

A variety of MR probes have been developed to study various biological processes (e.g., thrombosis, angiogenesis, inflammation, and neoplasia) and diseases (e.g., cancer, cardiovascular disease, stroke, and diabetes) by targeting a spectrum of molecular markers as summarized in Table 2. Most of these agents are in the preclinical stage, with only a few approved for clinical use. The table provides examples of contrast agents under development for investigational and diagnostic purposes with potential application for monitoring of cardiovascular therapies. Of note, probes being developed for use in one context may well have applicability for other disease processes. For example, markers of

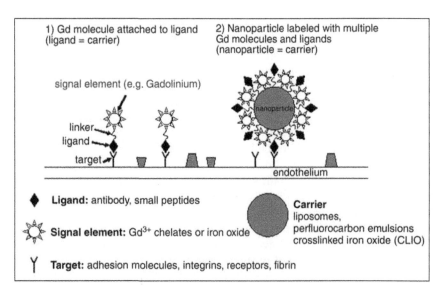

Figure 9 Schematic of molecular contrast agents targeted against endothelial activation. The basic components of each molecular contrast agent consist of a ligand that binds to a specific target and a signal element, which in case of MR, is made of a Gd³⁺ chelate or an iron oxide. These two basic components can be directly (1) linked to each other or may be attached to or incorporated within a larger nanoparticle (=carrier) as demonstrated in (2).

Table 2 Examples of MR Molecular Imaging Probes

Biological processes	Targets	Ligand	Carrier	Signal generating component	Size/ weight	Relaxivities mM⁻¹s⁻¹	Disease	References
Thrombosis	Fibrin	Anti-fibrin F(ab)' fragment	Perfluoro-carbon nanoparticle	10,000–50,000 Gd³⁺	~250 nm	r1 = 0.18–0.54 mL x s⁻¹ x pmol⁻¹ /nanoparticle	CVD	47
	Fibrin	Peptide	Peptide	4 Gd³⁺	~4000 kDa	r1= 21/Gd³; r1= 84 /molecule	CVD	8
	Platelets	RGD-peptide	USPIO nanoparticle	USPIO			CVD	56
Angiogenesis	αvβ3	Peptidomimetic vitronectin antagonist	Nanoparticle	~90,000 Gd³⁺	~270 nm	18/25 (/Gd³⁺) 1.7×10⁶/2.4×10⁶ (/nanoparticle)	CVD, cancer	49,50
	E-selectin	anti-human E-selectin F(ab')₂ fragment	CLIO nanoparticle	CLIO	~40 nm	0.3–0.6mg Fe/ml T2 = 29–40 ms bound T2 ≅ 1500 ms Unbound	CVD, cancer	57
Apoptosis	Phosphatidyl serine	Annexin-V	CLIO Nanoparticle	CLIO	~40 nm	0.3–0.6mg Fe/ml T2 = 29–40 ms bound T2 ≅ 1500 ms Unbound	CVD, cancer	58
Vascular inflammation	E-selectin	anti-human E-selectin F(ab')₂ fragment	CLIO nanoparticle	CLIO	~40 nm	0.3–0.6mg Fe/ml T2 = 29–40 ms bound T2 ≅ 1500 ms unbound	CVD, cancer	57
Neoplasia	macrophage		USPIO nanoparticle	USPIO	~20–30 nm	r1 = 7 r2 = 81	CVD, CNS	59,60

angiogenesis may be important for study of neoplasia and atherosclerotic lesions, as well as to study therapies aimed to stimulate new vessel growth. Later chapters of this book are devoted to molecular MRI of atherosclerosis and angiogenesis so they will be only briefly mentioned here.

MRI of Thrombosis

Imaging of fibrin has potential clinical applications for diagnosis of several significant medical conditions including acute coronary syndromes, deep venous thrombosis, and pulmonary emboli. In recent studies by Yu (46), Flacke (47), and by our group (7,8), Gd labeled fibrin-avid nanoparticles and small peptides have been successfully used for imaging of thrombus in the jugular vein (47), aorta (7), the pulmonary (48), and coronary arteries (8,48). Figure 10 demonstrates imaging of acute thrombus in an animal model of plaque rupture with EP-1873. Gadolinium concentrations as low as ~50 μM (r1 ≅ 21 mM⁻¹s⁻¹ per Gd) were sufficient for ready visualization of mural and lumen encroaching thrombus. A similar compound, EP-2104R,

enabled imaging of coronary in-stent thrombosis (Fig. 11) (8) and pulmonary embolism (48) in an experimental animal model. The administered dose was 4–7.5 μmol/kg, much lower than that for conventional non-targeted Gadolinium based contrast agents (typically ~0.1 mmol Kg⁻¹).

MRI of Integrins

Integrins, such as αvβ3 are over-expressed in activated neovascular endothelial cells, which are believed to play an integral role in tumor growth and the initiation and development of atherosclerosis. Wickline, Lanza and co-workers have developed perfluoronanoparticles that can carry as many as 90,000 paramagnetic Gd chelates per particle and which can be targeted against various biomarkers by attaching appropriate ligands. In a recent study, they directed such nanoparticles to the αvβ3-integrin by attaching a peptidomimetic vitronectin antagonist. Using this approach, they were able to image angiogenesis in nascent Vx-2 rabbit tumors (49) and in early stage atherosclerosis (50). This approach will be discussed

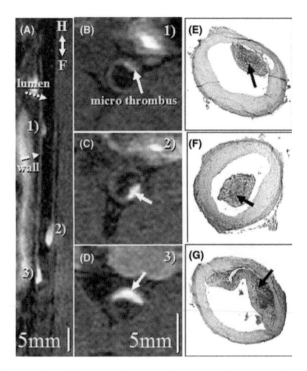

Figure 10 (**A**) Reformatted view of a coronal 3D data set showing the sub-renal aorta approximately 20 hours post EP-1873 administration. Three well delineated mural thrombi (*arrows*) can be observed with good contrast between thrombus (numbered), arterial blood (*dotted arrow*), and the vessel wall (*dashed arrow*). The in-plane view of the aorta allows simultaneous display of all thrombi showing head, tail, length, and relative location. (**B–D**) Corresponding cross-sectional views show good agreement with histopathology (**E–G**). *Source*: From Ref. 7.

in detail in later chapters on MRI of angiogenesis and atherosclerosis.

MRI of Endothelial Activation

Vascular inflammation and associated endothelial activation is believed to play an integral role in initiation and progression of atherosclerosis. Endothelial activation is characterized by the upregulation of leukocyte adhesion molecules such as E- and P-selectin, which facilitate adhesion and migration of monocytes. Differentiation of monocytes into macrophages and subsequent digestion of lipoproteins by macrophages occur in a later stage and eventually lead to the accumulation of lipid filled macrophages, which are believed to be a precursor of rupture-prone vulnerable plaque.

In a recent studies by Sibson et al. (51) and Barber et al. (52), early endothelial activation was observed in focal ischemia in mice brains (Fig. 12) (52) and in brain inflammation in rats (after IL-1β and TNF-α induced E- and P-selectin upregulation) using a novel MR contrast agent. This novel Gd labeled contrast agent, Gd-DTPA-B(sLe^x)A (53), consists of the Sialyl Lewis^x (sLe^x) carbohydrate, which interacts with both E- and P-selectin. The relaxivity was measured as 3.5 mM^{-1}s^{-1}at 1.5T and thus is similar to Gd-DTPA. These promising results encourage further studies and other potential applications including potential assessment of E- and P-selection upregulation in early atherosclerotic lesions.

MRI of Stem Cell Therapy

Stem cell therapy holds enormous therapeutic potential for a broad spectrum of diseases and thus is being investigated at major research institutes world-wide. Non-invasive monitoring of stem cell delivery, migration and therapy are essential to optimize various delivery approaches (systemic vs. local delivery) and to assess the potential impact of cell engraftment and differentiation on various disease processes.

With the recent demonstration of stem cells successfully labeled with MR contrast agents, non-invasive imaging of stem cell therapy has become feasible. Labeling is performed by incubation of stem cells with ferumoxides injectable solution (25 µg Fe/mL, Feridex, Berlex Laboratories, Montville, New Jersey) in culture

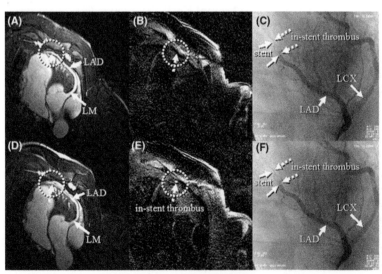

Figure 11 In-vivo MR molecular imaging of coronary in-stent thrombosis. Bright blood images of the left main and left anterior descending coronary arteries before (**A**) and after (**D**) stent placement and injection of a fibrin-binding Gadolinium-labeled contrast agent, EP-2104R. No apparent thrombus and no stent artifacts are visible on the post stent placement and post EP-2104R bright blood images (**D**). Black blood IR images before (**B**) and after stent placement and EP-2104R (**E**). A bright spot (*arrow*) is visible after intra-coronary injection of EP-2104R (**E**). Thrombus was subsequently confirmed by X-ray angiography (**C, F**). *Source*: From Ref. 8.

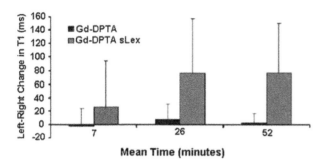

Figure 12 The graph demonstrates the mean pre–post contrast T1 differences between hemispheres within the region of interest (the stroke lesion) measured within three mean time periods (*p < 0.05, different between groups). *Source*: From Ref. 52.

medium for 24–48 hours with 375 ng/mL poly-L-lysine (54,55). Stem cell viability after magnetic labeling was found to be >95% as determined trypan blue exclusion. In an early study using this approach, magnetically labeled stem cells were delivered via direct injection to the site of injured myocardium in a swine model of myocardial infarction. Subsequent imaging using a gradient echo imaging sequence allowed determination of the size and location of each injection (Fig. 13) (55). Ongoing studies are investigating the impact of systemic and local stem cell injection on improvement in cardiac function and in better defining the underlying biology of stem cell differentiation. Although this field is still at a very early stage, further advances in development of novel approaches with potential therapeutic application are expected.

CONCLUSION

Molecular MRI has great promise as a tool to improve understanding of biologic processes and to aid in clinical diagnosis and monitoring of response to treatment.

Although progress in the field must overcome numerous technical challenges to optimize imaging methods and develop novel agents to detect molecular targets, a multidisciplinary approach to these problems will lead to further advances in this field.

ACKNOWLEGMENTS

We like to thank Drs. Phil Barber, PhD, Peter Caravan, PhD, Hannes Dahnke, PhD, Dara Kraitchman VMD, PhD, and Tobias Schaeffter, PhD for providing some of the material for the figures. Drs. Yeon, Spuentrup, and Botnar receive research grant support from EPIX Pharmaceuticals, Inc. Dr. Manning receives research grant support from Philips Medical Systems and EPIX Pharmaceuticals, Inc. He is also a consultant to EPIX Pharmaceuticals, Inc.

REFERENCES

1. Johnson GA, et al. Histology by magnetic resonance microscopy. Magn Reson Q 1993; 9:1–30.
2. Weinmann HJ, Brasch RC, Press WR, Wesbey GE. Characteristics of gadolinium-DTPA complex: a potential NMR contrast agent. AJR Am J Roentgenol 1984; 142:619–624.
3. Laniado M, Weinmann HJ, Schorner W, Felix R, Speck U. First use of GdDTPA/dimeglumine in man. Physiol Chem Phys Med NMR 1984; 16:157–165.
4. Caravan P, Ellison JJ, McMurry TJ, Lauffer RB. Gadolinium(III) chelates as MRI contrast agents: structure, dynamics, and applications. Chem Rev 1999; 99:2293–2352.
5. Weissleder R, et al. Ultrasmall superparamagnetic iron oxide: characterization of a new class of contrast agents for MR imaging. Radiology 1990; 175:489–493.
6. Fischer SE, Wickline SA, Lorenz CH. Novel real-time R-wave detection algorithm based on the vectorcardiogram for accurate gated magnetic resonance acquisitions. Magn Reson Med 1999; 42:361–370.

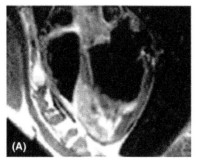 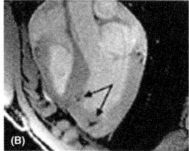 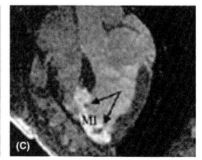

Figure 13 Representative hypointense lesions in double-inversion recovery fast spin echo (FSE) (**A**), fast gradient-recalled echo (FGRE) (**B**), and delayed (15 post contrast) enhancement (DE)-MRI (**C**) of ferumoxide labeled mesenchymal stem cells (MR-MSC) injection sites (*arrows*) within 24 hours of injection. MR-MSCs were injected into the myocardial infarction (MI, hyperintense region in **C**). *Source*: From Ref. 55.

7. Botnar RM, et al. In vivo molecular imaging of acute and subacute thrombosis using a fibrin-binding magnetic resonance imaging contrast agent. Circulation 2004; 109:2023–2029.

8. Botnar RM, et al. In vivo magnetic resonance imaging of coronary thrombosis using a fibrin-binding molecular magnetic resonance contrast agent. Circulation 2004; 110:1463–1466.

9. Niendorf T, Hardy CJ, Giaquinto RO, et al. Toward single breath-hold whole-heart coverage coronary MRA using highly accelerated parallel imaging with a 32-channel MR system. Magn Reson Med 2006; 56(1):167–176.

10. Look DC, Locker DR. Time saving in measurement of NMR and EPR relaxation times. Rev Sci Instrum 1970; 41:250–251.

11. Henderson E, McKinnon G, Lee TY, Rutt BK. A fast 3D look-locker method for volumetric T1 mapping. Magn Reson Imaging 1999; 17:1163–1171.

12. Messroghli DR, et al. Modified Look-Locker inversion recovery (MOLLI) for high-resolution T1 mapping of the heart. Magn Reson Med 2004; 52:141–146.

13. Reeder SB, Faranesh AZ, Boxerman JL, McVeigh ER. In vivo measurement of T^*2 and field inhomogeneity maps in the human heart at 1.5 T. Magn Reson Med 1998; 39:988–998.

14. Clare S, Francis S, Morris PG, Bowtell, R. Single-shot T2(*) measurement to establish optimum echo time for fMRI: studies of the visual, motor, and auditory cortices at 3.0 T. Magn Reson Med 2001; 45:930–933.

15. Dahnke H, Weiss S, Schaeffter T. Simultaneous T2* mapping and anatomical imaging using a fast radial multi-gradient-echo acquisition. in ISMRM 1745 (International Society for Magnetic Resonance Imaging, Kyoto, 2004).

16. Dahnke H, Schäffter, T. Limits of detection of SPIO at 3.0T using T2* relaxometry. Magn Reson Med 2005; in press.

17. Moody AR, et al. Characterization of complicated carotid plaque with magnetic resonance direct thrombus imaging in patients with cerebral ischemia. Circulation 2003; 107:3047–3052.

18. Murphy RE, et al. Prevalence of complicated carotid atheroma as detected by magnetic resonance direct thrombus imaging in patients with suspected carotid artery stenosis and previous acute cerebral ischemia. Circulation 2003; 107:3053–3058.

19. Toussaint JF, LaMuraglia GM, Southern JF, Fuster V, Kantor HL. Magnetic resonance images lipid, fibrous, calcified, hemorrhagic, and thrombotic components of human atherosclerosis in vivo. Circulation 1996; 94:932–938.

20. Yuan C, et al. In vitro and in situ magnetic resonance imaging signal features of atherosclerotic plaque-associated lipids. Arterioscler Thromb Vasc Biol 1997; 17:1496–1503.

21. Fayad ZA, et al. Noninvasive in vivo high-resolution magnetic resonance imaging of atherosclerotic lesions in genetically engineered mice. Circulation 1998; 1541–1547.

22. Moody AR. Direct imaging of deep-vein thrombosis with magnetic resonance imaging. Lancet 1997; 350:1073.

23. van Beek EJ, et al. MRI for the diagnosis of pulmonary embolism. J Magn Reson Imaging 2003; 18:627–640.

24. Yuan C, et al. In vivo accuracy of multispectral magnetic resonance imaging for identifying lipid-rich necrotic cores and intraplaque hemorrhage in advanced human carotid plaques. Circulation 2001; 104: 2051–2056.

25. Fayad ZA, et al. In vivo magnetic resonance evaluation of atherosclerotic plaques in the human thoracic aorta: a comparison with transesophageal echocardiography. Circulation 2000; 101:2503–2509.

26. Sirol M, et al. Lipid-rich atherosclerotic plaques detected by gadofluorine-enhanced in vivo magnetic resonance imaging. Circulation 2004; 109:2890–2896.

27. Bach-Gansmo T. Ferrimagnetic susceptibility contrast agents. Acta Radiol Suppl 1993; 387:1–30.

28. Nivorozhkin AL, et al. Enzyme-activated Gd(3+) magnetic resonance imaging contrast agents with a prominent receptor-induced magnetization enhancement. We thank Dr. Shrikumar Nair for helpful discussions. Angew Chem Int Ed Engl 2001; 40:2903–2906.

29. Aime S, et al. Insights into the use of paramagnetic Gd(III) complexes in MR-molecular imaging investigations. J Magn Reson Imaging 2002; 16:394–406.

30. Ferrucci JT, Stark DD. Iron oxide-enhanced MR imaging of the liver and spleen: review of the first 5 years. AJR Am J Roentgenol 1990; 155:943–950.

31. Small WC, Nelson RC, Bernardino ME. Dual contrast enhancement of both T1- and T2-weighted sequences using ultrasmall superparamagnetic iron oxide. Magn Reson Imaging 1993; 11:645–654.

32. Bousquet JC, et al. Gd-DOTA: characterization of a new paramagnetic complex. Radiology 1988; 166: 693–698.

33. Weinmann HJ, Press WR, Gries H. Tolerance of extracellular contrast agents for magnetic resonance imaging. Invest Radiol 1990; 25(suppl 1):S49–S50.

34. Bartolini ME, et al. An investigation of the toxicity of gadolinium based MRI contrast agents using neutron activation analysis. Magn Reson Imaging 2003; 21: 541–544.

35. Niendorf HP, Haustein J, Cornelius I, Alhassan A, Clauss W. Safety of gadolinium-DTPA: extended clinical experience. Magn Reson Med 1991; 22:222–228; discussion 229–32.

36. Pouliquen D, Le Jeune JJ, Perdrisot R, Ermias A, Jallet P. Iron oxide nanoparticles for use as an MRI contrast agent: pharmacokinetics and metabolism. Magn Reson Imaging 1991; 9:275–283.

37. Schmitz SA, et al. Superparamagnetic iron oxide-enhanced MRI of atherosclerotic plaques in Watanabe heritable hyperlipidemic rabbits. Invest Radiol 2000; 35:460–471.

38. Ruehm SG, Corot C, Vogt P, Kolb S, Debatin JF. Magnetic resonance imaging of atherosclerotic plaque with ultrasmall superparamagnetic particles of iron oxide in hyperlipidemic rabbits. Circulation 2001; 103:415–422.

39. Kooi ME, et al. Accumulation of ultrasmall superparamagnetic particles of iron oxide in human atherosclerotic plaques can be detected by in vivo magnetic resonance imaging. Circulation 2003; 107:2453–2458.

40. Pirko I, et al. In vivo magnetic resonance imaging of immune cells in the central nervous system with superparamagnetic antibodies. Faseb J 2004; 18:179–182.

41. Gohr-Rosenthal S, Schmitt-Willich H, Ebert W, Conrad J. The demonstration of human tumors on nude mice using gadolinium-labelled monoclonal antibodies for magnetic resonance imaging. Invest Radiol 1993; 28:789–795.

42. Artemov D. Molecular magnetic resonance imaging with targeted contrast agents. J Cell Biochem 2003; 90:518–524.

43. Yuan C, et al. Contrast-enhanced high resolution MRI for atherosclerotic carotid artery tissue characterization. J Magn Reson Imaging 2002; 15:62–67.

44. Kramer CM, et al. Magnetic resonance imaging identifies the fibrous cap in atherosclerotic abdominal aortic aneurysm. Circulation 2004; 109:1016–1021.

45. Weinmann HJ, Ebert W, Misselwitz B, Schmitt-Willich H. Tissue-specific MR contrast agents. Eur J Radiol 2003; 46:33–44.

46. Yu X, et al. High-resolution MRI characterization of human thrombus using a novel fibrin-targeted paramagnetic nanoparticle contrast agent. Magn Reson Med 2000; 44:867–872.

47. Flacke S, et al. Novel MRI contrast agent for molecular imaging of fibrin: implications for detecting vulnerable plaques. Circulation 2001; 104:1280–1285.

48. Spuentrup E, et al. Molecular magnetic resonance imaging of coronary thrombosis and pulmonary emboli with a novel fibrin-targeted contrast agent. Circulation 2005.

49. Winter PM, et al. Molecular imaging of angiogenesis in nascent Vx-2 rabbit tumors using a novel alpha(nu)beta3-targeted nanoparticle and 1.5 tesla magnetic resonance imaging. Cancer Res 2003; 63:5838–5843.

50. Winter PM, et al. Molecular imaging of angiogenesis in early-stage atherosclerosis with alpha(v)beta3-integrin-targeted nanoparticles. Circulation 2003; 108:2270–2274.

51. Sibson NR, et al. MRI detection of early endothelial activation in brain inflammation. Magn Reson Med 2004; 51:248–252.

52. Barber PA, et al. MR molecular imaging of early endothelial activation in focal ischemia. Ann Neurol 2004; 56:116–120.

53. Laurent S, Vander Elst L, Fu Y, Muller RN. Synthesis and physicochemical characterization of Gd-DTPA-B(sLex)A, a new MRI contrast agent targeted to inflammation. Bioconjug Chem 2004; 15:99–103.

54. Bulte JW, et al. Magnetodendrimers allow endosomal magnetic labeling and in vivo tracking of stem cells. Nat Biotechnol 2001; 19:1141–1147.

55. Kraitchman DL, et al. In vivo magnetic resonance imaging of mesenchymal stem cells in myocardial infarction. Circulation 2003; 107:2290–2293.

56. Johansson LO, Bjornerud A, Ahlstrom HK, Ladd DL, Fujii DK. A targeted contrast agent for magnetic resonance imaging of thrombus: implications of spatial resolution. J Magn Reson Imaging 2001; 13:615–618.

57. Kang HW, Josephson L, Petrovsky A, Weissleder R, Bogdanov A Jr. Magnetic resonance imaging of inducible E-selectin expression in human endothelial cell culture. Bioconjug Chem 2002; 13:122–127.

58. Schellenberger EA, Hogemann D, Josephson L, Weissleder R. Annexin V-CLIO: a nanoparticle for detecting apoptosis by MRI. Acad Radiol 2002; 9(suppl 2):S310–1.

59. Rogers J, Lewis J, Josephson L. Use of AMI-227 as an oral MR contrast agent. Magn Reson Imaging 1994; 12:631–639.

60. Dousset V, et al. In vivo macrophage activity imaging in the central nervous system detected by magnetic resonance. Magn Reson Med 1999; 41:329–333.

6

Molecular and Cellular Imaging with Targeted Contrast Ultrasound

Carolyn Z. Behm
University of Virginia, Charlottesville, Virginia, U.S.A.

Jonathan R. Lindner†
Cardiovascular Division, Oregon Health and Science University, Portland, Oregon, U.S.A.

Non-invasive medical imaging has become an essential part of the practice of cardiovascular medicine and traditionally relies on the ability to characterize disease-related anatomic and/or physiologic changes. In the past decade, there has been considerable progress in the development of new techniques for imaging the pathophysiologic molecular processes that are responsible for initiation and progression of diseases. This chapter focuses on recent advances that have allowed the imaging of molecular and cellular alterations of disease with targeted contrast enhanced ultrasound (CEU).

TARGETED ULTRASOUND CONTRAST AGENTS

In its most common form, site-targeted imaging relies on the detection of selectively retained contrast agent at a prespecified site of disease. Conventionally, CEU is performed by ultrasound detection of gas-containing microbubbles that have shells composed of protein, lipids, or biocompatible polymers. Agents that are in clinical use worldwide are generally several microns in diameter and are pure intravascular tracers that behave similar to red cells in the microcirculation (1,2). For the purposes of targeting, microbubbles can be retained in diseased tissue by virtue of ligands that are conjugated to the microbubble shell surface or, in a more simple fashion, by changing certain key chemical properties of the microbubble shell. When developing a targeting strategy for any modality, there are many considerations regarding the behavior of the contrast agent and the properties of the molecular target. Some of the issues that pertain to microbubbles are schematically

illustrated in Figure 1. One key issue is the biodistribution and biokinetics of microbubbles. Because of their localization to the vascular space, microbubbles have generally been targeted to antigens expressed on either endothelial cells or blood components such as leukocytes, platelets, or thrombi. The confinement of microbubble agents to the vascular compartment does limit the potential cell types to be targeted. However, this property can be advantageous in certain circumstances. For example, the constitutive or induced expression of a targeted molecule in nonvascular cell types would confound selective assessment of vascular phenotype for diffusible agents but not for pure intravascular tracers. Ultrasound contrast agents that are smaller than conventional microbubbles (submicron) have also been developed that are composed of either liposomes that contain a small amount of air (3), or perfluorocarbon emulsions (4). By virtue of their size and composition, some of these agents are capable of extravascular migration in regions of vascular injury or regions where vascular permeability is abnormally high (4).

The relative rate at which free tracer is cleared is an important issue and determines temporal resolution since targeted imaging is adversely affect by high concentrations of freely circulating or diffusing unbound tracer. In this regard, microbubble tracers have an advantage over most other types of tracers since they are for the most part cleared from the circulation within minutes after injection. Rapid clearance of free microbubbles provides for a low free-circulating background signal soon after their injection. However, because circulation time is limited, the in vivo affinity of microbubbles for target must be relatively high. Targeted microbubbles are multivalent particles and the determinants of their attachment in many ways reflects that of biologic systems such as the process of leukocyte recruitment. The likelihood of attachment of microbubbles in the face of vascular shear stresses depends on the number of ligands on the surface, the

†Dr. Lindner is supported by grants from the National Institutes of Health (R01-DK063508, R01-HL074443, and R01-HL078610), Bethesda, Maryland, U.S.A.; and a Grant-in-Aid from the American Heart Association Mid-Atlantic Affiliate, Baltimore, Maryland, U.S.A.

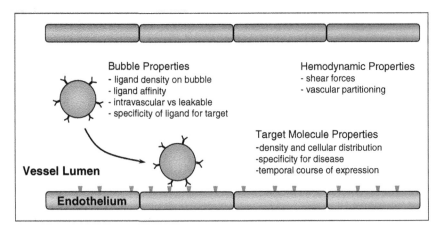

Figure 1 Considerations for the development of targeting with microbubble ultrasound contrast agents.

on-rates and off-rates of the ligand for its target, and the density of target expression (Fig. 1).

The properties of the imaging detectors must also be considered when developing a method for molecular assessment of disease. Ideally, an imaging modality should be sensitive with high signal-to-noise ratio, have a good spatial resolution, and require only a brief time for image acquisition. For conventional forms of non-invasive contrast-enhanced imaging, there tends to be an inverse relationship between sensitivity and resolution. CEU neither has the sensitivity of radionuclide imaging nor the spatial resolution of magnetic resonance (at conventional frequencies used for CEU). However, CEU is generally regarded to be the most balanced in terms of these two considerations. CEU is superior to all modalities in terms of its temporal resolution requiring only minutes for tracer injection, clearance of free tracer, and image acquisition.

MICROBUBBLE-ULTRASOUND INTERACTIONS

Expanding on the founding principles of Lord Raleigh and Plesset, Noltingk and Nepiras described in theory the behaviour of gas bubbles exposed to an acoustic field in the mid 20th century (5,6). One of the most important considerations is the relative compressibility of the gas-containing microbubbles in relation to that of the surrounding medium (5–7). The signal intensity for a gas body increases non-linearly with the microbubble radius. Hence, microbubbles must be sufficiently large to produce a robust signal but small enough to transit the microcirculation unimpeded. Optimization of size distribution and maintenance of this size after injection has been achieved by shell encapsulation and/or use of high-molecular weight gases that have low solubility and diffusivity.

The ability to distinguish the signal generated by microbubbles from that of the surrounding tissue during imaging is enhanced by the non-linear behaviour of microbubbles (8). Provided the transmission

frequency is appropriate, the radial oscillation in microbubble volume becomes exaggerated and non-linearly related to the acoustic pressure, thereby producing harmonics (8,9). It is now also thought that exaggerated radial oscillation can result in release of free gas bubbles that are free of the damping effects of a shell and can therefore generate very large signals at the fundamental (transmitted) and harmonic frequencies. Tissue in the same ultrasound field produces a significantly lower relative signal in the harmonic ranges. Improvement in bubble to tissue signal ratio can be achieved by receiving at the second harmonic, between the fundamental and second harmonic frequency, or beyond the second harmonic (Fig 2) (10). Filtering to receive non-linear microbubble signal during high-power imaging (MI ≥ 0.9) or even non-linear signal at the fundamental frequency has been used to detect retained targeted microbubbles during molecular imaging due to the need to maximize sensitivity for agent detection. Smaller perfluorocarbon nanoparticles and immunoliposomes have generally been imaged with fundamental imaging at higher frequencies than that used for microbubble agents.

TARGETED MICROBUBBLE DESIGN AND DETECTION

Acosutically active agents for molecular imaging have been developed using two different targeting strategies (Fig. 3). The first and more simple strategy involves the manipulation of the chemical or charge properties of the microbubble or microparticle shell. The shells of microbubbles, nanoparticles, or liposomes may also be modified by the surface conjugation of ligand molecules, such as monoclonal antibodies, peptides, glycoproteins, etc. Most formulations involve surface conjugation of ligand after microbubbles have been prepared since this strategy generally requires less ligand and avoids denaturing or other alteration in the ligand.

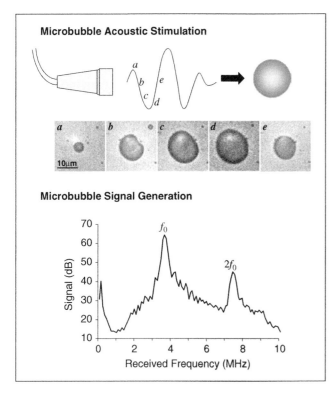

Figure 2 Schematic illustration of the acoustic properties of microbubble contrast agents. The microscopy images obtained 330 ns apart demonstrate volumetric oscillation of a microbubble during exposure to ultrasound (500 KHz) that occurs during high and low pressure phases. Microbubble images were recorded at a constant magnification. Frequency versus amplitude data from microbubbles demonstrating returning signal at the fundamental (f_0) and second harmonic ($2f_0$) frequencies, as well as between the harmonic peaks. *Source*: Courtesy of Postema M, de Jong N, Erasmus University (microbubble images); Burns P, University of Toronto (revised, frequency versus amplitude data).

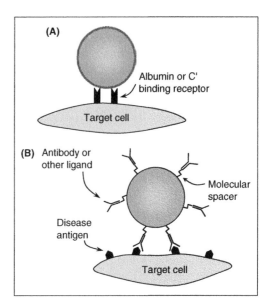

Figure 3 Strategies for targeting microbubbles to disease. Microbubbles or other microparticles can be targeted to regions of disease either by chemical or charge properties of the shell constituents that augment non-specific interactions with certain upregulated receptors, or by surface conjugation of specific ligands or antibodies that bind to disease-related antigens.

CEU imaging protocols that have been developed to assess tissue perfusion are not directly applicable to the imaging to retained contrast agents. One challenge for imaging retained site-targeted microbubbles is to adequately exclude the signal from circulating, non-retained tracer. The solution for microbubble agents has been to use a delayed imaging protocol for molecular imaging with contrast-enhanced ultrasound, similar to that employed with most other imaging methods (Fig. 4) (11). The tracer is administered as a single intravenous bolus injection. Attachment to the target molecule is greatest early after injection when the freely circulating concentration is the highest. Imaging of retained tracer is delayed in order to allow freely circulating or diffusible tracer to be cleared. The relative heights and time intervals of the curves illustrated in Figure 4 depend largely on the tracer used. As mentioned previously, targeted imaging with microbubbles can be performed at 8–15 minutes due to their rapid clearance by the reticuloendothelial system.

Image processing algorithms have been developed to derive signal only from retained or attached microbubbles (Fig. 4) (11,12). The first frame of ultrasound obtained after resumption of imaging is acquired, and contains signal both retained and the few remaining circulating microbubbles in the tissue. The ultrasound

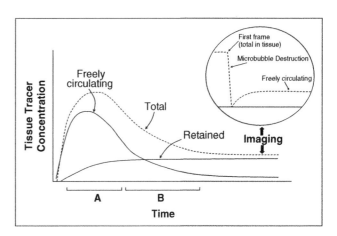

Figure 4 Schematic representation of the protocol for imaging retained microbubbles. After a bolus injection of microbubbles, most of the signal in tissue is attributable to free tracer in the blood pool. The rate of accumulation at the target will be the greatest when free tracer concentration is at its highest (interval A). Time is then allowed for free tracer to be cleared (interval B). Imaging is delayed until a time when most remaining tracer in a diseased tissue is from targeted attachment. At the time of imaging, the relative signal from retained microbubbles can be determined by the difference between the initial signal and the signal obtained late after microbubble destruction.

beam is rapidly cleared of all microbubbles applying high-power imaging that disrupts all microbubbles. After this destruction phase, imaging is then repeated in a fashion to obtain the signal from freely circulating microbubbles. This latter signal from the low concentration of circulating microbubbles can be digitally subtracted from that initially obtained upon resumption of imaging to yield a single image reflecting retained microbubbles alone.

SPECIFIC APPLICATIONS

Inflammation

The inflammatory response is a key feature in many cardiovascular diseases. A key part of this response is the recruitment of circulating leukocytes to the vascular endothelium through leukocyte capture, rolling, adhesion, and transmigration. These events are mediated by the expression of both leukocyte adhesion molecules and endothelial cell adhesion molecules (ECAMs) (13). Families of ECAMs have been defined and their roles in inflammatory cell recruitment are increasingly well understood. Initial leukocyte capture and rolling on the post-capillary venular endothelium is largely mediated by the selectin family (14). Rolling leukocytes become progressively activated, and are capable of firm adhesion mediated by interactions between activated integrins on the leukocyte surface with ECAM's (ICAM-1 and VCAM-1) expressed by activated endothelial cells (13). Adherent activated leukocytes then can undergo transendothelial migration. The increased understanding of the molecular processes involved in leukocyte recruitment has provided a variety of possible targets for molecular imaging strategies.

Microbubble retention by activated leukocytes adherent to the microvascular endothelium in regions of local inflammation forms the basis of one strategy for imaging inflammation. Both albumin and lipid-shelled microbubbles can interact with activated leukocytes. Attachment of albumin-shelled microbubbles is mediated mostly by the β_2-integrin Mac-1, whereas lipid microbubbles are retained due to opsonization with serum complement (15). Microbubbles can even be phagocytosed soon after attachment to the activated leukocytes (12).

The avidity of microbubbles for activated leukocytes has been augmented by shell alterations that increase the likelihood for complement deposition such as the addition of lipid moieties that carry a net ionic charge at physiologic pH. For example, the addition of a small amount of phosphatidylserine (PS) to the lipid shell has been shown to increase complement deposition on the microbubble surface, thereby increasing the retention of microbubbles by activated

leukocytes on the vascular endothelium (11). This strategy has been used to image the severity and spatial extent of inflammation following ischemia reperfusion injury in the kidney and the myocardium (Fig. 5) (11,16). Another strategy that can increase microbubble-cell interactions is the absence of a protective polymeric layer [usually polyethyleneglycol (PEG)] that is normally present on many lipid microbubble preparations. Witihout a PEG coat, there is greater surface deposition of complement and, more importantly, greater ability of microbubbles to interact with cells, including even normal endothelial cells (17). Negatively charged microbubbles lacking a PEG coat have been shown to enhance the myocardium in the setting of ischemia-reperfusion injury and cardiac transplant rejection, although the exact mechanism for microbubble retention has yet to be definitively shown (18,19).

An alternative strategy for imaging inflammation relies on microbubble surface conjugation of ligands for ECAMs. The ligands investigated thus far have been monoclonal antibodies, including those to P-selectin (20), ICAM-1 (21,22), VCAM-1 (23,24), and MAdCAM-1 (25). Flow cytometry and quantitative fluorometry have demonstrated that >60000 ligands can be conjugated to the surface of each individual microbubble (20,26). The capability of CEU and

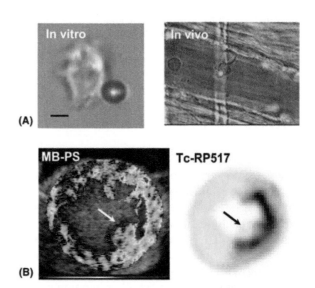

(A)

(B)

Figure 5 Targeted CEU imaging of post-ischemic myocardial injury. (**A**) Microscopy illustrating non-specific interaction between lipid microbubbles and activated leukocytes in vitro, and by intravital microscopy in vivo where a fluorescent microbubble is attached to a leukocyte adherent in a TNF-α-stimulated venule. (**B**) Myocardial short axis images obtained in vivo with CEU and phosphatidylserine-containing leukocyte-targeted microbubbles (MB-PS) demonstrating signal enhancement from the inflammatory response following ischemia-reperfusion injury of the left circumflex artery territory, and corresponding region of inflammation defined by radionuclide labeling of the leukotriene-B4 receptor of leukocytes (Tc-RP517). *Source*: From Refs. 11, 16.

ligand-targeted microbubbles to detect early inflammatory responses in vivo after intravenous injection was first demonstrated using P-selectin-targeted microbubbles. Preferential attachment of the P-selectin-targeted microbubbles to inflamed endothelium was observed by intravital microscopy, and CEU imaging demonstrated marked signal enhancement following intravenous injection of P-selectin-targeted microbubbles has been found in the kidney and heart following ischemia-reperfusion injury (20). For the purposes of detecting heterotopic heart transplant rejection, CEU with ICAM-1-targeted microbubbles has been used in a rat model of strain mismatch (Fig. 6) (22).

Inflammation plays a central role in the genesis and progression of atherosclerosis, and there is considerable interest in non-invasive imaging strategies for assessing the vulnerable plaque and the early endothelial manifestations of plaque formation. Both VCAM-1 and ICAM-1 have been investigated as target ligands for CEU in the imaging of large-vessel atherosclerotic disease. Atherosclerosis imaging was first described with acoustically-active sub-micron liposomes with antibody against ICAM-1 conjugated to the membrane. The in vivo binding capacity of this agent has been evaluated in an inflammatory atherosclerotic model in pig carotid arteries (27). After direct intra-arterial injection, anti-ICAM-1 liposomes were observed to attach to the endothelium overlying the atherosclerotic plaque, as evidenced by focal regions of acoustic enhancement with high-frequency ultrasound imaging (Fig. 7A). More recently, phenotypic characterization of lesions according to ECAM expression, presence of tissue factor, and surface microthrombi has been performed with intra-arterial injection of immunoliposomes (23).

Ultrasound contrast agents can also be targeted to regions of inflamed plaque using an intravenous route of contrast administration. Perfluorocarbon-filled lipid microbubbles targeted against VCAM-1 have been shown to adhere preferentially to inflamed atherosclerotic plaques in the aortas of Apo-E-deficient mice (Fig. 7B) (24). These preliminary findings are being further investigated to assess the utility of imaging arterial inflammation for early detection of disease, risk assessment, and evaluation of the response to drug-therapy.

Angiogenesis

There is a rapidly growing interest in the use of molecular imaging of neovascularization, including angiogenesis and arteriogenesis. For oncology applications, imaging of angiogenesis or arteriogenesis may be useful for diagnosing neoplasms, detecting metastases, and assessing the response to tumoricidal therapies including angiogenesis inhibitors. For cadiovascular applications, imaging angiogenesis in ischemic tissue may be useful for studying endogenous neovascularization, the response to pro-angiogenic therapies designed to treat severe coronary artery disease and peripheral vascular disease, and pathologic process during atherogenesis.

The endothelial cells in angiogenic vessels are characterized by the expression of a variety of adhesion molecules, growth factor receptors, and other cell

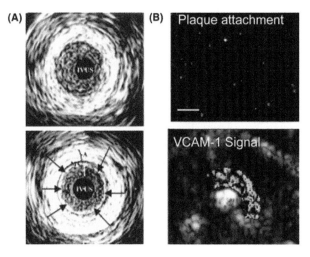

Figure 7 CEU molecular imaging of VCAM-1 expression in atherosclerosis. (**A**) Intravascular ultrasound images of a carotid artery in an atherosclerotic miniswine demonstrating no enhancement after control saline injection (*top*), and enhancement (*arrows*) after intra-arterial injection of VCAM-1-targeted immunoliposomes (*bottom*). (**B**) Ex vivo, en face image of an atherosclerotic plaque in the aortic arch in an ApoE-deficient mouse where fluorescently-labeled VCAM-1-targeted microbubbles have attached 10 min after their intravenous injection (*top*); and transthoracic ultrasound enhancement of the aortic arch in vivo after intravenous injection of VCAM-1-targeted microbubbles. *Source*: From Refs. 23 and 24.

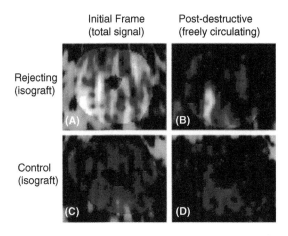

Figure 6 Short-axis CEU images of cardiac allografts after injection of ICAM-1-targeted microbubbles demonstrating strong signal enhancement on the initial frame in the setting of transplant rejection (*top*) and low signal enhancement when rejection is absent (*bottom*). Signal intensity from subsequent post-destructive frames was low, indicating the presence of few freely circulating microbubbles. *Source*: From Ref. 22.

surface proteins that can be targeted by intravascular agents such as microbubbles. The best investigated ligands for targeting are the α-integrins—$\alpha_v\beta_3$ and $\alpha_v\beta_5$—which are differentially expressed in animal and in vitro models of angiogenesis (28,29). Microbubbles have been targeted to angiogenesis by surface conjugation of peptides such as disintegrins that bind to $\alpha_v\beta_3$ and, to a lesser degree, other integrins expressed in angiogenic vessels (such as $\alpha_5\beta_1$) (30). The feasibility of imaging the in vivo angiogenic response to ischemia and to growth factor (FGF-2) therapy has been demonstrated in a rat chronic ischemic hindlimb model (Fig. 8) (31). These studies have demonstrated that the molecular signal of vascular remodeling provides information on neovascularization even prior to any detectable increases in blood flow.

Thrombus

Thrombus-targeted microbubbles have been developed to improve the diagnostic accuracy of ultrasound for detecting vascular or intracardiac thrombi. One strategy for has been to target microbubbles to the platelet glycoprotein IIb/IIIa receptor that is expressed at a very high density on the surface of platelets upon

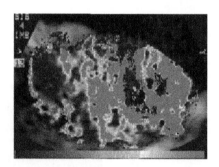

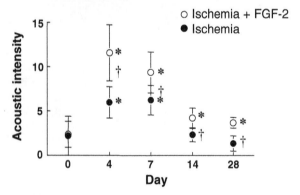

Figure 8 Molecular imaging of ischemia and growth factor-mediated angiogenesis with targeted microbubbles. (*Top*) Signal enhancement in ischemic hindlimb skeletal muscle in rats treated with FGF-210 min after intravenous injection of microbubbles targeted to neointegrins. (*Bottom*) The time course and extent of neointegrin expression from molecular imaging in ischemic limbs (*solid circles*) and ischemic limbs treated with FGF-2 (*open circles*). Signal from control limbs was non-detectable (*data not shown*). *Source*: From Ref. 31.

activation and is necessary platelet aggregation. Lipid-shelled microbubbles bearing a cyclic oligopeptide containing the RGD sequence recognized by the active binding site of platelet IIb/IIIa receptors have been observed to adhere to the surface of newly formed thrombus in vitro and in vivo (32,33).

Fibrin targeting of nanoparticle emulsion agents has also been used to enhance vascular thrombi on ultrasound imaging (34). Intracardiac thrombi in animals have also been detected with intravenous administration of fibrinogen-targeted echogenic immunoliposomes (35). The potential to better identify intracardiac thrombi is a particularly timely issue given the expanded use of echocardiography to identify blood clots in patients undergoing cardiac procedures and the desire to monitor microembolization during treatment of patients with acute myocardial infarction.

UNIQUE ASPECTS OF CEU

The relative advantages and disadvantages for each molecular imaging modality must be considered when contemplating their use in the research, or in the future, clinical setting. As mentioned previously, disadvantages of CEU for molecular imaging include the restriction to molecular process that can be accessed by a pure intravascular tracer, and the use of a complex multivalent particle rather a single labeled targeting molecule. However, there are some distinct advantages of CEU that should be considered as well. For example, CEU is well suited for the characterization of disease pathophysiology in animal models of disease and in genetically manipulated animal lines. For this application, CEU molecular imaging meets many of the key requirements such as high spatial resolution for tissue or organ-specific localization in small animals, high-throughput, and ability to be performed sequentially in the same subject. It is also well suited for assessing response at the molecular level to therapies in animal models and patients since it is quantitative and has excellent temporal resolution which can be important for evaluating acute responses to therapy. Finally, there are also practical considerations when introducing any new imaging technology to clinical practice. Practical advantages of CEU include ease of use, rapidity of imaging protocols, ability to be performed at the bedside or in an outpatient setting, and cost.

SUMMARY AND FUTURE DIRECTIONS

Molecular imaging is a rapidly developing field that is already beginning to play an important role in the research setting, but also has great potential in the case of patients with cardiovascular disease. Feasibility for imaging inflammation, angiogenesis, and thrombus formation with targeted CEU has clearly been demonstrated; and current efforts are underway to improve both the targeting efficiency of microbubbles and the ultrasound detection methods. Given some of the practical advantages of the technique in terms of cost, time, and ease of use, CEU will likely play an important role in the future of non-invasive cardiac imaging.

REFERENCES

1. Lindner JR, Song J, Jayaweera AR, Sklenar J, Kaul S. Microvascular rheology of definity microbubbles following intra-arterial and intravenous administration. J Am Soc Echocardiogr 2002; 15:396–403.
2. Skyba DM, Camarano G, Goodman NC, Price RJ, Skalak TC, Kaul S. Hemodynamic characteristics, myocardial kinetics and microvascular rheology of FS-069, a second generation echocardiographic contrast agent capable of producing myocardial opacification from a venous injection. J Am Coll Cardiol 1996; 28:1292–1300.
3. Huang SL, Hamilton AJ, Pozharski E, et al. Physical correlates of the ultrasonic reflectivity of lipid dispersions suitable as diagnostic contrast agents. Ultrasound Med Biol 2002; 28;339–348.
4. Lanza GM, Abendschein DR, Hall CS, et al. In vivo molecular imaging of stretch-induced tissue factor in carotid arteries with ligand-targeted nanoparticles. J Am Soc Echocardiogr 2000; 13:608–614.
5. Noltingk BE, Neppiras EA. Cavitation produced by ultrasonics. Proc Phys Soc 1950; B63:674–685.
6. Epstein PS, Plesset MS. On the stability of gas bubbles in liquid–gas solutions. J Chem Phys 1950; 18: 1505–1509.
7. de Jong N, Hoff L, Skotland T, Bom N. Absorption and scatter of encapsulated gas-filled microspheres: Theroetical considerations and some measurements. Ultrasonics 1992; 30:95–103.
8. de Jong N, Cornet R, Lancee CT. Higher harmonics of vibrating gas-filled microspheres. Ultrasonics 1994; 32:455–459.
9. Hoff L, Sontum PC, Hoff B. Acoustic properties of shell-encapsulated, gas-filled ultrasound contrast agents. Proceedings of the IEEE Ultrasonics symposium 1996; 1441–1444.
10. Burns PN, Powers JE, Simpson DH, Uhlendorf V, Fritzsche T. Harmonic imaging: Principles and preliminary results. Clin Radiol 1996; 51(suppl):50–55.
11. Lindner JR, Song J, Xu F, et al. Noninvasive ultrasound imaging of inflammation using microbubbles targeted to activated leukocytes. Circulation 2000; 102: 2745–2750.
12. Lindner JR, Dayton PA, Coggins MP, et al. Noninvasive imaging of inflammation by ultrasound detection of phagocytosed microbubbles. Circulation 2000; 102: 531–538.
13. Ley K. Molecular mechanisms of leukocyte recruitment in the inflammatory process. Cardiovasc Res 1996; 32:733–742.
14. Ley K, Tedder TF. Leukocyte interactions with vascular endothelium. J Immunol 1995; 155:525–528.
15. Lindner JR, Coggins MP, Kaul S, Klibanov AL, Brandenburger GH, Ley K. Microbubble persistence in the microcirculation during ischemia/reperfusion and inflammation is caused by integrin- and complement-mediated adherence to activated leukocytes. Circulation 2000; 101:668–675.
16. Christiansen JP, Leong-Poi H, Klibanov AL, Kaul S, Lindner JR. Noninvasive imaging of myocardial reperfusion injury using leukocyte-targeted contrast echocardiography. Circulation 2002; 105:1764–1767.
17. Fisher NG, Christiansen JP, Klibanov AL, Taylor RP, Kaul S, Lindner JR. Influence of surface charge on capillary transit and myocardial contrast enhancement. J Am Coll Cardiol 2002; 40:811–819.
18. Kondo I, Ohmori K, Oshita A, Takeuchi H, et al. Leukocyte-targeted myocardial contrast echocardiography can assess the degree of acute allograft rejection in a rat cardiac transplantation model. Circulation 2004; 109:1056–1061.
19. Kunichika H, Peters B, Cotto B, et al. Visualization of risk-area myocardium as a high-intensity, hyperenhanced "hot spot" by myocardial contrast echocardiography following coronary reperfusion: Quantitative analysis. J Am Coll Cardiol 2003; 42:52–557.
20. Lindner JR, Song J, Christiansen JP, Klibanov AL, Xu F, Ley K. Ultrasound assessment of inflammation and renal tissue injury with microbubbles targeted to P-selectin. Circulation 2001; 104:2107–2112.
21. Villanueva FS, Jankowski RJ, Klibanov AL, Brandenburger GH, Wagner WR. Microbubble targeted to intercellular adhesion molecule-1 bind to activated coronary endothelial cells. Circulation 1998; 98:1–5.
22. Weller GE, Lu E, Csikari MM, et al. Ultrasound imaging of acute cardiac transplant rejection with microbubbles targeted to intercellular adhesion molecule-1. Circulation 2003; 108:218–224.
23. Hamilton AJ, Huang SL, Warnick D, et al. Intravascular ultrasound molecular imaging of atheroma components in vivo. J Am Coll Cardiol 2004; 43:453–460.
24. Kaufmann B, Sanders JM, Davis C, et al. Molecular imaging of inflammation in atherosclerosis with targeted ultrasound detection of vascular cell adhesion molecule-I. Circulation 2007 (in press).
25. Bachmann C, Klibanov AL, Olson T, et al. A novel diagnostic tool for the non-invasive evaluation of intestinal inflammation. Gastroent 2006; 130:8–16.
26. Takalkar AM, Klibanov AL, Rychak JJ, Lindner JR, Ley K. Binding and detachment of microbubbles targeted to p-selectin under controlled shear flow. J Contr Rel 2004; 96:473–482.
27. Demos SM, Alkan-Onyuksel H, Kane BJ, et al. In vivo targeting of acoustically reflective liposomes for intravascular and transvalvular ultrasonic enhancement. J Am Coll Cardiol 1999; 33:867–875.
28. Brooks PC, Clark RAF, Cheresh DA. Requirement of vascular integrin $\alpha v \beta 3$ for angiogenesis. Science 1994; 264:569–571.
29. Friedlander M, Brooks PC, Shaffer R, Cheresh DA. Definition of two angiogenic pathways by distinct αv integrins. Science 1995; 270:1500–1503.

30. Leong-Poi H, Christiansen JP, Klibanov AL, Kaul S, Lindner JR. Non-invasive assessment of angiogenesis by ultrasound and microbubbles targeted to α_v integrins. Circulation 2003; 107:455–460.

31. Leong-Poi H, Christiansen JP, Heppner P, et al. Assessment of endogenous and therapeutic arteriogenesis by integrin imaging with targeted contrast ultrasound. Circulation 2005; 111:3248–3254.

32. Unger E, McCreery TP, Sweitzer RH, Shen D, Wu G, In vitro studies of a new thrombus-specific ultrasound contrast agent. Am J Cardiol 1998; 81:58G–61G.

33. Schumann PA, Christiansen JP, Quigley RM, et al. Targeted microbubble binding selectively to GPIIbIIIa receptors on platelet thrombi. Invest Radiol 2002; 37:587–593.

34. Lanza GM, Abendschein D, Hall C, et al. Molecular imaging of stretch-induced tissue factor expression in carotid arteries with intravascular ultrasound. Invest Radiol 2000; 35:227–234.

35. Hamilton A, Huang SL, Warnick D, et al. Left ventricular thrombus enhancement after intravenous injection of echogenic immunoliposomes. Circulation 2002; 105:2772–2778.

7

Intravascular Radiation Detectors to Detect Vulnerable Atheroma in the Coronary Arteries

Daniel Pryma

Department of Radiology, Memorial Sloan Kettering Cancer Center, New York, New York, U.S.A.

Bradley E. Patt

Gamma Medica Inc., Northridge, California, U.S.A.

H. William Strauss

Department of Radiology, Memorial Sloan Kettering Cancer Center, New York, New York, U.S.A.

INTRODUCTION

Atheroma

Atherosclerosis is a common arterial disease characterized by subendothelial areas of cellular infiltration and degeneration in regions of lipid and cholesterol deposits. Autopsies on premature newborns, particularly those with hyperlipidemic mothers (1), demonstrated fatty streaks, suggesting that atheroma starts in the prenatal environment. Atherosclerosis is a panvascular disease affecting multiple vessels at numerous sites. Individual atheromas evolve at different rates from subendothelial fatty streaks to a ruptured plaque causing thrombosis. The multiplicity of lesions is important because, although there is often just one culprit lesion causing a clinical event, there are many near "culprit" lesions, ready to cause subsequent events. This progression is typically clinically silent for decades, presenting clinically as a catastrophic stroke or myocardial infarction. Detecting atheroma, especially lesions with histologic characteristics suggesting a high likelihood of rupture (vulnerable plaque), can be difficult. Most lesions have repetitive cycles of lipid deposition, inflammation, and healing (often associated with calcification and luminal narrowing). However, about 50% of lipid laden lesions are associated with eccentric dilatation of the vessel.

Eccentric Dilatation

This phenomenon, described by Glagov and his colleagues in 1987 (2), can compensate for atheroma occupying up to 40% of the vessel area. Under this circumstance, the vessel maintains the size of the lumen by accommodating the lesion. Although there is no impingement on the lumen, eccentric arterial dilatation causes an increase in thickness of the vessel in the region of the lipid collection. Computed tomography (CT) and magnatic resonance imaging (MRI) have sufficient resolution to identify these 1–3 mm focal increases in vessel thickness, but the subtle changes are difficult to identify. Progression of atheroma is associated with additional lipid deposition which is irritating to the tissue, leading to an inflammatory response. Inflammation results in attraction of inflammatory cells to the lesion site, further enlargement of the lesion and stimulation of additional vasa vasorum (3), and thinning of the cap covering the lesion. Continued progression exhausts the ability of eccentric dilatation to compensate and the lesion begins to encroach on the lumen. Although most coronary interventions are performed on vessels with significant narrowing, in almost half of all infarcts, lesions in the infarct related artery were not associated with angiographically significant stenoses (i.e., <50% narrowing) prior to the acute event (4). This observation suggests that new methods and new criteria will be required to detect atheroma during the phase of eccentric compensation.

External Imaging

Lesions calcify during the healing phase of atheroma, possibly as a result of macrophage apoptosis (5). As a result, vascular calcification does not imply metabolically active atheroma (6). However, the presence of vascular calcification provides evidence of the presence of atheroma. External imaging with multidetector computed tomography (MDCT) without intravenous contrast provides information about calcification in plaque. MDCT with injection of intravenous contrast depicts the vessel lumen, indicating the presence of late stage disease based on luminal narrowing. Although the resolution of MDCT can define vessel thickness, the technology cannot identify the composition of a lesion. Similarly, MR provides information about the vessel lumen, thickness of the arterial wall, and may indicate the lipid content of the lesion, but cannot define the cellular content or metabolic activity. Although radionuclide techniques cannot provide the spatial resolution of MR or CT, the contrast resolution of radionuclide techniques makes up for this deficiency. Radionuclide techniques can specifically label plaque contents (7) [LDL—using radiolabeled LDL, monocytes, and macrophages—using radiolabeled monocyte chemoattractant peptide (MCP-1)], define the metabolic activity of the cells in the lesion (using fluorodeoxyglucose), or indicate the expression of inflammatory indicators, such as up-regulation of $\alpha_v\beta_3$ receptors (using radiolabeled peptides that bind to these integrins). In the coronaries, the majority of lesions occur within the proximal third of the vessel (8). Eccentric atheromas are often only a few millimeters in length, <1 mm thick, and may involve only a portion of the vessels circumference. The small lesion size makes detection particularly challenging with external imaging. On the other hand, a catheter based system may readily identify these lesions. A probe-like catheter device has the advantage of very high sensitivity for small lesions because of geometric proximity to the lesion and the nearly 4Π geometry of lesions that partially surround the detector. Interrogating these lesions to determine the specific inflammatory components would be particularly helpful to develop a specific course of local or systemic therapy.

Radionuclide Approaches

In patients experiencing a major acute cardiac event (MACE), urgent catheterization defines the location and extent of the problem and delivers local therapy. Although the event is usually caused by a culprit lesion, where an occlusive thrombus formed on a ruptured plaque, coronary angioscopy has demonstrated there are typically a large number of lesions at other sites in the coronary tree (9) that have characteristics similar to the culprit lesion prior its acute rupture. This observation suggests a pancoronary arteritis (10). These lesions remain clinically silent, unless the thrombus propagates, leading to occlusion of the vessel. Since these lesions are difficult or impossible to identify angiographically, other approaches are under development. The metabolic rate of atheroma varies, but lesions with large amounts of lipid, especially oxidized low density lipoprotein, have a high metabolic rate. The high metabolic rate leads to increased vasa vasorum. The vasa vasorum are thin walled, and can rupture into the plaque, markedly increasing intralesional pressure, with subsequent loss of integrity of the cap of the plaque, and rupture of the lesion. Alternatively, some lesions erode (lose their endothelium, most likely due to apoptosis of the overlying endothelial cells), exposing the underlying thrombogenic collagen to blood, resulting in formation of thrombus with subsequent progression of stenosis or complete occlusion of the vessel.

Many atheromas are not visible angiographically and so alternative approaches have been advocated to detect these lesions. Because a key feature of progressive atheroma is inflammation, a number of radionuclide approaches have focused on identifying specific attributes of inflammation. Cells in the lesion are activated and have markedly increased metabolic rates. Glucose is their major metabolic substrate (especially for macrophages), which is obtained from exogenous glucose in the extracellular fluid. Laboratory studies demonstrate that these cells concentrate the glucose analog ^{18}F-fluorodeoxyglucose (^{18}FDG). Macrophages attracted to the lesion have increased expression of receptors on their surface. Radiolabeled peptides localizing in these receptors, such as MCP-1, are found in higher concentration at sites of inflammation. In a similar fashion, there is increased expression of integrin receptors ($\alpha_2\beta_3$) at the lesion site, which can be identified by peptides that recognize these receptors. Inflammation damages and frequently leads to the death of cells responding to the noxious stimulus, often by apoptosis. Therefore, markers of apoptosis, such as radiolabeled annexin V, localize in regions of inflammation.

Laboratory studies have demonstrated the feasibility of identifying experimental atheroma with each of these approaches using external imaging. However, the animal studies were usually performed in hyperlipidemic animals, with relatively extensive lesions. Many atheromas are small, extending about 2–10 mm in length, 1–2 mm in depth, and frequently occupying less than the full circumference of the vessel. The small size is often below the resolution limit of external imaging devices. To detect very small lesions the target to background ratio must increase in an exponential fashion. Although these lesions may be impossible to see with external imaging, a catheter based detector, placed in close proximity, could identify the lesion.

Fabricating the radiation detector from materials optimized to sense charged particles rather than gamma or X-radiation makes the catheter insensitive to radioactivity concentrating outside the vessel. Charged particle radiation is emitted as a byproduct of nearly all radioactive decay, but is typically most abundant in radionuclides that decay by beta emission (either positrons or negatrons). Prototype catheters using a plastic scintillator mated to an optical fiber have been tested in the laboratory using the positron emitting radiopharmaceutical ^{18}FDG. The catheter had sufficient sensitivity to detect lesions concentrating ~0.000001% of the typical 15 mCi dose of ^{18}FDG. To increase the sensitivity and specificity of vulnerable plaque localization, multi-sensor catheters, capable of measuring several parameters simultaneously, are in development. Contemporaneous measurement of endothelial temperature (with a thermister), fluorescence emitted in proportion to local apoptosis (using optically labeled annexin), and glucose utilization (with [18]FDG) with the same device, would enhance the likelihood of correct lesion characterization.

Patients with symptomatic MACE often have many lesions that meet the histologic criteria for vulnerable plaque. The stimuli leading to rupture of the culprit lesion also act on the remaining lesions, contributing to the high incidence of recurrent events that often occur during recovery. Managing patients with MACE must address both the acute event (usually with a percutaneous intervention) and reduce the likelihood of another event during convalescence. To reduce the possibility of another event, aggressive pharmacologic therapies are advocated to reduce inflammation and stabilize the remaining vulnerable lesions. An alternative approach may be the immediate stabilization of these vulnerable non-culprit lesions using a percutaneous catheter-based intervention applied at the time of the emergency percutaneous intervention therapy for the patient's acute coronary event. A major challenge to this therapeutic approach is identifying and localizing these lesions with certainty. Since many of these lesions are not visible in coronary angiograms, and are too small for external imaging, a catheter-based detection technique may be most useful.

Catheter Detector

A major question is what should the catheter be capable of detecting? There are number of characteristics of vulnerable plaque (Table 1) which could be evaluated. Specific devices have been advocated to measure each of these parameters. Cap thickness can be measured with intravascular magnetic resonance imaging (11); a multiwire basket catheter carrying multiple thermisters can detect increased temperature (12) (Fig. 1); the

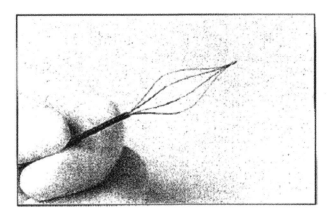

Figure 1 Photograph of the wire basket thermal catheter. (as depicted in patent 6,763,261 in Ref. 12). Multiple thermistors are placed on the surface of the wire. The thermistors are brought in contact with the endothelium of the vessel to take the temperature of the vessel.

size and characteristics (protein/lipid ratio) of the lipid pool can be measured with near infrared imaging (13); local acidosis can be identified with optical imaging using dyes that change colors at various pH values; and the metabolic activity (14) and chemotactic receptor expression (15) of inflammatory cells can be measured with radionuclide techniques.

The biologic rationale for the changes is the striking concentration of inflammatory cells (macrophages, granulocytes, and lymphocytes) in these lesions. These activated cells have markedly elevated metabolic activity, causing the lesion to have an elevated temperature. Local acidosis is likely caused by the limited oxygen supply in the lesion, causing the macrophages to function in a relatively anaerobic environment. Energy for the metabolic activity of these cells is provided by exogenous glucose. Areas of vascular inflammation appear as regions of increased tracer localization on positron emission tomograms in patients imaged with [18]FDG (20,21).

CONSIDERATIONS IN RADIATION CATHETER DESIGN

Catheter design must consider: diameter, flexibility, safety, and performance. The following section will review the potential detector materials that could be used in the intravascular environment: (*i*) Gas-filled detectors (e.g., ionization chambers); (*ii*) Semiconductors (e.g., cadmium telluride, lithium-drifted silicon, germanium), and (*iii*) Scintillators (including organic liquid scintillators, solid crystals such as sodium iodide or bismuth germinate, and plastics). Most nuclear medicine imaging devices use solid crystal scintillators for their detector material because of the penetrating nature of

Table 1 Attributes of Vulnerable Plaque

Attribute	Comments
Inflammation	Macrophages, granulocytes, and lymphocytes are abundant (16)
Increased vasa vasorum	Vasa vasorum increased with the intensity of inflammation. Thallium localizes in areas of increased vasa vasorum.
Increased temperature	Inflammatory cell metabolism increases local temperature > 0.1°C (17)
Local acidosis	pH in the lesion is reduced by 0.2 to 0.3 pH units (18)
Lipid pool	Vulnerable plague has a large extracellular lipid pool (19)
Cap thickness	The thickness of the cap is <65 μM in vulnerable lesions (19)

gamma rays and X-rays. Solid crystal scintillators have sufficient density to have a high probability of interaction between the gamma ray or X-ray and the crystalline structure of the detector. If the detector can be brought close to the source of radiation, however, it may be preferable to detect beta radiation (or very low energy conversion electrons) than gamma radiation. A major advantage of this approach is the elimination of the need for shielding and collimation because of the very short range of charged particles (<2 mm) compared with that of gamma rays (10s of cm to m) in tissue. Limiting the sensitivity of a beta detector for gamma rays minimizes the likelihood that radiation arising at some distance from the detector will be sensed. Most beta detectors have beta/gamma sensitivity ratios of greater than 100 to 1 favoring beta radiation. This ratio is achieved by having a relatively low density material for the sensor. As a result, the probability of interaction of gamma or X-ray radiation in the detector is minimized. Based on the geometry of emitted radiation from the source and the short path length of charged particle radiation, a charged particle detector has the advantage of low background and a significant increase in count rate as the source is approached.

The beta particles sensed by the detector can be positively or negatively charged. Positively charged particles, positrons, are emitted as part of positron emitting radionuclide decay (nuclides such as fluorine-18) while negatively charged particles are emitted in the course of conventional (negatron) beta decay (iodine-131, or phosphorus-32). The mean path length of both positron and negatron particles depends on the energy of emitted radiation, and is typically less than 3 mm in water. This physical phenomena permits localization of beta radiation based solely on proximity of the source to the detector. For example, ^{18}Fluorine (97% incidence of mean energy 0.25 MeV beta +) has an average range in water of 0.64 mm (very similar to the beta particle from ^{131}Iodine), while ^{82}Rubidium (83% incidence of mean energy 1.5 MeV beta +) still has an average range of only 4.29 mm.[a]

In addition to preferential detection of beta particles, an intravascular detector must be relatively flexible with a very small diameter (on the order of 1 mm) in order to traverse irregular atherosclerotic lesions that increase the tortuosity of the vessels. While the probability that a particle will deposit its energy in the detector is related to the size of the detector and geometry of the lesion, this must be balanced by flexibility and size constraints. Whatever detector is selected, the material must be biocompatible. Because of both size and path length considerations, there is

very little space for sealing the detector. Therefore, toxic materials cannot be used, nor can materials that would be damaged by the intravascular environment (e.g., hygroscopic detector materials). As a result, the detector must be "covered" with a thin layer of biocompatible material.

Most intravascular detector concepts incorporate a plastic scintillator coupled to an optical fiber. An alternative to plastic scintillators is a gas-filled detector, such as an ionization chamber or Geiger–Müller counter. Although these detectors are very sensitive for charged particle radiation, they have the disadvantage that the sensitive portion of the detector must be rigid, because a constant distance needs to be maintained between the anode and cathode. An additional disadvantage is the requirement for a high voltage across the chamber in order to detect ionizations, which causes safety concerns.

Semiconductor detectors work in a similar fashion to gas-filled ionization chambers. However, instead of gas, they employ a solid that can conduct ionization energy when a potential is applied across the semiconductor. This improves the stopping power since semiconductors are orders of magnitude denser than gas. Furthermore, semiconductors produce a much higher electrical signal per ionization than gas-filled detectors, improving signal strength. A major drawback hampering the use of semiconductors has been the difficulty in manufacturing pure crystals. Exceptionally pure substrate is still quite expensive. Pure crystals are required to minimize the background detector noise. Some of the newer detector materials can be operated at low voltages (e.g., 25–50 volts) at safe currents (5–10 picoamps), which makes them a viable alternative to plastic scintillators. The engineering challenge of achieving the necessary electrical shielding and flexibility of the detector material have not yet been met. This leaves scintillators as the best choice for intravascular detectors.

This leaves us with scintillating materials. Organic scintillating liquids are often used for detecting beta particles since the isotope containing material can be directly mixed with the scintillating liquid. This clearly cannot be done in an intravascular detector. However, one could fill a small probe with a scintillating liquid. It is easy to imagine that such a probe would have excellent flexibility and could be trivially made in any size desired. However, liquid scintillators have relatively low light output compared to other scintillators. Additionally, the primary scintillator often produces a wavelength of light that is not ideal for photomultiplier tubes, so a second, wave-shifting scintillator must be used. Finally, because these organic liquids are highly toxic and could enter the circulation upon any damage to the catheter, it would be difficult to safely use such a design.

---

[a]http://ie.lbl.gov/decay.html

Non-hygroscopic scintillating crystals, such as bismuth germinate (BGO), lutetium oxyorthosilicate (LSO), and germanium oxyorthosilicate (GSO) have properties that are attractive for intravascular detector design. LSO and GSO have relatively short decay times and excellent photon yield (LSO having the highest yield). These scintillators also have a photon wavelength that is very well suited to amplification with photomultiplier tubes. However, these materials are very dense, which will increase their stopping power (and thus the relative detection of background gamma radiation). Furthermore, they are not trivial to machine to the necessary dimensions for intravascular probes and are relatively delicate.

Finally, we come to plastic scintillators. Many plastic scintillators are, in essence, organic scintillators in solid form. Therefore, they are easy to manufacture in the exact size and shape desired and they are generally biologically inert and can be very safely deployed in humans. Plastic scintillators are not nearly as dense as most semiconductors or scintillating crystals, improving their ratio of beta to gamma detection. Furthermore, they are exceptionally fast scintillators so they can accommodate very high relative activities. Their light output is similar to that of BGO or GSO crystals, which is quite adequate. Furthermore, plastics can be designed and tuned for a desired output wavelength. They can also be successfully coupled to optical fibers for good photon detection. For these reasons, plastic scintillators have become the most widely studied for intravascular probe design. The utility of the catheter for detecting the lesions will depend on both the physical properties of the radionuclide and the biologic behavior of the selected tracer. Although external imaging utilizes gamma and X-radiation, this is not desirable for intravascular detection (22). Since gamma radiation is very penetrating, activity outside the vessel (e.g., in the myocardium) could create a potentially overwhelming background, making it difficult if not impossible to see the relatively weak signal from the vulnerable plaque. Charged particle radiation (either positron, negatron, or conversion electrons), on the other hand, is not very penetrating in blood (with path lengths ranging from <1 mm to a few millimeters). As a result, using a detector with a high sensitivity for charged particle radiation, one can be certain that the detected charged particle radiation is emitted from sites in close proximity to the detector. An additional advantage of detecting charged particle radiation is elimination of the requirement for conventional collimation. On the other hand, the limited tissue penetration of the charged particle requires the detector to be close to the lesion. The energy of the particle must be sufficient to have a high probability of traversing the lesion, endothelium, adjacent blood, and catheter sheath to deposit its energy in the radiation detector.

Radionuclides that preferentially decay with emission of a charged particle, such as fluorine-18 (β^+ max = 633.5 keV; β^+ average = 242.8 keV; average path length ~1.5 mm), are particularly well suited for this task.

The catheter must be extremely sensitive for the detection of beta radiation. There are two elements that contribute to sensitivity. One is the stopping power of the detector, that is, the likelihood that a charged particle of a specific energy will deposit energy in the detector. In the case of a plastic scintillator (a material that is not hygroscopic and is easy to work with in a biocompatible environment), a 300 keV charged particle is likely to deposit all of its energy in a 1 mm thick detector. In the case of the positron emitted by fluorine-18, sufficient light can be generated in a detector between 0.3–1 mm thick. The second is the geometric sensitivity of the device. Atheroma typically surround a portion of the vessel. Particles which are discharged into the vessel are likely to be detected, while those emitted toward the basement membrane of the vessel (traveling away from the detector) will not be detected. This gives an overall geometric sensitivity of ~50%. Since decay of a radionuclide is a random event, it is necessary to collect a sufficient number of counts (1,000 counts yield ~3% uncertainty) to reliably distinguish a lesion from background. To detect this number of events in 2 s, assuming the detector will have a scintillation for 100% of the interacting charged particles and a geometric efficiency of 50%, the lesion must contain about 0.00003 mCi.[b]

TESTING PROTOTYPE RADIATION SENSITIVE CATHETERS

Experimental studies in apo e −/− mice and New Zealand white rabbits fed a hyperlipidemic diet with catheter induced lesions in the aorta, suggest that the lesions only concentrate 1/10,000 of the administered dose. Using FDG as an example, a typical clinical FDG imaging study is performed with an injected dose of 15 millicuries. One ten-thousandth of the dose is 0.0015 mCi. After one half-life of radioactive decay, 0.00075 mCi would be resident in the lesion. Measurements made using a single plastic scintillator mated to an optical fiber (Fig. 2A,B) (22,23) demonstrated that this was 25 fold greater than the detection limit of the catheter. These calculations suggest that measurements could be recorded during a catheter pullback with a velocity of ~1–2 mm/second.

---

[b]This is in the absence of gamma radiation. The detector is much more sensitive for beta than gamma radiation, but some gamma radiation will cause a scintillation in the detector. Typically, the ratio of beta to gamma sensitivity is 20:1. However, as the detector is made thinner the ratio increases. In experiments with the prototype device, Janacek (22) measured a ratio of 1100:1.

(A)

(B)

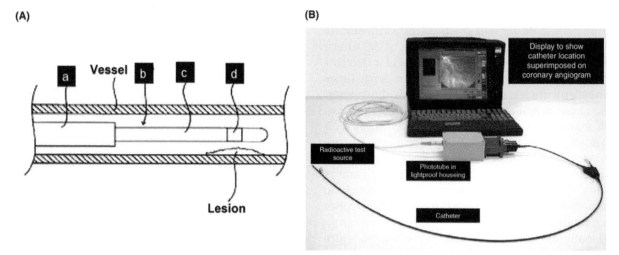

Figure 2   (**A**) Diagram of radiation sensitive intravascular catheter (as depicted in patent 6,782,289 in Ref. 23) using a plastic scintillator (d), an optical fiber to conduct the light (c), a lightproof sheath (a), and the vessel lumen (b). (**B**) A photograph of the prototype catheter undergoing testing with a small beta radiation source. The catheter connects to a phototube in a lightproof housing, which changes the individual light pulses to an electrical signal for display on the readout. The catheter position is superimposed on the coronary arteriogram and the radiation intensity is displayed as a false color signal on the arteriogram.

In addition to FDG, there are a number of radiolabeled tracers that could be used for the detection of vulnerable plaque. Table 2 lists some of these agents, their radiolabels, the biologic principle underlying their localization in atheroma, and the major radiation that can be useful for lesion detection.

The catheter must also have sufficient flexibility and be of small enough diameter to navigate tight lesions. In an effort to determine the radial location of the plaque we constructed several multifiber catheters (Fig. 3). However, multifiber catheters have marked limitations of flexibility, especially at the tip, because of the stiffness of the coupling point of the scintillator and the optical fiber. Another design requirement is to

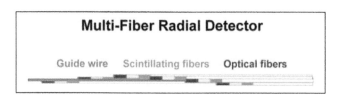

Figure 3   Diagram of a multiple scintillating fiber catheter capable of providing radial information about plaque location.

make the final device <4 French (1.3 mm external diameter), or if possible, about the diameter of a guidewire. Other investigators, such as Wahl et al. (27), Zhu et al. (28), and Mukai et al. (29) have suggested designs for radiation sensitive catheters. Most of these devices

Table 2   Radiopharmaceuticals for Catheter Detection of Vulnerable Plaque

Agent	Biologic rationale	Major detectable radiation	Comment
Tc-99m low density lipoprotein (LDL)	Lipid incorporation in plaque	Conversion electrons	Charged particles resulting from conversion electrons, are about 10% abundant, have very low energy, and a path length <<1 mm. These characteristics make it challenging to detect lesions with the charged particle principle with technetium-99m as the radiolabel
In-111 Z2D3 (24)	Antibody recognizing proliferating vascular smooth muscle	Cadmium X-rays	Average energy of ~23 keV
F-18 FDG	Metabolic substrate for inflammatory cells	$\beta+$ (100%)	Average path length 1–2 mm. This path length is acceptable for detection in a vessel with a diameter of ~3 mm
Tc-99m Annexin (25)	Apoptosis of inflammatory and smooth muscle cells	Conversion electrons	See above comments for Tc-99m. Annexin can be labeled with other nuclides, such as F-18 (26)
Ga-68 Annexin	Apoptosis of inflammatory and smooth muscle cells	$\beta+$ (100%)	Charged particle $\beta+$ max = 1.899keV, $\beta+$ average = 740 keV, average path length 3 mm

had prototypes that were 2–3 mm diameter (6–9 French), which limits the utility of the design.

POTENTIAL FUTURE APPROACH

To assure the correct identification of vulnerable lesions, it would be helpful to measure multiple parameters simultaneously. This can be accomplished by adding a small thermistor to the tip of radiation sensitive catheter to take the temperature of the lesion during the pullback. In addition, an optical sensor would allow determination of lesion pH, or possibly the extent of apoptosis (using an agent such as optically labeled annexin).

CONCLUSION

Radiation sensitive catheters have been tested in animal models of atheroma. These devices can provide key information about the location and metabolic status of the plaque. With further technical refinement, these catheters may be useful to interrogate vessels of patients with MACE for the presence of other highly vulnerable lesions, which may benefit from a local intervention.

REFERENCES

1. Napoli C, Pignalosa O, de Nigris F, Sica V. Childhood infection and endothelial dysfunction: A potential link in atherosclerosis (editorial). Circulation 2005; 111:1568–1570.
2. Glagov S, Weisenberg E, Zarins CK, Stankunavicius R, Kolettis GJ. Compensatory enlargement of human atherosclerotic coronary arteries. N Engl J Med. 1987; 316:1371–1375.
3. Carlier S, Kakadiaris IA, Dib N, et al. Vasa Vasorum Imaging: a new window to the clinical detection of vulnerable atherosclerotic plaques. Current Atherosclerosis Reports 2005; 7:164–169.
4. Ambrose JA, Tannenbaum MA, Alexopoulos D, et al. Angiographic progression of coronary artery disease and the development of myocardial infarction. J Am Coll Cardiol 1988; 12:56–62.
5. Mody N, Tintut Y, Radcliff K, Demer LL. Vascular calcification and its relation to bone calcification: Possible underlying mechanisms. J Nucl Cardiol 2003; 10: 177–83.
6. Dunphy MPS, Freiman A, Larson SM, Strauss HW. Association of vascular fdg uptake with vascular calcification. J Nucl Med 2005; 46:1278–1284.
7. Narula J, Strauss HW. Imaging of unstable atherosclerotic lesions. Eur J Nucl Med Mol Imag 2005; 32(1):1–5.
8. Schoen FJ. The heart, Chap 12. In: Kumar V, Robbins SL, Cotran RS, eds. Pathologic Basis of Disease, 7th Edition. Elsevier 2005; 555–618.
9. Asakura M, Ueda Y, Yamaguchi O, et al. Extensive development of vulnerable plaques as a pan-coronary process in patients with myocardial infarction: an angioscopic study. J Am Coll Cardiol 2001; apr 37(5): 1284–1288.
10. Rioufol G, Ernei G, Ginon J, et al. Multiple atherosclerotic plaque rupture in acute coronary syndrome: a three-vessel intravascular study. Circulation 2002.
11. Worthley SG, Helft G, Fuster V, et al. A novel nonobstructive intravascular MRI coil: In vivo imaging of experimental atherosclerosis. Arterioscler Thromb Vasc Biol 2003; 23:346–350.
12. Casscells SW, Willerson JT, Naghavi M, Guo B. Patent 6,763,261, issued Jul 13, 2004.
13. Wang J, Geng YJ, Guo B, et al. Near-infrared spectroscopic characterization of human advanced atherosclerotic plaques. J Am Coll Cardiol 2002; 39(8):1305–1313.
14. Ogawa M, Ishino S, Mukai T, et al. (18)F-FDG accumulation in atherosclerotic plaques: Immunohistochemical and PET imaging study. J Nucl Med 2004; 45(7):1245–1250.
15. Ohtsuki K, Hayase M, Akashi K, Kopiwoda S, Strauss HW. Detection of monocyte chemoatteractant peptide-1 receptor expression in experimental atherosclerotic lesions: An autoradiographic study. Circulation 2001; 104:203–208.
16. Burke AP, Virmani R, Galis Z, Haudenschild CC, Muller JE. 34th Bethesda Conference: Task force #2—What is the pathologic basis for new atherosclerosis imaging techniques? J Am Coll Cardiol 2003; 41(11):1874–1878.
17. Courtney BK, Nakamura M, Tsugita R, et al. Validation of a thermographic guidewire for endoluminal mapping of atherosclerotic disease: An in vitro study. Catheter Cardiovasc Interv 2004; 62(2):221–229.
18. Khan T, Soller B, Naghavi M, Casscells W. Tissue pH determination for the detection of metabolically active, inflamed vulnerable plaques using near-infrared spectroscopy: An in-vitro feasibility study. Cardiology 2005; 103(1):10–16.
19. Davies MJ, Richardson PD, Woolf N, Katz DR, Mann J. Risk of thrombosis in human atherosclerotic plaques: Role of extracellular lipid, macrophage, and smooth muscle cell content. Br Heart J 1993; 69:377–381.
20. Rudd JH, Warburton EA, Fryer TD, et al. Imaging atherosclerotic plaque inflammation with [18F]-fluoro-deoxyglucose positron emission tomography. Circulation 2002; 105(23):2708–2711.
21. Yun M, Jang S, Cucchiara A, Newberg AB, Alavi A. 18F FDG uptake in the large arteries: A correlation study with the atherogenic risk factors. Semin Nucl Med 2002; 32(1):70–76.
22. Janecek M, Patt BE, Iwanczyk JS, et al. Intravascular probe for detection of vulnerable plaque. Mol Imag Biol 2004; 6:131–138.
23. Strauss HW. Patent 6,782,289, issued Aug 24, 2004.
24. Carrio I, Pieri PL, Narula J, et al. Noninvasive localization of human atherosclerotic lesions with indium 111-labeled monoclonal Z2D3 antibody specific for

proliferating smooth muscle cells. J Nucl Cardiol 1998; 5:551–557.

25. Kolodgie FD, Petrov A, Virmani R, et al. Targeting of apoptotic macrophages and experimental atheroma with radiolabeled annexin V: A technique with potential for noninvasive imaging of vulnerable plaque. Circulation 2003; 108:3134–3139.

26. Grierson JR, Yagle KJ, Eary JF, et al. Production of [F-18]fluoroannexin for imaging apoptosis with PET. Bioconjug Chem 2004; 15:373–379.

27. Wahl RL, Lederman RJ. Patent 6,295,6L80, issued Oct 2, 2001.

28. Zhu Q, Piao D, Sadeghi MM, Sinusas AJ. Simultaneous optical coherence tomography imaging and beta particle detection. Opt Lett 2003; 28: 704–1706.

29. Mukai T, Nohara R, Ogawa M, et al. A catheter-based radiation detector for endovascular detection of atheromatous plaques. Eur J Nucl Med Mol Imaging 2004; 31:1299–1303.

8

Intravascular Ultrasound for Molecular Imaging

Matthew O'Donnell
Department of Biomedical Engineering, University of Michigan, Ann Arbor, Michigan, U.S.A.

INTRODUCTION

Contrast angiography remains the most common method to guide endovascular therapies. In the last decade, intravascular ultrasound (IVUS) has emerged as a valuable adjunct, overcoming many of angiography's limitations for a number of clinical applications. In particular, the tomographic character of IVUS helps visualize the entire vessel circumference rather than just a projection.

IVUS uses small ultrasound transducer(s) integrated into an intraluminal catheter. The typical IVUS catheter is only about 0.9–1.2 mm in diameter. Two different approaches have been developed, one based on mechanical rotation of a single element transducer and the other using multi-element electronic arrays. The single element catheter operates at about 30 MHz with frame rates up to 50 Hz. Although high spatial resolution is possible, images are often corrupted by artifacts due to non-uniform rotational distortion (NURD) (1). In contrast, full electronic images produced with catheter-borne arrays do not exhibit the spatial resolution of mechanical scanners, but are free of rotational artifacts and are easier to manipulate since they can be delivered over the same guidewire used for conventional therapeutic catheters (2).

Images produced from IVUS are cross-sections of the artery perpendicular to the flow direction. IVUS has several advantages over other modalities. It provides real-time images from inside the vessel at very high resolution (50–75 μm) with large penetration depths of 5–7.5 mm (i.e., 10–15 mm field of view). Furthermore, intravascular ultrasound enables not only anatomical imaging of vessels but also functional imaging. Currently, the primary disadvantages of IVUS include being minimally invasive (the catheter is placed inside the vasculature via the femoral artery), imaging one cross-section of the vessel at a time, and requiring additional time and cost in the catheterization lab. Overall, IVUS has been shown to change clinical outcomes in assessing diffusely diseased vessel segments, ostial or bifurcation stenoses, eccentric plaques, and angiographically foreshortened vessels (3–4). Future developments in IVUS could further increase its role in the catheterization lab, especially by increasing the imaging capabilities of the basic catheter to include prospective (i.e., forward-looking) three-dimensional (3D) imaging and linking it with specific probes for molecular imaging in cardiovascular applications.

CLINICAL PROBLEM

Cardiovascular disease remains the primary killer in the United States, leading to nearly 1 million deaths a year. The American Heart Association estimates that nearly 60 million Americans have some form of cardiovascular disease, with a total economic burden of hospitalization, treatment, and lost work time amounting to $330 billion annually (5).

The most dangerous form of cardiovascular disease is atherosclerotic coronary artery disease, which leads to narrowing and stiffening of arteries supplying blood to the heart. Atherosclerosis arises from repeated injury, subsequent inflammation, and repair of the vascular wall. Endothelial injury leads to an inflammatory response and lipid accumulation mostly from low-density lipoprotein cholesterol in the blood. Endothelial cells, along with macrophages, promote rapid lipid accumulation leading to fatty streaks or foam cells. At this stage, atherosclerosis is mostly reversible with appropriate changes to risk factors (smoking, exercise, healthy diet, etc.). If atherosclerosis continues, a plaque (atheroma) forms around a lipid core. A stable plaque (fibroatheroma) is characterized by a thick fibrous cap composed of smooth muscles cells and a small lipid core. In contrast, rupture-prone plaque is often characterized by a large lipid core and a thin fibrous cap. They are sometimes referred to as vulnerable plaques. Note that the lipid core is highly thrombolytic. Upon rupture, lipids interact with the blood to induce thrombosis, artery blockage, and ultimately myocardial infarction. If the atheroma stabilizes, smooth muscle cell proliferation can prevent rupture but still lead to calcification and severe fibrosis. Continued plaque growth can lead to severe stenosis and myocardial ischemia. Recent studies have shown that plaque rupture, not vessel occlusion, is the leading cause of these acute events (6–7).

CLINICAL TREATMENTS

The most common interventional procedure to treat atherosclerosis is percutaneous transluminal coronary angioplasty (PTCA). In PTCA, a balloon catheter is inserted into the arterial vasculature via the femoral artery over a smaller guide wire. Once the balloon catheter is placed at the coronary lesion to be dilated, the balloon is pressurized with fluid to expand the arterial lumen. The first balloon angioplasty procedure was performed in 1977 (8) and is most effective for softer plaques. Although angioplasty proved effective in many cases, restenosis (subsequent occlusion of the artery lumen after the procedure) rates range from 25–35% for most catheterization labs (9–10).

To enhance interventional outcomes, stents were developed to maintain lumen diameter after the balloon intervention. Stents are meshed scaffold tubes implanted in a vessel and delivered over an angioplasty balloon. Since gaining FDA approval in 1994, they have become the primary interventional procedure accounting for a majority of angioplasty procedures in catheterization labs (11). Stents effectively prop open the occluded vessel by pushing on the plaque and arterial wall to expand lumen diameter. Clinical complications related to stent deployment are underdeployment and overdeployment. Underdeployment can lead to stent collapse if it is not in complete contact with the vessel lumen. Furthermore, blood can collect in the gap between the arterial tissues, leading to thrombosis.

To avoid these complications, stents are usually overdeployed or overpressurized to prevent stent collapse. However, this practice can lead to vessel rupture, edge dissections of the vessel, or distal embolism. Furthermore, stent deployment often leads to restenosis as over-pressurization triggers an injury response from the vessel. The inflammatory response from endothelial damage initiates neointimal hyperplasia (smooth muscle cell proliferation into the lumen leading to restenosis) in the very vessel the stent was intended to open. Most recently, drug-coated stents have been developed to deliver chemicals inhibiting neointimal proliferation. The recent study by Sousa et al. shows promising initial results in human patients with no in-stent or edge restenosis or major clinical events 8 months after a stent procedure (12). While more recent studies with a larger number of patients and longer follow-up evaluation have yielded less promising results (13), drug-coated stents appear to significantly reduce restenosis rates.

Other common interventional procedures include atherectomy, brachytherapy, and coronary bypass surgery. Atherectomy uses a small rotating blade on the tip of a catheter to disrupt plaque and collect debris with a suction tube. It is usually reserved for calcified plaques resistant to angioplasty or stent procedures (14), but it has the risk of vessel rupture or injury. Another interventional treatment is brachytherapy. It delivers focused radiation at the lesion site using a catheter, preferentially killing plaque cells. In more severe cases in which angioplasty or stent procedures are ineffective, coronary bypass surgery is required. Vessels are usually grafted from extremities to provide collateral flow around coronary occlusions. This surgical procedure requires a thorocotomy to gain access to the heart.

For all these therapies, pharmaceuticals play a significant and ever growing role in the treatment of cardiovascular disease. During the procedure, anticoagulants such as heparin are usually delivered to help minimize thrombosis in the treated artery. Furthermore, cholesterol-lowering drugs have been shown to be valuable adjuvants to lifestyle changes (stopping smoking, regular exercise, and healthy diet) to help manage coronary artery disease (1,15). Research is focusing on anti-inflammatory drugs to minimize the inflammatory response in the treated region. Also, chemicals that inhibit smooth muscle growth factor [i.e., vascular endothelial growth factor (VEGF)] are used in drug-coated stents (12,16).

To optimize drug treatment, especially of anti-inflammatories, there is significant clinical need to localize and quantify the inflammatory response post-intervention. Molecular imaging with IVUS may be able to provide this information. Current molecular imaging agents for ultrasound use the intrinsic mechanical properties of the agents themselves (i.e., density and compressibility) to provide the means for detection. Microbubbles are the dominant contrast agent (17,18). They represent efficient scatterers with nonlinear acoustic characteristics enabling sensitive detection at even relatively low imaging frequencies. At the higher frequencies used in IVUS, single microbubbles can be detected. Recently, specific targeting of these agents has been explored for molecular imaging and therapeutics (19,20). Because of their large size compared to most biomolecules, however, it is very difficult to transport them across cell membranes other than by phagocytosis (21,22). Consequently, most targeting studies have focused on applications where extracellular accumulation provides clinically significant information.

INTRAVASCULAR ULTRASOUND IMAGING

To help diagnose and treat cardiovascular disease, intravascular ultrasound imaging was developed to visualize arterial pathologies from the vessel lumen.

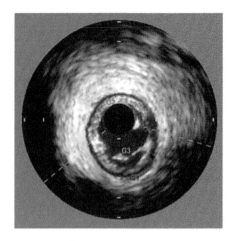

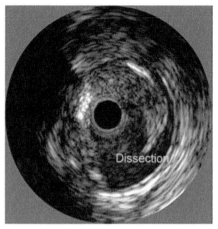

**IVUS Image of Diseased
Coronary Artery with
Intimal Flap**

**IVUS Image of
Dissection post Balloon
Angioplasty**

Figure 1 Two typical B-Mode clinical images obtained with a 3.5F imaging catheter. (*Left*) A diseased artery with an intimal flap. (*Right*) Arterial dissection post angioplasty on the right.

IVUS uses small ultrasound transducer(s) integrated into an intraluminal catheter. B-mode (Brightness mode) cross-sectional images of the vessel are acquired and displayed in real-time for the clinician, as illustrated in Figure 1. The typical IVUS catheter is only about 1 mm in diameter and is inserted into the patient's vasculature via the femoral artery. The two most common IVUS catheters are a 64-element circumferential array transducer and a single-element rotating (element or mirror) transducer. Figure 2 compares the two designs. The single element catheter operates at a center frequency of 30 MHz at frame rates up to 50 Hz. It can produce image artifacts from NURD due to frictional forces when the catheter is inside the vessel (14).

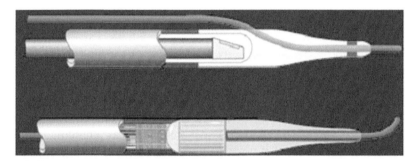

Figure 2 Two types of IVUS catheters. The top drawing compares a single, rotating ultrasound element transducer with a solid state ultrasound array transducer. The bottom images present the size and layout of a modern solid state ultrasound catheter. *Abbreviation*: IVUS, intravascular ultrasound.

The solid-state circumferential array transducer operates at a center frequency of 20 MHz using synthetic aperture processing to form images at frame rates of about 30 Hz. Its advantages are that there are no mechanical parts, elements are controlled electronically, and its smaller footprint allows it to be integrated into combined devices.

IVUS is primarily used as a diagnostic tool to measure lumen diameter and degree of stenosis pre- and post-intervention (angioplasty or stent deployment). In addition, IVUS is also used to measure blood flow to determine if normal physiological flow has been restored after an interventional procedure.

The main advantages of IVUS are real-time imaging from inside the vessel, high-resolution images (50–75 μm), large penetration depths of 5–7.5 mm, anatomical imaging of vessels and plaques, and functional imaging (e.g., flow and strain). The primary disadvantages are that it is minimally invasive, requires extra time during interventional procedures, adds cost, and images only one cross-section of the vessel at a time.

OTHER IMAGING MODALITIES

To better understand IVUS, it is important to compare it to other cardiovascular imaging. X-ray angiography remains the gold standard for coronary arteries. Its main advantages are high spatial resolution (0.15 mm) and relatively low cost compared to other modalities (23). The main drawbacks are the radiation exposure for the patient during an intervention as multiple injections of radioactive contrast agent are delivered. Also, angiography is a minimally invasive procedure using a catheter to deliver the radioactive contrast agent. Furthermore, planar X-ray imaging captures a projection of the vasculature and can often misdiagnose asymmetric lesions (23).

An extension of angiography is computed tomography (CT). Its main advantage is non-invasive, cross-sectional imaging of vessels to diagnose vascular occlusions and atherosclerotic calcified lesions. In addition, fast multi-slice CT produces 3-dimensional volume data acquisition and detection of vessels larger than 1 mm in diameter. CT can also be used to image vessels after an interventional procedure and during non-invasive follow-up studies. Unfortunately, CT, like angiography, exposes the patient to radiation and iodonized contrast agents are often used during the procedure.

Another non-invasive modality is magnetic resonance imaging (MRI). MRI uses a high-field, whole-body magnet, and radio-frequency coils to manipulate nuclear spins in the body for functional and anatomical imaging (24). MRI can identify large lipid regions in a plaque, which is useful to identify vulnerable plaques.

Furthermore, like CT, MRI collects 3D data so that any slice can be viewed after data acquisition. Its drawbacks are long data acquisition times and the inherent tradeoff between spatial resolution and frame rate.

Nuclear imaging uses radiopharmaceuticals injected intravenously in small doses to study physiological parameters (as opposed to imaging anatomical features) by collecting images over time. As different regions of the body selectively absorb these agents, gamma rays are emitted and detected. Although nuclear imaging provides important functional information, both spatial and temporal resolutions are lower than other modalities and the patient must be injected with radioactive isotopes to achieve contrast.

Optical Coherence Tomography (OCT) imaging is a relatively new modality developed in 1991 (25). It operates much like a single element ultrasound system but uses broadband optics to transmit and receive 1-dimensional signals, which are then combined to form an image. Its primary advantage is high resolution (about 10 μm) at frame rates around 10 Hz, providing histological image quality to help characterize plaques inside a vessel. The major obstacle to vascular imaging with OCT is the high scattering of transmitted light by blood. Continuous saline flushes are need for real-time imaging (26).

CLINICAL APPLICATIONS

Conventional IVUS images vessel cross-section to a depth of 5–8 mm from the catheter. Quantitative lumen measurements are made from these images to diagnose vessel stenosis. With the growth of vascular interventional procedures such as angioplasties and stents, comparing pre- and post-procedure lumen measurements are especially useful in evaluating an intervention. Thus, a clinician can immediately determine if further interventions are necessary.

In addition to lumen morphology, IVUS images can also assess coronary blood flow during diagnostic or post-therapeutic interventions, as illustrated in Figure 3. One problem in measuring flow with IVUS is that blood moves normal to the side-looking image plane (vessel cross-section). Therefore, Doppler processing cannot be used to image the flow field. Recent developments in IVUS technology have enabled real-time blood speed imaging using a filter bank approach (27). With quantitative flow measurements, the degree of stenosis can be evaluated before and after an intervention.

FUTURE DIRECTIONS

There has been a growing interest in measuring mechanical properties of arterial tissues and pathologies

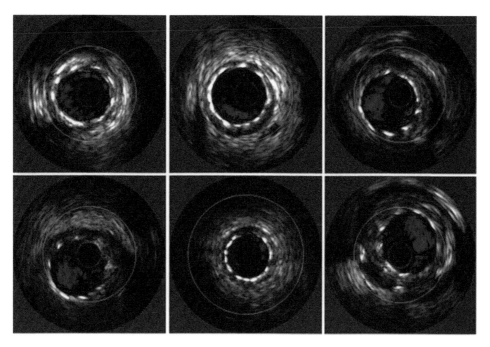

Figure 3 Non-Doppler flow images along the length of a coronary artery post-stenting. The free lumen is clearly identified in these flow images and there is no flow between stent struts and the arterial wall, demonstrating full deployment.

using strain imaging and IVUS (28–30). If vulnerable plaques can be identified and their mechanical properties quantified, interventional procedures can be optimized and plaque rupture can be better understood. Intravascular ultrasound has been used for elasticity imaging (with deformations applied by a balloon catheter or the cardiac pulse) to differentiate soft (lipid) and hard (fibrotic) regions of a plaque. Tissue displacements are accurately measured with speckle tracking algorithms applied to real-time image frames as an artery or lesion is deformed. Displacement data are accumulated and processed to create strain and elasticity images of vascular tissues and pathologies. Clinical evidence has shown that distensibility of atherosclerotic coronary lesions may vary widely, from undilatable hard lesions to soft, highly compliant ones (31). In particular, the elastic moduli of a normal arterial wall and a coronary plaque can vary by over an order of magnitude (32).

Another exciting future direction for IVUS is a forward-viewing catheter to image in front of the device (33). Using a ring-annular array of ultrasound elements, 3-dimensional data can be acquired to image tissue and flow in front of the catheter, as illustrated in Figure 4. Multi-view images can help guide the IVUS catheter to the lesion of interest and analyze the plaque without disturbing it. A C-mode image in this geometry (Fig. 5) images the vessel cross-section, similar to the conventional B-mode image in the side-looking IVUS catheter. However, the image plane is in

front of the catheter, an ideal position for guidance. The B-mode image in this geometry is orthogonal to this plane and can image along the length of the vessel. Using a forward-viewing geometry, Doppler flow measurements would also be possible in the region distal to the IVUS array. This forward-viewing array can be fabricated by modifying the current solid state IVUS catheter to operate in both side-viewing (conventional) and forward-viewing modes for imaging at the location of the catheter as well as the region in front of it. A prototype device is presented in the lower right panel of Figure 4.

Forward-Looking Catheter

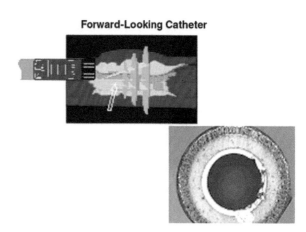

Figure 4 Forward-viewing IVUS array drawing and simulated images (*top*). The lower right shows a photograph of a prototype array. *Abbreviation*: IVUS, intravascular ultrasound.

Multiple Views

C-Mode

B-Mode

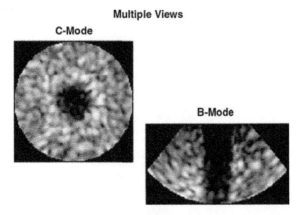

Figure 5 (*Left*) C-scan at a depth of 4.2 mm from the catheter surface over approximately a 6.8 mm x 6.8 mm area for a simulated coronary artery with a 2.4 mm diameter lumen. (*Right*) Precisely the same simulation as the left panel but for the case of a B-Scan display perpendicular to the C-Scan display plane.

Integrated balloon-ultrasound catheter with stent

IVUS imaging array   semi-compliant balloon

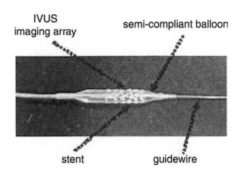

stent   guidewire

Figure 6 Integrated balloon ultrasound catheter prototype with a stent placed on top of the balloon.

Conventional IVUS imaging can be extended to provide 3D images by acquiring beams in real-time over a forward-looking cone, segmenting the data, and rendering reconstructed images to view the vessel and pathology from any angle. The physician could rotate the 3D image and also view different imaging planes (or slices) to better visualize plaque and vessel morphology. Three-dimensional IVUS imaging would be particularly useful for unusual vessel or plaque shapes, suspicious plaques, and regions near vessel branches.

Perhaps the greatest clinical limitation in balloon angioplasty and stent deployment procedures is the lack of quantitative feedback during the procedure. Even pre-procedure estimates of lumen diameter and occlusion can lead to significant errors due to planar projections obtained by X-ray angiography (23). Without image guidance, stents are often underdeployed, leading to thrombosis or stent collapse, or overdeployed, resulting in greater intimal injury and higher restenosis rates. However, current IVUS uses are still limited to imaging before and after the procedure but not *during* it. Furthermore, imaging pre- and post-procedure requires additional catheters and lengthens the total intervention time. Both of these factors can significantly add to the total cost of patient treatment.

To address these important clinical problems, an integrated balloon ultrasound catheter has been proposed to combine IVUS imaging and therapy in real-time during an interventional procedure (28,34). Figure 6 shows a photo of a prototype integrated device. The clinical opportunity and significance of an integrated device are numerous. By imaging in real-time as the balloon "probes" the tissue, lesions could be characterized before a complete angioplasty procedure. Even during stent procedures, strain imaging could prevent overdeployment. For balloon

procedures such as angioplasty and stent deployment, the integrated balloon ultrasound catheter could provide both real-time image guidance and quantitative feedback during the interventional procedures. Both visual and quantitative feedback would help clinicians achieve optimal stent deployment. After a procedure is performed, the balloon can be deflated and moved in pullback mode along the length of the lesion to verify that the stent has been fully deployed. Lastly, current IVUS limitations can be overcome with a single catheter device for balloon interventions, potentially reducing total intervention time while providing significant feedback during the procedure.

The ultimate goal of an integrated device is to significantly change clinical outcomes by reducing intimal injury and subsequent restenosis rates. By verifying stent deployment with the same catheter after the procedure, complete deployment can be easily validated in pullback mode. If any regions are underdeployed, the balloon can easily be re-aligned and re-inflated using the same metric and real-time image guidance. Ultimately, the integrated catheter could transform cardiovascular interventions into a single catheter procedure. By reducing the number of catheters, the total procedure time would decrease for the patient and lead to increased patient throughput for a catheterization lab.

New transducer technologies on the horizon can greatly improve the overall image quality of IVUS systems, especially forward-looking 3D scanners. Catheters with capacitive micromachined ultrasound transducers provide arrays of capacitive transducers with integrated electronics (35.36). These systems exhibit high transduction sensitivity and very broad bandwidths. Guidewires containing arrays of opto-acoustic transducers have been proposed (37). Such devices can provide real-time high-frequency imaging to every intervention using a guidewire. Both of these advanced systems have the potential for greatly improved spatial resolution and sensitivity compared to current IVUS systems. This can be especially valuable for molecular imaging.

Finally, several research groups are investigating ultrasound mediated gene therapy and molecular imaging applications with high frequency ultrasound. One group is studying the effects of intravascular ultrasound therapy to enhance transgenic expression (38) and reduce neointimal hyperplasia. Recent work by another group has explored several targeted contrast agents for intravascular ultrasound to image evolving thrombi and the inflammatory response following PCTA (39,40). To overcome some of the limitations of site-targeted microbubbles, molecular contrast agents based on perfluorocarbons have been developed (41,42). These nanosystems augment ultrasonic reflectivity from fibrin thrombi by two orders of magnitude (40,43–45). Additionally, vascular epitopes, such as growth factors induced in smooth muscle cells after angioplasty, can be targeted because particles can penetrate through microfissures into the vascular media (45). Ultrasonic backscatter from a single perfluorocarbon nanoparticle does not compare to that from a single microbubble—high concentrations are required for ultrasonic detection. Nevertheless, agents such as this combined with high resolution IVUS hold the promise for molecular imaging of the inflammatory response following an intervention.

CONCLUSION

Intravascular ultrasound is a valuable imaging modality in the catheterization lab used to diagnose cardiovascular diseases. Recent developments and current research show promise for IVUS gaining a larger role in guiding interventional procedures. With the development of a forward-viewing IVUS array and an integrated balloon ultrasound catheter, IVUS can be used as a single interventional device to not only diagnose but guide therapies and provide real-time feedback to the clinician. Further advances in IVUS catheter and guidewire imaging systems will lead to higher resolution images and smaller devices. Finally, emerging applications such as 3D rendering, strain processing, and molecular imaging will continue to make IVUS an important imaging tool in treating cardiovascular disease.

ACKNOWLEDGMENTS

I would like to thank the group in the Biomedical Ultrasonics Lab at the University of Michigan over the last decade who have worked on IVUS systems, including Ben Shapo, John Crowe, Charles Choi, Andrei Skovoroda, Stas Emelianov, Javier deAna, and Yan Shi. I would also like to thank my pals at the former Endosonics who helped to develop the In-Visions system, including Mike Eberle, Doug Stephens, Jerry Litzza, Dave Bleam, and Ignacio Cespedes. Finally, long-term support from NHLBI under grant HL47401 is gratefully acknowledged.

REFERENCES

1. Mikhailidi DP, Ganotakis ES, Spyropoulos KA, Jagroop IA, Byrne DJ, Winder AF. Prothrombotic and lipoprotein variables in patients attending a cardiovascular risk management clinic: Response to cipofibrate or lifestyle advice. Int Angiol 1998; 17:225–233.

2. Schulze-Clewing J, Eberle MJ, Stephens DN. Miniaturized circular array. Proceedings of the 2000 IEEE International Ultrasonics Symposium 2000; 1253–1254.

3. Nissen SE, Yock P. Intravascular ultrasound: Novel pathophysiological insights and current clinical applications. Circulation 2001; 103:604–616.

4. Sharma S, Gulati G. Current trends and future applications of intravascular ultrasound. InD J Radiol Imag 2003; 13:53–60.

5. American Heart Association. 2002 Heart and Stroke Statistical Update. Dallas, Texas: American Heart Association, 2001.

6. Peterson M, Dangas G, Fuster V. Atherosclerosis and thrombosis. In: Safian RD, Freed MS, eds. The Manual of Interventional Cardiology, 3rd ed. Royal Oak, MI: Physicians' Press, 2001; 959–973.

7. Lee RT, Libby P. The unstable atheroma. Atheroscler Thromb Vasc Biol 1997; 17:1859–1867.

8. Norell MS. History, development and current activity. In: Norell MS, Perrins EJ, eds. Essential Interventional Cardiology. London: Harcourt Publishers Limited, 2001; 3–12.

9. Schiele F, Vuillemenot A, Meneveau N, Pales-Espinosa D, Gupta S, Bassand J-P. Effects of increasing balloon pressure on mechanism and results of balloon angioplasty for treatment of restenosis after palmaz-schatz stent implantation: An angiographic and intravascular ultrasound study. Catheter Cardiovas Interven 1999; 46:314–331.

10. Shiran A, Mintz GS, Waksman R, et al. Early lumen loss after treatment of in-stent restenosis: An intravascular ultrasound study. Circulation 1998; 98:200–203.

11. Lee SF, Brooks MR, Livingston DL, Squire JC. Smart catheter for stent placement. IEEE Potentials 2003; 22:19–22.

12. Sousa JE, Costa MA, Abizaid A, et al. Lack of neointimal proliferation after implantation of sirolimus-coated stents in human coronary arteries: A QCA and 3-D IVUS study. *Circulation* 2001; 103:192–195.

13. Liistro F, Stankovic G, Di Mario C, et al. First clinical experience with a paclitaxel derivative-eluting polymer stent system implantation for in-stent restenosis. Circulation 2002; 105:1883–1886.

14. Atar S, Siegel RJ. Intravascular echocardiography. In: Pohost GM, O'Rourke RA, Shah PM eds. Imaging in Cardiovascular Disease. Philadelphia: Lippincott Williams and Wilkins 2000; 97–113, 2000.

15. Dupis J, Tardif JC, Cernack P, Theroux P. Cholesterol reduction rapidly improves endothelial function after acute coronary syndromes. The RECIFE (reduction of cholesterol in ischemia and function of the endothelium) trial. Circulation 1999; 99:3227–3233.

16. Grube E, Gerckens, Rowold S, Yeung AC, Stertzer SH. Inhibition of in-stent restenosis by a drug-eluting polymer stent: Pilot trial with 18-month follow-up. Circulation 2000; 102:II-554.

17. Frinking PA, deJong N, Cespedes EE. Scattering properties of encapsulated gas bubbles at high ultrasound pressures. J Acoust Soc Am 1999; 105:1989–1996.

18. Becher H, Burns PN. Handbook of Contrast Echocardiography: LV Function and Myocardial Perfusion. New York: Springer Verlag, 2000.

19. Dayton PA, Chomas JE, Lum AFH, et al. Optical and acoustical dynamics of microbubble contrast agents inside neutrophils. Biophysical Journal 2001; 80:1547.

20. Dayton PA, Ferrara K. Targeted imaging using ultrasound. J Mag Reson Imaging 2002; 16:362.

21. Lindner JR, Coggins MP, Kaul S, et al. Microbubble persistence in the microcirculation during ischemia/reperfusion and inflammation is caused by integrin- and complement-mediated adherence to activated leukocytes. Circulation 2000; 101:668.

22. Lindner JR, Dayton PA, Coggins MP, et al. Noninvasive imaging of inflammation by ultrasound detection of phagocytosed microbubbles. Circulation 2000; 102:531.

23. Ohnesorge BM, Becker CR, Flohr TG, Reiser MF. Multislice CT in Cardiac Imaging: Technical Principles, Clinical Application and Future Developments. New York: Springer, 2002.

24. Nishimura DG. Principles of Magnetic Resonance Imaging. Stanford, CA: Nishimura, 1996.

25. Huang D, Swanson EA, Lin CP, et al. Optical coherence tomography. Science 1991; 254:1178–1181.

26. Brezenski ME, Fujimoto JG. Optical coherence tomography: High resolution imaging in nontransparent tissue. IEEE J Select Top Quan Elect 1999; 5:1185–1192.

27. Crowe JR, O'Donnell M. Quantitative blood speed imaging with intravascular ultrasound. IEEE Trans on Ultrasonics, Ferroelectrics and Frequency Control UFFC 2001; 48:477–487.

28. Shapo BM, Crowe JR, Skovoroda AR, Eberle MJ, Cohn NA, O'Donnell M. Displacement and strain imaging of coronary arteries with intraluminal ultrasound. IEEE Trans on Ultrasonics, Ferroelectrics and Frequency Control 1996; 43:234–246.

29. O'Donnell M, Eberle MJ, Stephens DN, et al. Catheter arrays: Can intravascular ultrasound make a difference in managing coronary artery disease? Proceedings of the 1997 IEEE Ultrasonics Symposium 1997; 1447–1456.

30. de Korte CL, van der Steen AFW, Cespedes EI, Pasterkamp G. Intravascular ultrasound elastography in human arteries: Initial experience in vivo. Ultrasound Med Biol 1998; 24:401–408.

31. Cespedes EI, de Korte CL, van der Steen AFW. Intraluminal ultrasonic palpation: Assessment of local and cross-sectional tissue stiffness. Ultrasound Med Biol 2000; 26:285–296.

32. Falk E. Why do plaques rupture? Circulation 1992; 86(suppl III):III-30–III-42.

33. Wang Y, Stephens DN, O'Donnell M. Optimizing the beam pattern of a forward-viewing ring-annular ultrasound array for intravascular imaging. IEEE Trans on Ultrasonics, Ferroelectrics and Frequency Control 2002; 49(12):1652–1664.

34. Choi CD, Skovoroda AR, Emelianov SY, O'Donnell M. An integrated compliant balloon ultrasound catheter for intravascular strain imaging. IEEE Trans on Ultrasonics, Ferroelectrics and Frequency Control 2002; 49(11):1552–1560.

35. Oralkan O, Ergun S, Johnson JA, et al. Capacitive micromachined ultrasonic transducers: Next generation arrays for acoustic imaging? IEEE Trans on Ultrasonics, Ferroelectrics and Frequency Control 2002; 49:1596–1610.

36. Oralkan O, Hansen ST, Bayram B, Yaralioglu GG, Ergun AS, Khuri-Yakub BT. CMUT ring arrays for forward-looking intravascular imaging. Proc IEEE Ultrason Symp, 2004; 403–406.

37. Buma T, Hamilton JD, Spisar M, O'Donnell M. High frequency ultrasonic imaging using optoacoustic arrays. Proc IEEE Ultrason Symp 2002; 553–562.

38. Lawrie A, Brisken AF, Francis SE, et al. Ultrasound enhances reporter gene expression after transfection of vascular cells in vitro. Circulation 1999; 99:2617–2620.

39. Lanza GM, Wallace KD, Scott MJ, et al. A novel site-targeted ultrasonic contrast agent with broad biomedical application. Circulation 1998; 94:3334–3340.

40. Lanza GM, Trousil RL, Wallace KD, et al. In vitro characterization of a novel, tissue-targeted ultrasound contrast system with acoustic microscopy. J Acous Soc Am 1999; 104:3665–3672.

41. Wickline SA, Lanza GM. Molecular imaging, targeted therapeutics, and nanoscience. J Cellular Biochem Supp 2002; 39:90–97.

42. Wickline SA, Lanza GM. Nanotechnology for molecular imaging and targeted therapy. Circulation 2003; 107: 1092–1095.

43. Lanza GM, Wallace KD, Scott MJ, et al. A novel site targeted contrast agent with broad biomedical application. Circulation 1996; 95:3334–3340.

44. Lanza GM, Wallace KD, Fisher SE, et al. High frequency ultrasonic detection of thrombi with a targeted contrast agent. Ultrasound Med Biol 1997; 23:863–870.

45. Lanza GM, Abendschein DR, Hall CH, et al. In vivo molecular imaging of stretched-induced tissue factor in carotid arteries with ligand-targeted nanoparticles. J Am Soc Echocardiography 2000; 13:608–614.

9

Optical Coherence Tomography for Imaging the Vulnerable Plaque

Guillermo J. Tearney

Wellman Center for Photomedicine, Department of Pathology, Harvard Medical School, Massachusetts General Hospital, Boston, Massachusetts, U.S.A.

Ik-Kyung Jang

Cardiology Division, Harvard Medical School, Massachusetts General Hospital, Boston, Massachusetts, U.S.A.

Brett E. Bouma

Wellman Center for Photomedicine, Harvard Medical School, Massachusetts General Hospital, Boston, Massachusetts, U.S.A.

INTRODUCTION

Acute myocardial infarction (AMI) is the leading cause of death in the United States and industrialized countries (1,2). Research conducted over the past 15 years has demonstrated that several types of minimally or modestly stenotic atherosclerotic plaques, termed vulnerable plaques, are precursors to coronary thrombosis, myocardial ischemia, and sudden cardiac death. Post-mortem studies have identified one type of vulnerable plaque, the thin-cap fibroatheroma (TCFA), as the culprit lesion in approximately 80% of sudden cardiac deaths (3–7). Over 90% of TCFAs are found within the most proximal 5.0 cm segment of each of the main coronary arteries (left anterior descending—LAD; left circumflex—LCx; and right coronary artery—RCA) (3,5). The TCFA is typically a minimally occlusive plaque characterized histologically by the following features: (*i*) thin fibrous cap (<65 μm), (*ii*) large lipid pool, and (*iii*) activated macrophages near or within the fibrous cap (3,5,7–9). It is hypothesized that these features predispose TCFAs to rupture in response to biomechanical stresses (10,11). Following rupture and the release of procoagulant proteins, such as tissue factor, a substrate for thrombus formation is created, leading to an acute coronary event (12,13). While TCFAs are associated with the majority of AMIs, recent autopsy studies have shown that coronary plaques with erosions or superficial calcified nodules may also precipitate thrombosis and sudden occlusion of a coronary artery (3,5,14,15).

Although autopsy studies have been valuable in determining features of culprit plaques, the retrospective nature of these studies limits their ability to quantify the risk of an individual plaque for causing acute coronary thrombosis. For instance, TCFAs are a frequent autopsy finding in asymptomatic or stable patients and are found with equal frequency in culprit and nonculprit arteries in acute coronary syndromes (16). Moreover, disrupted TCFAs have been found in 10% of noncardiac deaths (16). Recent findings of multiple ruptured plaques (17) and increased systemic inflammation in acute patients (18) has challenged the notion of a single vulnerable plaque as the precursor for AMI (19–21). An improved understanding of the natural history and clinical significance of these lesions would accelerate progress in diagnosis, treatment, and prevention of coronary artery disease (CAD).

An attractive approach to studying the evolution of vulnerable plaques is noninvasive or intracoronary imaging of individual lesions at multiple time points. Unfortunately, the microscopic features that characterize vulnerable plaque are not reliably identified by conventional imaging technologies, such as intravascular ultrasound (IVUS) (22–27), CT (28–31), and MRI (31–34). While experimental intracoronary imaging modalities, such as integrated backscatter IVUS (35,36), elastography (31,37), angioscopy (38–42), near-infrared spectroscopy (43), Raman spectroscopy (44), and thermography (45,46) have been investigated for the detection of vulnerable plaque, no method to date has been shown to reliably identify all of the characteristic features of these lesions.

While our understanding of vulnerable coronary plaque is still at an early stage, the concept that certain types of plaques predispose patients to developing an AMI and sudden cardiac death continues to be at the forefront of cardiology research to improve the mortality of CAD. Intracoronary optical coherence tomography (OCT) has been developed to both identify and study these lesions due to its distinct resolution advantage over other imaging modalities. In this chapter, we summarize clinical research conducted at the Massachusetts General Hospital over the past decade to develop, validate, and utilize this technology to improve our understanding of vulnerable plaque. Our results show that intracoronary OCT may be safely conducted in patients and that it provides abundant information regarding plaque microscopic morphology, which is essential to the identification and study of high-risk lesions. Even though many basic biological, clinical, and technological challenges must be addressed prior to widespread use of this technology, the unique capabilities of OCT ensure that it will have a prominent role in shaping the future of cardiology.

OPTICAL COHERENCE TOMOGRAPHY

Intracoronary OCT is a microscopic imaging technology that has been developed for the identification of vulnerable plaque. OCT acquires cross-sectional images of tissue reflectance and, since it may be implemented through an optical fiber probe, it is readily adaptable to coronary catheters (47) for circumferential imaging of arterial pathology. The first investigation of vascular optical coherence tomography ex vivo demonstrated the potential of this technique to identify arterial microstructure (48). Subsequent development of OCT technology enabled image acquisition at rates sufficient for intracoronary imaging in human patients (49,50,51). In this chapter, we will describe studies conducted with this technology over the past decade at the Massachusetts General Hospital. Results from these studies show that a wide variety of microscopic features, including those associated with TCFAs, can be identified by OCT imaging both ex vivo and in living human patients. These findings suggest that this technology will play an important role in improving our understanding of CAD, guiding local therapy, and decreasing the mortality of AMI.

OPTICAL COHERENCE TOMOGRAPHY SYSTEM

A schematic of the OCT system is shown in Figure 1 (52). Briefly, the system consisted of a polarization-diverse fiber-optic nonreciprocal interferometer that operated in the time domain. The light source was centered at 1300 nm and had a Gaussian spectral full-width-at-half-maximum of 70 nm, providing an axial resolution of approximately 8 μm in tissue. The transverse resolution, determined by the focal spot size produced by the probe, was 25 μm. Group delay scanning at a rate of 2 kHz was conducted by utilizing a phase-control rapidly scanning optical delay (RSOD) in the reference arm (53). Images (500 pixels transverse

× 250 pixels axial) were obtained at 4 frames per second and stored digitally. A custom-built fiber-optic rotary junction was utilized for catheter-based circumferential imaging and a galvanometer mirror was used for the free-space experiments (54). Catheters were constructed by modifying a commercially available 3.0 F (950 μm diameter) IVUS catheter to incorporate a central fiber, a distal gradient index (GRIN) lens, and a deflecting prism, which were rotated to construct a circumferential image (54).

EX VIVO STUDIES

Plaque Characterization
The first step in validating this imaging modality was to establish objective image criteria for discrimination of atherosclerotic plaque types ex vivo. In order to accomplish this task, 50 cadaver plaques were imaged by OCT and correlated with histology obtained at the imaging site (55). Fibrous plaques were characterized by homogeneous, signal-rich regions, fibrocalcific plaques by signal-poor regions with sharp borders, and lipid-rich plaques by signal-poor regions with diffuse borders (Fig. 2). The accuracy of OCT for characterizing plaque type was then determined by prospectively applying these criteria to images of 307 plaques, with histopathologic diagnosis as the gold standard. These criteria yielded a sensitivity and specificity of 94% and 92% respectively for lipid-rich plaques (overall agreement, $\kappa = 0.84$), demonstrating that objective OCT criteria are highly sensitive and specific for differentiating lipid-rich plaques from other plaque types.

Quantification of Macrophage Content
Macrophages are central to the etiology of coronary artery disease (12,56–58). Due to the high quantity of intracellular phagolysosomes containing lipid and other cellular debris, the authors hypothesized that the refractive index contrast provided by the cytoplasm of macrophages would result in a strong optical signal from these cells (Fig. 3). Furthermore, since macrophages are typically heterogeneously distributed in atherosclerotic tissue, the spatial variance of OCT signal in plaques with a high macrophage content should also be elevated. In order to test this hypothesis, cap macrophage and smooth muscle densities of 27 necrotic core fibroatheromas were quantified morphometrically by immunoperoxidase staining with CD68 and smooth muscle actin. Measurements of cell density were then compared to the normalized standard deviation (NSD) of the OCT signal intensity at corresponding locations (59). We found a high degree of positive correlation between OCT and histologic measurements of fibrous cap macrophage density ($r = 0.84$, $p < 0.0001$) and a negative correlation between OCT

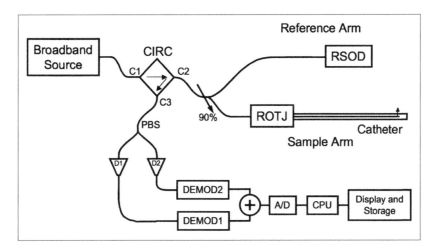

Figure 1 Schematic of the time-domain intracoronary optical coherence tomography system. Polarized, broad bandwidth light passes through a circulator (CIRC, port C1 to port C2) and is split into reference and sample arms via a 10/90 fiber optic beam splitter. The optical path length (group delay) of the reference arm is scanned by translating the galvanometer of the rapidly scanning optical delay (RSOD) line. Sample arm light is coupled into the catheter by a rotating optical junction (ROTJ). Light returned from the reference and sample arms is combined at the splitter and transmitted back to the circulator at port C2. The circulator then passes this light through port C3 to a polarizing beam splitter (PBS). The two orthogonal polarization states are detected separately by photodiodes D1 and D2. The two signals are demodulated and summed to create the final output signal, which is digitized (A/D) and transferred to the CPU. Detection of the fringe patterns created by sample and reference arm interference allows one radial scan (A-line) to be constructed that maps tissue reflectivity to a given axial or depth location. A cross-sectional image is generated by repeating this process at successive transverse locations on the sample while the ROTJ rotates the internal components of the catheter.

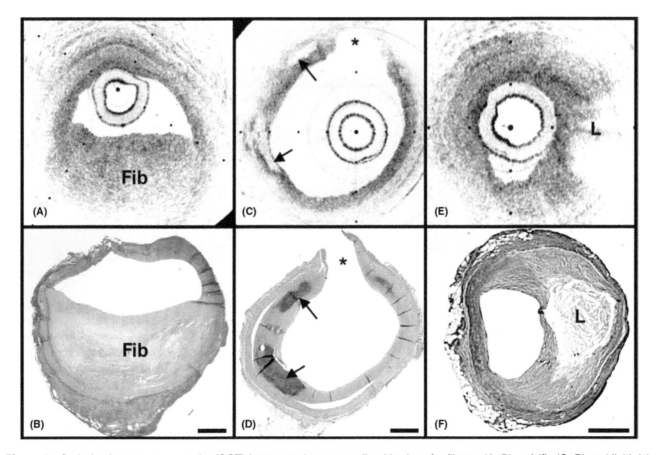

Figure 2 Optical coherence tomography (OCT) images and corresponding histology for fibrous (**A, B**), calcific (**C, D**) and lipid-rich (**E, F**) plaque types. In fibrous plaques, the OCT signal (Fib) is observed to be strong and homogeneous. In comparison, both calcific (*arrows*) and lipid-rich regions (L) appear as signal poor regions within the vessel wall. Lipid-rich plaques have diffuse or poorly demarcated borders, while the borders of calcific nodules are sharply delineated. (**B, D**) Hematoxylin and eosin; (**F**) Masson's trichome, original magnification 40×. Scale bars, tick marks, 500 μm.

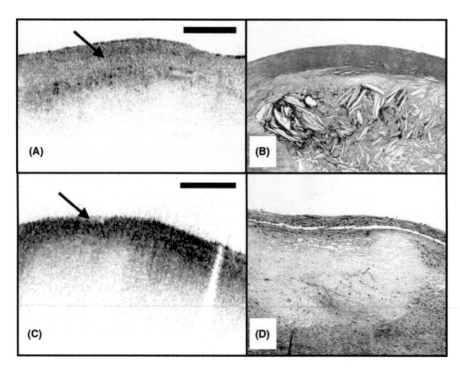

Figure 3 (**A**) Optical coherence tomography (OCT) image of a fibroatheroma with a low density of macrophages within the fibrous cap (*arrow*). (**C**) OCT image of a fibroatheroma with a high density of macrophages within the fibrous cap (*arrow*). (**B, D**) Histology corresponding to (**A**) and (**C**), respectively; Masson's trichome, original magnification 40×. Scale bars, 500 μm.

and histologic measurements of smooth muscle actin density ($r = -0.56$, $p < 0.005$). A range of NSD thresholds (6.15–6.35%) yielded greater than 90% sensitivity and specificity for identifying caps containing >10% CD68 staining. This study demonstrated that the high contrast and resolution of OCT enables the quantification of macrophages within fibrous caps of atherosclerotic plaques. The impact of this finding was significant, as prior to this work there was no established tool that could permit the investigation of these crucial inflammatory cells in patients.

Other Plaque Features
While the previously described studies demonstrated accurate characterization of features associated with TCFAs, OCT is capable of identifying additional plaque components that may be associated with acute coronary events. These features are described subsequently.

Calcific Nodules
Calcific nodules have been associated with plaque thrombosis in a minority of cases (3,5). Calcium by OCT appears as a signal poor region with a sharp delineation between the nodule and the surrounding tissue (Fig. 2C). In our histopathologic study of plaque characterization, we were able to use this criterion to diagnose calcific nodules with 96% sensitivity and 97% specificity (55).

Thrombus
Differentiation of thrombus from plaque is critical for accurate plaque characterization. In addition, there is evidence to suggest that thrombus type, platelet-rich versus red blood cell-rich, is an indicator of the flow conditions associated with the thrombus formation and could be an important predictor of the efficacy of thrombolytic therapy (60). We have conducted a study to correlate histopathologic sections of thrombus with OCT. Our preliminary results suggest that a "red" thrombus (red blood cell-rich) rapidly attenuates the signal in a manner similar to whole blood (Fig. 4A). In contrast, a "white" thrombus (platelet-rich) appears to exhibit a homogeneous moderate-strong signal with significantly less attenuation (Fig. 4B) (51). In our experience, both forms of thrombus are usually easily distinguished from the arterial wall proper (61).

Macrophages at the Cap-Lipid Pool Border
High-resolution cross-sectional optical imaging affords the unique opportunity to study the location of macrophage accumulations within a given plaque cross-section. Using our macrophage data set (59), we found that 84% of OCT images demonstrated a high signal at the junction between the cap and the lipid pool (Fig. 5A, B) when CD68 staining at this interface was greater than 50%.

Giant Cells
Multinucleated macrophages, or giant cells, are an inflammatory response to a foreign body (e.g., cholesterol crystals) within atherosclerotic plaques. Giant cells can be visualized by OCT as large highly reflecting regions within the cap and at the cap-lipid pool interface (Fig. 5C, D).

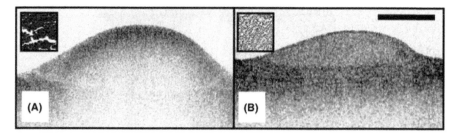

Figure 4   Optical coherence tomography (OCT) images of red blood cell-rich (**A**) and platelet-rich (**B**) thrombi. The red blood cell-rich thrombus demonstrates high OCT signal attenuation, whereas the platelet rich thrombus shows a homogeneous scattering signal with relatively little attenuation. Insets depict corresponding histology sections from each thrombus; hematoxylin and eosin, original magnification 40×. Scale bar, 500 μm.

Cholesterol Crystals

Research investigating the biomechanical properties of atherosclerotic plaques has shown that the presence of cholesterol crystals increases the stiffness of lipid pools, and as a result, may decrease the likelihood of plaque rupture (62). Images of cholesterol crystals demonstrate oriented, linear, highly reflecting structures within the plaques (Fig. 5E, F) (63).

CLINICAL STUDIES

Histopathologic validation of qualitative and quantitative image criteria ex vivo provided a foundation for interpreting data obtained from living human patients. Between January 2000 and September 2003, a total of 83 patients undergoing routine percutaneous transluminal coronary intervention (PTCI) were enrolled in a study at the Massachusetts General Hospital (Boston,

Massachusetts, U.S.A.) to investigate the feasibility of intracoronary OCT. Imaging was performed in culprit lesions and remote sites with IVUS and OCT pre- and postcoronary intervention (51,64). Clear visualization of the arterial wall was accomplished by use of intermittent saline flushes (8–10 cc) through the guide catheter. All patients tolerated the procedure well without complication. Images demonstrating detailed arterial microstructure were successfully obtained in all patients studied and in all three major coronary arteries (51,64). Summaries of the results from our clinical study are described subsequently.

OCT images obtained in living patients contained the same image features as those obtained ex vivo. All of the characteristics of macrophage-rich TCFAs as well as the additional plaque features described in the previous section were observed in vivo; there were no image features in the clinical data that had not been observed ex vivo (Fig. 6) (51). These observations

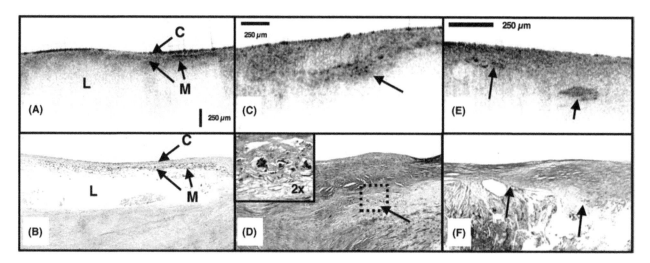

Figure 5   (**A**) Optical coherence tomography (OCT) image of a fibroatheroma with macrophages (M) present at the cap (C) lipid-pool (L) interface. (**C**) OCT image demonstrating giant cells (*arrow*). (**E**) Cholesterol crystals (*arrows*) appear as signal-rich linear structures. (**B, D,** and **F**) Histology corresponding to (**A**), (**C**), and (**E**); (**B**) CD68; (**D, F**) Masson's trichrome, original magnification 40×. Scale bars, 250 μm.

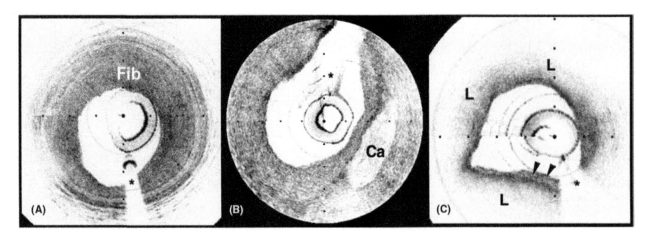

Figure 6 Optical coherence tomography (OCT) images of coronary plaques acquired from living human patients. (**A**) Fibrous plaque (Fib); (**B**) calcific nodule (Ca); (**C**) TCFA with circumferential lipid pool (L) and a region consistent with a platelet-rich thrombus (*arrowheads*). *Note*: ⋆, guidewire artifacts. Tick marks, 500 μm.

suggest that image interpretation criteria and algorithms validated ex vivo can also be applied to images obtained from patients.

OCT observations were consistent with IVUS, the current gold standard for intracoronary imaging (51). Although IVUS is unable to resolve microstructural features associated with vulnerable plaque, it can identify nonatherosclerotic (normal) vessels, large thrombi, calcific deposits, and pronounced arterial disruptions. Although unconfirmed, indirect evidence suggests that IVUS may detect large lipid deposits (23). In all cases where IVUS identified these characteristics, blinded OCT observations were consistent. OCT detected additional cases of intimal hyperplasia, thrombus, intimal disruption, and lipid pool not identified by IVUS (51).

Imaging of culprit lesions demonstrated a higher prevalence of TCFA in patients with acute coronary disease than patients with stable coronary disease (65). In

this analysis, TCFA was defined as a plaque with lipid area more than 2 quadrants and cap thickness < 65 μm. From a total of 57 patients, 20 presented with AMI, 20 with an acute coronary syndrome (ACS), and 17 with stable angina pectoris (SAP). TCFAs identified by OCT were found in 13 AMI patients (65%), 9 with ACS (45%), and 3 with SAP (18%) (65). TCFA were more prevalent in acute presentations (AMI and ACS) of CAD (55% vs. 18%, $p = 0.012$) (65). Plaque disruption was found more frequently in acute CAD (20% vs. 12%, $p = 0.053$) and calcifications were more frequent in stable disease (41% vs. 12%, $p = 0.049$) (65). These findings represent the first observation of presentation-dependent plaque morphology in living human patients and confirm our current knowledge of the relationship between morphology and patient outcome that has been obtained in previous autopsy studies (3–7).

Macrophage content was significantly higher in fibroatheroma caps in patients with acute presentations

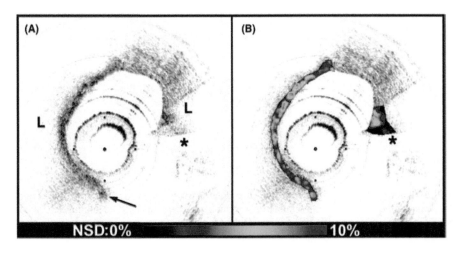

Figure 7 (**A**) Conventional optical coherence tomography (OCT) image of a ruptured thin-cap fibroatheroma obtained from a patient with acute myocardial infarction. (**B**) In this image, macrophage density data from the fibrous cap is displayed using a color look up table. L, lipid pool; arrow, intimal disruption; ⋆, guidewire artifact. Tick marks, 500 μm.

of CAD, versus caps in patients with SAP (66). For this study, we created NSD (macrophage density) images (Fig. 7) and analyzed the macrophage distributions in both culprit and remote plaques (n=225) of 49 patients with different clinical syndromes. Macrophage density was found to be significantly higher in the AMI (5.54±1.48%) and ACS (5.86±2.01%) groups compared to the SAP group (4.14±1.81 %) ($p < 0.003$) (66).

Both focal and multi-focal elevations of macrophage density were associated with severity of clinical presentation (66). There is great interest in understanding the role of focal versus multi-focal plaque features in the pathogenesis of acute coronary thrombosis. The unique ability of OCT to quantify macrophage content and observe the spatial distribution of plaque macrophages provides a valuable tool for investigating this critical question. Supporting the concept of focal risk, sites of plaque rupture demonstrated a greater macrophage density than nonrupture sites (6.95 ± 0.48 %, $5.71 \pm 0.37\%$; $p = 0.01$) (66). In addition, macrophage density was significantly higher at the surface (first 50 µm) of culprit plaques in comparison to remote lesions, indicating that superficial macrophages confer a higher risk for developing an acute thrombus (66). This finding represents a new understanding of the role of macrophages in the pathogenesis of acute coronary disease and may provide an additional parameter to assess individual plaque risk.

We also found evidence in support of the multi-focal hypothesis. Macrophage densities at remote sites were correlated with measurements at culprit sites within the same patient ($r = 0.66$; $p = 0.01$) (66). Fibrous plaques that are not considered to be high-risk lesions, had a higher macrophage content in patients with acute disease, compared with stable patients ($p = 0.025$) (66). Taken together, these results suggest that both focal and generalized macrophage distributions play important roles governing the severity of CAD.

CURRENT TECHNOLOGY CHALLENGES

Studies conducted with OCT clearly demonstrate the potential of this technology for impacting the management of CAD. However, limitations of current technology may preclude widespread use. Most importantly, removal of blood interposed between the catheter and the arterial wall is required in order to obtain a clear, unobstructed view. Saline purging adequately removes blood from the field, but at current imaging frame rates (4–8 fps), intracoronary OCT with saline flushing reduces to a single cross-sectional measurement. As a result, large area screening of vessel pathology required for widespread adoption is untenable with this paradigm. Balloon occlusion with saline purging, such as is commonly employed during angioscopy, remains a

viable option for certain interventional communities (41,67). However, this technique is not favored in the United States. Strategies such as reducing the scattering of blood by administering an index matching fluid or administration of a transparent oxygen-carrying blood substitute have been proposed to reduce the blood-attenuation problem (68,69). More research needs to be conducted in these areas to determine the clinical viability of these methods. Another line of attack for overcoming the blood attenuation problem is to increase the frame rate significantly, which can commensurately increase the information yield per purge. While conventional time-domain OCT systems have a frame rate of approximately 4–8 frames per second, new frequency-domain OCT technology, spectral-domain OCT (SD-OCT) and optical frequency domain imaging (OFDI), enable more than an order of magnitude higher frame rates (70,71).

One common misconception regarding OCT is that it can replace intravascular ultrasound. In fact, each technology excels at different tasks. IVUS obtains images through blood with approximately 100 µm resolution and a depth of penetration of approximately 1.0 cm. OCT on the other hand obtains images with much higher resolution, but cannot penetrate as deeply (~2 mm). As a result, OCT is particularly well suited to investigating microscopic features at the surface of the arterial wall, which are characteristic of high-risk vulnerable plaques. However, the infrared light utilized by OCT does not reach the back wall of thick atherosclerotic lesions and therefore this technology is probably not appropriate for evaluating pathology, such as remodeling that is manifested by alterations of the elastic laminae.

Separate from the depth of penetration issue, conventional time-domain technology also has a limited depth range over which signals can be obtained. This limited ranging depth presents difficulties visualizing the entire arterial circumference, notably when the catheter is eccentrically placed within a large diameter vessel. Auto-ranging that adaptively adjusts the RSOD galvanometer to follow the lumen of the arterial wall is one viable solution to this problem (72). Additionally, second generation OFDI systems inherently provide up to 7.0 mm of ranging depth, which is sufficient to visualize the entire artery even at the coronary ostia (70).

CONCLUSION

To date, optical coherence tomography has had a tangible impact on the quest to understand and identify the vulnerable plaque. It is the only method demonstrated to be capable of measuring all of the microscopic features associated with TCFAs. Our knowledge of the morphology of plaques associated with AMI, previously predicated on retrospective autopsy studies, has now been confirmed in living human patients with this technology. New information regarding macrophage distributions in patients with severe CAD and insight into the focal and diffuse nature of the inflammatory atherosclerotic process has been uncovered.

The promise of intracoronary OCT is great, yet important implementation challenges still remain. Much needs to be learned about CAD in order to ascertain the eventual role of this or any other intravascular modality in clinical practice. The prevalence, incidence, and natural history of high-risk lesions are presently poorly understood. Studies are currently underway using OCT and other imaging modalities to investigate the evolution of individual coronary lesions (73). Tests capable of screening a large population to determine patients that require more detailed intracoronary evaluation must be developed. A variety of local treatment strategies have been proposed, including lesion stabilization with drug-eluting stents (74,75), and photodynamic therapy (76,77), but the long-term clinical viability of these approaches is currently unknown. While there is still much to be done, we anticipate that the unique capabilities of OCT as an investigational tool for high-risk lesions will serve the cardiology community well as it advances to understand, identify, and ultimately treat the vulnerable plaque.

ACKNOWLEDGMENTS

This chapter is reproduced from Tearney et al. Journal of Biomedical Optics 2006; 11(2):021001-1-10. Studies by the authors described in this chapter were funded in part by the Center for Integration of Medicine and Innovative Technology (development of the imaging platform), Guidant Corporation, and the National Institutes of Health (grants R01-HL70039 and R01-HL76398).

REFERENCES

1. American Heart Association: Heart and Stroke Facts: 1996, Statistical Supplement.
2. American Heart Association: Heart Disease and Stroke Statistics—2003 Update.
3. Kolodgie FD, Burke AP, Farb A, et al. The thin-cap fibroatheroma: a type of vulnerable plaque: the major precursor lesion to acute coronary syndromes. Curr Opin Cardiol 2001; 16(5):285–292.
4. Virmani R, Kolodgie FD, Burke AP, Farb A, Schwartz SM. Lesions from sudden coronary death: a comprehensive morphological classification scheme for atherosclerotic lesions. Arterioscler Thromb Vasc Biol 2000; 20: 1262–1275.
5. Virmani R, Burke AP, Farb A, Kolodgie FD. Pathology of the unstable plaque. Prog Cardiovasc Dis 2002; 44(5):349–356.
6. Davies MJ. Stability and instability: two faces of coronary atherosclerosis. The Paul Dudley White Lecture 1995. Circulation 1996; 94:2013–2020.
7. Falk E. Why do plaques rupture? Circulation 1992; 86(suppl 6):30–42.
8. Davies MJ. Detecting vulnerable coronary plaques. Lancet 1996; 347:1422–1423.
9. Sukhova GK, Schonbeck U, Rabkin E, et al. Evidence for increased collagenolysis by interstitial collagenases-1 and -3 in vulnerable human atheromatous plaques. Circulation 1999; 99(19):2503–2509.
10. Cheng GC, Loree HM, Kamm RD, Fishbein MC, Lee RT. Distribution of circumferential stress in ruptured and stable atherosclerotic lesions. A structural analysis with histopathological correlation. Circulation 1993; 87(4):1179–1187.
11. Lee RT, Grodzinsky AJ, Frank EH, Kamm RD, Schoen FJ. Structure dependent dynamic mechanical behavior of fibrous caps from human atherosclerotic plaques. Circulation 1991; 83:1764–1770.
12. Moreno PR, Bernardi VH, Lopez-Cuellar J, et al. Macrophages, smooth muscle cells, and tissue factor in unstable angina. Implications for cell-mediated thrombogenicity in acute coronary syndromes. Circulation 1996; 94:3090–3097.
13. Zaman AG, Helft G, Worthley SG, Badimon JJ. The role of plaque rupture and thrombosis in coronary artery disease. Atherosclerosis 2000; 149(2):251–266.
14. van der Wal AC, Becker AE, van der Loos CM, Das PK. Site of intimal rupture or erosion of thrombosed coronary atherosclerotic plaques is characterized by an inflammatory process irrespective of the dominant plaque morphology. Circulation 1994; 89:36–44.
15. Farb A, Burke AP, Tang AL, et al. Coronary plaque erosion without rupture into a lipid core. A frequent cause of coronary thrombosis in sudden coronary death. Circulation 1996; 93:1354–1363.
16. Arbustini E, Grasso M, Diegoli M, et al. Coronary atherosclerotic plaques with and without thrombus in ischemic heart syndromes: a morphologic, immunohistochemical, and biochemical study. Am J Cardiol 1991; 68(7):36B-50B.
17. Rioufol G, Finet G, Ginon I, et al. Multiple atherosclerotic plaque rupture in acute coronary syndrome: a three-vessel intravascular ultrasound study. Circulation 2002; 106(7):804–808.
18. Biasucci LM, Liuzzo G, Colizzi C, Maseri A. The role of cytokines in unstable angina. Expert Opin Investig Drugs 1998; 7(10):1667–1672.
19. Maseri A, Fuster V. Is there a vulnerable plaque? Circulation 2003; 107:2068–2071.
20. Cascells W, Naghavi M, Willerson JT. Vulnerable atherosclerotic plaque: a multfocal disease. Circulation 2003; 107:2072–2075.
21. Kereiakes DJ. The emperor's clothes: in search of the vulnerable plaque. Circulation 2003; 107:2076–2077.

22. Yock PG, Fitzgerald PJ. Intravascular ultrasound: state of the art and future directions. Am J Cardiol 1998; 81(7A):27E-32E.

23. Yamaguchi M, Terashima M, Awano K, et al. Morphology of vulnerable coronary plaques: insights from follow-up of patients examined by intravascular ultrasound before an acute coronary event. J Am Coll Cardiol 2000; 35(1):106–111.

24. Martin AJ, Ryan LK, Gotlieb AI, Henkelman RM, Foster FS. Arterial imaging: comparison of high-resolution US and MR imaging with histologic correlation. Radiographics 1997; 17(1):189–202.

25. Schoenhagen P, Nissen SE. Understanding coronary artery disease: tomographic imaging with intravascular ultrasound. Heart 2002; 88:91–96.

26. Tobis JM, Mallery J, Mahon D, et al. Intravascular ultrasound imaging of human cornary arteries in vivo: analysis of tissue characterizations with comparison to in vitro histological specimens. Circulation 1991; 83:913–926.

27. Prati F, Arbustini E, Labellarte A, et al. Correlation between high frequency intravascular ultrasound and histomorphology in human coronary arteries. Heart 2001; 85(5):567–570.

28. Rumberger JA, Behrenbeck T, Breen JFSheedy PF, 2nd. Coronary calcification by electron beam computed tomography and obstructive coronary artery disease: a model for costs and effectiveness of diagnosis as compared with conventional cardiac testing methods. J Am Coll Cardiol 1999; 33(2):453–462.

29. Wong ND, Vo A, Abrahamson D, Tobis JM, Eisenberg H, Detrano RC. Detection of coronary artery calcium by ultrafast computed tomography and its relation to clinical evidence or coronary artery disease. Am J Cardiol 1994; 73(4):223–227.

30. Budoff MJ, Brundage BH. Electron beam computed tomography: screening for coronary artery disease. Clin Cardiol 1999; 22(9):554–558.

31. Naghavi M, Madjid M, Khan MR, Mohammadi RM, Willerson JT, Casscells SW. New developments in the detection of vulnerable plaque. Curr Atheroscler Rep 2001; 3(2):125–135.

32. Baer FM, Theissen P, Crnac J, Schmidt M, Jochims M, Schicha H. MRI assessment of coronary artery disease. Rays 1999; 24(1):46–59.

33. Toussaint JF, LaMuraglia GM, Southern JF, Fuster V, Kantor HL. Magnetic resonance images lipid, fibrous, calcified, hemorrhagic, and thrombotic components of human atherosclerosis in vivo. Circulation 1996; 94:932–938.

34. Helft G, Worthley SG, Fuster V, et al. Progression and regression of atherosclerotic lesions: monitoring with serial noninvasive magnetic resonance imaging. Circulation 2002; 105(8):993–998.

35. Machado JC, Foster FS. Ultrasonic integrated backscatter coefficient profiling of human coronary arteries in vitro. IEEE Trans Ultrason Ferroelectr Freq Control 2001; 48(1):17–27.

36. Urbani MP, Picano E, Parenti G, et al. In vivo radiofrequency-based ultrasonic tissue characterization of the atherosclerotic plaque. Stroke 1996; 24(10):1507–1512.

37. Korte CL, van der Steen AFW, Cespedes EI, et al. Characterization of plaque components and vulnerability with intravascular ultrasound elastography. Phys Med Biol 2000; 45:1465–1475.

38. Ueda Y, Asakura M, Yamaguchi O, Hirayama A, Hori M, Kodama K. The healing process of infarct-related plaques. J Am Coll Cardiol 2001; 38(7):1916–1922.

39. Asakura M, Ueda Y, Yamaguchi O, et al. Extensive development of vulnerable plaques as a pan-coronary process in patients with myocardial infarction: an angioscopic study. J Am Coll Cardiol 2001; 37(5): 1284–1288.

40. Kodama K, Hirayama A, Ueda Y. Usefulness of coronary angioscopy for the evaluation of hyperlipidemia. Nippon Rinsho - Japanese Journal of Clinical Medicine 2002; 60(5):927–932.

41. Mizuno K, Nakamura H. Percutaneous coronary angioscopy: present role and future direction. Ann Med 1993; 25(1):1–2.

42. Waxman S. Characterization of the unstable lesion by angiography, angioscopy, and intravascular ultrasound. Cardiol Clin 1999; 17(2):295–305.

43. Moreno PR, Lodder RA, Purushothaman KR, Charash WE, O'Connor WN, Muller JE. Detection of lipid pool, thin fibrous cap, and inflammatory cells in human aortic atherosclerotic plaques by near-infrared spectroscopy. Circulation 2002; 105(8):923–927.

44. Romer TJ, Brennan JF, 3rd, Fitzmaurice M, et al. Histopathology of human coronary atherosclerosis by quantifying its chemical composition with Raman spectroscopy. Circulation 1998; 97(9):878–885.

45. Casscells W, Hathorn B, David M, et al. Thermal detection of cellular infiltrates in living atherosclerotic plaques: possible implications for plaque rupture and thrombosis. Lancet 1996; 347(9013):1447–1451.

46. Stefanadis C, Toutouzas K, Tsiamis E, et al. Increased local temperature in human coronary atherosclerotic plaques: an independent predictor of clinical outcome in patients undergoing a percutaneous coronary intervention. J Am Coll Cardiol 2001; 37(5):1277–1283.

47. Tearney GJ, Boppart SA, Bouma BE, et al. Scanning single-mode fiber optic catheter-endoscope for optical coherence tomography. Optics Letters 1996; 21.

48. Brezinski ME, Tearney GJ, Bouma BE, et al. Optical coherence tomography for optical biopsy: properties and demonstration of vascular pathology. Circulation 1996; 93(6):1206–1213.

49. Tearney GJ, Brezinski ME, Bouma BE, et al. In vivo endoscopic optical biopsy with optical coherence tomography. Science 1997; 276(5321):2037–2039.

50. Bouma BE, Tearney GJ. Power-efficient nonreciprocal interferometer and linear-scanning fiber-optic catheter for optical coherence tomography. Optics Letters 1999; 24(8):531–533.

51. Jang IK, Bouma BE, Kang DH, et al. Visualization of coronary atherosclerotic plaques in patients using optical coherence tomography. J Am Coll Cardiol 2002; 39:604–609.

52. Bouma BE, Tearney GJ. Power-efficient nonreciprocal interferometer and linear-scanning fiber optic catheter

for optical coherence tomography. Optics Letters 1999; 24(8):531–533.

53. Tearney GJ, Bouma BE, Fujimoto JG. Phase and group delay relationships for the phase control rapid-scanning optical delay line. Optics Letters 1997; 22:1811–1813.

54. Shishkov M, Bouma BE, Jang IK, et al. Optical coherence tomography of coronary arteries in vitro using a new catheter. In: Optical Society of America Biomedical Topical Meetings 2000; Miami, FL, 2000; p. SuC4–1, 35–7.

55. Yabushita H, Bouma BE, Houser SL, et al. Characterization of human atherosclerosis by optical coherence tomography. Circulation 2002; 106:1640–1645.

56. Moreno PR, Falk E, Palacios IF, Newell JB, Fuster V, Fallon JT. Macrophage infiltration in acute coronary syndromes: implications for plaque rupture. Circulation 1994; 90:775–778.

57. Lendon CL, Davies MJ, Born GV, Richardson PD. Atherosclerotic plaque caps are locally weakened when macrophage density is increased. Atheroosclerosis 1991; 87:87–90.

58. Davies MJ, Richardson PD, Woolf N, Katz DR, Mann J. Risk of thrombosis in human atherosclerotic plaques: role of extracellular lipid, macrophage, and smooth muscle cell content. Br Heart J 1993; 69(5):377–381.

59. Tearney GJ, Yabushita H, Houser SL, et al. Quantification of macrophage content in atherosclerotic plaques by optical coherence tomography. Circulation 2003; 107:113–119.

60. Goto S, Handa S. Coronary thrombosis. Effects of blood flow on the mechanism of thrombus formation. Jpn Heart J 1998; 39(5):579–596.

61. Jang IK, Hursting MJ. When heparins promote thrombosis: review of heparin induced thrombocytopenia. Circulation 2005; 111(20):2671–2683.

62. Loree HM, Grodzinsky AJ, Park SY, Gibson LJ, Lee RT. Static circumferential tangential modulus of human atherosclerotic tissue. J Biomechanic 1994; 27(2): 195–204.

63. Tearney GJ, Jang IK, Bouma BE. Evidence of cholesterol crystals in atherosclerotic plaque by optical coherence tomographic (OCT) imaging. J Am Coll Cardiol 2004; 44:972–979.

64. Jang IK, Tearney G, Bouma B. Visualization of tissue prolapse between coronary stent struts by optical coherence tomography: comparison with intravascular ultrasound. Circulation 2001; 104(22):2754.

65. Jang IK, Tearney GJ, MacNeill B, et al. In vivo characterization of coronary atherosclerotic plaque by use of optical coherence tomography. Circulation 2005; 111(12):1551–1555.

66. BD MacNeill, Jang IK, Bouma BE, Halpern EF, Tearney GJ. Coronary plaque macrophage distributions in living patients. Eur Heart J 2003; 24:1462.

67. Uchida Y, Fumitaka N, Tomaru T, et al. Prediction of acute coronary syndromes by percutaneous coronary angioscopy in patients with stable angina. Am Heart J 1995; 130(2):195–203.

68. Brezinski M, Saunders K, Jesser C, Li X, Fujimoto J. Index matching to improve optical coherence tomography imaging through blood. Circulation 2001; 103(15):1999–2003.

69. Villard JW, Feldman MD, Kim J, Milner TE, Freeman GL. Use of a blood substitute to determine instantaneous murine right ventricular thickening with optical coherence tomography. Circulation 2002; 105(15):1843–1849.

70. Yun SH, Tearney GJ, de Boer JF, Iftimia N, Bouma BE. High-speed optical frequency domain imaging. Opt Express 2003; 11:2953–2963.

71. Yun SH, Tearney GJ, Bouma BE, Park BH, de Boer JF. High-speed spectral-domain optical coherence tomography at 1.3 m wavelength. Opt Express 2003; 26(11):3598–3604.

72. Iftima N, Bouma BE, de Boer JF, Park BH, Cense B, Tearney GJ. Adaptive ranging for optical coherence tomography. Optics Express 2004; 12(17):4025–4034.

73. Wald JD, Boylan J, Wang C. The status of the vulnerable plaque opportunity: a report for the 3rd annual conference on vulnerable plaque. Heathcare industry note: Edwards; 2005 June 21.

74. Valgimigli M, van Mieghem CA, Ong AT, et al. Short- and long-term clinical outcome after drug-eluting stent implantation for the percutaneous treatment of left main coronary artery disease: insights from the Rapamycin-Eluting and Taxus Stent Evaluated At Rotterdam Cardiology Hospital registries (RESEARCH and T-SEARCH). Circulation 2005; 111(11):1383–1389.

75. Chieffo A, Stankovic G, Bonizzoni E, et al. Early and mid-term results of drugeluting stent implantation in unprotected left main. Circulation 2005; 111(6): 791–795.

76. Demidova TN, Hamblin MR. Macrophage-targeted photodynamic therapy. Int J Immunopathol Pharmacol 2004; 17(2):117–126.

77. Kereiakes DJ, Szyniszewski AM, Wahr D, et al. Phase I drug and light doseescalation trial of motexafin lutetium and far red light activation (phototherapy) in subjects with coronary artery disease undergoing percutaneous coronary intervention and stent deployment: procedural and long-term results. Circulation 2003; 108(11):1310–1315.

10

Cardiovascular Molecular Imaging: Overview of Cardiac Reporter Gene Imaging

Joseph C. Wu

Division of Cardiology, Department of Medicine, and Division of Nuclear Medicine, Department of Radiology, Stanford University School of Medicine, Stanford, California, U.S.A.

Sanjiv S. Gambhir

Molecular Imaging Program at Stanford (MIPS), Division of Nuclear Medicine, Department of Radiology, Bioengineering, and Bio-X Program, Stanford University School of Medicine, Stanford, California, U.S.A.

INTRODUCTION

Significant advances have been made over the past three decades in imaging technologies such as computed tomography (CT), magnetic resonance imaging (MRI), single photon emission computed tomography (SPECT), positron emission tomography (PET), and ultrasound. While these diagnostic modalities largely focus on gathering structural or physiological information, the advent of the human genomic project dictates a need to develop novel assays capable of imaging changes at the genetic and biochemical level (1). Driven by this need, a new discipline called "molecular imaging" has evolved over the past few years. Broadly, it can be defined as the visual representation, characterization, and quantification of biological processes at the cellular and subcellular levels within intact living organisms (2). Molecular imaging is a unique field in that it requires the integration of cell biology, molecular biology, synthetic chemistry, imaging sciences, and translational sciences. Since 1998, the National Cancer Institute has identified molecular imaging as one of six extraordinary opportunities for development (3). While the National Heart Lung and Blood Institute has not yet adopted such measures, the field of cardiovascular molecular imaging has slowly but steadily forged ahead (4). This chapter is intended as an overview of cardiac reporter gene imaging. The chapter is divided into four major sections: (*i*) basic concept of reporter gene imaging, (*ii*) different reporter gene imaging modalities, (*iii*) PET reporter genes and reporter probes, and (*iv*) overview of recent studies on cardiac reporter gene imaging.

BACKGROUND OF REPORTER GENE IMAGING

Traditionally, molecular biologists can monitor gene expression by using reporter genes such as β-galactosidase (5) and chloramphenicol-acetyl transferse (6), which require invasive biopsy or postmortem tissue sampling for analysis. Molecular imaging offers distinct advantages by allowing for noninvasive, quantitative, and repetitive imaging of targeted macromolecules and biological processes in living organisms (7). Although a vast array of molecular imaging techniques is available, they all require two fundamental elements: (*i*) a molecular probe which would signal gene expression by detecting a protein or messenger ribonucleic acid (mRNA) transcripts, and (ii) a method to monitor these probes or events (8). To date, the two most widely used strategies are direct and indirect imaging.

Direct molecular imaging involves direct probe-target interaction. The target can be a receptor, enzyme, or mRNA. For probe-receptor imaging, radiolabeled monoclonal antibodies binding to tumor cell-specific surface antigens has been used for the past two decades (9). Recent example of cardiac application involves imaging $\alpha v \beta 3$ integrin receptor expressed in angiogenic vessels after myocardial infarction by using ^{111}In-RP748, a radiolabeled quinolone targeted at $\alpha v \beta 3$ (10). For probe-enzyme imaging, the most well known cardiac application is 18F-fluorodeoxyguclose ($[^{18}F]$-FDG) to assess for tissue viability after myocardial infarction. The $[^{18}F]$-labeled glucose analog is transported across the intact cell membrane, undergoes phosphorylation by hexokinase, and is retained in the cell in proportion to the rate of cellular glycolysis (11). The radioactive $[^{18}F]$ undergoes positron annihilation into two high energy γ rays (511 keV) which can be detected as coincidence signals by PET. For probe-mRNA imaging, radiolabeled antisense oligonucleotide (RASON) probes can be used (12,13). These radiolabeled probes are typically 12–35 nucleotides long and are complementary to a small segment of the target mRNA. However, the RASON approach is less practical at this point because of (*i*) low number of target mRNA (~1000 copies) per cell compared to proteins (>10,000 copies); (*ii*) limited tracer penetration across cell membrane; (*iii*) poor intracellular stability; (*iv*) slow washout of unbound oligonucleotide probes; and (*v*) low target/background ratios (14). Despite these obvious complexities, antisense imaging is making steady progress as recent data by Hnatowich and colleagues have further addressed the issues of delivery and targeting in cell cultures and small animal models (15,16). For similar reasons, direct imaging of the DNA (2 copies) within the nuclear membrane is extremely difficult and not yet feasible. In addition, knowing the *activity* of gene expression (as reflected in mRNA transcripts or protein levels) rather than the *number* of DNA copies is more relevant for biological research. For the direct imaging approach, the main disadvantage is that it requires synthesizing a customized probe for the product (e.g., receptor, enzyme, or mRNA) of every therapeutic gene of interest, which

can be time consuming and lack the generalizability that is needed for most applications.

Indirect molecular imaging using reporter genes have only been recently validated. The concept of imaging reporter gene expression is illustrated in Figure 1. A reporter gene is first introduced into target tissues by viral or non-viral vectors. Using molecular biology techniques, the promoter or regulatory regions of genes can be cloned into different vectors to drive the transcription of a reporter gene into mRNA. The promoter activity can be "constitutive" (always on), "inducible" (turned on or off), or "tissue-specific" (expressed only in the heart, liver, or other organs). Translation of the mRNA leads to a reporter protein that can interact with the reporter probe. This interaction may be enzyme-based or receptor-based (discussed in a later section). The signals can then be

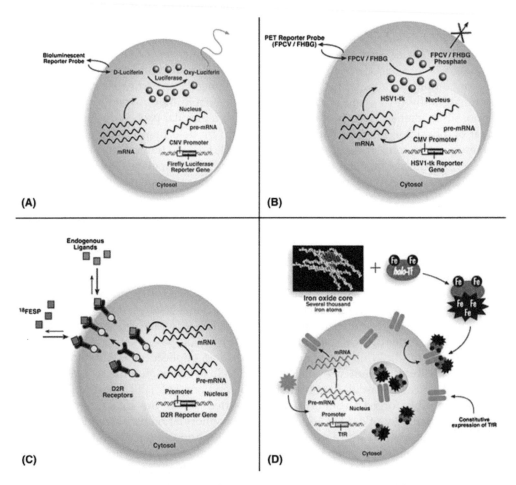

Figure 1   Four different strategies of imaging reporter gene/reporter probe. (**A**) Enzyme-based bioluminescence imaging. Expression of the firefly luciferase (Fluc) reporter gene leads to the firefly luciferase reporter enzyme (FL), which catalyzes the reporter probe (D-Luciferin) that results in a photochemical reaction. This yields low levels of photons that can be detected, collected, and quantified by a charge-coupled device (CCD) camera. (**B**) Enzyme-based PET imaging. Expression of the herpes simplex virus type 1 thymidine kinase (HSV1-tk) reporter gene leads to the thymidine kinase reporter enzyme (HSV1-tk), which phosphorylates and traps the reporter probe ([^{18}F]-FHBG) intracellularly. Radioactive decay of [^{18}F] isotopes can be detected using PET. (**C**) Receptor-based PET imaging. The [^{18}F]-FESP is a reporter probe that interacts with the dopamine 2 receptor (D2R) to result in probe trapping on or in cells expressing the D2R gene. (**D**) Receptor-based MRI imaging. Overexpression of engineered transferrin receptors (TfR) results in increased cell uptake of the transferrin-monocrystalline iron oxide nanoparticles (Tf-MION). These changes result in a detectable contrast change on MRI. *Source*: From Ref. 2.

detected by various imaging modalities such as optical charged coupled device (CCD) camera, PET, or MRI. Clearly, the main advantage of reporter gene system is its flexibility and multiplexing capacity. By altering various components, the reporter gene can provide information about the regulation of DNA by upstream promoter, the fate of intracellular protein trafficking, and the efficiency of vector transduction into cells. Likewise, the reporter probe itself does not have to be changed if one wishes to study a new biological process, which saves valuable time needed to synthesize, test, and validate new radiotracer agents. However, the main disadvantage of indirect imaging is that it is a surrogate marker of the physiologic process of interest rather than a direct measurement of the receptor density, intracellular enzyme, or mRNA copies as discussed above, which are likely to be more clinically relevant.

With both the direct and indirect molecular imaging methods, the ideal reporter gene and/or reporter probe should have the following characteristics: (*i*) The chromosomal integration or episomal expression of reporter gene should not adversely affect the cellular metabolism or physiology. (*ii*) The reporter gene product should not elicit a host immune response. (*iii*) The size of the promoter/enhancer elements and reporter gene should be small enough to fit into a delivery vehicle. (*iv*) Transfection (e.g., plasmid) or transduction (e.g., lentivirus or adeno-associated virus) using the delivery vector should not be cytotoxic to the cells. (*v*) The reporter probe should be stable in vivo and reach the target site despite natural biological barriers (e.g., blood vessel or blood brain barrier). (*vi*) The reporter probe should only accumulate within cells that express the reporter gene to yield high signal to background ratio. (*vii*) Afterwards, the reporter probe should clear rapidly from the circulation to allow repetitive imaging within the same living subject. (*viii*) The reporter probe or its metabolites should not be cytotoxic to the cells. (*ix*) The image signals should correlate well with true levels of reporter gene mRNA and protein in vivo. (*x*) The reporter gene and reporter probe should be applicable for human imaging in the future (14). Currently, no single reporter gene and reporter probe assay meets all of these stringent criteria listed above. Therefore, the choice of the reporter gene and reporter probe assay to use will depend on the particular application, organ system, and imaging modality available at a given institution.

MOLECULAR IMAGING TECHNOLOGY

Medical discoveries often start with a clinical observation, followed by years of in vitro testing and small animal validation before returning to clinical application.

With this paradigm, there is a great need for novel devices that can image biological and biochemical processes in small living animals. Indeed, considerable effort has already been made toward the development of miniaturized imaging systems such as bioluminescence or fluorescence CCD camera, small animal ultrasound, small animal SPECT, small animal PET, or small animal MRI (Fig. 2). With this extensive armamentarium, it is worthwhile to discuss and compare their advantages and disadvantages.

Optical Imaging

Number of different optical imaging approaches have been described recently. These techniques include bioluminescence imaging (17), intravital microscopy of green fluorescence protein (GFP) (18), and near-infrared fluorescence (19). In bioluminescence imaging, luciferase genes have been cloned from different organisms, such as firefly (*Photinus pyralis*), jellyfish (*Aequorea*), coral (*Renilla*), and dinoflagellates (*Gonyaulx*). For firefly, the luciferase enzyme uses energy from adenosine triphosphate to convert its substrate (D-Luciferin) to oxyluciferin, which emit low levels of photons (2–3 eV) that can be detected and counted by an ultra-sensitive CCD camera (e.g., Xenogen IVIS system) (17). In fluorescence imaging, the GFP does not need the injection of a reporter substrate but requires an excitation wavelength followed by an emission wavelength that can be captured (20). Recent technological advances (e.g., eXplore Optix system) has allowed the measurement, quantification, and visualization of fluorescence intensity in small living animals using the time-domain approach (21). In general, the main advantages of optical based imaging are relatively low cost of instrumentation (i.e., typically $100–200k versus $500–2000k for small animal PET and small animal MRI systems) and capacity for high throughput studies (i.e., several mice can be scanned once). But both techniques suffer from photon attenuation and scattering within deep tissues and inability to extrapolate to clinical usage (22). In the future, novel intravascular devices capable of detecting either bioluminescence or fluorescence signals within the coronary arteries may be possible.

Magnetic Resonance Imaging

In contrast to optical imaging, MRI has the advantages of a very high spatial resolution (25–100 μm) and the ability to measure more than one physiological parameter at once using different radiofrequency pulse sequences (23). These features make MR very attractive for imaging reporter gene expression. The imaging signal is generated as a result of spin relaxation effects, which can be altered by atoms with high magnetic moments (e.g., gadolinium and iron). One particularly useful MR imaging signal amplification system is

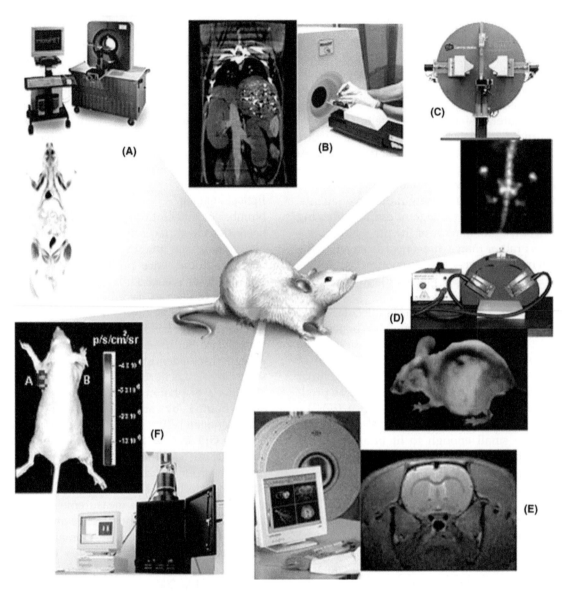

Figure 2 Multiple imaging modalities are available for small-animal molecular imaging. (**A**) Small animal PET whole-body coronal image of a rat injected with [^{18}F]-FDG, showing uptake of tracer in tissues including muscles, heart, brain, and bladder due to renal clearance. (**B**) MicroCT coronal image of a mouse abdomen after injection of intravenous iodinated contrast medium. (**C**) MicroSPECT coronal image of a mouse abdomen and pelvis regions after injection of [99mTc]-methylene diphosphonate, showing spine, pelvis, tail, vertebrae, and femurs due to accumulation of tracer in bone. (**D**) Optical fluorescence image of a mouse showing GFP-expressing tumors that have spread to the liver, abdomen, spine, and brain. (**E**) MicroMRI coronal T2-weighted image of a mouse brain. (**F**) Optical bioluminescence image of a mouse with a subcutaneous xenograft expressing renilla luciferase (Rluc) in the left shoulder region, after tail-vein injection of the reporter substrate coelentrazine. Images were obtained using a cooled CCD camera. The color image of visible light is superimposed on a photographic image of the mouse with a scale in photons per second per square centimeter per steradian (sr). *Source*: From Ref. 2.

based on the cellular internalization of superparamagnetic probes such as monocrystalline iron oxide nanoparticles (MION) (24). In this study, Weissleder et al. over-expressed an engineered transferring receptor (T*f*R) as a reporter gene in gliosarcoma cells. Next the cells were implanted into shoulders of nude mice and animals were injected with MION conjugated with human holo-transferrin (T*f*-MION). Binding of T*f*-MION to T*f*R in vivo was visualized by MR imaging (1.5 T; imaging time, 3–7 min per sequence; voxel

resolution, $300 \times 300 \times 700$ µm). In a follow up study by the same investigators, the T*f*R was probed for with superparamagnetic transferrin crossed linked iron oxides (T*f*-CLIO) probes while using a herpes simplex virus-based amplicon vector system (25). These studies hold promise for in vivo imaging in humans because of the availability of clinical MR scanners, the superparamagentic particles are relatively nontoxic when administered intravenously, the iron oxide core is biodegradable, and similar preparations are already in clinical use (26). However, the issues of persisting signals from the superparamagnetic particles may hinder its capacity for quantitative and repetitive imaging. MR is also several log of orders less sensitive (10^{-3} to 10^{-5} molar) for detection of reporter probes compared to optical bioluminescence imaging (10^{-15} to 10^{-17}) and PET (10^{-11} to 10^{-12}) imaging (2). Therefore, further strategies for robust signal amplification will be necessary before this modality can be of practical use for imaging cardiac gene expression.

Radionuclide Imaging

In many ways, molecular imaging has its roots in nuclear medicine since many of the techniques rely on radionuclide probes. PET, SPECT, and planar scintigraphy have been used to detect radionuclide-labeled probes. PET offers several advantages over other imaging modalities. First, PET is more sensitive compared to SPECT and MRI for detection of probe activity as discussed above. This may allow monitoring of gene delivery by vectors with relatively weak promoters (e.g., tissue-specific) or low transfection efficiency (e.g., plasmid). Second, PET imaging (unlike MRI) is more quantitative, which allows dynamic imaging with tracer kinetic modeling for analysis of the rate constants of the underlying biochemical processes (27). Third, because of the short half-lives of many PET isotopes (~110 min for [^{18}F]), daily repetitive imaging of tracer retention by targeted tissues is possible. Fourth, PET imaging (unlike optical imaging) is tomographic and so a relatively precise location of gene expression can be identified. The current generations of small animal PET scanners have a resolution of $1^3 - 2^3$ mm^3 compared to ~6^3 mm^3 for clinical PET scanners (28). Finally, using the same basic principles, studies performed in a small animal PET scanner can be scaled upward to patients using a clinical PET scanner in the future (29).

PET REPORTER GENES AND PROBES

Of these three techniques, radionuclide imaging represents the most promising approach for clinical imaging of reporter gene expression. It has several desirable characteristics, including robust detection sensitivity, quantitative capacity, tomographic resolution, and clinical available scanners. Significant progress has been made over the past few years. The first phase I/II clinical trial was published in 2001 (30). This study involved five patients (age range 49–67 years) with recurrent glioblastomas who were infused intratumourally with liposome complex containing HSV1-tk. After vector administration, only one of five patients had specific [^{124}I]-FIAU-associated radioactivity observed within the infused tumor. No specific

FIAU-accumulation was observed in the other four patients, whose histology showed a significantly lower number of proliferating tumor cells. These data indicate that a certain critical number of the thymidine kinase-gene transduced tumor cells per voxel (threshold) have to be present for accumulation of FIAU and detection by PET. Complications of the blood-brain-barrier and clearance of tracer somewhat limited the findings of this study. Recently, a more detailed clinical trial involving PET imaging was completed (31). In this study, seven patients (age range 51–78) with hepatocellular carcinomas underwent intratumoral injection of recombinant adenovirus carrying the cytomegalovirus promoter driving herpes simplex virus type 1 thymidine kinase (Ad-CMV-tk). Successful PET imaging was visualized in patients using 9-(4-fluoro-3-hydroxymethylbutyl)guanine ([^{18}F]-FHBG) as the PET reporter probe with very good signal to background. Repeated imaging was also possible in this study because of the relatively short half-life of Fluorine-18. The HSV1-tk, in addition to serving as a reporter gene, can be used to destroy tumor cells by its ability to convert a nontoxic prodrug such as valganciclovir into a phosphorylated cytotoxic compound. Thus, both these studies demonstrate the principle of using molecular imaging to assess novel therapeutic strategies and monitor clinical responses in living subjects (Fig. 3A). Finally, the PET reporter probe [^{18}F]-FHBG has also been shown to have favorable characteristics such as in vivo stability, rapid blood clearance, low background signal, and acceptable radiation dosimetry in humans (Fig. 3B) (32). Recently we have obtained an investigational new drug approval for [^{18}F]-FHBG from the FDA. It is hoped that with further validation, similar studies can be performed in patients with cardiovascular diseases using a variety of PET reporter genes and PET reporter probes discussed below.

The most common *enzyme-based* reporter gene is the herpes simplex virus type-1 thymidine kinase (HSV1-tk) (33,34). Unlike the mammalian thymidine kinase enzyme, the HSV1-tk has very broad substrate specificity. The HSV1-tk can phosphorylate thymidine analogue probes such as FIAU and guanosine analogue probes such as FHBG. Once phosphorylated, the reporter probes are trapped intracellularly and the radioisotope ([^{124}I] or [^{18}F]) attached to the compounds can be detected by PET imaging (Fig. 1). The main advantage of [^{18}F]FHBG versus [^{124}I]FIAU is its shorter half-life ($T_{1/2}$ 110 min versus 4.2 days), which allows for repetitive imaging on a daily basis. The second version of HSV1-tk is a mutant form (HSV1-sr39tk) that differs from HSV1-tk by seven nucleotide substitutions leading to five different nonpolar amino acids (35). The HSV1-sr39tk can phosphorylate acycloguanosine derivatives (e.g,. fluorinated ganciclovir

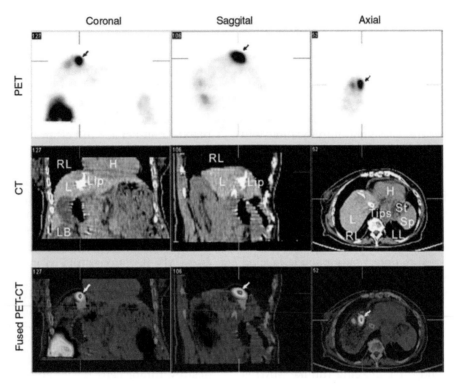

Figure 3 PET-CT imaging of HSV1-tk transgene expression in humans. Columns 1 to 3 show the 5 mm thick coronal, sagittal and transaxial slices respectively, from a [^{18}F]FHBG-PET-CT study in patient. All sections are centered on the treated tumor lesion (dotted lines in the CT images), and show [^{18}F]FHBG accumulation at the tumor site (*arrows*). Anatomic and metabolic correlation can be obtained by fused PET-CT imaging. The white spots on the liver seen on the CT images correspond to lipiodol (*arrowheads*) retention after trans-arterial embolization of the tumor and a transjugular intrahepatic portosystemic shunt (St). Tracer signal can be seen in the treated lesion (*arrows*), while no specific accumulation of the tracer can be seen in necrotic, lipiodol-retaining regions around it. *Abbreviations*: H, heart; L, liver; LB, large bowel; RL, right lung; Sp, spleen. *Source*: From Ref. 31.

and penciclovir as reporter probes) effectively while minimizing interaction with endogenous thymidine kinase (36). Dynamic imaging in animal tumor models using the HSV1-sr39tk/[^{18}F]-FHBG combination also appears to be more sensitive compared to HSV1-tk/[^{14}C]FIAU (37). Whereas the [^{18}F]FHBG is well retained in HSV1-sr39tk expressing cells (C6-stb-sr39tk+) during 4 hour of injection, the [^{14}C]FIAU is rapidly cleared from HSV1-tk expressing cells (MH3924A-stb-tk+). A more recent study suggest that [^{3}H]2′-fluoro-2′-deoxyarabinofuranosyl-5-ethyluracil ([^{3}H]FEAU) may exhibit the most robust affinity to either HSV1-tk or HSV1-sr39tk reporter genes, although further works are clearly needed to assess the optimal combination of reporter probe/reporter gene (38). The third version of HSV1-tk gene has a deletion of the first 135 base pair that contains a nuclear localization signal and a cryptic testis-specific transcriptional start point (39). This deletion leads to more cytoplasmic localization of TK enzyme, which result in better image signal activity based on higher interaction rates between TK and FHBG reporter probe (40). The 135 base pair deletion mutant also lacks the cryptic testis-specific transcriptional start point and overcomes the problem of male sterility in transgenic mice carrying the HSV1-tk

gene (41). Finally, a "humanized" version of HSV1-tk with more robust TK activity and less host immune response is being evaluated for suicide and reporter gene purposes. In general, the main advantage for enzyme-based reporter genes is that the imaging sensitivity can be greatly enhanced by the enzymatic amplification of the reporter probe signal.

Several receptor-based PET reporter genes have also been described. The first is the dopamine 2 receptor (D2R), a 415 amino acid protein with a seven transmembrane domain, found in substantial levels primarily in the striatum (42). The D2R gene has been used as a PET reporter gene, both in an adenoviral delivery vector and in stably transfected tumor cell xenografts (43). The location, magnitude, and duration of D2R reporter gene expression can be monitored by PET detection of D2R-dependent sequestration of injected 3-(2′-[^{18}F]-fluoro-ethyl)-spiperone ([^{18}F]-FESP) probe, a high-affinity D2R ligand. The second is the somatostatin type 2 receptor (SSTr2), which is expressed primarily in the pituitary gland. When used as a reporter gene, the location of SSTr2 expression can be monitored by systemic injection of a ^{99m}Tc labeled SSTr2 peptide probe ([^{99m}Tc]-P829) and subsequent imaging with a conventional gamma camera (44). Another promising approach involves the human sodium/iodide symporter (hNIS) gene, which is expressed in the thyroid gland. The accumulation of radioactive isotopes iodine ([^{123}I] and [^{131}I]) has been used in nuclear medicine for the diagnosis and targeted therapy of thyroid pathology (45). In general, the main advantage for receptor-based reporter genes is that the D2R, SSTr2, and hNIS are cloned from endogenous genes (striatum, pituitary, and thyroid gland, respectively), which are less likely to evoke a host immune response (46).

CARDIAC REPORTER GENE IMAGING STUDIES

While most molecular imaging studies have focused on cancer biology (33,34), only recently has reporter gene imaging of cardiac transgene expression been investigated. In the following section, a summary of some of the most relevant studies published over the past years will be discussed.

In the first proof-of-principle study involving cardiac optical bioluminescence imaging, adenovirus with a constitutive cytomegalovirus (CMV) promoter driving firefly luciferase reporter gene (Ad-CMV-Fluc; 1×10^9 pfu) was injected into the rat myocardium via aseptic lateral thoracotomy (47). The reporter probe D-Luciferin (125 mg/kg body weight) was injected intraperitoneally before imaging. Cardiac transgene expression was assessed over 2 weeks (Fig. 4). The in vivo imaging results correlated well with in vitro enzyme assays (r^2 = 0.92), indicating that bioluminescence imaging can be used in parallel or in lieu of traditional biochemical assays. Cardiac firefly luciferase activity peaked within the first 3–5 days but declined rapidly thereafter due to host cellular immune response against the adenoviral vector (48). Leakage into the systemic circulation allowed the adenovirus to bind to coxackie-adenovirus receptors on hepatocytes, leading to expression at an unintended target site (49). The kinetics of cardiac transgene expression varied significantly among different animals, suggesting that the efficiency of gene transfer varies due to inter-individual response. The main drawback of bioluminescence imaging is that the exact location of cardiac transgene location could not be determined as discussed previously.

For these reasons, human cardiac imaging would most likely require PET based reporter genes and reporter probes. The cellular concept of imaging HSV1-tk and [^{125}I]FIAU in cardiomyocytes was demonstrated initially by Bengel et al. using postmortem autoradiography (50). In explanted heart tissues, they showed that transgene expression within the Ad-CMV-tk injection site to have 3.4 ± 2.2-fold higher level of radioactivity compared with the remote myocardium. Subsequently, in the first proof-of-principle study using in vivo cardiac PET imaging, adenovirus carrying

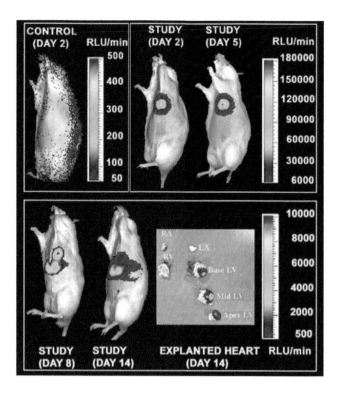

Figure 4 Optical imaging of cardiac reporter gene expression. Control rat transduced with Ad-CMV-HSV1-sr39tk (1×10^9 pfu) shows background signal. Study rat transduced with Ad-CMV-Fluc (1×10^9 pfu) emits significant cardiac firefly luciferase (FL) activity at day 2, day 5, day 8, and day 14 (P<0.05 vs control). Significant hepatic FL activity is seen, starting at day 8. The same rat with heart explanted and sliced into 3 sections at day 14. FL activity is localized at anterolateral wall of left ventricle along the site of virus injection. Note the bioluminescence scales are different for control rat, study rat days 2 to 5, and study rat days 8 to 14 to account for the wide range of cardiac FL activity observed. *Abbreviations*: RA, indicates right atrium; RV, right ventricle; LV, left ventricle. *Source*: From Ref. 47.

mutant thymidine kinase reporter gene (Ad-CMV-HSV1-sr39tk) was injected intramyocardially (51). Similar to the bioluminescence study, adenoviral transduction led to robust but transient (~2 weeks) gene expression in the myocardium as assessed by the PET reporter probe [^{18}F]-FHBG. With PET imaging, the transgene expression could be localized tomographically at the anterolateral wall as shown on short, vertical, and horizontal axis (Fig. 5). In a follow-up study, Inubushi et al. examined the quantitative aspects of the HSV1-sr39tk and [^{18}F]-FHBG system in detail (52). Myocardial [^{18}F]-FHBG accumulation was visualized with viral titers down to 1×10^7 particle forming units (pfu) but not with 1×10^6 pfu of HSV1-sr39tk. More recently, gamma camera imaging of cardiac transgene expression using adenoviral mediated expression of human sodium iodide symporter (Ad-CMV-hNIS) and ^{123}Iodide or ^{99m}Technetium as reporter probes was shown to be a practical and

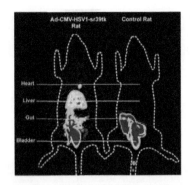 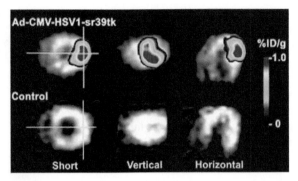

Figure 5 Small animal PET imaging of cardiac reporter gene expression. (*Left*) At day 4, whole-body small animal PET image of a rat shows focal cardiac [^{18}F]-FHBG activity at the site of intramyocardial Ad-CMV-HSV1-sr39tk injection. Liver [^{18}F]-FHBG activity is also seen due to systemic adenoviral leakage with transduction of hepatocytes. Control rat injected with Ad-CMV-Fluc shows no [^{18}F]-FHBG activity in both cardiac and hepatic regions. Significant gut and bladder activities are seen for both study and control rats due to route of [^{18}F]-FHBG clearance. (*Right*) Tomographic views of cardiac small animal PET images. The [^{13}N]-NH$_3$ (gray scale) images of perfusion are superimposed on [^{18}F]-FHBG images (color scale) demonstrating HSV1-sr39tk reporter gene expression. [^{18}F]-FHBG activity is seen in the anterolateral wall for study rat compared to background signal in control rat. *Note:* Perpendicular lines represent the axis for vertical and horizontal cuts. Color scale is expressed as %ID/g. *Source:* From Ref. 51.

effective alternative to HSV1-tk and HSV1-sr39tk imaging (53). In most of these studies, however, transient cardiac gene expression and unintended target site expression within the liver were noticed. These observations suggest that ischemic heart patients who underwent intracoronary infusion of adenovirus carrying either fibroblast growth factor (FGF) (54) or vascular endothelial growth factor (VEGF) (55) in clinical trials would likely have encountered these two issues.

An ideal vector system therefore should demonstrate high infectivity of cardiac muscle cells, minimal infectivity of unintended target site, and no elicitation of host immune response (56). The issue of transient gene expression may be addressed by substituting the adenovirus with less immunogenic vectors such as the adeno-associated virus (57) or gutless adenovirus (58). The issue of unintended target site expression may be addressed by using cardiac tissue specific promoters such as myosin light chain kinase that drive the reporter genes instead of the constitutive CMV promoter (59). However, tissue specific promoters are inherently less robust in driving gene expression compared to constitutive promoters. In the oncology literature, the two step transcriptional activation system has been used for amplification of gene expression from a weak tissue specific promoter such as the prostate specific antigen (60). Similar approaches should be feasible for amplifying transgene expression mediated by cardiac specific promoters. Other alternatives involve using viruses with high affinity to specific tissue type or modification of the viral particle surface with cell specific ligands (61). Despite these potential options, it remains a formidable challenge to translate results from small animals to large animals and eventually to human gene therapy trials.

The feasibility of transferring imaging results from a small animal PET scanner to a clinical PET scanner has been demonstrated in a porcine model (62). Bengel et al. showed that myocardial tissues infected with adenovirus expressing HSV1-tk had significantly higher [^{124}I]FIAU retention during the first 30 minutes after injection. The FIAU uptake correlated with ex vivo images, autoradiography, and immunohistochemistry for reporter gene product after euthanasia. However, the signal to background ratio at the site of HSV1-tk injection was only ~1.25 within the first 30 minutes post. Afterwards, there was significant washout from 45–120 min post injection and the [124I]-FIAU retention became similar to control myocardial regions. Thus, the combination of HSV1-tk/FIAU may be less ideal compared to HSV1-sr39tk/FHBG for imaging cardiac transgene expression (63), consistent with data from the oncology literature (37).

One of the goals of cardiac reporter gene imaging is to help guide cardiac gene therapy. Conceptually, this can be achieved by linking a PET reporter gene to a therapeutic gene of interest. To first demonstrate that the expression of two linked genes would have high fidelity, Chen et al. constructed an bicistronic adenoviral vector whereby a CMV driven receptor-based PET receptor gene (a mutant D2R, or D2R80a) was coupled to an enzyme-based PET reporter gene (HSV1-sr39tk) using an internal ribosomal entry site (Ad-CMV-D2R80a-IRES-HSV1-sr39tk) (Fig. 6) (64). After injection into the rat myocardium, longitudinal imaging with PET reporter probes [^{18}F]-FESP and [^{18}F]-FHBG, respectively, revealed a good correlation between the two linked PET reporter genes (r^2 = 0.73; P < 0.001). These results suggest that if one of the PET reporter genes is replaced with a therapeutic gene, a correlated

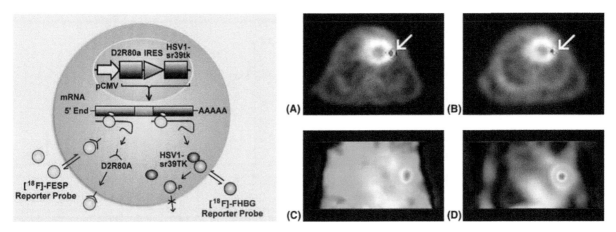

Figure 6 Small animal PET imaging of bicistronic PET reporter genes. (*Left*) Schematic of Ad-CMV-D2R80a-IRES-HSV1-sr39tk medi-ated gene expression. A CMV promoter drives the expression of a single mRNA containing D2R80a and HSV1-sr39tk cistrons, sepa-rated by an encephalomyocarditis virus (EMCV) internal ribosomal entry site (IRES). The translated product of D2R80a (D2R80A) binds to [^{18}F]-FESP mainly on the cell membrane, whereas the translated product of HSV1-sr39tk (HSV1-sr39TK) phosphorylates [^{18}F]-FHBG and traps it intracellularly. (*Right*) [^{18}F]-FESP and [^{18}F]-FHBG accumulations in rat myocardium. An athymic rat was intramyocardially injected with 2x10^9 pfu of Ad-CMV-D2R80a-IRES-HSV1-sr39tk and scanned (**A**) on day 2 for D2R80a-dependent accumulation of [^{18}F]-FESP and (**B**) on day 3 for HSV1-sr39tk-dependent sequestration of [^{18}F]-FHBG. The transaxial [^{18}F]-FESP and [^{18}F]-FHBG images (color) are individually superimposed on an [^{13}N]-ammonia perfusion scan (gray scale) for easier localization of tracer accumula-tion relative to myocardium. Distinct tracer accumulation (*arrow*) is seen in the anterolateral wall of myocardium, corresponding to the site of viral injection. Coronal (**C**) [^{18}F]-FESP and (**D**) [^{18}F]-FHBG images of the thorax, individually overlaid with an [^{13}N]-NH$_3$ perfusion scan, show that the lung activity is notably higher in the [^{18}F]-FESP than in the [^{18}F]-FHBG image, likely due to a slower clearance of [^{18}F]-FESP from the lung parenchyma. *Source*: From Ref. 64.

expression would likely exist and allow for indirect monitoring of the therapeutic gene. Besides the IRES system, other techniques of linking the therapeutic gene to the reporter gene are also available, including using two separate delivery vectors (65), fusion of reporter genes (40), bi-directional transcription (66), and dual promoter approach as described below.

In the first proof of principle study using PET imaging to track therapeutic gene expression, Wu et al. constructed an adenovirus with a CMV promoter driving a VEGF$_{121}$ therapeutic gene and a second CMV promoter driving HSV1-sr39tk reporter gene (67). The two expression cassettes are separated by poly adenine sequences (Ad-CMV-VEGF$_{121}$-polyA-CMV-HSV1-sr39tk-polyA). The adenovirus was injected into the myocardium of adult rats with myocardial infarc-tion by ligation of the left anterior descending artery. Control animals received adenovirus without an expression cassette (Ad-null) instead. Reporter gene expression persisted for approximately two weeks due to host cellular immune repsponse. Repeat injection of Ad-CMV-VEGF$_{121}$-polyA-CMV-HSV1-sr39tk-polyA into the same ischemic territory at two months did not induce any reporter gene expression due to host humoral immune response (Fig. 7) (48). At 10 weeks, the mean densities of capillaries (747 ± 104 versus 450 ± 101 per mm^2) and small blood vessels (8.1 ± 0.8 versus 5.1 ± 1.2 per mm^2) were significantly higher in the VEGF-treated study group compared with the

control group as assessed by histological staining for anti-CD31 and anti-smooth muscle actin antibodies (P < 0.05 for both). For functional studies, left ventricu-lar ejection fraction showed mild improvement in the VEGF-treated study group (43.4 ± 8.1% at baseline to 47.3 ± 12.5% at week 10) compared with the control group (47.5 ± 9.3% to 45.2 ± 8.4%), but this did not reach statistical significance. The VEGF-treated study animals also showed an encouraging trend toward lower [^{13}N]-NH$_3$ perfusion defects (15.2 ± 3.1% to 13.8 ± 2.6%) and [^{18}F]-FDG metabolism deficits (12.7 ± 4.3% to 11.5 ± 4.6%), but the changes were also not statistically significant. As expected, the control group did not show any significant changes in perfu-sion (14.0 ± 4.0% to 15.3 ± 4.1%) or metabolism (13.4 ± 2.3% to 15.1 ± 3.0%) scores (P=NS) (Fig. 8). Taken together, these results suggest that the micro-scopic level of neovasculature induced by VEGF did not translate into significant changes in clinically relevant physiological parameters such as myocardial contractility, perfusion, and metabolism under the study conditions tested.

These findings may shed light into interpretation of recent cardiac gene therapy trials. Historically, the field of cardiac angiogenesis attracted much attention from the cardiovascular community in the late 1990s as most animal studies and phase 1 nonrandomized trials uniformly showed positive results (68). However, recent phase 2 randomized trials such as the VIVA

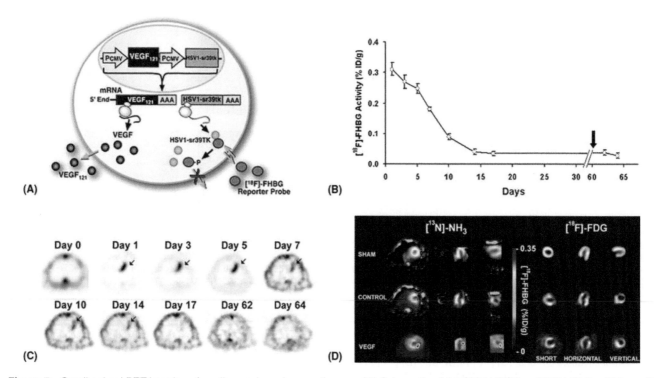

Figure 7 Small animal PET imaging of cardiac angiogenic gene therapy. (**A**) Schematic of Ad-CMV-VEGF$_{121}$-CMV-HSV1-sr39tk medi-ated gene expression. Two separate gene cassettes with CMV promoters driving the expression of a VEGF$_{121}$ therapeutic gene and an HSV1-sr39tk reporter gene separated by polyA tails. The translated product of VEGF$_{121}$ is soluble and excreted extracellularly, whereas the translated product of HSV1-sr39tk traps [^{18}F]-FHBG intracellularly by phosphorylation. (**B**) Noninvasive imaging of the kinetics of cardiac transgene expression. Gene expression peaked at day 1 and rapidly decreased thereafter. A second injection (*arrow*) of Ad-CMV-VEGF$_{121}$-CMV-HSV-sr39tk at day 60 yielded no detectable signal on day 62 and day 64. Error bars represent mean±S.E.M. (**C**) A representative rat scanned longitudinally with transaxial [^{18}F]-FHBG PET images shown at similar slice levels of the chest cavity. The gray scale is normalized to the individual peak activity of each image. In this rat, myocardial [^{18}F]-FHBG accumulation was visualized at the anterolateral wall (*arrow*) from day 1 to day 14 but not day 17, 62, or 64. (**D**) In vivo gene, perfusion, and metabolism imaging with PET. At day 2, representative images showing normal perfusion ([^{13}N]-NH$_3$) and metabolism ([^{18}F]-FDG) in a sham rat, anterolateral inf-arction in a control rat (Ad-null), and anterolateral infarction in a VEGF-treated study rat (Ad-CMV-VEGF$_{121}$-CMV-HSV1-sr39tk) in short, vertical, and horizontal axis (gray scale). The color scale is expressed as % ID/g for [^{18}F]-FHBG uptake. As expected, both the sham and control animals had background [^{18}F]-FHBG signal only (blue color) that outlined the shape of the chest cavity. In contrast, the study rat showed robust HSV1-sr39tk reporter gene activity near the site of injection. *Source*: From Ref. 67.

(Vascular Endothelial Growth Factor in Ischemia for Vascular Angiogenesis) (69), FIRST (FGF Initiating Revascularization Trial) (70), AGENT (Adenovirus Fibroblast Growth Factor Angiogenic Gene Therapy) (54), and KAT (Kuopio Angiogensis Trial) (55) have demonstrated neither consistent nor substantial effi-cacy. In retrospect, several lessons can be learned from these trials. They showed that angiogenesis is a com-plex process regulated by the interaction of various growth factors and likely cannot be duplicated by use of a single protein or gene injection for a brief period of time (71). The ideal injection method, delivery vector, and patient population remain to be determined. Because there was no available method of assessing gene expression in vivo, the investigators were blinded as to whether the lack of symptomatic improvement was a result of transient gene expression, poor delivery technique, or host inflammatory response. Thus, for

cardiac gene therapy to be successful, biological issues related to pharmacokinetics, functional, and physio-logical effects of gene expression will need to be fully understood before expecting clinical efficacy (72). It is exactly this reason that molecular imaging can and should be used for monitoring gene transfer.

In recent years, stem cell therapy has replaced gene therapy as the most promising treatment avenue for ischemic heart disease. Several phase 1 clinical studies have shown that the implantation of skeletal myoblasts (73), endothelial progenitor cells (74), or bone marrow stem cells (75) into the infarcted myocardium can result in improved function. The mechanisms may be related to stem cells secreting paracrine factors, providing a mechanical scaffold, or recruiting other peripheral or resident cardiac stem cells (76). However, the analysis of stem cells, just like gene therapy, often relies on postmortem histology to

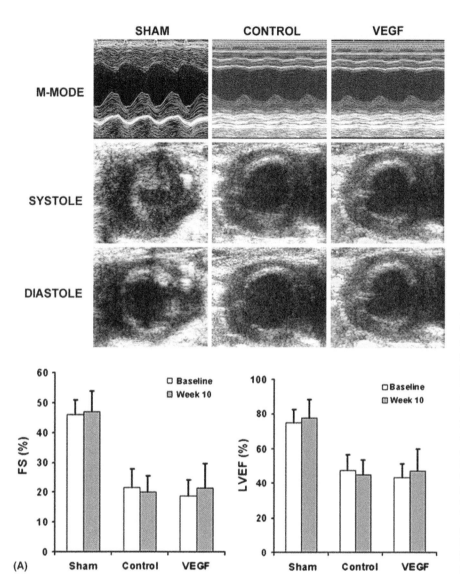

Figure 8 Evaluation of myocardial functional improvement after VEGF$_{121}$ gene therapy. (**A**) Ten weeks after myocardial infarction, echocardiography was performed to evaluate cardiac function by M-mode (*top row*) and short-axis views during systole (*middle row*) and diastole (*bottom row*). The left ventricular ejection fraction (LVEF) and fractional shortening (FS) were compared for all three groups. As expected, sham animals had normal LVEF and FS that remained unchanged. Control animals (Ad-null) showed slightly decreased LVEF and FS from baseline to week 10 while study animals (Ad-CMV-VEGF$_{121}$-CMV-HSV1-sr39tk) showed slightly increased LVEF and FS from baseline to week 10. However, both trends were not statistically significant ($n = 10$; $P =$ N.S.). (*Continued*)

identify their presence. The ability to study stem cell survival and proliferation in the context of the intact living body rather than postmortem histology would yield better insight into stem cell biology and physiology.

In the first proof of principle study using reporter genes to track cardiac cell survival, Wu et al. transduced embryonic rat H9c2 cardiomyoblasts with firefly luciferase or HSV1-sr39tk reporter genes before injecting into the rat myocardium. Cell survival was monitored noninvasively by optical bioluminescence or small animal PET imaging (Fig. 9) (77). Cell signal activity was quantified in units of photons per second per square centimeter per steradian (photons/sec/cm^2/sr) or percentage injected dose of [^{18}F]-FHBG per gram tissue (%ID/g), respectively. In both cases, drastic reductions in signal activity were seen within the first 1–4 days due to acute donor cell death from inflammation, ischemia, and apoptosis. Interestingly,

this pattern of cell death was consistent with other reports using traditional ex vivo assays such as TUNEL apoptosis (78), histologic stainings (79), and TaqMan polymerase chain reaction that require large numbers of animals to be sacrificed at different time points (80). Given that several types of reporter genes and reporter probes are available, it should be possible in the future to perform multi-modality imaging of stem cell transplantation. For example, recent study has demonstrated the feasibility of imaging murine embryonic stem cells transplanted into the rat myocardium (81). These cells were genetically manipulated to express a triple fusion reporter that consists of green fluorescence protein for FACS analysis and single cell fluorescence microscopy, firefly luciferase for high throughput bioluminescence imaging, and thymidine kinase for small animal PET imaging. Finally, reporter gene imaging may be helpful for evaluating other issues relevant

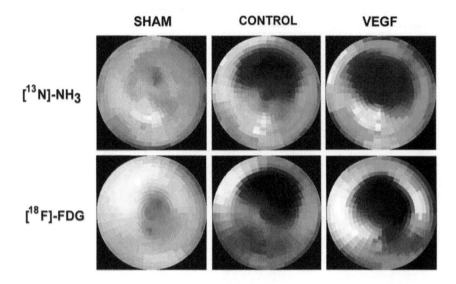

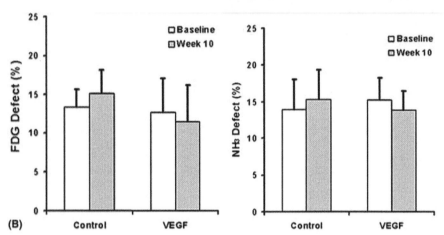

Figure 8 (*Continued*) (**B**) Evaluation of changes in myocardial perfusion and metabolism in relation to VEGF$_{121}$ gene therapy. Ten weeks after myocardial infarction, [^{13}N]-NH$_3$ perfusion and [^{18}F]-FDG metabolism imaging were performed again for sham, control (Ad-null), and study (Ad-CMV-VEGF$_{121}$-CMV-HSV1-sr39tk) animals. A representative sham animal showed normal myocardial perfusion and metabolism while control and VEGF study animals had anterior and anterolateral wall defects on polar maps. Compared to baseline, the extent of perfusion and metabolism improved slightly for VEGF study animals but did not reach statistical significance ($n = 10$; P = N.S.).

to stem cell biology, such as imaging stem cell survival, proliferation, and differentiation.

CONCLUSION

Different approaches to image cardiac reporter gene expression have been discussed. For imaging modalities, optical bioluminescence or fluorescence imaging has excellent temporal resolution but limited depth penetration. MRI provides spectacular image resolution but lower detection sensitivity of reporter genes. PET has both good detection sensitivity and good spatial resolution. The choice of the imaging modality will depend on the biological question asked. Regardless, improvement in detection sensitivity will be needed for all modalities as we move from imaging of organs and tissues to cells and genes. For reporter genes, many critical issues remain to be investigated such as the safety of stable chromosomal integration or transient episomal expression in cellular system, the potential

for host immune response against reporter genes, and the possibility of reporter gene silencing in stable cell lines (82,83). For cardiac studies, there needs to be a concerted effort to educate current cardiovascular investigators to be aware of the potentials of molecular imaging for analyzing gene transfer, cell survival, endogenous transcriptional regulation, transgenic animal phenotypes, and drug discovery (2,4,7). In summary, reporter gene imaging is an exciting field. It combines the disciplines of cell biology, molecular biology, synthetic chemistry, medical physics, and translational sciences into a new research paradigm. Continuation of this active collaboration will be needed as the field moves forward from animal studies to human cardiac imaging.

DISCLOSURE

The authors have indicated they have no financial conflicts of interest.

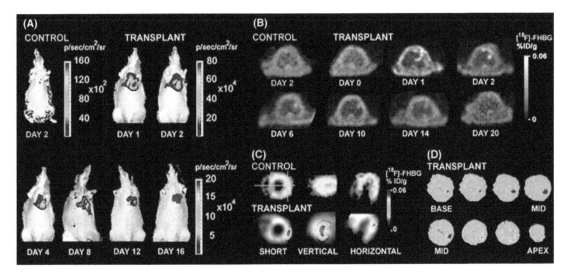

Figure 9 Optical bioluminescence and small animal PET imaging of cardiac cell transplantation in living animals. (**A**) Study rat transplanted with embryonic H9c2 cardiomyoblasts (3×10^6) emits significant cardiac bioluminescence activity at day 1, 2, 4, 8, 12, and 16 (P<0.05 vs. control). Control rat shows background signal only. (**B**) The location, magnitude, and duration of cell survival are determined by longitudinal imaging of [^{18}F]-FHBG activity (gray scale) within the same rat. (**C**) Tomographic views of cardiac PET images shown in short, vertical, and horizontal axis. At day 2, study rat transplanted with embryonic cardiomyoblasts expressing HSV1-sr39tk shows significant [^{18}F]-FHBG uptake (color scale) superimposed on [^{13}N]-NH$_3$ images (gray scale). Control rat shows homogeneous [^{13}N]-NH$_3$ perfusion but background [^{18}F]-FHBG uptake. (**D**) Autoradiography of the same study rat at day 2 confirms trapping of [^{18}F] by transplanted cells at the lateral wall at finer spatial resolution (~50 μm). *Source*: From Ref. 77.

REFERENCES

1. Collins FS, Green ED, Guttmacher AE, Guyer MS. A vision for the future of genomics research. Nature 2003; 422:835–847.
2. Massoud TF, Gambhir SS. Molecular imaging in living subjects: Seeing fundamental biological processes in a new light. Genes Dev 2003; 17:545–580.
3. National Cancer Institute. The nation's investment in Cancer Research for Fiscal 2004. (http://plan.cancer.gov/discovery/imaging.htm)). 2004.
4. Wu JC, Tseng JR, Gambhir SS. Molecular imaging of cardiovascular gene products. J Nucl Cardiol 2004; 11:491–505.
5. Schröder G, Risch K, Nizze H, et al. Immune response after adenoviral gene transfer in syngeneic heart transplants: Effects of anti-CD4 monoclonal antibody therapy. Transplantation 2000; 70:191–198.
6. Kass-Eisler A, Falck-Pedersen E, Alvira M, et al. Quantitative determination of adenovirus-mediated gene delivery to rat cardiac myocytes in vitro and in vivo. Proceedings of the National Academy of Sciences of the United States of America. 1993; 90: 11498–11502.
7. Blasberg RG, Tjuvajev JG. Molecular-genetic imaging: Current and future perspectives. J Clin Invest 2003; 111:1620–1629.
8. Herschman HR. Molecular imaging: Looking at problems, seeing solutions. Science 2003; 302:605–608.
9. Verel I, Visser GW, van Dongen GA. The promise of immuno-PET in radioimmunotherapy. J Nucl Med 2005; 46(suppl 1):164S–171S.
10. Meoli DF, Sadeghi MM, Krassilnikova S, et al. Noninvasive imaging of myocardial angiogenesis following experimental myocardial infarction. J Clin Invest 2004; 113:1684–1691.
11. Schelbert HR. 18F-deoxyglucose and the assessment of myocardial viability. Semin Nucl Med 2002; 32: 60–69.
12. Shi N, Boado RJ, Pardridge WM. Antisense imaging of gene expression in the brain in vivo. Proc Natl Acad Sci U S A. 2000; 97: 14709–14714.
13. Tavitian B, Terrazzino S, Kuhnast B, et al. In vivo imaging of oligonucleotides with positron emission tomography. Nat Med 1998; 4:467–471.
14. Gambhir SS, Barrio JR, Herschman HR, Phelps ME. Imaging gene expression: Principles and assays. J Nucl Cardiol 1999; 6:219–233.
15. Nakamura K, Fan C, Liu G, et al. Evidence of antisense tumor targeting in mice. Bioconjug Chem 2004; 15:1475–1480.
16. Hnatowich DJ, Nakamura K. Antisense targeting in cell culture with radiolabeled DNAs—a brief review of recent progress. Ann Nucl Med 2004; 18:363–368.
17. Contag PR, Olomu IN, Stevenson DK, Contag CH. Bioluminescent indicators in living mammals. Nat Med 1998; 4:245–247.
18. Potter SM, Wang CM, Garrity PA, Fraser SE. Intravital imaging of green fluorescent protein using two-photon laser-scanning microscopy. Gene 1996; 173:25–31.
19. Weissleder R, Tung CH, Mahmood U, Bogdanov A, Jr. In vivo imaging of tumors with protease-activated near-infrared fluorescent probes. Nat Biotechnol 1999; 17:375–378.
20. Lippincott-Schwartz J, Patterson GH. Development and use of fluorescent protein markers in living cells. Science 2003; 300:87–91.

21. Ramjiawan B, Ariano RE, Mantsch HH, Maiti P, Jackson M. Immunofluorescence imaging as a tool for studying the pharmacokinetics of a human monoclonal single chain fragment antibody. IEEE Trans Med Imaging 2002; 21:1317–23.

22. Wu JC, Sundaresan G, Iyer M, Gambhir SS. Noninvasive optical imaging of firefly luciferase reporter gene expression in skeletal muscles of living mice. Mol Ther 2001; 4:297–306.

23. Bogdanov A, Weissleder R. In vivo imaging of gene delivery and expression. Trends in Biotechnology 2002; 20:S11–S18.

24. Weissleder R, Moore A, Mahmood U, et al. In vivo magnetic resonance imaging of transgene expression. Nat Med 2000; 6:351–355.

25. Ichikawa T, Hogemann D, Saeki Y, et al. MRI of transgene expression: Correlation to therapeutic gene expression. Neoplasia 2002; 4:523–530.

26. Bulte JW, Kraitchman DL. Iron oxide MR contrast agents for molecular and cellular imaging. NMR Biomed 2004; 17:484–499.

27. Schelbert HR, Inubushi M, Ross RS. PET imaging in small animals. J Nucl Cardiol 2003; 10:513–520.

28. Tai YC, Chatziioannou AF, Yang Y, et al. MicroPET II: Design, development and initial performance of an improved microPET scanner for small-animal imaging. Phys Med Biol 2003; 48:1519–1537.

29. Phelps ME. Inaugural article: Positron emission tomography provides molecular imaging of biological processes. Proc Natl Acad Sci U S A. 2000; 97:9226–9233.

30. Jacobs A, Voges J, Reszka R, et al. Positron-emission tomography of vector-mediated gene expression in gene therapy for gliomas. Lancet 2001; 358:727–729.

31. Penuelas I, Mazzolini G, Boan JF, et al. Positron emission tomography imaging of adenoviral-mediated transgene expression in liver cancer patients. Gastroenterology 2005; 128:1787–1795.

32. Yaghoubi S, Barrio JR, Dahlbom M, et al. Human pharmacokinetic and dosimetry studies of [(18)F]FHBG: A reporter probe for imaging herpes simplex virus type-1 thymidine kinase reporter gene expression. J Nucl Med 2001; 42:1225–1234.

33. Tjuvajev JG, Finn R, Watanabe K, et al. Noninvasive imaging of herpes virus thymidine kinase gene transfer and expression: A potential method for monitoring clinical gene therapy. Cancer Res 1996; 56:4087–4095.

34. Gambhir SS, Barrio JR, Wu L, et al. Imaging of adenoviral-directed herpes simplex virus type 1 thymidine kinase reporter gene expression in mice with radiolabeled ganciclovir. J Nucl Med 1998; 39:2003–2011.

35. Black ME, Newcomb TG, Wilson HM, Loeb LA. Creation of drug-specific herpes simplex virus type 1 thymidine kinase mutants for gene therapy. Proc Natl Acad Sci USA 1996; 93:3525–3529.

36. Gambhir SS, Bauer E, Black ME, et al. A mutant herpes simplex virus type 1 thymidine kinase reporter gene shows improved sensitivity for imaging reporter gene expression with positron emission tomography. Proc Natl Acad Sci USA 2000; 97:2785–2790.

37. Min JJ, Iyer M, Gambhir SS. Comparison of [18F]FHBG and [14C]FIAU for imaging of HSV1-tk reporter gene expression: adenoviral infection vs stable transfection. Eur J Nucl Med Mol Imaging 2003; 30:1547–1560.

38. Kang KW, Min JJ, Chen X, Gambhir SS. Comparison of [14C]FMAU, [3H]FEAu, [14C]FIAU, and [3H]PCV for monitoring reporter gene expression of wild type and mutant herpes simplex virus type 1 thymidine kinase in cell culture. Eur J Nucl Med Mol Imaging 2005; 7(4):296–303.

39. Degreve B, Johansson M, De Clercq E, Karlsson A, Balzarini J. Differential intracellular compartmentalization of herpetic thymidine kinases (TKs) in TK gene-transfected tumor cells: Molecular characterization of the nuclear localization signal of herpes simplex virus type 1 TK. J Virol 1998; 72:9535–9543.

40. Ray P, De A, Min JJ, Tsien RY, Gambhir SS. Imaging tri-fusion multimodality reporter gene expression in living subjects. Cancer Res 2004; 64:1323–1330.

41. Cohen JL, Boyer O, Salomon B, et al. Fertile homozygous transgenic mice expressing a functional truncated herpes simplex thymidine kinase delta TK gene. Transgenic Res 1998; 7:321–330.

42. Missale C, Nash SR, Robinson SW, Jaber M, Caron MG. Dopamine receptors: From structure to function. Physiol Rev 1998; 78:189–225.

43. MacLaren DC, Gambhir SS, Satyamurthy N, et al. Repetitive, non-invasive imaging of the dopamine D2 receptor as a reporter gene in living animals. Gene Ther 1999; 6:785–791.

44. Zinn KR, Buchsbaum DJ, Chaudhuri TR, Mountz JM, Grizzle WE, Rogers BE. Noninvasive monitoring of gene transfer using a reporter receptor imaged with a high-affinity peptide radiolabeled with 99mTc or 188Re. J Nucl Med 2000; 41:887–895.

45. Kaminsky SM, Levy O, Salvador C, Dai G, Carrasco N. The Na+/I– symporter of the thyroid gland. Soc Gen Physiol Ser 1993; 48:251–262.

46. Min JJ, Gambhir SS. Gene therapy progress and prospects: Noninvasive imaging of gene therapy in living subjects. Gene Ther 2004; 11:115–125.

47. Wu JC, Inubushi M, Sundaresan G, Schelbert HR, Gambhir SS. Optical imaging of cardiac reporter gene expression in living rats. Circulation 2002; 105: 1631–1634.

48. Yang Y, Li Q, Ertl HC, Wilson JM. Cellular and humoral immune responses to viral antigens create barriers to lung-directed gene therapy with recombinant adenoviruses. J Virol 1995; 69:2004–2015.

49. Guzman RJ, Lemarchand P, Crystal RG, Epstein SE, Finkel T. Efficient gene transfer into myocardium by direct injection of adenovirus vectors. Circ Res 1993; 73:1202–1207.

50. Bengel FM, Anton M, Avril N, et al. Uptake of radiolabeled 2'-fluoro-2'-deoxy-5-iodo-1-beta-D-arabinofuranosyluracil in cardiac cells after adenoviral transfer of the herpesvirus thymidine kinase gene: The cellular basis for cardiac gene imaging. Circulation 2000; 102:948–950.

51. Wu JC, Inubushi M, Sundaresan G, Schelbert HR, Gambhir SS. Positron emission tomography imaging of cardiac reporter gene expression in living rats. Circulation 2002; 106:180–183.

52. Inubushi M, Wu JC, Gambhir SS, et al. Positron-emission tomography reporter gene expression imaging in rat myocardium. Circulation 2003; 107:326–332.

53. Miyagawa M, Beyer M, Wagner B, et al. Cardiac reporter gene imaging using the human sodium/iodide symporter gene. Cardiovasc Res 2005; 65:195–202.

54. Grines CL, Watkins MW, Helmer G, et al. Angiogenic Gene Therapy (AGENT) trial in patients with stable angina pectoris. Circulation 2002; 105:1291–1297.

55. Hedman M, Hartikainen J, Syvanne M, et al. Safety and feasibility of catheter-based local intracoronary vascular endothelial growth factor gene transfer in the prevention of postangioplasty and in-stent restenosis and in the treatment of chronic myocardial ischemia: Phase II results of the Kuopio Angiogenesis Trial (KAT). Circulation 2003; 107:2677–2683.

56. Wang Z, Zhu T, Qiao C, et al. Adeno-associated virus serotype 8 efficiently delivers genes to muscle and heart. Nat Biotechnol 2005; 23:321–328.

57. Chu D, Sullivan CC, Weitzman MD, et al. Direct comparison of efficiency and stability of gene transfer into the mammalian heart using adeno-associated virus versus adenovirus vectors. J Thorac Cardiovasc Surg 2003; 126:671–679.

58. Fleury S, Driscoll R, Simeoni E, et al. Helper-dependent adenovirus vectors devoid of all viral genes cause less myocardial inflammation compared with first-generation adenovirus vectors. Basic Res Cardiol 2004; 99:247–256.

59. Boecker W, Bernecker OY, Wu JC, et al. Cardiac-specific gene expression facilitated by an enhanced myosin light chain promoter. Mol Imaging 2004; 3:69–75.

60. Iyer M, Salazar FB, Lewis X, et al. Noninvasive imaging of enhanced prostate-specific gene expression using a two-step transcriptional amplification-based lentivirus vector. Mol Ther 2004; 10:545–552.

61. Du L, Kido M, Lee DV, et al. Differential myocardial gene delivery by recombinant serotype-specific adeno-associated viral vectors. Mol Ther 2004; 10:604–608.

62. Bengel FM, Anton M, Richter T, et al. Noninvasive imaging of transgene expression by use of positron emission tomography in a pig model of myocardial gene transfer. Circulation 2003; 108:2127–2133.

63. Miyagawa M, Anton M, Haubner R, et al. PET of cardiac transgene expression: Comparison of 2 approaches based on herpesviral thymidine kinase reporter gene. J Nucl Med 2004; 45:1917–1923.

64. Chen IY, Wu JC, Min JJ, et al. Micro-positron emission tomography imaging of cardiac gene expression in rats using bicistronic adenoviral vector-mediated gene delivery. Circulation 2004; 109:1415–1420.

65. Yaghoubi SS, Wu L, Liang Q, et al. Direct correlation between positron emission tomographic images of two reporter genes delivered by two distinct adenoviral vectors. Gene Ther 2001; 8:1072–1080.

66. Sun X, Annala AJ, Yaghoubi SS, et al. Quantitative imaging of gene induction in living animals. Gene Ther 2001; 8:1572–1579.

67. Wu JC, Chen IY, Wang Y, et al. Molecular imaging of the kinetics of vascular endothelial growth factor gene expression in ischemic myocardium. Circulation 2004; 110:685–691.

68. Yla-Herttuala S, Alitalo K. Gene transfer as a tool to induce therapeutic vascular growth. Nat Med 2003; 9:694–701.

69. Henry TD, Annex BH, McKendall GR, et al. The VIVA trial: Vascular endothelial growth factor in Ischemia for Vascular Angiogenesis. Circulation 2003; 107:1359–1365.

70. Simons M, Annex BH, Laham RJ, et al. Pharmacological treatment of coronary artery disease with recombinant fibroblast growth factor-2: Double-blind, randomized, controlled clinical trial. Circulation 2002; 105:788–793.

71. Dor Y, Djonov V, Abramovitch R, et al. Conditional switching of VEGF provides new insights into adult neovascularization and pro-angiogenic therapy. Embo J 2002; 21:1939–1947.

72. Pislaru S, Janssens SP, Gersh BJ, Simari RD. Defining gene transfer before expecting gene therapy: Putting the horse before the cart. Circulation 2002; 106:631–636.

73. Menasche P, Hagege AA, Vilquin JT, et al. Autologous skeletal myoblast transplantation for severe postinfarction left ventricular dysfunction. J Am Coll Cardiol 2003; 41:1078–1083.

74. Assmus B, Schachinger V, Teupe C, et al. Transplantation of progenitor cells and regeneration enhancement in acute myocardial infarction (TOPCARE-AMI). Circulation 2002; 106:3009–3017.

75. Wollert KC, Meyer GP, Lotz J, et al. Intracoronary autologous bone-marrow cell transfer after myocardial infarction: the BOOST randomised controlled clinical trial. Lancet 2004; 364:141–148.

76. Wollert KC, Drexler H. Clinical applications of stem cells for the heart. Circ Res 2005; 96:151–163.

77. Wu JC, Chen IY, Sundaresan G, et al. Molecular imaging of cardiac cell transplantation in living animals using optical bioluminescence and positron emission tomography. Circulation 2003; 108:1302–1305.

78. Zhang M, Methot D, Poppa V, Fujio Y, Walsh K, Murry CE. Cardiomyocyte grafting for cardiac repair: Graft cell death and anti-death strategies. J Mol Cell Cardiol 2001; 33:907–921.

79. Murry CE, Wiseman RW, Schwartz SM, Hauschka SD. Skeletal myoblast transplantation for repair of myocardial necrosis. J Clin Invest 1996; 98:2512–2523.

80. Muller-Ehmsen J, Whittaker P, Kloner RA, et al. Survival and development of neonatal rat cardiomyocytes transplanted into adult myocardium. J Mol Cell Cardiol 2002; 34:107–116.

81. Cao F, Krishnan M, Dylla SJ, et al. Molecular imaging of embryonic stem cell transplantation, survival, and proliferation in living animals. International Society for Stem Cell Research 2005 (Abstract). Circulation 2006 113(7):1005–1114, Epub 2006 Feb 13.

82. Kim YH, Lee DS, Kang JH, et al. Reversing the silencing of reporter sodium/iodide symporter transgene for stem cell tracking. J Nucl Med 2005; 46:305–311.

83. Wu JC, Wang D, Chen IY, et al. Molecular mechanisms of reporter gene silencing and reversal in cell transplant imaging. Mol Imaging Biol 2005 (Abstract). FASEB J 20(1):106–108, Epub 2005 Oct 24.

11

In-Vivo Imaging of Transgene Expression Using the Herpesviral Thymidine Kinase Reporter Gene

Frank M. Bengel

Division of Nuclear Medicine, Russell H. Morgan Department of Radiology and Radiological Science, Johns Hopkins University School of Medicine, Baltimore, Maryland, U.S.A.

Advances in genomics and proteomics have continuously refined the understanding of molecular changes in heart disease, and have set the stage for development of therapeutic interventions at the cellular and sub-cellular level. It is the vision of molecular pharmacologists to target specific disease mechanisms and to modify them at their earliest biologic stages. Within recent years, a variety of gene therapeutic approaches, directed against targets involved in atherosclerosis, ischemia, heart failure, and arrhythmia, have been introduced and in cardiology evaluated in the experimental setting, and some have reached the stage of clinical trials (1). Results, however, were not always consistent, and consensus on some basic issues such as the most suitable approach for vector delivery, or the relationship between transgene expression and functional therapy effects, has not yet been obtained.

More recently, efforts have focused on the potential of stem cells and progenitor cells, either genetically modified or not, to regenerate myocardial and vascular integrity (2). An advantage of transferring non-differentiated cells is that not only a single gene, but rather an entire biologic system, which inherently provides plasticity and flexibility, is being used to modify disease-generated alterations. Initial clinical trials of cell therapy of the heart have been conducted, but similar questions as for gene therapy remain unanswered, and issues related to cell homing, viability and differentiation need to be addressed in more detail.

Noninvasive molecular imaging techniques based on reporter genes will significantly contribute to address these open issues noninvasively. The herpesviral thymidine kinase reporter gene (HSV1-tk) is at present best established for assessment of transgene expression using nuclear imaging. This chapter first summarizes the current status of gene and cell therapeutic cardiac interventions and highlights open issues which may be resolved by use of noninvasive imaging. Subsequently, the present experience with application of HSV1-tk reporter gene imaging in the heart is outlined. Finally, open questions regarding the usefulness of reporter gene imaging itself are presented along with potential solutions that may help to contribute to a further evolution of the technique.

HUMAN CARDIOVASCULAR GENE THERAPY

Current Status

Primary targets for cardiovascular gene therapy have been myocardial ischemia and heart failure, but applications in other diseases such as arrhythmia, restenosis, and transplant rejection are also being pursued. The most advanced field of cardiovascular gene therapy to date is induction of angiogenesis. The term "therapeutic angiogenesis" characterizes the promotion of revascularization of ischemic tissue through angiogenic growth factors such as vascular endothelial growth factor (VEGF) and fibroblast growth factor. Initial clinical application occurred in individuals with peripheral artery disease and critical limb ischemia, where gene transfer using plasmids encoding for VEGF resulted in improvement of symptoms (3). The same concept was also applied to myocardial ischemia in patients with advanced coronary artery disease. Early trials used direct intramyocardial injection of plasmid (4) or adenovirus (5) carrying the VEGF gene during thoracotomy, and demonstrated improvement in anginal pain and exercise capacity. Catheter-based myocardial vector delivery was also introduced, and intracoronary administration as an adjunct to percutaneous intervention was employed (6,7). While symptoms almost always improved in those trials, results of myocardial perfusion imaging did not always show improvement. Similarly, results of recent randomized trials using recombinant growth factors did not show significant advantages of treatment most likely due to the observation of improvement in symptoms even with placebo application (8,9).

Treatment of heart failure is another scenario for gene therapy, where it is close to entering clinical trials (10). Molecular targets for restitution of contractile function in the failing heart include interventions in calcium homeostasis [e.g., overexpression of sarcoplasmic reticulum Ca-ATPase (SERCA) to increase contraction and relaxation velocity, or inhibition of phospholamban to increase SERCA levels (11,12)], in β-adrenergic signal transduction [e.g., β-receptor overexpression to increase contractility, or inhibition of β-adrenoceptor kinases to reduce receptor desensitization (13)] and in the cascade of apoptotic cell death [e.g., overexpression of the anti-apoptotic factor Bcl-2 (14)].

Gene therapy in other areas of cardiovascular disease has not yet reached clinical application, but a wide variety of therapeutic genes is intensively investigated in experimental settings and the field is moving forward rapidly.

Open Questions

Several key issues concerning cardiac gene therapy remain unresolved at present (Table 1). This starts with the basic methodologic question about which technique for gene transfer should be applied in which situation. Successful cardiac gene therapy requires a suitable technique for local delivery to the heart, and an appropriate vector for delivering and expressing gene(s) in the desired cell type (15). Viral vectors include retrovirus, adenovirus, adeno-associated virus, and lentivirus. Most frequently applied non-viral gene therapy methods include the use of liposomes and injection of vector-free DNA (naked DNA). Vector delivery techniques include direct epi- or endocardial intramyocardial injection, pericardial application, and arterial or retrovenous perfusion. Consensus on the best approach for a given situation has not yet been obtained.

Additionally and even more importantly, as demonstrated in the case of therapeutic angiogenesis, questions with regard to effectiveness in the clinical setting remain to be resolved. The success of gene transfer in a given therapeutic setting, its reproducibility, its direct linkage with improvements in functional alterations, and the influence of the individual pathobiologic milieu on efficiency of gene transfer and therapeutic effectiveness all are issues which require detailed investigation before clinical application of an approach.

Table 1 Open Questions in Cardiac Gene Therapy that Can Be Answered by Imaging

What is the best technique for gene delivery?
Was gene transfer successful in an individual?
Location, magnitude, and time course of transgene expression?
Relationship between transgene expression and functional therapy effects?
Contamination of extracardiac structures?

At present, most of these questions are addressed by molecular ex vivo tissue analysis in experimental studies. These post-mortem analyses, however, remain limited to a single time point in the individual course of therapy. The difficulty to transfer such data from experimental work to a clinical therapeutic setting is emphasized by the variable success of clinical gene therapy. A noninvasive approach, which allows one to repetitively monitor the expression of transferred genes along with their functional effects is thus highly desirable to refine the understanding of current gene therapeutic approaches.

HUMAN CARDIOVASCULAR CELL-BASED THERAPY

Current Status

Based on observations that stem cells are able to differentiate into cardiac cells of adult phenotype, interest in the role of cell transplantation for regeneration of damaged myocardium is rapidly and enormously increasing (2). Cell types which are being studied comprise embryonal stem cells, skeletal myoblasts, bone-marrow derived hematopoetic and mesenchymal stem cells, and endothelial progenitor cells from peripheral circulation (16). The potential of these cell types to improve ventricular performance in animal models of myocardial damage has been demonstrated. Furthermore, cells can be genetically engineered ex vivo to enhance their in vivo regenerative potential. When enhanced by overexpression of VEGF or telomerase, the renewal and revascularization capacity of endothelial progenitor cells, for example, increased significantly (17).

Cell therapy offers the advantage over gene therapy that an entire, flexible system is installed which may dynamically adapt itself to the environment and replace lost function of damaged tissue. This potential advantage explains why current developments in cardiac moelcular therapy seem to favor the application of cells over modification of genes.

Clinical trials of cell therapy have been performed in small patient groups. Initial human studies using skeletal myoblasts observed improvement in contractility (18), but sustained ventricular arrhythmias requiring defibrillator implantation occurred in some patients (19) and reduced enthusiasm. Bone-marrow derived cells and endothelial-progenitor cells have also been applied to humans and safety of the approaches was demonstrated (20–22).

Methods for direct cardiac delivery of stem cells are largely the same as those for vector delivery in gene therapy. Clinical delivery of cells has been performed using direct surgical engraftment (19) or intraarterial application (21). Transendocardial catheter-based delivery is also under investigation. These concepts of direct delivery have been challenged, however, by observations that cells can home to the myocardium for differentiation in vivo (2). For example, Y-chromosome containing cells with cardiocyte properties have been observed in specimens of male recipients of a female donor heart (23). Although not understood in detail, the homing process may support local cardiac accumulation of intravenously administered cells and alleviate the need for direct myocardial injection.

Further experimental work has substantiated a considerable intrinsic regenerative potential through resident or remote adult cardiomyogenic stem cells which may be stimulated by cytokines, thereby questioning the need for exogenous cell transplantation (24). Also, the occurrence of cell fusion after transplantation has been reported as a confounding factor and challenged the initial assumptions of a transdifferentiation potential of stem cells (25). Other reports, which could not reproduce transdifferentiation of hematopoetic stem cells to cardiac cells (26,27), further emphasized the need for a better understanding of the biologic mechanisms involved in cell-based therapies of the heart.

Open Questions

Compared to gene therapy, there are at least as many unknown issues concerning cell transplantation for cardiac repair (Table 2). No consensus is yet obtained about the most suitable cell type, the most suitable method of delivery and the suitability of systemic administration and efficiency of subsequent biologic homing. Furthermore, the mechanisms through which poorly differentiated cells contribute to functional improvement in the long term are still not fully understood, and the benefit of this approach of cardiac repair over conventional cardiovascular therapies have not been fully elucidated.

There are growing concerns about the application of cardiac cell-based therapies in clinical studies in humans without understanding the biological mechanisms involved (28). This emphasizes not only the need for a closer cooperation between basic scientists and clinicians. It also points out the necessity of biologic imaging techniques which can be applied in experimental animals, but also—and in a similar fashion—in clinical studies.

Table 2 Open Questions in Cardiac Cell Therapy that Can Be Answered by Imaging

What is the best technique for cell delivery?
Did cells migrate or home to the target area?
Are cells still viable?
Do cells differentiate (and express new endogenous genes)?
Relationship between cell viability/differentiation and functional
    therapy effects?

REPORTER GENE IMAGING IN THE HEART USING HSV1-TK

Basic Principle

Detection of the level of reporter gene product activity through accumulation of a reporter probe provides indirect information about the level of reporter gene expression (29,30). Reporter gene imaging was initially developed for postmortem tissue analysis (e.g., ß-gal/ LacZ assay), but several studies have now established reporter genes for noninvasive imaging. The gene product of suitable reporter genes is usually not present in target tissue and has little effects or interaction with tissue function. The HSV1-tk has been used as tumoricidal suicide gene when combined with antiviral drug treatment, but because of its absence in normal tissue and its potential to phosphorylate and thus accumulate nucleosides, this gene has also emerged as the currently best investigated imaging reporter gene (Fig. 1). The largest experimental body of evidence for the usefulness of reporter gene imaging continues to be in oncology, and initial application in a clinical setting of brain tumors has been reported (31).

Application in the Heart

In a first study in cardiac cells, it was demonstrated that HSV1-tk can be expressed in a manner similar to tumor cells using an adenoviral vector, and specific accumulation of the radioiodinated pyrimidine derivative FIAU (2'-fluoro-2'-deoxy-5-iodo-1-ß-D-arabinofuranosyluracil) in the area of in vivo myocardial gene transfer was demonstrated at autoradiography (32). Using a mutant HSV1-tk (HSV1-sr39tk) and another nucleoside, the acycloguanosine derivative FHBG (9-(4-^{18}F-fluoro-3-hydroxy-methylbutyl)guanine), in vivo images of cardiac reporter gene expression were obtained for the first time using micro positron emission tomography (PET) in rats (33). Further studies showed that adenoviral titers as low as $1*10^7$ produced enough signal to be identified, and that the reporter gene signal peaked at 3–5 days after gene transfer but was no longer detectable after 10–17 days (34). More recently, it has been shown that results obtained in rats using dedicated small field of view micro PET can be transferred to a clinical setting using conventional PET scanners (35) (Fig. 2). Areas injected with adenovirus expressing HSV1-tk were specifically visualized in the pig heart using ^{124}I-FIAU, and were validated versus ex vivo count rates and immunohistochemistry. Furthermore, co-analysis of myocardial blood flow showed a small but significant increase of flow, likely due to adenovirus-induced inflammatory reaction. This emphasizes a strength of noninvasive imaging, namely that information on transgene expression can be co-registered with physiologic parameters such as

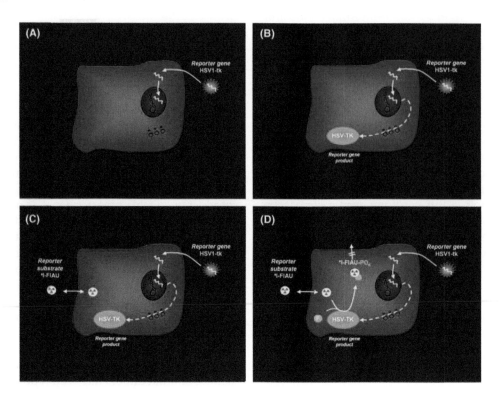

Figure 1 Using HSV1-tk as reporter gene. (**A**) A vector (viral or non-viral) is used for gene transfer. (**B**) The reporter gene is transcribed and translated to a reporter gene product (the herpesviral thymidine kinase enzyme). (**C**) A radiolabelled reporter probe (e.g., the nucleoside FIAU) is administered intravenously and passes the cell membrane via nucleoside transporters. (**D**) The probe is phosphorylated and accumulated only in presence of the reporter gene product.

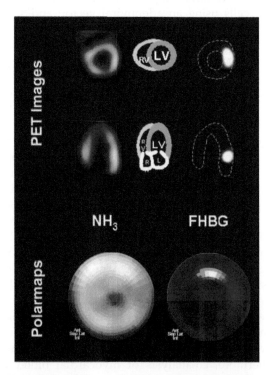

Figure 2 PET study of a pig two days after regional myocardial gene transfer using an adenovirus expressing HSV1-sr39tk reporter gene. N-13 ammonia (NH_3) determined perfusion is homogeneous and images using the reporter probe F-18 FHBG show significant regional accumulation only in the myocardial area of gene transfer.

perfusion and metabolism in the same imaging session to determine functional effects of gene transfer. Washout of FIAU following initial specific uptake in the HSV1-tk transfected myocardium was identified as a limitation for the use of this technique (35). The mechanisms of this washout are not fully understood yet, but they are not due to systemic tracer degradation and may be related to myocardium-specific dephosphorylation of FIAU-phosphate or to transport mechanisms, as indicated in studies using isolated heart perfusion (36). A direct comparison of FIAU as tracer for wild-type HSV1-tk with FHBG as tracer for mutant HSV1-sr39tk in small and large animals revealed an advantage for FHBG and mutant SV1-sr39tk because tracer washout is not observed. Instead, continuous accumulation of the reporter probe was found over time, yielding high contrast (37).

How to Answer Unresolved Issues in Molecular Therapy

Because reporter gene expression can be quantified volumetrically by noninvasive means, it will allow for comparing different techniques of gene/cell transfer with regard to the achieved pattern and magnitude of gene expression/cell accumulation, and thus help identifying the most suitable technique of delivery in a given therapeutic setting. The final goal for monitoring

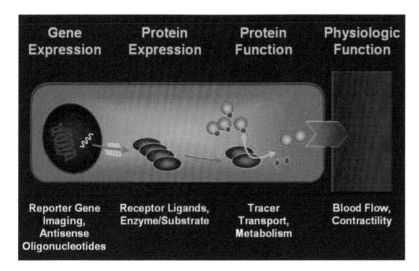

Figure 3 Schematic display of molecular targets amenable for imaging. Addition of reporter gene imaging will allow for characterization from genotype to phenotype.

of gene therapy is to coexpress a reporter gene with a therapeutic gene so that the expression of the therapeutic gene is indirectly monitored via reporter gene imaging. Further biologic variables such as blood flow, metabolism and contractility can be obtained in the same imaging session using additional tracers. This will allow for linkage of the reporter gene signal with functional effects of gene therapy, and is expected to refine the understanding of mechanisms of therapy (Fig. 3).

Approaches for coexpression of two genes include, for example, bicistronic vectors in which the two genes are linked by an internal ribosomal entry site (IRES), vectors which encode for fusion proteins, or vectors which express the two transgenes under control of separate promoters. Using an IRES-based approach, two imaging reporter genes (HSV1-sr39tk and the dopamine D2 receptor) were recently successfully coexpressed within rat myocardium, and both were imaged using microPET (38), suggesting the usefulness for

future monitoring of a therapeutic gene. An adenoviral vector which coexpresses the angiogenic gene VEGF together with HSV1-sr39tk in separate expression cassettes has also been introduced recently and was validated in vitro (39) and in vivo in a rat model of myocardial infarction (40). The technology thus is close to an application in a true therapeutic setting such as angiogenesis induction in chronic myocardial ischemia or augmentation of contractility in heart failure.

To monitor the fate of stem cells after transplantation, approaches for direct labeling with radioisotopes (41,42) or magnetic particles (43) have been introduced. Those provide a stable signal for imaging, but have the disadvantages that the signal is not directly linked with cell viability. It may also be lost in case of cell division, and it will be lost upon radioactive decay or magnetic particle exocytosis. Ex vivo transfer of a reporter gene to stem cells prior to administration would allow for almost unlimited noninvasive monitoring of cell fate in vivo (Fig. 4). Reporter gene transfer to therapeutic cells

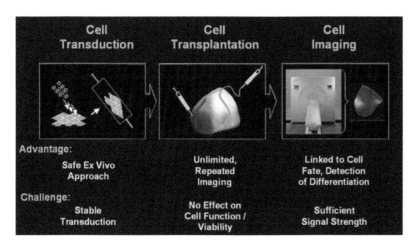

Figure 4 Schematic display of steps necessary for using reporter genes for cell imaging.

can be performed in vitro using various available vector techniques, and cells can be evaluated with regard to succesful reporter gene expression prior to in vivo administration. Initial proof of principle for such an approach was obtained in a recent study, which transiently transfected rat cardiomyoblasts ex vivo with adenovirus carrying HSV1-sr39tk and luciferase genes, and demonstrated that cell-specific in vivo optical and micro PET imaging is feasible for up to two weeks following intramyocardial cell injection (44). But in vivo studies using cells with a therapeutic potential, which were stably transfected with a reporter gene, have not yet been performed.

Meanwhile, however, reporter gene imaging technology continues to advance. A variety of reporter genes other than HSV1-tk have been identified and successfully applied. Some of them, like the sodium-iodide symporter gene, provide the advantage that a species-specific protein is expressed which is nonimmunogenic (45). Studies performed mainly in tumor models have introduced several even more specific approaches which may soon also be applied for cardiac molecular-genetic imaging (30). For example, endogenous biologic processes can be monitored if the reporter gene is controlled by a specific response element rather than a constitutional, nonspecific promoter. Such is the case of the p53-response element, which was demonstrated to promote expression of reporter genes only after DNA-damage induced upregulation of p53 (46). Additionally, methods have been introduced to augment the signal derived from reporter genes expressed under control of weak, tissue specific promoters (47). Two-step transcriptional transactivation was shown to increase promoter strength for imaging while maintaining tissue specificity. When transferred to the heart, such techniques could allow for obtaining specific reporter gene information from various tissue types (e.g., myocardium, endothelium, and fibroblasts), for obtaining a readout that would be specific for disease-related molecular events in target tissue (e.g., identification of upregulation of hypoxia-inducible factor during ischemia and/or angiogenesis by expressing reporter genes under control of the hypoxia-responsive element), or for identifying differentiation steps during maturation of transplanted immature cells.

WHAT LIES AHEAD AND WHAT NEEDS TO BE DONE?

Proof of principle has been obtained to date for several applications of reporter genes for imaging in the cardiovascular system. Similar to molecular cardiovascular therapy, the field of reporter gene imaging is rapidly advancing. The greatest challenge for reporter gene imaging, however, will be to translate the technique from proof of principle to practical application in molecular therapy and from experimental to clinical application. Imaging scientists will need to get closer together with therapists to unite forces and advance molecular medicine in general. Open questions in cardiac gene and cell therapy may be answered in the future by reporter gene imaging approaches, which will be incorporated into multimodality, multi-tracer imaging strategies for serial quantitative monitoring of gene and cell therapy in the heart (Fig. 5).

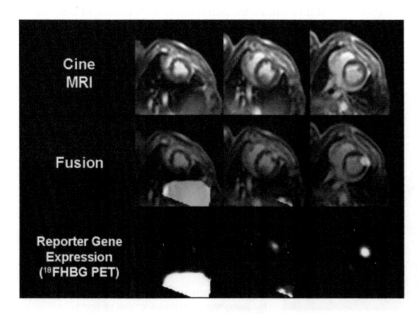

Figure 5 Fusion of cine magnetic resonance imaging and PET reporter gene imaging with HSV1-sr39tk and F-18 FHBG, allowing for combined assessment of gene transfer and its functional effects.

REFERENCES

1. Isner JM. Myocardial gene therapy. Nature 2002; 415:234–239.
2. Forrester JS, Price MJ, Makkar RR. Stem cell repair of infarcted myocardium: an overview for clinicians. Circulation 2003; 108:1139–1145.
3. Isner JM, Walsh K, Symes J, et al. arterial gene therapy for therapeutic angiogenesis in patients with peripheral artery disease. Circulation 1995; 91:2687–2692.
4. Losordo DW, Vale PR, Symes JF, et al. Gene therapy for myocardial angiogenesis: Initial clinical results with direct myocardial injection of phVEGF165 as sole therapy for myocardial ischemia. Circulation 1998; 98:2800–2804.
5. Rosengart TK, Lee LY, Patel SR, et al. Angiogenesis gene therapy: Phase I assessment of direct intramyocardial administration of an adenovirus vector expressing VEGF121 cDNA to individuals with clinically significant severe coronary artery disease. Circulation 1999; 100:468–474.
6. Losordo DW, Vale PR, Hendel RC, et al. Phase 1/2 placebo-controlled, double-blind, dose-escalating trial of myocardial vascular endothelial growth factor 2 gene transfer by catheter delivery in patients with chronic myocardial ischemia. Circulation 2002; 105: 2012–2018.
7. Hedman M, Hartikainen J, Syvanne M, et al. Safety and feasibility of catheter-based local intracoronary vascular endothelial growth factor gene transfer in the prevention of postangioplasty and in-stent restenosis and in the treatment of chronic myocardial ischemia: Phase II results of the Kuopio Angiogenesis Trial (KAT). Circulation 2003; 107:2677–2683.
8. Henry TD, Annex BH, McKendall GR, et al. The VIVA trial: Vascular endothelial growth factor in Ischemia for Vascular Angiogenesis. Circulation 2003; 107: 1359–1365.
9. Simons M, Annex BH, Laham RJ, et al. Pharmacological treatment of coronary artery disease with recombinant fibroblast growth factor-2: Double-blind, randomized, controlled clinical trial. Circulation 2002; 105:788–793.
10. Williams ML, Hata JA, Schroder J, et al. Targeted beta-adrenergic receptor kinase (betaARK1) inhibition by gene transfer in failing human hearts. Circulation 2004; 109:1590–1593.
11. Hasenfuss G. Calcium pump overexpression and myocardial function. Implications for gene therapy of myocardial failure. Circ Res 1998; 83:966–968.
12. He H, Meyer M, Martin JL, et al. Effects of mutant and antisense RNA of phospholamban on SR Ca(2+)-ATPase activity and cardiac myocyte contractility. Circulation 1999; 100:974–980.
13. Maurice JP, Koch WJ. Potential future therapies for heart failure: Gene transfer of beta-adrenergic signaling components. Coron Artery Dis 1999; 10:401–405.
14. Chatterjee S, Stewart AS, Bish LT, et al. Viral gene transfer of the antiapoptotic factor Bcl-2 protects against chronic postischemic heart failure. Circulation 2002; 106:I212–217.
15. Avril N, Bengel FM. Defining the success of cardiac gene therapy: How can nuclear imaging contribute? Eur J Nucl Med Mol Imaging 2003; 30:757–771.
16. Hassink RJ, Dowell JD, Brutel de la Riviere A, Doevendans PA, Field LJ. Stem cell therapy for ischemic heart disease. Trends Mol Med 2003; 9:436–441.
17. Iwaguro H, Yamaguchi J, Kalka C, et al. Endothelial progenitor cell vascular endothelial growth factor gene transfer for vascular regeneration. Circulation 2002; 105:732–738.
18. Menasche P, Hagege AA, Scorsin M, et al. Myoblast transplantation for heart failure. Lancet 2001; 357: 279–280.
19. Menasche P, Hagege AA, Vilquin JT, et al. Autologous skeletal myoblast transplantation for severe postinfarction left ventricular dysfunction. J Am Coll Cardiol 2003; 41:1078–1083.
20. Assmus B, Schachinger V, Teupe C, et al. Transplantation of Progenitor Cells and Regeneration Enhancement in Acute Myocardial Infarction (TOPCARE-AMI). Circulation 2002; 106:3009–3017.
21. Strauer BE, Brehm M, Zeus T, et al. Repair of infarcted myocardium by autologous intracoronary mononuclear bone marrow cell transplantation in humans. Circulation 2002; 106:1913–1918.
22. Wollert KC, Meyer GP, Lotz J, et al. Intracoronary autologous bone-marrow cell transfer after myocardial infarction: The BOOST randomised controlled clinical trial. Lancet 2004; 364:141–148.
23. Quaini F, Urbanek K, Beltrami AP, et al. Chimerism of the transplanted heart. N Engl J Med 2002; 346:5–15.
24. Beltrami AP, Barlucchi L, Torella D, et al. Adult cardiac stem cells are multipotent and support myocardial regeneration. Cell 2003; 114:763–776.
25. Alvarez-Dolado M, Pardal R, Garcia-Verdugo JM, et al. Fusion of bone-marrow-derived cells with Purkinje neurons, cardiomyocytes and hepatocytes. Nature 2003.
26. Murry CE, Soonpaa MH, Reinecke H, et al. Haematopoietic stem cells do not transdifferentiate into cardiac myocytes in myocardial infarcts. Nature 2004; 428: 664–668.
27. Balsam LB, Wagers AJ, Christensen JL, Kofidis T, Weissman IL, Robbins RC. Haematopoietic stem cells adopt mature haematopoietic fates in ischaemic myocardium. Nature 2004; 428:668–673.
28. Mathur A, Martin JF. Stem cells and repair of the heart. Lancet 2004; 364:183–192.
29. Gambhir SS, Herschman HR, Cherry SR, et al. Imaging transgene expression with radionuclide imaging technologies. Neoplasia 2000; 2:118–138.
30. Blasberg RG, Tjuvajev JG. Molecular-genetic imaging: Current and future perspectives. J Clin Invest 2003; 111:1620–1629.
31. Jacobs A, Voges J, Reszka R, et al. Positron-emission tomography of vector-mediated gene expression in gene therapy for gliomas. Lancet 2001; 358:727–729.
32. Bengel FM, Anton M, Avril N, et al. Uptake of radiolabeled 2'-fluoro-2'-deoxy-5-iodo-1-beta-D-arabinofuranosyluracil in cardiac cells after adenoviral transfer of the herpesvirus thymidine kinase gene: The cellular

basis for cardiac gene imaging. Circulation 2000; 102: 948–950.

33. Wu JC, Inubushi M, Sundaresan G, Schelbert HR, Gambhir SS. Positron emission tomography imaging of cardiac reporter gene expression in living rats. Circulation 2002; 106:180–183.

34. Inubushi M, Wu JC, Gambhir SS, et al. Positron-emission tomography reporter gene expression imaging in rat myocardium. Circulation 2003; 107:326–332.

35. Bengel FM, Anton M, Richter T, et al. Noninvasive imaging of transgene expression by use of positron emission tomography in a pig model of myocardial gene transfer. Circulation 2003; 108:2127–2133.

36. Simoes MV, Miyagawa M, Reder S, et al. Myocardial kinetics of the reporter probe [124I]-FIAU in isolated perfused rat hearts following in vivo adenoviral transfer of the herpesviral thymidine kinase reporter gene. J Nucl Med 2005; (in press).

37. Miyagawa M, Anton M, Haubner R, et al. PET imaging of cardiac transgene expression—Comparison of two approaches based on herpesviral thymidine kinase reporter gene. J Nucl Med 2004; 45:1917–1923.

38. Chen IY, Wu JC, Min JJ, et al. Micro-positron emission tomography imaging of cardiac gene expression in rats using bicistronic adenoviral vector-mediated gene delivery. Circulation 2004; 109:1415–1420.

39. Anton M, Wittermann C, Haubner R, et al. Coexpression of herpesviral thymidine kinase reporter gene and VEGF gene for noninvasive monitoring of therapeutic gene transfer—an in vitro-evaluation. J Nucl Med 2004; 45:1743–1746.

40. Wu JC, Chen IY, Wang Y, et al. Molecular imaging of the kinetics of vascular endothelial growth factor gene expression in ischemic myocardium. Circulation 2004; 110:685–691.

41. Barbash IM, Chouraqui P, Baron J, et al. Systemic delivery of bone marrow-derived mesenchymal stem cells to the infarcted myocardium: Feasibility, cell migration, and body distribution. Circulation 2003; 108:863–868.

42. Aicher A, Brenner W, Zuhayra M, et al. Assessment of the tissue distribution of transplanted human endothelial progenitor cells by radioactive labeling. Circulation 2003; 107:2134–2139.

43. Hill JM, Dick AJ, Raman VK, et al. Serial cardiac magnetic resonance imaging of injected mesenchymal stem cells. Circulation 2003; 108:1009–1014.

44. Wu JC, Chen IY, Sundaresan G, et al. Molecular imaging of cardiac cell transplantation in living animals using optical bioluminescence and positron emission tomography. Circulation 2003; 108:1302–1305.

45. Miyagawa M, Beyer M, Wagner B, et al. Cardiac reporter gene imaging using the human sodium/iodide symporter gene. Nature 2003; 425(6961):968–973.

46. Doubrovin M, Ponomarev V, Beresten T, et al. Imaging transcriptional regulation of p53-dependent genes with positron emission tomography in vivo. Proc Natl Acad Sci U S A 2001; 98:9300–9305.

47. Iyer M, Wu L, Carey M, Wang Y, Smallwood A, Gambhir SS. Two-step transcriptional amplification as a method for imaging reporter gene expression using weak promoters. Proc Natl Acad Sci U S A. 2001; 98:14595–14600.

12

Optical Cardiovascular Imaging

Crystal M. Ripplinger
Department of Biomedical Engineering, Washington University School of Medicine, St. Louis, Missouri, U.S.A.

Guy Salama
Department of Cell Biology and Physiology, University of Pittsburgh, Pittsburgh, Pennsylvania, U.S.A.

Igor R. Efimov
Department of Biomedical Engineering, Washington University School of Medicine, St. Louis, Missouri, U.S.A.

INTRODUCTION

Mammalian physiology has an ingrained hierarchy with molecular and cellular physiology at its base followed by the interactions of large populations of cells, organ systems, and the integration of multiple organ functions of an entire animal. Although there have been great strides in advancing our understanding of molecular and cellular mechanisms, there is a growing realization that organs such as the heart are comprised of several types of interacting cells with significant heterogeneities of properties, cell-to-cell coupling, and function within each group. Thus, an understanding of molecular and cellular mechanisms must still be integrated to explain the more complex organ system while taking into account spatial and temporal heterogeneities of cell functions throughout the organ. Unfortunately, experimental methodologies available for studies at the organ level are not as abundant as at the cellular scale. However, optical modes of imaging in combination with parameter-sensitive probes have demonstrated their ability to overcome the problem of spatio-temporal resolution in two dimensions for a wide range of applications from single molecular events to in vivo whole animal physiology.

Fluorescence has been used to measure a wide range of physiological parameters in cells and tissues. One such example is the use of fluorescence for monitoring cellular metabolism by the intrinsic fluorescence changes of NADH or flavoproteins (1), the differential absorption changes of mitochondrial cytochromes, or the oxygen content of blood and cardiac muscle through the absorption changes of oxy to deoxy hemoglobin and myoglobin, respectively (2,3). Recently, probes have been developed that can selectively measure functional parameters such as membrane potential, intracellular concentrations of free calcium, magnesium, sodium and potassium, pH, nitric oxide, oxygen tension, and sulfhydryl redox sate. Of these, probes of membrane potential and intracellular free calcium ($[Ca]_i$) have had the most impact in cardiovascular physiology.

The development of optical recordings of membrane potential was driven by the need to overcome many obstacles in electrophysiology and the promise of a technology "for measuring membrane potential in systems where, for reasons of scale, topology, or complexity, the use of electrodes is inconvenient or impossible" (4). Based on our current experience in cardiac electrophysiology, this list needs to be extended to recordings of action potentials in the presence of external electric fields during stimulation and defibrillation; an impossible task with both extra- and intracellular electrodes due to the large electrical artifacts caused by external fields. Optical mapping techniques and potentiometric probes have now made major contributions to our understanding of cardiac electrophysiology in ways that could not have been accomplished with other approaches.

GENERAL PRINCIPLES OF FLUORESCENT RECORDINGS

Over 30 years ago, investigators discovered molecular probes that bind to the plasma membrane of neuronal (5) and cardiac cells (6) and exhibit changes in fluorescence and/or absorption that mimic changes in transmembrane potential. Several optical properties of membrane-bound dyes can be used to measure membrane potential changes namely: fluorescence, absorption, dichroism, birefringence, fluorescence resonance energy transfer, nonlinear second harmonic generation, and resonance Raman absorption. However, most studies in cardiac cells or tissues have relied on fluorescence which tends to yield higher fractional changes in signal magnitude compared to the other modes and because fluorescence signals tend to be considerably less sensitive to movement artifacts generated by muscle contractions (6,7).

The general principle of fluorescence involves the absorption of photons of a certain energy by a fluorescent compound. The compound is then excited from the ground state E_0 to an unstable energy-rich state E_2 (Fig. 1). Then the energy of E_2 is partially dissipated, resulting in the relaxed excited state E_1 from which fluorescence emission originates. When the compound falls back to the ground state E_0, the compound fluoresces by emitting a photon of a lower energy than the exciting photon. The wavelength of a photon is a function of its energy, therefore emitted light always has a longer wavelength than the exciting light. This principle is called Stoke's shift.

Voltage-dependent changes in fluorescence of dye molecules are a consequence of interactions of the electric field with the dye molecules resulting in intra- and extra-molecular rearrangements of the dye in the membrane. Voltage-sensitive dyes are classified into two groups (8), fast and slow dyes, based on their response times and presumed molecular mechanisms. Only the fast probes are used in cardiac electrophysiology, due to their ability to respond to voltage changes in a matter of microseconds (9). The precise mechanisms underlying the voltage-dependent spectroscopic properties of fast voltage-sensitive dyes are still not fully understood. The *electrochromic* theory (10) states that a dye will be voltage sensitive if (*i*) the photon-produced excitation of the chromophore is accompanied by a shift in electric charge and (*ii*) the vector of intramolecular charge movement is parallel with the electric field gradient. Therefore, if charge movement in a dye molecule occurs perpendicular to the cell membrane, the dye's fluorescence will be sensitive to changes in transmembrane potential. An alternative theory is the *solvatochromic* theory (11) which contends that dye molecules experience a change in the polarity of the lipid environment during reorientation produced by the voltage gradient. This dependency causes the spectral voltage-dependence of the chromophore.

Several useful classes of chromophores have emerged over the last 30 years, including merocyanine, oxonol, and styryl dyes. However, styryl dyes represent the most popular family of dyes, with RH-421 and di-4-ANEPPS being the most important members of this family. The spectroscopic properties of these dyes have been shown to have a linear response to transmembrane potential changes in the normal physiological range (7,12,13).

NECESSITY OF OPTICAL MAPPING

Characterizing the spread of electrical activity is essential for our understanding of the mechanisms responsible for normal cardiac rhythm and for the initiation and maintenance of arrhythmias. While intracellular microelectrode recordings have increased our knowledge of the ionic basis of the cardiac action potential, these single cell impalements cannot practically be used to simultaneously record action potentials from hundreds of sites. Therefore, surface electrograms measured with arrays of electrodes are generally used to map activation and repolarization (14,15). While surface electrodes can describe the spread of excitation and repolarization, interpretation of data in some cases is uncertain (14,16). For instance, activation sequences are difficult to interpret during rapid synchronous depolarization, as after electric shock application, and during slowly changing low level depolarization, as in ischemia. Repolarization measured with an electrogram often does not coincide with the actual repolarization at the recording site (16,17). Therefore, optical mapping with potentiometric dyes has made great strides in furthering our understanding of these often complex and dynamic activation and repolarization sequences.

COMPENSATION FOR MOTION ARTIFACT

A significant limitation of optical mapping of the heart is motion artifact introduced by muscle contractions.

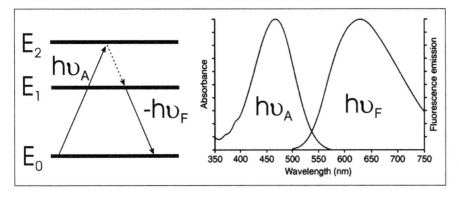

Figure 1 Quantum mechanism of the difference between the absorbance and fluorescent emission spectra (see text for detail).

These "movement" artifacts distort optical action potentials by altering the fluorescence intensity. When the tissue contracts, it can move relative to both the sensor and light source, causing artificial changes in fluorescence. Since muscle contraction begins immediately after the action potential upstroke, motion artifacts are most pronounced during the plateau and repolarization phases. Several methods have been used in the past to minimize the effect of motion artifact. Mechanical restriction of the movement can successfully limit the artifact without affecting the physiology of the heart (18). This method works particularly well with small hearts such as mice, rats, and guinea pigs. A popular alternative is the use of various pharmacological agents, such as calcium channel blockers (19), 2,3-butanedione monoxime (BDM) (20,21) or cytochalasin D (22). However, all of these agents may have effects on the electrical activity of the heart. Calcium channel blockers are often avoided due to the many calcium-dependant cellular processes (23,24). BDM has an effect on a variety of ion channels and may alter action potential duration in a number of species (25,26). Therefore, BDM may not be appropriate for studies of repolarization. Cytochalasin D may provide a promising alternative for some species (22,27). Therefore, the effects of any pharmaceutical agents used need to be taken into consideration for an appropriately designed experiment.

Other approaches for reducing motion artifact involve more complex signal processing techniques. One such approach is a ratiometric technique (28,29). This method is based on simultaneous measurements of fluorescent signals at two different wavelength ranges: one in which the dye exhibits a large voltage-dependent response and the other at a wavelength where the potentiometric dye exhibits an inverted or no voltage-dependent response. The ratio or difference of these two signals yields an optical signal free from, or with significantly reduced, motion artifacts (28–30). Unfortunately, this method works well only with relatively low intensity contractions (31). One of the newest techniques employed is the use of image registration to effectively "track" the motion from frame to frame and register each image with the use of a reference frame (32). However, this method has difficulty with image sequences that have a large activation signal-to-noise ratio. In these cases, the image registration algorithm may confound image features with the activation signal itself (32).

CALIBRATION OF OPTICAL RECORDINGS

A subtlety of fluorescence recordings is that the signals themselves do not provide an absolute measurement of transmembrane potential. Instead, they linearly track the changes in transmembrane potential with high temporal fidelity. Initial studies with fluorescence recordings calibrated optical action potentials by simultaneously recording fluorescence and microelectrode signals from the same region of the heart (6). More recent studies have concentrated on the application of ratiometry approaches for quantitative measurements of transmembrane potential with newer and better dyes. Early experiments with this technique were performed on neuroblastoma cells (33). However, these experiments have not been repeated to confirm the linearity of these dyes in cardiac cells. Simultaneous ratiometric optical and microelectrode recordings of transmembrane potential in perfused hearts confirmed excellent correlation and linearity of optical recordings (30).

OPTICAL SENSORS

Currently, several technologies exist for fast image acquisition, including photomultipliers (PMT), laser scanning (34), charge-coupled device (CCD) cameras, and photodiode arrays (PDA). Optical imaging of the heart is usually performed with CCD cameras and PDA detectors (18,35–38). Complementary metal-oxide semiconductor (CMOS) cameras represent another emerging candidate. CCD technology has a seemingly significant advantage of higher spatial resolution due to the large number of pixels on a CCD sensor. However, the rate of data acquisition is usually significantly lower. The rate can be increased by pixel binning, yet this defeats the major advantage of CCD technology since binning effectively reduces the spatial resolution. Theoretically, CCD technology should also provide significant advantages with respect to the signal-to-noise ratio, particularly at low-light applications. However, it is important to emphasize that an optical mapping experiment on a whole heart or tissue preparation tends not to be light-limited, rather the greater the fluorescence intensity from the preparation, the greater the absolute amplitude of the fractional fluorescence change. The dynamic range of CCD cameras is constrained by the accuracy of analog-to-digital conversion and by the saturation of the sensor at light levels readily detected from the heart. Therefore, a dynamic range of 10^3 is not easily achieved. Thus, for practical purposes the majority of single cell and cell culture optical mapping studies have been done using PDAs (39–41).

PDAs have been used for optical mapping studies in neurophysiology and cardiology since 1981 (42). With a quantum efficiency above 0.8, photodiodes are the most sensitive sensors for medium to large light intensities. Their main drawback is a large dark current, which may limit their usefulness at very low light intensities as in neurophysiological applications. Photodiodes are packaged in arrays of 100, 144, 256 or more. Each recording channel can be independently signal conditioned before analog-to-digital conversion. Therefore, PDAs have become the optical sensor of choice for many optical mapping laboratories.

OPTICAL MAPPING SYSTEM

Any optical imaging system must consist of a two-dimensional optical sensor and a stable light source, such as a laser or DC-powered tungsten-halogen lamp, mercury source, or light emitting diodes. Figure 2 describes a typical design. It consists of a 16 × 16 PDA, 256-channel signal conditioner/amplifier, analog-to-digital converter, and a computer. Excitation light passes through a 520 ± 45 nm filter, is reflected by a 585 nm dichroic mirror, passes through a 50 mm 1.2f lens and illuminates the preparation in the tissue chamber. The fluorescence emitted from the preparation is collected by the same lens, passes through the dichroic mirror, a 610 nm long-pass filter, and is collected by the 16 × 16 PDA. The signal from the PDA is then amplified, digitized, and stored on the computer.

MAPPING OF ACTIVATION AND REPOLARIZATION

Figure 3 shows typical activation and repolarization maps constructed from optical signals recorded from the anterior surface of a Langendorff-perfused guinea pig heart. An optical trace at the right illustrates the high quality of optical fluorescence signals and their ability to faithfully record epicardial action potentials. When movement artifacts are a concern, activation and repolarization times can still be determined by calculating the maximum first derivative of the action potential upstroke and the maximum second derivative of the downstroke, respectively (18) (Fig. 3). If motion artifact is not a concern, activation and repolarization times can be determined from the time point at which the upstroke reaches 50% or recovers to 90% of its maximum amplitude (43,44), respectively. Activation and repolarization data are usually presented as isophasic or isochronal maps (Fig. 3) (3,18,38,45–47), allowing quantitative assessment of the spread of activation and repolarization in highly anisotropic heart muscle under normal and pathological conditions (48–50).

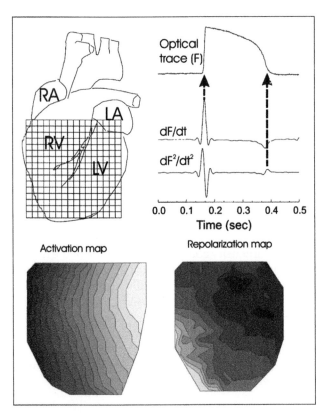

Figure 3 Maps of optical action potentials and methods of analysis. (*Left to right, top to bottom*): Schematic diagram of anterior epicardium of the guinea pig heart and the imaging field of view. Detection of activation and repolarization time points using first (dF/dt) and second (d^2F/dt^2) derivatives of fluorescence, F. Maps of activation and repolarization.

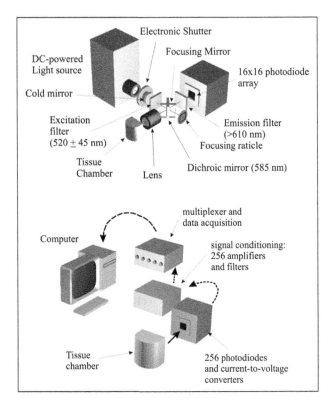

Figure 2 Schematics of a typical optical mapping system. *Source*: From Ref. 77.

Conventional high-resolution multielectrode mapping was successful in assessing activation sequences. However, repolarization was generally beyond the capabilities of electrode mapping (17). Optical mapping provides a tool with the unique ability to faithfully reconstruct maps of repolarization. In particular, optical mapping has made a significant contribution to our understanding of the role of intrinsic myocardial heterogeneities in repolarization of the normal heart (18,48); of dynamic concordant and discordant alternans in the onset of ventricular tachyarrhythmias (51,52); and the role of the slope of the restitution curve in the transition from ventricular tachycardia to fibrillation (53–55).

MAPPING OF STIMULATION AND DEFIBRILLATION

Optical recordings of electrical activity are the only methods which resolve questions related to the interaction of electric fields with excitable cells, as conventional electrode recordings are often distorted by large-amplitude artifacts. Over the past 20 years,

optical methods have provided a vast body of knowledge about transmembrane potential changes during stimulation and defibrillation that was previously undiscovered using conventional electrode recordings. Dillon (19) demonstrated the prolongation of action potential duration by strong electric shocks applied during the refractory period. Optical mapping explained the mechanisms of epicardial unipolar (56) and bipolar pacing (57). Optical mapping also provided the experimental basis for the new theory of stimulus-induced arrhythmogenesis and the related theory of success and failure of defibrillation based on virtual electrode effects (35,58,59), which are also known as secondary source effects (40,60).

Figure 4D presents an example of optically recorded changes in transmembrane potential before, during, and after monophasic shock application, which was applied at the plateau phase of a propagated action potential in the rabbit heart. Figure 4A shows the preparation photograph with the field of view of the optical mapping system. Analysis of data from all 256 channels allowed for reconstruction of the surface map of transmembrane polarization at the end

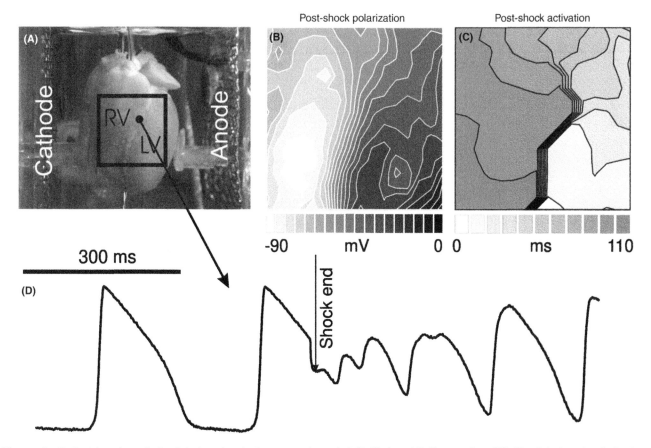

Figure 4 Optical imaging of shock-induced arrhythmogenesis and defibrillation. (**A**) Preparation. (**B**) Shock-induced polarization. (**C**) Shock-induced conduction pattern. (**D**) Optical recording of transmembrane potential during normal action potential, T-wave shock and shock-induced arrhythmia. *Source*: From Ref. 103.

of the shock (Fig. 4B) as well as the activation map of the subsequent arrhythmia genesis (Fig. 4C). Only optical imaging techniques allow the merging of several important parameters which are responsible for arrhythmogenesis during shock-induced arrhythmia or failed defibrillation shocks: shock-induced dispersion of transmembrane polarization, followed by dispersion of repolarization and new wavefront generation, followed by formation of phase singularities and reentry (59).

MAPPING OF THE CONDUCTION SYSTEM OF THE HEART

Optical mapping has made a significant contribution to functional studies of impulse propagation in the conduction system of the heart. The structural complexity of the right atrial preparation, especially in the nodal areas, makes it difficult to apply conventional electrode mapping techniques. While in unipolar electrograms, the distinction between local and distant events is often impossible, bipolar electrograms represent complex spatial derivatives of the underlying spatio-temporal substrate of electrical activity (17). Since the precise field of view of both types of electrodes is unknown, interpretation of the electrograms remains challenging and requires verification by means of simultaneous microelectrode recordings.

Like electrograms, optical recordings represent the integrated electrical response of cells from a volume

of tissue. However, the two-dimensional extent of this volume is precisely defined, even though the exact depth of the recording may be unknown. Because of this well-defined field of view and high spatial resolution, optical mapping has been instrumental in investigating the genesis of spontaneous electrical activity in the embryonic heart (61–64), as well as in the identification of the nonradial spread of activation via preferential pathways from the SA node toward the AV node (65,66). Our groups applied optical mapping to studies of AV nodal conduction (21,66,67), AV nodal reentrant arrhythmias (Fig. 5) (68,69), and AV junctional rhythm (70). Figure 5 demonstrates the construction of 3D stack plots (Fig. 5A) of AV nodal reentry from the first derivative of the fluorescence signal (dF/dt) compared with typical electrode recordings (Fig. 5D).

MAPPING OF VENTRICULAR TACHYARRHYTHMIAS

Optical mapping techniques presented a unique opportunity to study the mechanisms of supraventricular and ventricular arrhythmias, due to unprecedented spatio-temporal resolution as well as the ability to map all phases of electrical activity, including activation and repolarization. Jalife's group pioneered the application of optical mapping to study arrhythmogenesis (71–74) and made numerous significant contributions to our understanding of the mechanisms of both atrial (73) and ventricular arrhythmias (74,75). Since that time, many groups have presented optical mapping

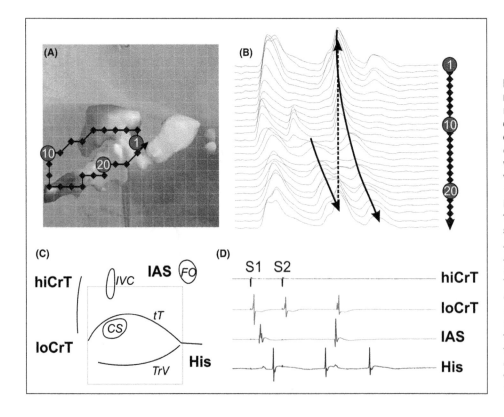

Figure 5 Reconstruction techniques for visualizing AV nodal reentry. (**A–B**) Reconstruction of conduction by 3D stack plot of optical signal derivatives and graph of overlapping signal traces from the whole reentrant circuit. Signals were low-pass filtered at 50 Hz. (**C**) Preparation schematic: loCrT and hiCrT indicate low and high crista terminalis areas: IAS, interatrial septum; His, bundle of His; and AVN, compact AV node area. Anatomical landmarks: CS, coronary sinus; tT, tendon of Todaro; TrV, tricuspid valve; FO, fossa ovalis; IVC, inferior vena cava. (**D**) Bipolar electrograms from locations noted in C. S1 and S2 correspond to the pacing artifact recorded from two sensing leads of a quadrupole electrode when the two other leads were used for pacing. *Source*: From Ref. 69.

data supporting the reentrant nature of ventricular tachycardia (59,71,76,77) and fibrillation (75,78,79). Furthermore, optical mapping presented evidence for the three-dimensional nature of ventricular reentry, which is sustained by scroll waves (77,80).

Figure 6 shows theoretical (panel A) and experimental (panels B and C) recordings of breakthrough activity of a 3D scroll wave. Electrotonic coupling between cells located across a line of block produces low-amplitude passive responses (81). Therefore, action potentials recorded near the line of block exhibit a dual-hump morphology. There is good agreement between bidomain simulations of this activity (82) and experimental results obtained with fluorescent

imaging (77), proving that optical techniques are well-suited for characterizing ventricular reentry as they can faithfully record the dual-hump phenomena.

MULTIPARAMETRIC OPTICAL MAPPING: IMAGING OF TRANSMEMBRANE POTENTIAL AND INTRACELLULAR CALCIUM

Calcium cycling is the most important component of cardiac excitation-contraction coupling. Normally, depolarization triggers intracellular Ca^{2+} ($[Ca^{2+}]_i$) transients but in pathological conditions, abnormalities in $[Ca^{2+}]_i$ handling may activate Ca^{2+} dependent currents

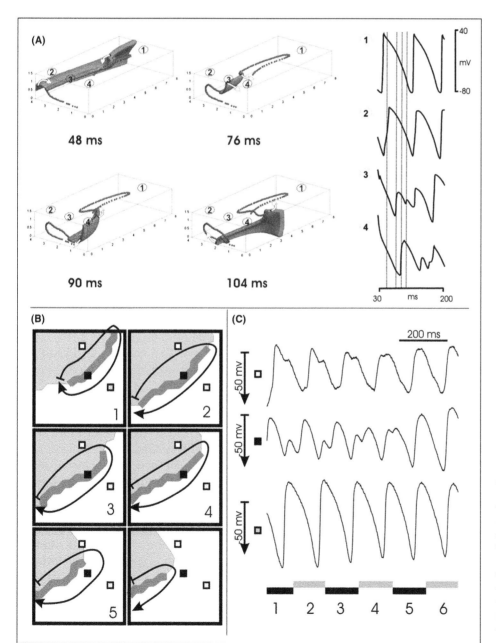

Figure 6 Simulated and experimental optical recordings of breakthrough activity of a 3D scroll wave. (**A**) Bidomain simulation of a 3D scroll wave with corresponding electrical activity observed on the top of the 3D tissue structure. Numbers on the individual traces correspond to locations indicated on the top of the tissue structure. (**B–C**) Experimental results obtained with optical imaging methods. *Source*: Panels (**B**) and (**C**) from Ref. 77.

that influence the time course of the AP and trigger a spontaneous membrane depolarization (83,84). Abnormalities in $[Ca^{2+}]_i$ handling have been implicated as the underlying mechanism in a number of pathologies that promote arrhythmias such as ischemia-reperfusion arrhythmias, the generation of early and delayed afterdepolarizations, and *torsades de pointes*. $[Ca^{2+}]_i$ overload has been implicated in triggering electromechanical alternans and in increasing the steepness of APD restitution curves, which are both associated with the promotion of arrhythmias. Thus, one cannot overstate the importance of simultaneous measurements of APs and $[Ca^{2+}]_i$ transients in intact hearts to address fundamental questions regarding the spatio-temporal relationship of transmembrane potential and $[Ca^{2+}]_i$ and their interplay in arrhythmias.

Simultaneous $[Ca^{2+}]_i$ and transmembrane potential imaging of the heart has been achieved by co-staining with RH421 and Rhod-2 (85), RH237 and Rhod-2 (52,86), RH237 and Fluo-4/Oregon Green BARTA 1 (29), di-4-ANEPPS and Fluor3/4 (87) and with di-4-ANEPPS and Indo-1 (88). Figure 7A shows a typical optical system used for simultaneous recordings of transmembrane potential and $[Ca^{2+}]_i$ signals from perfused hearts stained with RH237 and Rhod-2 (52). These particular dyes can be excited at the same wavelength but fluoresce at different wavelengths, allowing for separation of the V_m and $[Ca^{2+}]_i$ signals. Therefore, two PDAs are required in this system; one for recording V_m, and the other for $[Ca^{2+}]_i$. Rhod-2 was found to be an excellent Ca^{2+} indicator for perfused hearts because of its rapid association and dissociation with Ca^{2+}, fast loading into myocytes of perfused hearts, and long-term stability or low levels of exocytosis (52). Superimposed recordings of V_m and $[Ca^{2+}]_i$ are illustrated in Figure 7B. These signals were simultaneously recorded from the same region of the myocardium. In this experiment, the $[Ca^{2+}]_i$ transient occurred 10.4 ± 0.4 ms after the AP upstroke.

Multiparametric optical mapping is still in its infancy. Other probes are being synthesized and evaluated for simultaneous recordings from the same tissue, stained with multiple fluorescent probes. This exciting progress is likely to bring about a better understanding of cellular physiology at both the tissue and organ level.

OPTICAL MAPPING TO REVEAL 3D CELLULAR STRUCTURE

Action potentials optically recorded from the triangle of Koch have been shown to have two distinct components that appear during anterograde as well as

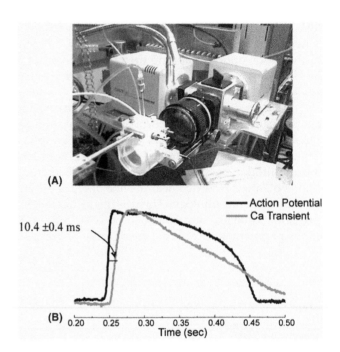

Figure 7 Multiparametric optical imaging: simultaneous optical recordings of transmembrane potential and intracellular calcium transients. (**A**) Experimental setup with two 16×16 photodiode arrays. Hearts were stained with voltage-sensitive dye RH 237 and Ca^{2+} indicator Rhod-2. (**B**) Superposition of an AP and $[Ca^{2+}]_i$ transient recorded from the same region of myocardium. AP and Ca_i were simultaneously recorded at a sampling rate of 4000 frames/sec for each array. The AP upstroke preceded the rise of $[Ca^{2+}]_i$ by 10.4 ± 0.4 ms.

retrograde conduction (21,66). This phenomenon is due to activation of the atrial-transitional layer of cells followed by activation of nodal cells. During retrograde conduction, the order of activation is reversed. This activation sequence has been verified with microelectrode recordings (21); however, optical imaging experiments using simple diffusion of potentiometric dye have also confirmed these findings and revealed the cellular origin of the two components (89).

In these experiments, right atrial preparations were stained by superfusion with di-4-ANEPPS at a concentration of 1 μmol/L. Due to the slow diffusion of the dye, superficial layers of tissue were stained first, followed by deeper cellular structures. During retrograde conduction, the first component of the optical trace, representing the deeper nodal structure, appeared after 20 minutes of staining. In contrast, the second component, representing the atrial-transitional layer, appeared after only 5–10 minutes. Figure 8 summarizes the results of these experiments. Although integration of electrical activity from a volume of tissue may in fact be a disadvantage in some optical imaging experiments, appropriately designed experiments can exploit this perceived shortcoming and instead use it to reveal 3D cellular structures.

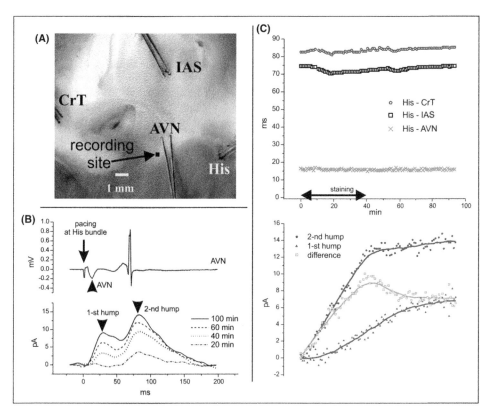

Figure 8   Experimental preparation and optical signals during staining. (**A**) Superfused right atrial preparation shows the location of the pacing (His) and recording (crista terminalis [CrT] and interatrial septum [IAS]) electrodes and the location of the optical recording site illustrated in panels **B** and **C**. The site was 375×375 mm². (**B**) Upper graph shows a bipolar electrogram recorded by the AVN electrode at the location shown in panel **A**. It contains a stimulus artifact (pacing at the His bundle), a slow response of the AVN and/or posterior nodal extension, and a fast response, reflecting atrial activation. Lower graph shows optical recordings collected during and after staining at the site shown in panel **A**. The amplitudes of the 1st and the 2nd humps are analyzed in panel **C**. (**C**) (*Top*) Stability of conduction delays between the bundle of His and CrT (upper line), the bundle of His and the IAS (middle line), and the bundle of His and the distal AVN (lower line). (*Bottom*) Kinetics of the amplitudes of the 1st and the 2nd humps of optical signals and their difference. *Source*: From Ref. 89.

EMERGING TECHNOLOGY: OPTICAL COHERENCE TOMOGRAPHY

Imaging with voltage-sensitive probes has several limitations. Potentiometric dyes are associated with phototoxicity (90) and increased contractility (91). Yet, the major limitation of tissue and organ-level optical imaging is its restricted depth of penetration. Depth restriction is due to light absorption and scattering on intrinsic differences of tissue optical properties (endogenous contrast). Absorption occurs primarily at endogenous chromospheres of hemoglobin, melanin, fat, and water. Scattering is usually due to refractive index differences of extra- and intracellular structures. These properties are strongly wavelength-dependent.

Optical coherence tomography (OCT) is a non-invasive technology that uses back-reflected infrared light (1,310 nm) to perform 3D imaging (92). Infrared light falls within the range known as the "therapeutic window." Due to the low intrinsic tissue absorbance, high scattering of photons dominates this range of wavelengths. OCT exploits this high degree of scattering and,

at present, can reconstruct cardiac tissue 1–3 mm in depth with up to 1 µm resolution. These characteristics make tomography a method of choice for in situ analysis of embryonic heart morphology (93,94), AV nodal multilayer structures (95), and the complex three-dimensional geometry of trabeculated structures of myocardium and the Purkinje network (96,97) (Fig. 9). Presently available OCT technology is limited to structural imaging only. However, fluorescent imaging with potentiometric probes combined with OCT, as well as the emerging second harmonic OCT technique (98,99), could provide powerful new tools for structure-function studies at the cellular and tissue levels.

CONCLUSION

Fluorescent imaging has been developed and further refined in a number of laboratories (19,39,47,56,66,71, 100–102) and has emerged as a powerful new investigative approach, making countless contributions to the area of cardiac electrophysiology over the last two

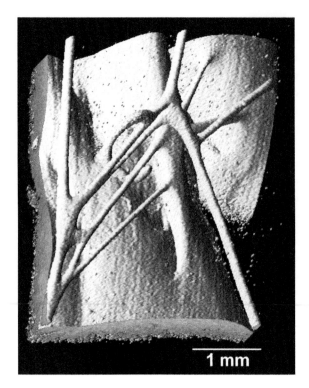

Figure 9 Optical coherence tomography of the purkinje network. The 3D OCT volume is $2.8 \times 3.6 \times 4.3$ mm³. *Source:* From Ref. 97.

decades. Exciting new developments in biophotonics suggest that the best is yet to come. The next decade is likely to yield: (*i*) novel optical molecular probes for multiparametric optical sensing of various biological parameters, processes, molecules, proteins, and their functional states; (*ii*) novel optical imaging modalities for three-dimensional optical interrogation of molecular probes with precise anatomical localization of the signal origin with sub-cellular spatial resolution.

ACKNOWLEDGMENTS

This work was supported by grant awards from the National Heart Lung and Blood Institute HL057929 (G.S.), HL70722 (G.S.), HL69097 (G.S.), HL074283 (I. R.E.), and from the American Heart Association 0515562Z (C.M.R.).

REFERENCES

1. Chance B, Salkovitz IA, Kovach AG. Kinetics of mitochondrial flavoprotein and pyridine nucleotide in perfused heart. Am J Physiol 1972; 223:207–218.
2. Chance B. On the mechanism of the reaction of cytochrome oxidase with oxygen. Ann N Y Acad Sci 1975; 244:163–173.
3. Salama G, Lombardi R, Elson J. Maps of optical action potentials and NADH fluorescence in intact working hearts. Am J Physiol 1987; 252:H384–H394.
4. Cohen LB, Lesher S. Optical monitoring of membrane potential: methods of multisite optical measurement. In: De Weer P, Salzberg BM, eds. Optical Methods in Cell Physiology. New York: Wiley-Interscience, 1986.
5. Davila HV, Salzberg BM, Cohen LB, Waggoner AS. A large change in axon fluorescence that provides a promising method for measuring membrane potential. Nat New Biol 1973; 241:159–160.
6. Salama G, Morad M. Merocyanine 540 as an optical probe of transmembrane electrical activity in the heart. Science 1976; 191:485–487.
7. Morad M, Salama G. Optical probes of membrane potential in heart muscle. J Physiol (Lond). 1979; 292:267–295.
8. Cohen LB, Salzberg BM. Optical measurement of membrane potential. Rev Physiol Biochem Pharmacol. 1978; 83:35–88.
9. Lev-Ram V, Grinvald A. Ca²⁺- and K⁺-dependent communication between central nervous system myelinated axons and oligodendrocytes revealed by voltage- sensitive dyes. Proc Natl Acad Sci U S A. 1986; 83:6651–6655.
10. Loew LM. Spectroscopic Membrane Probes. Boca Raton, Florida: CRC Press, 1988.
11. Clarke RJ, Zouni A, Holzwarth JF. Voltage sensitivity of the fluorescent probe RH421 in a model membrane system. Biophys J 1995; 68:1406–1415.
12. Ross WN, Salzberg BM, Cohen LB, et al. Changes in absorption, fluorescence, dichroism, and birefringence in stained giant axons: optical measurement of membrane potential. J Membr Biol 1977; 33: 141–183.
13. Salama G. Optical Measurements of Transmembrane Potential in Heart. In: Loew LM, ed. Spectroscopic Membrane Probes. Boca Raton, Florida: CRC Press, 1988.
14. Ideker RE, Smith WM, Blanchard SM, et al. The assumptions of isochronal cardiac mapping. Pacing Clin Electrophysiol 1989; 12:456–478.
15. Josephson ME, Horowitz LN, Farshidi A. Continuous local electrical activity. A mechanism of recurrent ventricular tachycardia. Circulation 1978; 57:659–665.
16. Haws CW, Lux RL. Correlation between in vivo transmembrane action potential durations and activation-recovery intervals from electrograms. Effects of interventions that alter repolarization time. Circulation 1990; 81:281–288.
17. Steinhaus BM. Estimating cardiac transmembrane activation and recovery times from unipolar and bipolar extracellular electrograms: a simulation study. Circ Res 1989; 64:449–462.
18. Efimov IR, Huang DT, Rendt JM, Salama G. Optical mapping of repolarization and refractoriness from intact hearts. Circulation 1994; 90:1469–1480.
19. Dillon SM. Optical recordings in the rabbit heart show that defibrillation strength shocks prolong the duration of depolarization and the refractory period. Circ Res 1991; 69:842–856.

20. Li T, Sperelakis N, Teneick RE, Solaro RJ. Effects of diacetyl monoxime on cardiac excitation-contraction coupling. J Pharmacol Exp Ther 1985; 232:688–695.

21. Efimov IR, Mazgalev TN. High-resolution three-dimensional fluorescent imaging reveals multilayer conduction pattern in the atrioventricular node. Circulation 1998; 98:54–57.

22. Wu J, Biermann M, Rubart M, Zipes DP. Cytochalasin D as excitation-contraction uncoupler for optically mapping action potentials in wedges of ventricular myocardium. J Cardiovasc Electrophysiol 1998; 9:1336–1347.

23. Girouard SD, Laurita KR, Rosenbaum DS. Unique properties of cardiac action potentials recorded with voltage- sensitive dyes. J Cardiovasc Electrophysiol 1996; 7:1024–38.

24. Niedergerke R, Orkand RK. The dual effect of calcium on the action potential of the frog's heart. J Physiol (Lond) 1966; 184:291–311.

25. Liu Y, Cabo C, Salomonsz R, Delmar M, Davidenko J, Jalife J. Effects of diacetyl monoxime on the electrical properties of sheep and guinea pig ventricular muscle. Cardiovascular Research 1993; 27:1991–1997.

26. Coulombe A, Lefevre IA, Deroubaix E, Thuringer D, Coraboeuf E. Effect of 2,3-butanedione 2-monoxime on slow inward and transient outward currents in rat ventricular myocytes. J Mol Cell Cardiol 1990; 22:921–932.

27. Biermann M, Rubart M, Wu J, Moreno A, Josiah-Durant A, Zipes DP. Effects of cytochalasin D and 2,3-butanedione monoxime on isometric twitch force and transmembrane action potentials in isolated canine right ventricular trabecular fibers. J Cardiovasc Electrophysiol 1998; 9:1348–1357.

28. Rohr S, Kucera JP. Optical recording system based on a fiber optic image conduit: assessment of microscopic activation patterns in cardiac tissue. Biophys J 1998; 75:1062–1075.

29. Kong W, Walcott GP, Smith WM, Johnson PL, Knisley SB. Emission ratiometry for simultaneous calcium and action potential measurements with coloaded dyes in rabbit hearts: reduction of motion and drift. J Cardiovasc Electrophysiol 2003; 14:76–82.

30. Knisley SB, Justice RK, Kong W, Johnson PL. Ratiometry of transmembrane voltage-sensitive fluorescent dye emission in hearts. Am J Physiol 2000; 279: H1421–H1433.

31. Salama, G. PhD thesis. 1977. University of Pennsylvania.

32. Rohde GK, Dawant BM, Lin SF. Correction of motion artifact in cardiac optical mapping using image registration. IEEE Trans Biomed Eng 2005; 52:338–341.

33. Zhang J, Davidson RM, Wei MD, Loew LM. Membrane electric properties by combined patch clamp and fluorescence ratio imaging in single neurons. Biophys J 1998; 74:48–53.

34. Dillon S, Morad M. A new laser scanning system for measuring action potential propagation in the heart. Science 1981; 214:453–456.

35. Banville I, Gray RA, Ideker RE, Smith WM. Shock-induced figure-of-eight reentry in the isolated rabbit heart. Circ Res 1999; 85:742–752.

36. Zaitsev AV, Guha PK, Sarmast F, et al. Wavebreak formation during ventricular fibrillation in the isolated, regionally ischemic pig heart. Circ Res 2003; 92: 546–553.

37. Sidorov VY, Woods MC, Wikswo JP. Effects of elevated extracellular potassium on the stimulation mechanism of diastolic cardiac tissue. Biophys J 2003; 84:3470–3479.

38. Laurita KR, Rosenbaum DS. Implications of ion channel diversity to ventricular repolarization and arrhythmogenesis: insights from high resolution optical mapping. Can J Cardiol 1997; 13:1069–1076.

39. Rohr S, Salzberg BM. Multiple site optical recording of transmembrane voltage (MSORTV) in patterned growth heart cell cultures: assessing electrical behavior, with microsecond resolution, on a cellular and subcellular scale. Biophys J 1994; 67:1301–1315.

40. Fast VG, Rohr S, Gillis AM, Kleber AG. Activation of cardiac tissue by extracellular electrical shocks: formation of "secondary sources" at intercellular clefts in monolayers of cultured myocytes. Circ Res 1998; 82:375–385.

41. Windisch H, Ahammer H, Schaffer P, Muller W, Platzer D. Optical multisite monitoring of cell excitation phenomena in isolated cardiomyocytes. Pflugers Archiv—European Journal of Physiology 1995; 430:508–518.

42. Grinvald A, Cohen LB, Lesher S, Boyle MB. Simultaneous optical monitoring of activity of many neurons in invertebrate ganglia using a 124-element photodiode array. J Neurophysiol 1981; 45:829–840.

43. Gray RA, Jalife J, Panfilov A, et al. Nonstationary vortexlike reentrant activity as a mechanism of polymorphic ventricular tachycardia in the isolated rabbit heart. Circulation 1995; 91:2454–2469.

44. Fast VG, Kleber AG. Block of impulse propagation at an abrupt tissue expansion: evaluation of the critical strand diameter in 2- and 3- dimensional computer models. Cardiovascular Research 1995; 30: 449–459.

45. Efimov IR, Ermentrout B, Huang DT, Salama G. Activation and repolarization patterns are governed by different structural characteristics of ventricular myocardium: experimental study with voltage-sensitive dyes and numerical simulations. J Cardiovasc Electrophysiol 1996; 7:512–530.

46. Choi BR, Burton F, Salama G. Cytosolic Ca^{2+} triggers early afterdepolarizations and Torsade de Pointes in rabbit hearts with type 2 long QT syndrome. J Physiol 2002; 543:615–631.

47. Rosenbaum DS, Kaplan DT, Kanai A, et al. Repolarization inhomogeneities in ventricular myocardium change dynamically with abrupt cycle length shortening. Circulation 1991; 84:1333–1345.

48. Kanai A, Salama G. Optical mapping reveals that repolarization spreads anisotropically and is guided by fiber orientation in guinea pig hearts. Circ Res 1995; 77:784–802.

49. Morley GE, Vaidya D, Samie FH, Lo C, Delmar M, Jalife J. Characterization of conduction in the ventricles of normal and heterozygous Cx43 knockout mice

using optical mapping. J Cardiovasc Electrophysiol 1999; 10:1361–1375.

50. Tamaddon HS, Vaidya D, Simon AM, Paul DL, Jalife J, Morley GE. High-resolution optical mapping of the right bundle branch in connexin40 knockout mice reveals slow conduction in the specialized conduction system. Circ Res 2000; 87:929–936.

51. Pastore JM, Girouard SD, Laurita KR, Akar FG, Rosenbaum DS. Mechanism linking T-wave alternans to the genesis of cardiac fibrillation. Circulation 1999; 99:1385–1394.

52. Choi BR, Salama G. Simultaneous maps of optical action potentials and calcium transients in guinea-pig hearts: mechanisms underlying concordant alternans. J Physiol 2000; 529 Pt 1:171–188.

53. Laurita KR, Girouard SD, Rudy Y, Rosenbaum DS. Role of passive electrical properties during action potential restitution in intact heart [in process citation]. Am J Physiol 1997; 273:Pt 2):H1205–H1214.

54. Banville E, Gray RA. Effect of Cytochalasin D on Action Potential Duration Restitution. PACE 2000: 23(4-II), 609.

55. Weiss JN, Chen PS, Qu Z, Karagueuzian HS, Lin SF, Garfinkel A. Electrical restitution and cardiac fibrillation. J Cardiovasc Electrophysiol 2002; 13:292–295.

56. Wikswo JP, Lin SF, Abbas RA. Virtual electrodes in cardiac tissue: a common mechanism for anodal and cathodal stimulation. Biophys J 1995; 69:2195–2210.

57. Nikolski V, Efimov IR. Virtual electrode polarization of ventricular epicardium during bipolar stimulation. J Cardiovasc Electrophysiol 2000; 11:605.

58. Efimov IR, Cheng YN, Biermann M, Van Wagoner DR, Mazgalev T, Tchou PJ. Transmembrane voltage changes produced by real and virtual electrodes during monophasic defibrillation shock delivered by an implantable electrode. J Cardiovasc Electrophysiol 1997; 8:1031–1045.

59. Efimov IR, Cheng Y, Van Wagoner DR, Mazgalev T, Tchou PJ. Virtual electrode-induced phase singularity: a basic mechanism of failure to defibrillate. Circ Res 1998; 82:918–925.

60. Gillis AM, Fast VG, Rohr S, Kleber AG. Mechanism of Ventricular Defibrillation: The Role of Tissue Geometry in the Changes in Transmembrane Potential in Patterned Myocyte Cultures. Circulation 2000; 101:2438–2445.

61. Fujii S, Hirota A, Kamino K. Optical recording of development of electrical activity in embryonic chick heart during early phases of cardiogenesis. J Physiol (Lond) 1981; 311:147–160.

62. Hirota A, Fujii S, Kamino K. Optical monitoring of spontaneous electrical activity of 8-somite embryonic chick heart. Jpn J Physiol 1979; 29:635–639.

63. Hirota A, Sakai T, Fujii S, Kamino K. Initial development of conduction pattern of spontaneous action potential in early embryonic precontractile chick heart. Dev Biol 1983; 99:517–523.

64. Sawanobori T, Hirota A, Fujii S, Kamino K. Optical recording of conducted action potential in heart muscle using a voltage-sensitive dye. Jpn J Physiol 1981; 31:369–380.

65. Sakai T, Hirota A, Momose-Sato Y, Sato K, Kamino K. Optical mapping of conduction patterns of normal and tachycardia-like excitations in the rat atrium. Jpn J Physiol 1997; 47:179–188.

66. Efimov IR, Fahy GJ, Cheng YN, Van Wagoner DR, Tchou PJ, Mazgalev TN. High Resolution Fluorescent Imaging of Rabbit Heart Does Not Reveal A Distinct Atrioventricular Nodal Anterior Input Channel (Fast Pathway) During Sinus Rhythm. J Cardiovasc Electrophysiol 1997; 8:295–306.

67. Choi BR, Salama G. Optical mapping of atrioventricular node reveals a conduction barrier between atrial and nodal cells [see comments]. Am J Physiol 1998; 274:H829–H845.

68. Nikolski, V. and Efimov, I. R. Fluorescent imaging of a dual-pathway conduction system of the AV-node. Circulation 2000; 102(18-II):3.

69. Nikolski VP, Jones SA, Lancaster MK, Boyett MR, Efimov IR. Cx43 and the dual-pathway electrophysiology of the AV node and AV nodal reentry. Circ Res 2003; 92:469–475.

70. Dobrzynski H, Nikolski VP, Sambelashvili AT, et al. Site of origin and molecular substrate of atrioventricular junctional rhythm in the rabbit heart. Circ Res 2003; 93:1102–1110.

71. Davidenko JM, Kent PF, Chialvo DR, Michaels DC, Jalife J. Sustained vortex-like waves in normal isolated ventricular muscle. Proc Natl Acad Sci U S A 1990; 87:8785–8789.

72. Davidenko JM, Pertsov AV, Salomonsz R, Baxter W, Jalife J. Stationary and drifting spiral waves of excitation in isolated cardiac muscle. Nature 1992; 355:349–351.

73. Gray RA, Ayers G, Jalife J. Video imaging of atrial defibrillation in the sheep heart. Circulation 1997; 95:1038–47.

74. Gray RA, Jalife J, Panfilov AV, et al. Mechanisms of cardiac fibrillation. Science 1995; 270:1222–1223; discussion 1224–1225.

75. Gray RA, Pertsov AM, Jalife J. Spatial and temporal organization during cardiac fibrillation. Nature 1998; 392:75–78.

76. Girouard SD, Pastore JM, Laurita KR, Gregory KW, Rosenbaum DS. Optical mapping in a new guinea pig model of ventricular tachycardia reveals mechanisms of multiple wavelengths in a single reentrant circuit. Circulation 1996; 93:603–613.

77. Efimov IR, Sidorov VY, Cheng Y, Wollenzier B. Evidence of 3D Scroll Waves with Ribbon-Shaped Filament as a Mechanism of Ventricular Tachycardia in the Isolated Rabbit Heart. J Cardiovasc Electrophysiol 1999; 10:1452–1462.

78. Witkowski FX, Leon LJ, Penkoske PA, et al. Spatiotemporal evolution of ventricular fibrillation [see comments]. Nature 1998; 392:78–82.

79. Valderrabano M, Lee MH, Ohara T, et al. Dynamics of Intramural and Transmural Reentry During Ventricular Fibrillation in Isolated Swine Ventricles. Circ Res 2001; 88:839–848.

80. Bray MA, Wikswo JP. Examination of optical depth effects on fluorescence imaging of cardiac propagation. Biophys J 2003; 85:4134–4145.

81. Allessie MA, Bonke FI, Schopman FJ. Circus movement in rabbit atrial muscle as a mechanism of tachycardia. iii. the "leading circle" concept: a new model of circus movement in cardiac tissue without the involvement of an anatomical obstacle. Circ Res 1977; 41:9–18.

82. Sambelashvili AT, Efimov IR. Bidomain-assisted reconstruction of scroll-wave filament from optical signals. PACE 2003; 26:997.

83. Laflamme MA, Becker PL. Ca^{2+}-induced current oscillations in rabbit ventricular myocytes. Circ Res 1996; 78:707–716.

84. Lakatta EG, Guarnieri T. Spontaneous myocardial calcium oscillations: are they linked to ventricular fibrillation? J Cardiovasc Electrophysiol 1993; 4:473–489.

85. Efimov IR, Rendt JM, Salama G. Optical maps of intracellular $[Ca^{2+}]i$ transients and action potentials from the surface of perfused guinea pig hearts. Circulation 1994; 90(II):632.

86. London B, Baker LC, Lee JS, et al. Calcium-dependent arrhythmias in transgenic mice with heart failure. Am J Physiol Heart Circ Physiol 2003; 284:H431–H441.

87. Johnson PL, Smith W, Baynham TC, Knisley SB. Errors caused by combination of Di-4 ANEPPS and Fluo3/4 for simultaneous measurements of transmembrane potentials and intracellular calcium. Ann Biomed Eng 1999; 27:563–571.

88. Laurita KR, Singal A. Mapping action potentials and calcium transients simultaneously from the intact heart. Am J Physiol Heart Circ Physiol 2001; 280: H2053–H2060.

89. Nikolski V, Efimov I. Fluorescent imaging of a dual-pathway atrioventricular-nodal conduction system. Circ Res 2001; 88:E23–E30.

90. Schaffer P, Ahammer H, Muller W, Koidl B, Windisch H. Di-4-ANEPPS causes photodynamic damage to isolated cardiomyocytes. Pflugers Arch 1994; 426: 548–551.

91. Cheng YN, Mazgalev T, Van Wagoner DR, Tchou PJ, Efimov IR. Voltage-sensitive dye RH421 increases contractility of cardiac muscle. Can J Physiol Pharmacol 1998; 76:1146–1150.

92. Huang D, Swanson EA, Lin CP, et al. Optical coherence tomography. Science 1991; 254:1178–1181.

93. Boppart SA, Tearney GJ, Bouma BE, Southern JF, Brezinski ME, Fujimoto JG. Noninvasive assessment of the developing Xenopus cardiovascular system using optical coherence tomography. Proc Natl Acad Sci U S A 1997; 94:4256–4261.

94. Yelbuz TM, Choma MA, Thrane L, Kirby ML, Izatt JA. Optical coherence tomography: a new high-resolution imaging technology to study cardiac development in chick embryos. Circulation 2002; 106: 2771–2774.

95. Gupta M, Rollins AM, Izatt JA, Efimov IR. Imaging of the atrioventricular node using optical coherence tomography. J Cardiovasc Electrophysiol 2002; 13:95.

96. Jenkins MW, Pederson CJ, Wade RS, et al. Three-dimensional OCT imaging of endocardial architecture. Photonics West: Technical Summary Digest 2004; 61.

97. Jenkins M, Wade RS, Cheng Y, Rollins AM, Efimov IR. Optical Coherence Tomography Imaging of the Purkinje Network. J Cardiovasc Electrophysiol 2005; 16:1–2.

98. Millard AC, Jin L, Lewis A, Loew LM. Direct measurement of the voltage sensitivity of second-harmonic generation from a membrane dye in patch-clamped cells. Opt Lett 2003; 28:1221–1223.

99. Jiang Y, Tomov I, Wang Y, Chen Z. Second-harmonic optical coherence tomography. Opt Lett 2004; 29:1090–1092.

100. Neunlist M, Tung L. Spatial distribution of cardiac transmembrane potentials around an extracellular electrode: dependence on fiber orientation. Biophys J 1995; 68:2310–2322.

101. Knisley SB, Hill BC. Optical recordings of the effect of electrical stimulation on action potential repolarization and the induction of reentry in two-dimensional perfused rabbit epicardium. Circulation 1993; 88:2402–2414.

102. Fast VG, Kleber AG. Microscopic conduction in cultured strands of neonatal rat heart cells measured with voltage-sensitive dyes. Circ Res 1993; 73:914–925.

103. Efimov IR, Aguel F, Cheng Y, Wollenzier B, Trayanova N. Virtual electrode polarization in the far field: implications for external defibrillation. Am J Physiol Heart Circ Physiol 2000; 279:H1055–H1070.

13

Targeted Molecular Imaging in Cardiology

William C. Eckelman

Molecular Tracer, LLC, Bethesda, Maryland, U.S.A.

INTRODUCTION

Targeted molecular imaging is a powerful technique because it leads to measurement of the interaction of a radiolabeled probe with a low density binding site such as receptors and enzymes by non-invasive, external imaging. Experiments at the beginning of the 20th century led to the hypothesis that there must be a specific binding site for a specific biological action. This was applied to both enzymes and receptors. The lock and key analogy of specificity has been in the vernacular since the time of Fischer who in 1895 studied two different enzyme preparations, emulsin and maltase and examined their ability to hydrolyze synthetic glucose derivatives that had been prepared in his laboratory. When one derivative fit emulsin but not maltase and another fit maltase but not emulsin, the lock-and-key analogy was born (1).

The concept of receptor binding ligands and receptors has an equally long history. The interaction of targeted molecules with a specific binding site has been postulated for some time. As early as the 1st century, Lucretius addressed this issue of targeted sites in general by proposing that differences in taste and smell are related to differences in the "pores with some smaller than others, some triangular, some square, some round, and others of various polygonal shapes" (2). The discovery of the odorant receptors and the organization of the olfactory system led to the Nobel Prize in Medicine/Physiology for Axel and Buck in 2004. Modern chemotherapy has been dated to the work of Paul Ehrlich, who at the turn of the century discovered effective agents to treat trypanosomiasis and syphilis. He discovered *p*-rosaniline, which has antitrypanosomal effects, and arsphenamine, which is effective against syphilis. Ehrlich postulated that it would be possible to find chemicals that were selectively toxic for parasites but not toxic to humans. Ehrlich also realized that this same concept was operative in his study of the interaction of antigens with specific, complementary, preformed receptors. Agents that are specific for a particular site were described by him as "magic bullets" (3). In 1906, Langley postulated the existence of "receptive substances" on cell surfaces (4)

and from that point on, this concept defined the action of a minute quantity of substrate on a particular target organ. The use of radiolabeled ligands to measure receptors in vitro has a shorter history. Jensen and Jacobson identified estradiol receptors using high specific activity [^{3}H]estradiol. The use of radiolabeled ligands was slowed by the inability to produce high specific activity radioligands such that the low concentration of isolated receptor would not be saturated, but today there are receptor binding radiotracer for in vitro use with most of the known receptors and their subtypes.

Today, receptor theory in pharmacology has moved past the simple bimolecular reaction. The ternary complex model and the probability model have replaced the bimolecular reaction model. Constitutive activity and reverse agonists have been added to the traditional agonists and antagonists and some 85% of competitive antagonists are now classified as reverse agonists. Such concepts as spontaneously occurring active states, organ selective and ligand selective agonists, auxiliary coupling proteins, allosterism, and receptor dimerization, have been introduced to explain the varied actions of receptors (5).

The practical approach to monitor these low density targets by external imaging came much later as gamma and positron emitting radionuclides and improved imaging devices became available and has been based on the bimolecular model, which for high specific activity radioligands, reduces to a first order reaction. The common radionuclides for single photon emission computed tomography (SPECT) are ^{99m}Tc (half-life = 6 h) and ^{123}I (half-life = 13 h). Radionuclides for positron emission tomography (PET) emit a positron that annihilates to give two 511 keV gamma rays at approximately 180 degrees. The positron-emitting radionuclides (with their half-lives) used most frequently are: ^{15}O (2.07 min), ^{11}C (20.4 min), and ^{18}F (109.7 min). The specific activities (Ci/mmol) of these radionuclides are high because they are made through a nuclear transformation; that is, one element is converted into another so that, except for trace contaminants, they are carrier free. The actual specific activities for the most-used PET radionuclides, ^{18}F and ^{11}C, are of the order of 1000–5000 Ci/mmol at the end of the cyclotron bombardment due to contamination by fluoride and carbon dioxide, respectively. Therefore, these radioactive probes are injected at tracer levels (~2–10 nmol injected) for a 10 mCi dose. The ^{123}I radioisotope of iodine can be obtained at near the theoretical specific activity of 233,700 Ci per mmol and therefore involves the injection of ~50 pmol. The uniqueness of the nuclear medicine technique based on this tracer principle is in measuring biochemistry in vivo, especially the biochemistry of low density sites, such as receptors, by external imaging. PET emphasized neurochemistry in the first efforts in external imaging and therefore most of the early imaging probes for low density sites were targeted to neuroreceptors (Table 1).

WHAT EXPERIMENTS DETERMINE THAT THE MOLECULAR PROBE IS BINDING TO THE TARGET SITE?

We have proposed several paradigms over the years outlining the steps necessary to develop a new target specific radioligand (6–15). One such list contains multiple steps to develop an imaging agent. This systematic approach is based in pharmacology and moves from in vitro to in vivo experiments.

The Choice of a Low Density Binding Site
In the pregenomic era, this choice was made primarily on the autopsy data that show a change in saturable binding site as a function of disease or more indirectly based on the current class of drugs used to treat a specific disease. However, with the advent of genetic linkage studies and genetic screening, a new avenue of radiotracer development has been pursued whereby abnormal genes and abnormal proteins could define the target. The post-genomic technique that is closest to molecular imaging is proteomics. Proteomics is used to identify the protein expression product that changes as a function of disease, just as imaging can identify an altered protein expression. Another approach that has developed in the post-genomic era is systems biology (16). Systems Biology is a scientific discipline that endeavors to quantify all the molecular elements of a

biological system to assess their interactions and to integrate that information into network models that serve as predictive hypotheses to explain changes as a result of disease.

There are both single gene diseases and multiple gene diseases. There are 6000 monogenic diseases, mostly relatively rare, and therefore the proteomics approach is valid (17). For example, the recent success of the tyrosine kinase inhibitor, imatinib, also known as Glivec and Gleevec, is a prime example of the use of proteomics to identify biochemistry that is upregulated in tumors versus normal tissue (18). In that case, the unique BCR/ABL-fusion tyrosine kinase produced from a reciprocal chromosomal translocation event that results in human chronic myelogenous leukemia (CML) was targeted. The upregulation of this tyrosine kinase in CML and gastrointestinal stromal tumors presents a specific biochemical target for chemotherapy. Where imaging likely to be most useful is in dividing phenotypic variations into discrete, non-overlapping categories. However, it is not always straightforward to go from single gene abnormalities to a target for imaging. Balaban has recently set criteria for the special case where the genotype and the phenotype (imaging target) could be linked in a causal relationship (eugenics) (19) if (i) the phenotypic differences are sharply discrete (the area where imaging is mostly like to contribute), (ii) the phenotypic differences are correlated with not more than three gene differences, and (iii) there are no changes due to developmental conditions. The classic example of a human phenotype that does not vary much as a function of the environment is cystic fibrosis. Cystic fibrosis is an example of cell surface protein mislocalization where the cAMP-regulated chloride conductance channel protein is not expressed on the cell surface. One that is very dependent on the environment is phenylketonuria and, in fact, the pathological phenotype can be altered by reducing the intake of phenylalanine.

The systems biology approach looks at multiple components simultaneously. Gropler's group has taken an approach based on systems biology using external imaging (20). They have measured myocardial blood flow (^{15}O[H_2O]), myocardial oxygen consumption (1-[^{11}C]acetate), glucose uptake (1-[^{11}C]glucose), myocardial fatty acid metabolism (1-[^{11}C]palmitate) compared with body mass index in obese women. From these experiments, obesity is a predictor of increased myocardial oxygen consumption and decreased efficiency. Insulin resistance is a predictor of myocardial fatty acid uptake, utilization, and oxidation. The use of multiple imaging probes led to an understanding of the various biochemical parameters involved in obesity and ultimately cardiac performance.

Table 1 "First" Clinical Study Using Receptor Binding Radiotracers

DATE	PET	SPECT
1983	[^{11}C]N-MeSpiperone	[^{123}I]IQNB
1984	[^{18}F]Cyclofoxy	
1985	[^{11}C]Raclopride	[^{99m}Tc]NGA
1985	[^{11}C]Carfentanil	
1985	[^{11}C]Flumazenil	

Use of a Mathematical Model to Choose Potential Receptor-Specific Radiopharmaceuticals: The Estimation of the K_D and Receptor Concentration (B_{max}) for the Target Compound and the Maximal Bound to Free Ratio (B/F)

Having chosen a receptor system based on a disease state, the use of a mathematical model to choose potential receptor-specific radiopharmaceuticals is the key. A simple model has been put forth by Katzenellenbogen (21) and Eckelman (22) as a first approximation. At high specific activity, the maximal B/F ratio will be B_{max}/K_D. Certainly, distribution factors, protein binding, metabolism, and other interactions will decrease the maximal B/F ratio. Therefore, this criterion is necessary but not sufficient. This estimation is especially important, as targeting of receptor systems with pM concentration becomes more prevalent.

Use of 3H-Labeled Compounds to Determine Distribution In Vivo

With the number of ^{3}H-labeled compounds now available commercially, the use of ^{3}H labeled compounds is an efficient method to determine if a particular binding site ligand is suitable for radiolabeling with a gamma emitting radionuclide. At the maximum specific activity of 30 Ci/mmol for one tritium per molecule, the specific activity is sufficient to produce the maximum B/F ratio in cases where the binding site concentration is approximately in nM. The exact analog (fluorinated derivative and, especially, a technetium derivative) is usually not available in the tritiated form and that has to be taken into account. One can expect a weaker K_D with large perturbations in the structure of the parent compound.

The Preparation of the Nonradioactive Analog and Determination of In Vitro and In Vivo Stability

The preparation of the nonradioactive substituted ligand is often the next step. This avoids the problems with no-carrier-added synthesis until the substituted ligand is shown to be a true tracer for the unsubstituted receptor binding ligand, especially in those cases where the substitution of the radionuclide introduces a substantial perturbation. This also produces the necessary reference compounds for the radiolabeled compound. The preparation of the nonradioactive compound better defines the radioactive compound because classical analytical methods can be used, whereas with the high specific activity compound, chromatography is the only analytical tool usually available for identification and quantification although the increasing sensitivity of mass spectrometry is making it a necessity in the modern radiochemistry laboratory. In the current regulatory climate, the non-radioactive compound is also needed to carry out the necessary toxicology and pathology studies.

The Use of Cryopreserved Hepatocytes and Liquid Chromatography/Mass Spectrometry

Most compounds are either chemically unstable and/ or metabolized in vivo. Therefore, the stability of the new derivative should be determined in plasma and in liver. The most convenient model for the enzymatic activity of the liver is commercially available cryopreserved hepatocytes (23). One difficulty with using the nonradioactive compound for stability testing is caused by the large concentrations used, which will mostly likely result in a second order reaction whereas the high specific activity ligand will most likely be in a pseudo first order reaction. This problem has been solved with the use of a combination liquid chromatography/mass spectrometry (LC/MS) analysis where picograms of material can often be analyzed. Another benefit is that an extraction procedure that allows the isolation of unmetabolized parent compound can be developed from the knowledge of the metabolite structures. In summary, there are two advantages of using LC/MS to define the metabolites of a potential radiopharmaceutical: (*i*) the identity of the metabolites are known and therefore questions about the presence of metabolites in the target tissue can be answered, (*ii*) simple extraction procedures can be developed to determine the percentage of parent in the plasma. Such a method allows rapid determination of the parent fraction in plasma and does not require less sensitive, time-consuming chromatographic analysis.

The Evaluation of Various Physical Parameters of the Nonradioactive Derivative (Structure-Distribution Relationship)

Just as structure-activity relationships have been the foundation of classic drug design, structure-distribution studies are the backbone of radiopharmaceutical development. The use of both theoretical and experimental parameters can direct the choice of radioligand. Many of the earlier radiopharmaceuticals were water-soluble, polar compounds that are excreted by the kidneys. As radioligands were developed that cross cell membranes, the properties of these compounds were often correlated with a particular physicochemical property. Most often, the property is lipophilicity determined by either measuring the organic-aqueous partition coefficient or using a related technique such as reversed phase high pressure liquid chromatography. In the radiopharmaceutical context, the percentage dose/g tissue gives a value proportional to the permeability coefficient.

Much of the structure activity relationships are now modeled using various programs using homology modeling. The protein data base is growing rapidly, but is still at only 1–2% of the known proteins, whereas advances in sequence comparison, fold recognition, and protein-modeling algorithms have been instrumental in closing the "sequence-structure gap" (24). For the G-protein-coupled receptors, homology modeling approaches are still based on the bovine rhodopsin structure. These homology models have been able to identify agonists and antagonists of G-protein receptors. Another important aspect of homology modeling is the comparison of different species based on the differences in sequence. Homology modeling can also be used to determine whether the proposed ligands will be substrates for the cytochrome P450s involved in metabolism. Only ~10 hepatic P450s are responsible for 90% of the metabolism of known drugs. Although the crystal structure of all 10 are not known, by homology modeling, the metabolism of various drugs can be proposed.

In Vivo Displacement of the ^3H or ^{125}I Labeled Compound with the Nonradioactive Derivative

If a tritium labeled compound is available for the binding site in question, the ability of the nonradioactive derivative to compete for the receptor should predict the distribution of the radioactive form of the same compound.

Preparation of the Radioactive Derivative and Use of Preinjection, Coinjection, or Postinjection to Decrease Effective Specific Activity of the Radioactive Derivative

Animal distribution studies using the radiolabeled ligand are the most telling experiments. Besides the requirement that the radioactivity is present in the target organ with a target to nontarget ratio of ~2 for PET and higher for SPECT and planar imaging, radioligands for saturable binding sites must also show specific binding. In general, this is carried out by using pre-, co-, or post-injection of a known receptor binding biochemical or drug. Since the definition of a receptor and a receptor binding radiotracer is operational, the distribution of the radioligand must fit the criteria of high affinity, specificity including stereospecificity, saturability, and correlation with biological activity. There are situations where these criteria cannot be tested. This inability to test these criteria occurs if the distribution of the binding site is homogeneous throughout the gray matter or if there are no specific ligands for that binding site from a different chemical class. One solution to this problem of not being able to use the traditional operational definition of a receptor binding radiotracer is the use of stereoisomers or as we discuss in the next section, the use of gene-manipulated mice.

Use of Active and Inactive Stereomers of the Radioactive Analog

The use of an active and inactive stereomers is especially important in human studies where injecting high concentrations of nonradioactive compound either pre-, co-, or post-injection of the radioligand could cause pharmacologic effects. The active and inactive stereomers can both be injected at high specific activity thereby avoiding that problem. The key to use the stereomer pair is to show that indeed the "inactive" isomer represents the free and nonspecific pools only and has the same metabolic profile. Otherwise, it will not be a true control for the active isomer in vivo.

Measurement of Sensitivity of the Radioligand, Most Often in Non-Human Primates

Kinetic sensitivity has been defined as the ability of a physiochemical parameter to alter the time-activity data of a radio-indicator (25). In the context of radiotracers for high affinity sites, the time-activity curve in the target organ should be sensitive to a change in binding site density. As pharmacokinetic analyses search for more accessible methods to analyze time activity curves for reversibly bound ligands, e.g., the Logan plot, then the slopes of such a plot should be sensitive to a change in binding site density. In order to use the more practical single-scan technique (26), the change must be recorded at a single time point. This information is often normalized to the metabolite-corrected plasma concentration of the radiotracer or a reference organ shown to be equivalent, and either the original target organ concentration or the target to nontarget ratios are compared to the receptor concentration obtained by analyzing tissue samples in vitro. A 1991 Nuclear Medicine and Biology editorial stated that "There is a pressing need to study positron emitting- and ^{123}I-labeled single-photon emitting-radiotracers to determine their sensitivity to biochemical changes and match that with diseases that undergo a similar change. This step has been delayed by the incessant development of new ligands that bind specifically in vivo, but are validated no further. This proof of specific binding is a necessary but by no means sufficient step in the complete validation of a probe for a biochemical process. Although the increased sensitivity of PET and SPECT imaging devices and the rapid development of specific biochemical radiotracers is encouraging, only with increased attention to the validation of these biochemical radiotracers will the clinical importance of SPECT and PET be realized" (27).

One of the first biochemical probes, 2-[^{18}F] fluoro-2-deoxyglucose, set a high standard for the development of other probes to be used in vivo.

Using ^{14}C deoxyglucose in studies spanning three decades, Sokoloff and his many colleagues developed a method of measuring glucose utilization in differing behavioral states, seizures, during development, during sensory stimulation, and following administration of drugs. This brilliant integration of science started with the methods for the measurement of cerebral blood flow developed by Kety, including the quantitative autoradiographic procedure using inert diffusible tracers for measuring local metabolic rates. This laid the groundwork for the crucial development: the determination of glucose utilization as a measure of energy metabolism and functional activity. Since the use of radiolabeled glucose necessitated very short measurement times, the metabolically trapped 2-deoxyglucose was studied. This substrate, like glucose, was phosphorylated by hexokinase, but the product could not be converted to fructose-6-phosphate, the next step in the glycolytic pathway. Therefore, 2-deoxyglucose-6-phosphate accumulates in brain to high levels because it is a poor substrate for downstream enzymes present and because glucose-6-phosphatase activity is very low in brain. Thus, 2-deoxyglucose could be used as a tracer for glucose with the autoradiographic technique that had been devised for the local cerebral blood flow method. Drawing on his experience with enzyme kinetics, Sokoloff developed a quantitative model to measure the local glucose utilization. As a result new relationships were revealed and the full constellation and degree of participation of structures simultaneously activated or inhibited as a result of a given behavior or response to stimulation were demonstrated for the first time (28). Shortly thereafter, the method was adapted for use in human subjects and has become the most important technique for measuring brain function by external detection using PET (29).

ACCELERATING THE VALIDATION OF THE PROBE FOR LOW DENSITY SITES USING KNOCK-OUT MICE

Validation of a new radioligand using the pharmacologic definition of site specific binding can be a long process and is neither commensurate with the fast-paced developments in genomics and proteomics, nor does it lend itself to impacting the drug development process (30,31). There are also situations where the pharmacologic approach, outlined above, does not yield a conclusive answer to the question of binding site specificity. The Lasker Award was given to M. Capecchi, M. Evans, and O. Smithies in 2001 for constructing the first knock-out (KO) mice by gene targeting. This occurred only 13 years ago (32). Yet these animals have proven invaluable in validating new radiotracers as ligands for a specific receptor (33). The mice are prepared by replacing the coding sequence for the binding site with a neomycin-resistance cassette in embryonic stem (ES) cells. These ES cells are then microinjected into blastocysts to generate male chimeric offspring, which in turn are mated with female mice. The subsequent generations produced heterozygous (+/−) and homozygous (−/−) mutant mice. These mutant mice (knockout) mice are

then used in experiments where biodistribution of the proposed binding-site-specific radioligand is obtained in groups of wild–type and knockout mice. The difference in binding of the radioligand in wild-type and knockout mice can be attributed to the lack of the specific binding site, all other variables being equal. This has become an important validation of receptor binding radiotracers given that doses of drug to inhibit binding at the specific-binding-site are rarely specific, especially at the doses needed to saturate the binding site. Therefore, knockout mice present a solution to the validation of new radiotracers that is quantitative and requiring the minimal experimentation.

The post-genomic era could lead to 5000 to 10,000 new drug targets whereas, at present, the pharmaceutical companies have investigated ~500 targets, but <100 targets are responsible for the top 100 best-selling drugs in 2001. Despite these advances in the post-genomic era, the approval rate of site directed molecular imaging probes has been low, with the pharmaceutical industry only producing an average of 31 New Molecular Entities (NME) each year over the past nine years. Even fewer of these are breakthrough targets, such as COX2 (cyclooxygenase 2) and PDE5 (phosphodiesterase type 5).

One approach to fewer drug failures because of lack of target specificity is to use KO mice. And in fact, this approach of using KO mice to determine binding site specificity for pharmaceuticals has been used for many years, but just recently documented (34). Just as in the development of new radiotracers, speed is important in the development of pharmaceuticals as well and KO mice lead to faster validation of the binding site than the more traditional pharmacologic approach.

KNOCK-OUT MICE IN CARDIOLOGY

NME in cardiology have been directed toward hypertension, blood coagulation and thrombosis. Knock-out of either angiotensin-converting enzyme (ACE) or the angiotensin receptor AT_1 results in significant decreases in resting blood pressure in mice. These KO mice be used as a milestone for complete blockade and compared to the effect of ACE inhibitors and AT_1 receptor inhibitors (34). KO of the ADP receptor and Factor X have helped to identify the mechanism of action of anti-thrombotic drugs.

Example of the Validation of a Parasympathetic Agent

Gene function and physiology are well conserved between mice and men. The success of the use of KO mice to define the specificity of potential pharmaceuticals argues that with the possible exception of metabolic pathways, the mouse is a good model for predicting human efficacy (34). [^{18}F]FP-TZTP has been used in vivo in several species: rats (35), monkeys (36,37), and mice (37). Its preference for the M2 receptor, evident from the M2 knock out mouse studies (34), and its high brain uptake resulting from its very high first passage extraction (36,37) led to its use in paradigms where the neurotransmitter concentration changes. In the first example, administering physostigmine, an acetylcholinesterase inhibitor assessed the sensitivity of [^{18}F]FP-TZTP binding to changes in monkey brain acetylcholine, via i.v. infusion beginning 30 min before tracer injection. Physostigmine produced a 35% reduction in cortical specific binding, consistent with increased competition from acetylcholine. Decreases in acetylcholine concentration can also be monitored. Clinical studies with [^{18}F]FP-TZTP in combination with positron emission tomography (PET) show a global increase in uptake of radioactivity in elderly versus young control subjects (38). More recently, Cohen et al. (39) have found that those elderly control subjects (52 to 75 years) with the APOE4 gene also had a higher uptake of [^{18}F]FP-TZTP than those without the APOE4 gene. Physostigmine inhibits degradation of acetylcholine via acetylcholinesterase and decreases binding of [^{18}F]FP-TZTP. In those elderly subjects, especially with the APOE4 gene, the acetylcholine concentration is apparently decreased causing increased binding of [^{18}F]FP-TZTP. It is clear the [^{18}F]FP-TZTP is capable of measuring more than receptor number. Rather, it measures the complex interaction between neurotransmitter release and uptake, receptor binding, and metabolism. The application of a muscarinic agonist such as FP-TZTP may be closer to the Systems Biology approach than the Proteomics approach.

THE IMPACT OF IMAGING ON PATIENT OUTCOME

The techniques developed recently have lead to many high throughput paradigms. Whether it is combitorial chemistry or in vitro screening procedures, the capability to prepare and screen large numbers of NME is now available. Likewise, these same approaches can be applied to radiotracers (40). In imaging especially, there must be an emphasis on developing the appropriate molecular probe against the target, but only after it is clear that this approach is likely to affect the management of the patient or the monitoring of drug treatment (41). It is instructive to review the large number of radiotracers that have been developed over the years and decide which of these has had a clinical impact. The answer is relatively few; for drugs the number is estimated to be 1 in 50. Brown and Superti-Furga recently called for rediscovering the sweet spot in drug discovery and the same might be said for molecular imaging (17). They attribute the present

woes to poorly validated targets, failure in lead identification or optimization, or the lead proves to be toxic. Certainly, imaging can play a role in validating the target, optimizing the lead compound, and perhaps in detecting P450 affinity and therefore toxicity.

CONCLUSION

Molecular Imaging's major contribution appears to be in accelerating the process of drug development and in monitoring therapy, especially individualized-medicine approaches. Validation of a new radioligand using the pharmacologic definition of site specific binding can be a long process, and is neither commensurate with the fast-paced developments in genomics and proteomics, nor does it lend itself to impacting the drug development process. In drug development, the validation of the target and optimization of the lead compound can be achieved efficiently and rapidly using a radiolabeled probe and mice with the target protein knocked-out. But it is equally important that cardiovascular scientists concentrate on the choice of an imaging ligand that targets the crucial protein which is a sensitive indicator of a particular cardiovascular disease.

REFERENCES

1. Clardy J. Borrowing to make ends meet. Proc Natl Acad Sci U S A 1999; 96(5):1826–1827.
2. Lucretius. On the Nature of the Universe. Translated by Latham RE. London: Penguin Books, 1951:149–152.
3. Ehrlich P. In: Himmelweit F, Marquardt M, Dale H, eds. Ehrlich P. The assay of the activity of diptheria-curative serum and its theoretical basis. Volume II–IV. London: Pergamon Press, 1957.
4. Langley JM. On nerve endings and special excitable submstances in cells. Proc R Soc B 1906; 8:170.
5. Kenakin T. Principles: Receptor theory in pharmacology. TIPS 2004; 25:186–192.
6. Eckelman WC. Receptor Binding Radiotracers. Boca Raton, FL: CRC Press, Inc. 1982.
7. Eckelman WC. The design of cholinergic tracers in neuro sciences. In: Feindel WF, Frackowiak RSJ, Gadian D, Magistretti PL, Zalutsky MR, eds. Discussions in Neuro Sciences. Netherlands: Foundation for the Study of the Nervous System 1985; 11:60–71.
8. Eckelman WC. Potentials of receptor binding radiotracers. In: Cahn J, Lassen N, eds. New Brain Imaging Techniques in Cerebrovascular Diseases. Paris: John Libbey Euro Text 1985; 2:113–124.
9. Eckelman WC. Receptor Binding Radiotracers, Takeda Science Foundation Symposium on Bioscience. Biomedical Imaging. Hayaishi Torizuka K. Tokyo, Japan: Academic Press, Inc., 1986:357–374.
10. Eckelman WC, Gibson RE. The design of site directed radiopharmaceuticals for use in drug discovery. Boston, MA: Birkhauser Boston, Inc., 1992.

11. Eckelman WC, Gibson RE, Rzeszotarski WJ, et al. The design of receptor binding radiotracers. In: Colombetti L, ed. Principles of Radiopharmacology. New York: CRC Press, 1979; 1:251–274.

12. Eckelman WC, Grissom M, Conklin J, et al. In vivo competition studies with analogues of quinucidinyl benzilate. J Pharm Sci 1984; 73:529–533.

13. Eckelman WC, Reba RC, Gibson RE, et al. Receptor binding radiotracers: A class of potential radiopharmaceuticals. J Nucl Med 1979; 20:350–357.

14. Gibson RE, Burns HD, Eckelman WC. The potential uses of radiopharmaceuticals in the pharmaceutical industry. In: Burns HD, Gibson RE, Dannals RF, Siegl PK, eds. Nuclear Imaging in Drug Discovery, Development and Approval. Boston: Birkhauser, 1992:321–332.

15. Eckelman WC. Mechanism of Target Specific Uptake Using Examples of Muscarinic Receptor Binding Radiotracers. In: Welch MJ, Redvanly C, eds. Handbook of Radiopharmaceuticals: Radiochemistry and Applications. West Sussex: John Wiley & Sons, Ltd., 2002:487–500.

16. Hood L, Heath JR, Phelps ME, Lin B. Systems Biology and New Technologies Enable Predictive and Preventive Medicine. Science 2004; 306:640–643.

17. Brown D, Superti-Furga G. Rediscovery the sweet spot in drug discovery. DDT 2003; 3:1067–1077.

18. Demetri GD, von Mehren M, Blanke CD, et al. Efficacy and safety of imatinib mesylate in advanced gastrointestinal stromal tumors. N Engl J Med 2002; 347:472–480.

19. Balaban E. Eugenics and individual phenotypic variation: To what extent is biology a predictive science? Science 1998; 11:331–356.

20. Peterson LR, Herrero P, Schechtman KB, et al. Effect of obesity and insulin resistance on myocardial substrate metabolism and efficiency in young women. Circulation 2004; 109:2191–2196.

21. Katzenellenbogen JA, Heiman DF, Carlson KE, Lloyd JE. In vitro and in vivo steroid receptor assays in the design of estrogen radiopharmaceuticals. In: Eckelman WC, ed. Receptor Binding Radiotracers. Boca Raton, FL: CRC Press, 1982; 1:93–126.

22. Eckelman WC. Radiolabeled adrenergic and muscarinic blockers for in vivo studies. In: Eckelman WC, ed. Receptor Binding Radiotracers. Boca Raton, FL: CRC Press, 1982; 1:69–92.

23. Ma Y, Kiesewetter D, Lang L, Eckelman WC. Application of LC-MS to the analysis of new radiopharmaceuticals. Mol Imaging Biol 2003; 5(6):397–403.

24. Hillisch A, Pineda LF, Hilgenfeld R. Utility of homology models in the drug discovery process. DDT, 2004; 9: 659–669.

25. Vera DR, Krohn KA, Schiebe PO, Stadalnik RC. Identifiability analysis of an in vivo receptor-binding radiopharmaceutical system. IEEE Trans Biomed Eng 1985; 32:312–322.

26. Carson RE. Mathematical Modeling and Compartmental Analysis. In: Nuclear Medicine, Diagnosis and Therapy Harbert J, Eckelman WC, Neumann R, eds. New York: Thieme Medical Publishers, 1996:167–194.

27. Eckelman WC. The status of radiopharmaceutical research. Int J Rad Appl Instrum B 1991; 18(7):iii–vi.

28. McCulloch J. Mapping functional alterations in the CNS with [^{14}C]deoxyglucose. In: Handbook of Psycho- pharmacology. New York: Plenum press, 1982: 321–410.

29. Reivich M, Kuhl D, Wolf A, Greenberg J, Phelps M, Ido T, Casella V, Fowler, Galliagher B, Hoffman E, Alavi A, Sokoloff L. Measurement of local cerebral glucose metabolism in man with ^{18}F-2-fluoro-2-deoxy-D-glucose. Acta Neurol Scand Suppl 1977; 64:190–191.

30. Eckelman WC. Accelerating drug discovery and development through in vivo imaging. Nucl Med Biol 2002; 29:777–782.

31. Eckelman WC. The use of PET and knockout mice in the drug discovery process. Drug Discov Today 2003; 8:404.

32. Capecchi MR. The new mouse genetics: altering the genome by gene targeting. Trends Genet 1989; 5: 70–76.

33. Eckelman WC, Erba PA, Schwaiger M, Wagner HN Jr, Alberto R, Mazzi U. Postmeeting summary on the round table discussion at the Seventh International Symposium on Technetium in Chemistry and Nuclear Medicine held in Bressanone, Italy on Sept 6–9, 2006. Nucl Med Biol 2007; 34(1):1–4.

34. Zambrowicz BP, Sands AT. Nature Reviews Drug Discovery 2003; 2:38–51.

35. Kiesewetter DO, Lee J, Lang L, Park SG, Paik CH, Eckelman WC. Preparation of ^{18}F-labeled muscarinic agonist with M2 selectivity. J Med Chem 1995; 38(1): 5–8.

36. Carson RE, Kiesewetter DO, Jagoda E, Der MG, Herscovitch P, Eckelman WC. Muscarinic cholinergic receptor measurements with [^{18}F]FP-TZTP: control and competition studies. J Cereb Blood Flow Metab 1998; 18(10):1130–1142.

37. Kiesewetter DO, Carson RE, Jagoda EM, Herscovitch P, Eckelman WC. In vivo muscarinic binding of 3-(alkylthio)-3-thiadiazolyl tetrahydropyridines. Synapse 1999; 31(1):29–40.

38. Podruchny TA, Connolly C, Bokde A, et al. In vivo muscarinic 2 receptor imaging in cognitively normal young and older volunteers. Synapse 2003; 48(1):39–44.

39. Cohen RM, Podruchny TA, Bokde AL, et al. Higher in vivo muscarinic-2 receptor distribution volumes in aging subjects with an apolipoprotein E-epsilon4 allele. Synapse 2003; 49(3):150–156.

40. Bergstrom M, Awad R, Estrada S, et al. Autoradiography with positron emitting isotopes in positron emission tomography tracer discovery. Mol Imaging Biol 2003; 5(6):390–396.

41. Ng JN, Ilag LL. Streamlining drug discovery: finding the right drug against the right target to treat the right disease. DDT 2004; 9:59–60.

14

In-Vivo Physiologic Evaluation of Murine Cardiovascular Phenotypes

Carla J. Weinheimer, Attila Kovacs, Michael Courtois, and Carolyn Mansfield
Center for Cardiovascular Research, Washington University School of Medicine, St. Louis, Missouri, U.S.A.

INTRODUCTION

Although mortality rates due to cardiovascular disease have decreased substantially in the last 20 years, it continues to be the most common threat to life and overall health in the United States. An estimated 20 million people are affected by some form of cardiovascular disease (1), and it is predicted that by 2020 cardiovascular disease will be the world's leading cause of death and disability (2).

Extensive research and the use of new therapeutic techniques have largely contributed to the decline in cardiovascular disease related deaths in the last few years. Historically, laboratory animals have been the primary models for investigating cardiovasuclar disease, and in recent years, this research has largely shifted to mice. Major advances in genetics and molecular biology have opened new doors for understanding cardiovascular disease. It is now clear that similar genes and signaling pathways regulate the heart and vasculature in mice and humans (2). A growing number of murine models have been developed to model human diseases and allow for evaluating the pathophysiologic processes involved. Consequently, these achievements in genetic engineering have created an increased demand for specialized techniques to study the intact mouse (3–5). Specifically designed dissecting microscopes, microsurgical instruments, high resolution imaging devices, and micro-sensors with electrodes have made murine research a viable strategy. To this end, it has become important to have established phenotyping cores that can provide reliable, validated techniques for assessing murine cardiovascular structure and function.

The aim of this chapter is to provide a brief description of the essential equipment, techniques, and common surgical mouse models used in physiologic cardiovascular evaluations in vivo. Because assessing data from such small animals can be challenging, additional considerations related to murine research will be addressed, along with descriptions of techniques used to optimize the extraction of reproducible results.

EQUIPMENT

To perform rigorous physiologic evaluations of the mouse circulatory system in both normal and disease states, to characterize unique phenotypes, and to provide in-depth analysis of data, several state-of-the art surgical, imaging, and hemodynamic methods are routinely employed in mouse physiology laboratories. This section outlines the use and function of some of these modalities.

- A dedicated surgical area containing a stereotactic dissecting microscope, rodent ventilator, warming apparatus, and a variety of microsurgical instruments are essential. Murine cardiac catheterization and sterile surgery is performed in this area. Typically, techniques for survival surgery can be used on mice above 6 grams.
- A cardiac echocardiography machine with a 15 MHz linear transducer used to perform transthoracic imaging to characterize the structure and function of the heart and great vessels of the mouse is employed. Because the average heart rate of a conscious mouse is 550–650 beats/min an echocardiographic machine with a high temporal resolution of 8 ms allows for acquisition rates of up to 18 frames per heart cycle. High spatial resolution and Doppler capabilities are also necessary components for noninvasive assessment of cardiovascular function in mice, particularly with specific models of disease. We have recently defined a method for documenting the degree of pressure overload in mice using Doppler echocardiography. This method will be discussed later.
- A 1.4 French Millar catheter instrument system with customized programs is commonly used for obtaining hemodynamic measurements. This catheter is a delicate tool made of polyimide with a micro-transducer located at the tip. This catheter is routinely placed into the left ventricle of mice through the right carotid artery. Typically, mice as small as 16 grams can be successfully catheterized.

With careful catheter placement and maintenance of basal temperatures, one can obtain measurements of heart rate, left ventricular end diastolic pressure, developed pressure over time, and the index of isovolumic relaxation rate. Using this catheter system, in conjunction with infusions of drugs such as inotropes, provides an exceptional method for assessing systolic and diastolic function in murine models of disease. A 1.4 F catheter that includes, in addition to a micro-pressure transducer, two conductance electrodes for the acquisition of simultaneous LV volume, is also used in many murine physiology cores.

- A commercially available ambulatory, telemetric electrocardiographic system, is used to identify and characterize cardiac rhythm disturbances and blood pressure in the conscious mouse over long periods of time (6,7). This system allows for the surgical placement of a telemetric implant into the subcutaneous space in the back or neck of a mouse. The mouse is then placed back into the cage on a recording receiver, and after recovery, is allowed to move about freely while data is collected.

- A commercially available Oxymax treadmill system is used for evaluating responses to short-term and long-term exercise protocols, while also allowing for determination of changes in respiratory exchange gases (8). Mice are evaluated in both resting and exertion states.

SURGICAL MODELS

Currently, two surgical models of cardiovascular disease are widely used in murine research. These surgical models provide a means of examining the cardiovascular differences in response to a pathologic stress between normal, transgenic, and knockout mice.

Pressure Overload

Pressure overload models are meant to mimic the hypertensive condition in human heart disease and induce LV hypertrophy. Transverse aortic constriction (TAC) is a common method used to create left ventricular pressure overload (3,9,10). During this surgery, a ligation is placed on the aortic arch between the carotid arteries. TAC provides a chronic pressure head in adult mice and produces a 25–30% increase in LV mass. This procedure causes LV dilatation acutely, but after 2–3 days, the heart goes through a compensatory hypertrophy that is maximal at approximately 1–2 weeks. Another method for producing LV hypertrophy is the neonatal ascending aortic banding model (NAB). This procedure has been adapted from the original protocol of Lorrel, et al. (4) and involves creating an aortic constriction in weanling mice at day 20 of life, weighing

only 5–8 grams. To perform this procedure, a loose suture is tied around the ascending aorta. This method causes a slower developing and more clinically relevant hypertrophy as the young mouse grows into the constriction over time, analogous to a patient gradually developing hypertension. The LV hypertrophy produced from this method is more robust, resulting in a 30–60% increase in LV mass.

Both of the above surgical models are excellent for evaluating altered signaling pathways for hypertrophic growth. It is important to note that there is some controversy concerning whether these hypertrophy models typically progress to overt heart failure. This variance may be due to strain, age, and genetic background of the mice studied.

Myocardial Infarction

The condition of acute myocardial infarction in patients has been modeled in mice by ligation of the left coronary artery. This procedure has become an important tool for evaluating a variety of murine phenotypes related to diabetes, metabolic disturbances, and rhythm abnormalities and has been widely utilized (5,11,12). To create this surgical model, a 7–0 suture is passed under the left anterior descending artery and tied proximal to the first bifurcation. Using this procedure, the mid to apical portion of the heart becomes infarcted and begins to remodel over the ensuing weeks. Most mice develop a 20–50% infarcted zone; however, the area affected varies significantly with the constriction location as a result of substantial anatomical variation of the left anterior descending artery among individual mice. To account for this variability, vital dyes can be used to define the area-at-risk. In addition, echocardiography can be used to evaluate infarct size and to study various parameters of post-infarct remodeling (11).

THE USE OF SPECIALIZED PROCEDURES FOR EVALUATING SPECIFIC PHENOTYPES CREATED BY CARDIOVASCULAR SURGICAL MODELS

Hypertrophy

Aortic constriction is an important tool for examining the response of genetically altered mice to a physiologic stress. The time course and structural features in the development of hypertrophy are very different across the two previous outlined models of aortic constriction. Echocardiography performed after surgery has revealed that there is acute dilatation of the left ventricle with segmental and global hypokinesis during the first few days after TAC surgery. As the hypertrophic response develops, the heart begins to remodel and by one week there is normalization of chamber size and systolic function indicating a compensated state of left ventricular hypertrophy.

Neonatal ascending aortic constriction causes a more gradual left ventricular hypertrophy. In contrast to the TAC procedure, serial noninvasive echocardiograms from sham operated and NAB mice demonstrate gradual development of concentric left ventricular hypertrophy as the mouse grows and the constriction becomes tighter. In addition, there is no evidence of acute left ventricular dysfunction. By two months, NAB results in a more robust increase in LV mass and less procedural variability in the amount of hypertrophy achieved. As mentioned previously, the degree to which different techniques of aortic constriction produce LV hypertrophy and/or failure is controversial (13–15). It has been our experience that both TAC and NAB procedures typically result in a compensated, hypertrophic state that does not lead to failure (16). One can also quantify left ventricular performance in hypertrophy models using hemodynamic measurements obtained from catheterizing the left ventricle of mice, coupled with echocardiographic analysis of lightly anesthetized animals.

Defining the degree of aortic constriction: Assessing the degree of aortic constriction is essential for ensuring that the TAC operation was successful and that the LV pressure head is similar in different experimental groups of mice. Figure 1 depicts our method for this examination, using echocardiography. To document the degree of pressure overload imposed on the LV and account for the procedural variability of aortic constriction surgery, Doppler echocardiography can be used to estimate the pressure gradient. This assessment is done by measuring the velocity of blood flow at the site of the constriction and applying the modified Bernoulli

equation. Figure 1 shows blood flow velocity measured in the ascending aorta by continuous wave Doppler in sham operated and NAB mice. Corresponding M-mode echocardiograms show development of concentric LVH in the banded mouse. There is close correlation between the peak velocity of trans-constriction blood flow and the extent of LV hypertrophy in response to the pressure overload. Divergence from this relationship in control groups can be used to evaluate an altered hypertrophic response of the myocardium in various genetically modified mice.

Compensatory hypertrophy: Results of an extensive evaluation of 110 sham-operated C57 wild-type mice compared with 114 TAC-operated C57 wild-type mice are presented in Figure 2. The graphs depict overall chamber parameters measured with echocardiography and hemodynamic analysis 8 weeks after the TAC procedure. As expected, LV weight to body weight and LV peak pressure are significantly increased in the TAC group (34% and 50% increase, respectively) (Fig. 2 top panels). However, as shown in the lower four panels of Figure 2, no evidence of either systolic or diastolic dysfunction was noted in the TAC mice compared to the shams. In addition, there is evidence for a hyperfunctioning heart in TAC animals, in response to the demands of an increased afterload. Systolic function is significantly enhanced as evidenced by a 20% augmentation in peak positive dp/dt in the TAC group (Fig. 2, middle right panel), and there is even a slight increase in diastolic performance of the LV chamber evidenced by a significant 7% decrease in tau, the index of isovolumic relaxation rate (Fig. 2, lower right panel). In this study, the diastolic dimensions were not different

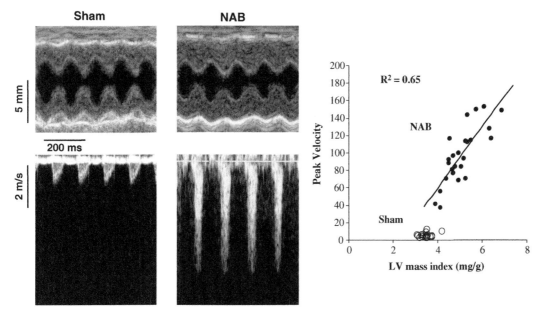

Figure 1   Doppler echocardiographic analysis of the degree of stenosis in FVB mice, receiving either an aortic constriction (*n*=32) or sham operation (*n*=25).   NAB mice show evidence of a linear relationship between hypertrophy and a non-invasively determined pressure gradient.

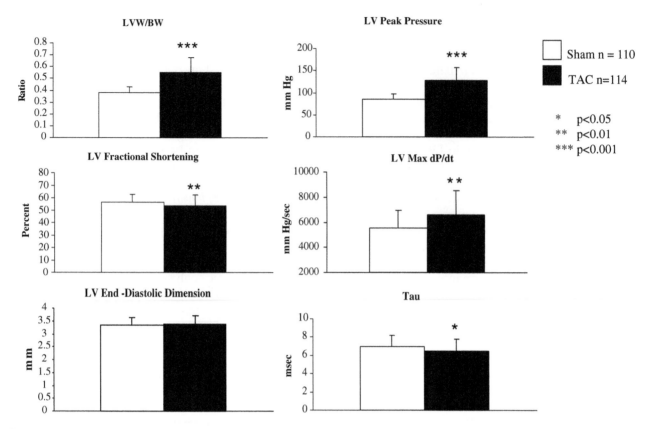

Figure 2 Evidence of compensatory hypertrophy in mice subjected to TAC vs. Sham operations. Echocardiographic data is presented in the left column and hemodynamic data on the right.

between the groups suggesting an absence of chamber dilation. However, the fractional shortening in this group is slightly lower in the TAC animals, probably due to a small number of mice that showed some degree of failure.

Systolic failure: Although LV failure is a rare occurrence in mice following aortic constriction at our institution, some laboratories have reported higher heart failure rates following pressure overload surgery (17,18). In our hands, both hemodynamic and echo-cardiographic data have shown that any tendency toward pressure overload LV failure has only been in small groups of C57 black females. LV weight to body weight and LV peak pressure are significantly increased in the TAC group of these mice (30% and 52%, respectively), and in contrast to compensatory hypertrophy, these small groups of females show trends consistent with systolic dysfunction (data not shown). Fractional shortening was significantly decreased by 23%, as was contractility assessed by peak positive dp/dt. A trend toward chamber dilatation is also noted in TAC mice in these groups. Finally, evidence of diastolic dysfunction is indicated by a significant increase in the value of tau. Taken as a whole, this data again emphasizes the need for rigorous control of strain, as well as gender in experimental designs and models utilizing mice.

Diastolic failure: We have also demonstrated that traditional hemodynamic and echocardiographic techniques applied to mice can categorize diastolic dysfunction independent of alterations in systolic function. Results of two groups of nine mice each of wild-type and transgenic animals are presented in Figure 3. The transgenic mice, genetically altered to over-express a fatty acid transport protein only in the heart, displays a pattern of LV function consistent with isolated diastolic dysfunction (19). The top two panels of Figure 3 indicate normal contractile function and contractility, using the echo parameter of fractional shortening and the hemodynamic index of peak positive dp/dt, respectively. In comparison, the four lower panels of Figure 3 all indicate impaired diastolic function. The Doppler parameters of transmittal flow deceleration time and the ratio of early to late filling flow are consistent with increased diastolic stiffness and impaired chamber compliance. The hemodynamic indices, tau and the ratio of maximum to minimum dp/dt, are consistent with impaired active relaxation of the left ventricular myocardium.

Evaluating cardiac function with pressure volume curves: The conductance catheter system is a recent important development which allows for simultaneous acquisition of left ventricular pressure and volume

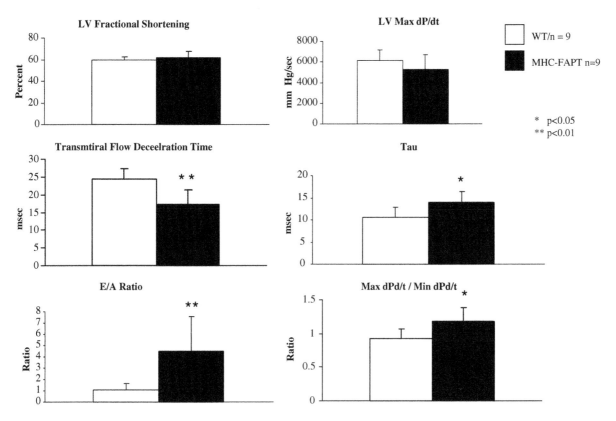

Figure 3 Evidence of impaired diastolic function in mice with preserved systolic function.

data (20). From these data, LV pressure–volume (P–V) loops can be constructed. P–V loops allow for the derivation of a variety of indices of LV chamber function. For example, the end-systolic P–V relation (Ees), shown in Figure 4A, is derived from data recorded during a transient inferior vena caval occlusion, and is considered a load–independent measure of LV contractility. The steepness of the slope of this relation varies with the LV contractile state. In the example, P–V data collected from a genetically altered mouse is compared to its wild-type control. The genetically manipulated animal demonstrates a dramatic increase in end-diastolic volume, reduced ejection fraction, and a decrease in Ees. In addition, measures of LV work, power, and passive diastolic chamber stiffness can be derived from P–V data.

The evaluation of LV contractile reserve can also be an important element in the proper assessment of certain mouse phenotypes. In our laboratory, we routinely evaluate adrenergic contractile reserve. This is accomplished with a graded increase in the infusion of dobutamine delivered intravenously, while assessing LV contractility by the maximum rate of LV pressure increase (peak + dp/dt) obtained from an indwelling LV micromanometric catheter. In Figure 4B, two curves, recorded from a genetically modified mouse and its wild-type control, are displayed. Although not differing

under baseline conditions, a steadily increasing infusion of dobutamine reveals significant differences in contractile reserve between the two animals.

Myocardial Infarction

Surgical ligation of the proximal left anterior descending (LAD) coronary artery has been performed in genetically engineered mice to delineate the role of specific molecular pathways in determining infarct size and post-infarct remodeling (12). Early studies using M-mode images showed that many features of this model closely resemble human myocardial infarction (21,22). Segmental involvement is the characteristic feature of myocardial infarction creating geometrically non-uniform distortion of LV shape. A single-dimensional image (M-mode) or even a 2-D image from 1 imaging plane cannot represent the full extent of involvement. This limitation of M-mode is well demonstrated by the example depicted in Figure 5.

Two-dimensional images using either orthogonal views of the ventricle or sequential parallel short-axis slices can accurately depict the extent and severity of segmental involvement (23) and subsequent changes in LV morphology and function (11). The accuracy of this detection can be improved by use of an intravenous echo-contrast agent, which improves endocardial visualization (24).

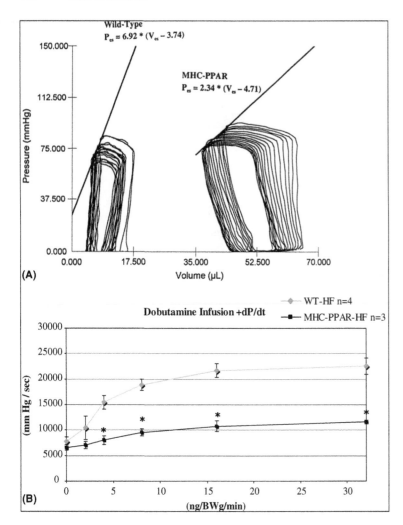

(A)

(B)

Figure 4 (**A**) Left ventricular pressure-volume loops recorded by conductance catheter during transient inferior vena caval occlusion in wild-type and MHC-PPAR mice. Severe LV chamber dilitation and depressed end-systolic P-V relation is evident in these transgenic mice. (**B**) Contractile response to dobutamine infusion assessed by LV peak +dP/dt in wild-type and MHC-PPAR mice both on high fat diet. Transgenic mice exhibit a pattern consistent with significant depression of contractile reserve related to adrenergic stimulation.

IMPORTANT CHALLENGES IN PHYSIOLOGIC MONITORING OF MICE

Accurate in-vivo evaluation of cardiovascular physiology in mice requires close attention to several procedural details. Two very important considerations involve choosing an appropriate anesthesia for the type of measurement being obtained, and determining a true resting condition for assessing changes after an intervention.

Anesthesia

Choosing appropriate anesthesia regimens to evaluate cardiovascular parameters is a crucial challenge to investigators. Several groups have done in-depth studies to determine the best anesthesia for mouse studies (25–27). Variables to consider include length of surgery, size and age of mouse, strain, and parameters to be measured. Maintaining heart rate is the single most important factor to control. Figure 6 is an echocardiographic comparison of three different anesthesia

regimens (avertin, pentothal, and ketamine/xylazine) on the same mouse, studied 1 week apart each time. Note the influence on heart rate and chamber diameter for each different drug. The best answer to the anesthesia dilemma is to choose a drug that has a minimal effect on heart rate and function, provides just enough anesthesia to accomplish the intervention, and is consistent within study groups. For our studies, we routinely use low dose Avertin for echocardiography, Ketamine/Xylazine for lengthy surgical procedures, and Pentothal for hemodynamic studies. Pentothal anesthesia typically yields heart rates of 450–600 beats per minute, which allows for accurate assessments of systolic and diastolic function while the mouse remains deep enough for catheterization.

Determining Resting Conditions

Figure 7 shows the importance of precisely defining the time point at which certain measurements, such as oxygen consumption are taken. In this protocol, mice are placed in a closed non-moving treadmill

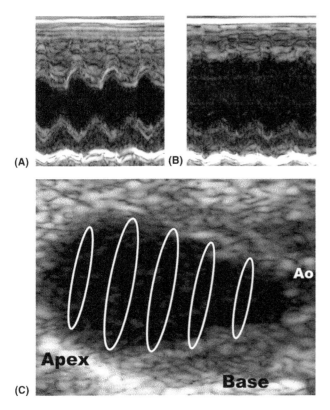

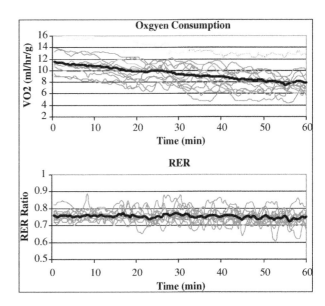

Figure 7   Measurements of oxygen consumption levels and respiratory exchange ratio (RER) in FVB mice (*n* = 12) using an Exer4-Oxymax. Data were obtained under resting conditions.

Figure 5   Echocardiograms of the left ventricle from a mouse 2 weeks after LAD ligation. The image obtained from the parasternal short-axis view by positioning the M-mode cursor immediately left to the LV midline (**A**) shows normal anterior wall motion. In contrast, the image acquired at the same short-axis level by sliding the M-mode line only slightly right to the midline shows markedly diminished anterior wall thickening (**B**). Serial short-axis images from base to apex allow accurate reconstruction of the 3-dimensional structure of the LV (**C**).

system and baseline (resting) measurements are taken to determine oxygen consumption levels and the respiratory exchange ratio (RER). It is difficult to determine resting conditions in a mouse. Figure 7

shows that while the RER is stable from time zero to one hour, the oxygen consumption continues to fall from roughly 11 ml/hr/g to under 8 ml/hr/g in this period. Specific parameters such as resting oxygen consumption are probably not achieved until mice are left undisturbed for over a 1 hr period. In addition, data from Figure 7 emphasizes the need for well-defined inclusion criteria. While 11 of the 12 mice in this study settled down after a 60 minute period and baseline values could be assessed, one mouse (top dotted line) remained agitated by visual assessment and clearly manifested elevated oxygen consumption levels. This is a mouse that should probably be excluded from the study.

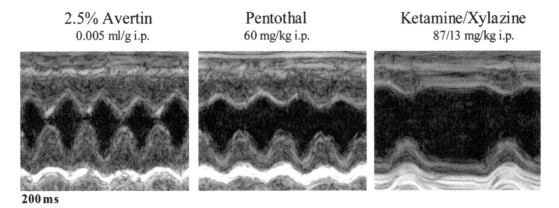

Figure 6   Echocardiographic M-Mode images representing different anesthesia effects in the same FVB mouse. There are significant changes in heart rate and chamber dimensions as the anesthesia regimen varies.s Ketamine/Xylazine causes profound bradycardia and a decrease in ejection fraction.

CONCLUSION

The goal in developing an in-vivo mouse phenotyping core is to generate and evaluate models that have reproducible results with little experimental variation. It is critical that relatively small, individual studies can still produce statistically significant results. The mouse cardiovascular phenotyping core should provide consistent, validated techniques for a systematic approach to study the murine cardiovascular system. This type of facility can optimize survival rates, be cost efficient, and reduce the number of animals required in each study. We have shown that several surgical models can be routinely used to mimic hypertension and myocardial infarction in humans. With highly specialized equipment developed solely for murine analysis, we have been able to examine and quantify left ventricular performance in both physiologic and non-physiologic conditions, using echocardiography and miniaturized catheterization techniques. We have also demonstrated that traditional hemodynamic and echo techniques applied to mice are capable of categorizing diastolic dysfunction independent of alterations in systolic function. While many advances have been made in studying genetically engineered mice, there are several challenges that must be overcome when making in-vivo physiology assessments. We have outlined a few of the common difficulties.

As the number of mouse models for common cardiovascular diseases continue to increase, there will be greater demands for more specialized techniques and procedures to evaluate cardiovascular physiology. Phenotyping cores provide a critical service to biomedical scientists interested in deciphering the molecular mechanisms of cardiovascular diseases, and also provide an important opportunity to validate new in-vivo techniques. These specialized techniques can provide an efficient way to examine a variety of different models to develop more effective therapies for treating patients with cardiovascular disease.

REFERENCES

1. Foster V, Alexander RW, O'Rourke RA. Hurst's the Heart, 10th ed. New York: McGraw Hill, 2001; 3–17.
2. Braunwald E, Zipes DP, Libby P. Heart Disease, 6th ed. Philadelphia: Saunders, 2001; 1–18, 114–1219, 1955–1976.
3. Rockman HA, Knowlton KU, Ross J, et al. In vivo murine cardiac hypertrophy. Circulation 1993; 2:87.
4. Ding B, Price RL, Goldsmith EC, et al. Left ventricular hypertrophy in ascending aortic stenosis mice. Circulation 2001; 101:2854–2862.
5. Nossuli TO, Lakshminarayanan V, Baumgarten G, et al. A chronic mouse model of myocardial ischemia-reperfusion essential in cytokine studies. Am J Physiol 2000; 278:H1049–H1055.
6. Brunet S, Aimond F, Li H, et al. Heterogeneous expression of repolarizing, voltage-gated K^+ currents in adult mouse ventricles. J Physiol 2004; 559:103–120.
7. Chen Y, Joaquim LF, Farah VM, et al. Cardiovascular autonomic control in mice lacking angiotensin AT1a receptors. Am J Physiol Regul Integr Comp Physiol 2004; 10.1152/ajpregu.00231.
8. Leone TC, Lehman JJ, Finck BN, et al. PGC-1alpha deficiency causes multi-system energy metabolic derangements: Muscle dysfunction, abnormal weight control, and hepatic steatosis. PLoS Biol 2005; 3(4):e101.
9. Barger PM, Brandt JM, Leone TC, et al. Deactivation of peroxisome proliferator-activated receptor-alpha during cardiac hypertrophic growth. J Clin Invest 2000; 105:1723–1730.
10. Hasegawa K, Lee SJ, Jobe SM, et al. Cis-acting Sequences that mediate induction of beta-myosin heavy chain gene expression during left ventricular hypertrophy due to aortic constriction. Circulation 1997; 96:3943–3953.
11. Kanno S, Lerner DL, Schuessler RB, Betsuyaku, et al. Echocardiographic evaluation of ventricular remodeling in a mouse model of myocardial infarction. J Am Soc Echocardiogr 2002; 15:601–609.
12. Michael LH, Entman ML, Hartley CJ, et al. Myocardial ischemia and reperfusion: A murine model. Am J Physiol Heart Circ Physiol 1995; 269:H2147–H2154m.
13. Hirota H, Chen J, Betz UA, et al. Loss of a gp130 cardiac muscle cell survival pathway is a critical event in the onset of heart failure during biomechanical stress. Cell 1999; 97(2):189–198.
14. Liao Y, Ishikura F, Beppu S, et al. Echocardiographic assessment of LV hypertrophy and function in aortic-banded mice: Necropsy validation. Am J Physiol Heart Circ Physio 2002; 282(5):H1703–1708.
15. Wang BH, Du XJ, Autelitano DJ, et al. Adverse effects of constitutively active alpha (1B)-adrenergic receptors after pressure overload in mouse hearts. Am J Physiol Heart Circ Physiol, 2000; 279(3):H1079–1086.
16. Zhang S, Weinheimer CJ, Courtois M, et al. The role of the Grb2-p38 MAPK signaling pathway in cardiac hypertrophy and fibrosis. J Clin. Invest 2003; 111:833–841.
17. Esposito G, Rapacciuolo A, Naga Prasad SV, et al. Genetic alterations that inhibit in vivo pressure-overload hypertrophy prevent cardiac dysfunction despite increased wall stress. Circulation 2002; 105(1):85–92.
18. Ito K, Yan X, Feng X, et al. Transgenic expression of sarcoplasmic reticulum Ca^{2+} ATPase modifies the transition from hypertrophy to early heart failure. Circ Res 2001; 89:422–429.
19. Chiu HC, Kovacs A, Courtois M, et al. Perturbation of cardiomyocyte lipid homeostasis causes diastolic dysfunction and electrophysiologic abnormalities. Circ Res 2005; 96:225–233.
20. Georgakopoulas D, Kass D. Assessment of cardiovascular function in the mouse using pressure-volume relationships. Acta Cardiol Sin 2002; 18:101–112.
21. Patten RD, Aronovitz MJ, Deras-Mejia L, et al. Ventricular remodeling in a mouse model of myocardial infarction. Am J Physiol 1998; 274:H1812–1820.
22. Gao XM, Dart AM, Dewar E, et al. Serial echocardiographic assessment of left ventricular dimensions and

function after myocardial infarction in mice. Cardiovasc Res 2000; 45:330–338.

23. Scherrer-Crosbie M, Steudel W, Hunziker PR, et al. Three-dimensional echocardiographic assessment of left ventricular wall motion abnormalities in mouse myocardial infarction. J Am Soc Echocardiogr 1999; 12:834–840.

24. Suehiro K, Takuma S, Cardinale C, et al. Assessment of segmental wall motion abnormalities using contrast two-dimensional echocardiography in awake mice. Am J Physiol Heart Circ Physiol 2001; 280:H1729–H1735.

25. Roth DM, Swaney JS, Dalton ND, et al. Impact of anesthesia on cardiac function during echocardiography in mice. Am J Physiol 2002; 282:H2134–H2140.

26. Zuurbier CJ, Emons VM, Ince C. Hemodynamics of anesthetized ventilated mouse models: Aspects of anesthetics, fluid support, and strain. Am J Physiol 2002; 282:H2099–H2105.

27. Kass DA, Hare JM, Georgakopoulos D. Murine cardiac function- a cautionary tail. Circ Research 1998; 82: 519–522.

15

Quantitative Cardiac PET in Small Animals: Technical Aspects

Michael J. Welch

Division of Radiological Sciences, Mallinckrodt Institute of Radiology, Washington University School of Medicine, St. Louis, Missouri, U.S.A.

INTRODUCTION

A major strength of positron emission tomography (PET) is the ability to noninvasively quantify biological or physiological processes of interest. In the heart, PET techniques have been developed to quantify myocardial perfusion, substrate metabolism, and neuronal function (1–4). These approaches have been used to provide key insights in a variety of normal and abnormal cardiac states such as normal aging, myocardial ischemia, dilated cardiomyopathy, and diabetic heart disease (5–8). To date, these studies have been limited to large animal models of disease and humans. The advent of small animal PET holds the promise of performing the same studies in rodents facilitating our understanding of alterations in gene expression to relevant phenotypes of human disease. However, to fulfill this promise of quantifying cardiac biological or physiological parameters in rats and mice will require the development of new techniques that will overcome the unique challenges of imaging the rodent heart in vivo. In this chapter, we describe various approaches that have been developed in this regard. For example, we show how these methods have been applied to quantify myocardial perfusion and substrate metabolism in the rodent heart.

DEVELOPMENT IN ANIMAL HANDLING TECHNIQUES

The first hurdle in obtaining quantitative data, in particular in mice, is the development of animal handling methods that result in minimal trauma to the animals. Furthermore, in PET studies of the heart, multiple tracer administrations to a single animal are frequently necessary to perform measurements of both myocardial perfusion and the metabolism of various substrates such as oxygen, glucose, and fatty acids. Due to the inherent rapid kinetics of flow and metabolic PET radiopharmaceuticals and the need for proper image alignment, the agents must be administered with the minimum amount of effort and animal movement. To conduct such involved studies, techniques have been developed that employ inhaled anesthetics, physiological monitoring, microsurgical techniques, and special restraining devices (Fig. 1) (9). The use of inhalant anesthetics (e.g., 1–2% Isoflurane) offers the advantages to perform imaging studies of extended duration (~1.5–2 hrs) with minimal manipulation of the animal and stable hemodynamics. Specially designed rodent anesthesia delivery systems have been devised. One such system consists of an induction chamber, which is fed by a dual station inhaled anesthetic system with a single Isoflurane vaporizer. The animal is placed in the induction chamber and allowed to breathe 1–1.5% Isoflurane until unconscious. Following this initial induction, anesthesia is delivered via nose-cone for the maintenance of anesthesia during subsequent microsurgical procedures and imaging sessions (Fig. 1A–D). Physiological monitoring of rodents includes measurements of core body temperature, blood oxygen saturation, hematocrit, heart rate, respiratory rate, and blood pressure using an instrumentation modified for small animals (Fig. 1E). Sterile microsurgery is performed for arterial and venous catheter placement for the purposes of radiopharmaceutical injection, drug administration and blood sampling. A micro-catheter (Harvard Apparatus, Inc.) is routinely used for intravenous injections of multiple radiopharmaceuticals in the jugular vein (Fig. 1F). Prototype micro-columns have been designed and several blood withdrawal techniques have been applied for serial blood sampling during imaging studies (Fig. 1G,H). These measurements are used for measuring plasma radiolabeled metabolites and substrates both of which are critical for applying kinetic models of PET metabolic radiotracers such as 1-^{11}C-glucose and 1-^{11}C-palmitate in the heart (3–7).

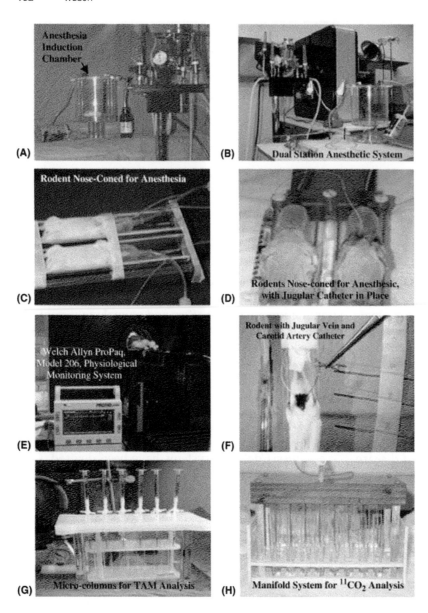

Figure 1 (**A**)The anesthetic setup: induction chamber; (**B**) dual-station inhaled anesthetic system with a single isoflurane vaporizer; (**C**) maintenance anesthesia via nose cone apparatus (**C** and **D**); (**E**) Welch Allyn ProPaq (Salt Lake City, Utal), Model 206, physiological monitoring system and (microPET-R4 Scanner, Concorde Microsystems Knoxville, TN); (**F**) placement of a microcatheter (Harvard Apparatus, Holleston, MA) in a rat jugular vein and carotid artery for blood withdraw; (**G**) prototype microcolumns (AG1-8X-formate resin) used to measure labeled acidic metabolites in blood; (**H**) manifold system used to measure $^{11}CO_2$ in blood. *Source*: From Ref. 9.

QUANTIFICATION OF TRACER KINETICS IN BLOOD AND MYOCARDIUM

Accurate quantification of blood and myocardial activity over time is critical for accurately measuring myocardial perfusion or metabolism using kinetic modeling. The small size and rapid heart rate of rodent hearts, particularly mouse heart, makes this task most challenging. The quantification of the tracer kinetics within the rodent heart utilizing small animal PET and the corrections applied due to confounding factors such as the partial volume and spillover effects have been extensively examined with the use of phantoms. One major challenge in the application of quantitative kinetic models is the measurement of blood time activity curves corrected for tracer metabolites (10). Discussed below are current approaches to perform these measurements as well measurements of myocardial time activity curves in-vivo.

Invasive Methods to Measure the Blood Activity

Beta-probe: Measurements of arterial PET radioactivity have been performed routinely in large animals in-vivo and ex-vivo hearts of rodents using beta probes. The goal of this technique is to detect directly the activity contained within the blood as it passes through the blood vessel. Because positrons emitted by PET radionuclides must travel a short distance in matter prior to annihilation it is possible to measure the positron before the annihilation event occurs using appropriately designed probes. However because of this short distance, the probe must be placed in direct contact with the blood or the blood vessel wall. The former approach offers the advantage of higher efficiency but requires complicated microsurgery that is both time consuming and more invasive. The latter method only requires exposure of the artery, but has much lower sensitivity due to the attenuation of signal by the blood vessel wall. An immediate benefit of using beta probes is the high temporal resolution and sampling method. However, there are numerous disadvantages to this technique. First, optical insulation is required to reduce beta particle signal contamination from ambient light. Second, contamination of the beta particle signal can occurs from background gamma rays (from the positron annihilation). Finally, the size of the probe, when used intravenously, has to be small relative to the blood vessel to prevent hemodynamic perturbation. This technique has been successfully applied in rats in a dynamic measurement of ^{18}F-fluorodexoyglucose in which the probe was inserted in the femoral artery (11). However, it has not been as successful for mouse studies because of the smaller blood vessel dimensions. An alternate design employs two scintillators in phoswich mode. A first fast plastic scintillator is used for beta-ray measurement, and a second slower scintillator such as bismuth germanate measures the gamma-ray component. The scintillation light from both scintillators can be collected using a single photo-multiplier tube and the particle type is determined from a pulse-shape discrimination technique (12). We have evaluated the utility of this approach in both rats and mice by placing the probe along the carotid artery. Although the initial peak of activity in the blood could be detected, gamma contamination was still significant limiting the accuracy of the method (10).

Arterial blood sampling: Rapid arterial blood sampling is typically considered the reference standard for measuring blood activity. In rodents, this approach is more challenging due to the small blood volume of these animals. Our group has developed techniques whereby blood samples as small as 5 μL can be withdrawn from the carotid artery of rodents. To measure the blood time activity curve, radiopharmaceutical is administered via a catheter secured in the jugular or femoral vein and rapid blood sampling at regular intervals occurs via catheter secured in the carotid artery. Blood samples are obtained in pre-weighted micropipette tubes. True counts are determined from the well counter measurements of the tubes and difference in weight of the tubes pre- and post-sampling. In addition, radiolabeled metabolite corrections can be performed on these samples (9,10). High quality curves can be obtained with this method (Fig. 2). The disadvantages of the method include its invasiveness and the need for numerous samples (typically ~20) which results in significant blood loss, particularly in mice. Consequently, this method typically cannot be used when multiple radiopharmaceuticals are to be administered or serial imaging is desired.

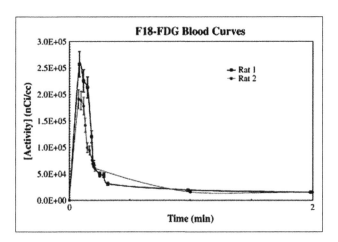

Figure 2 Blood time–activity of normal Sprague–Dawley rats with ^{18}F-fluorodeoxyglucose over 2 min (0.005 ml whole blood, $t = 1$ s, 4–17 s; 1 and 2 min after radiotracer injection). Error bars on the data consist of the statistical error of the counts and the error of blood sample weight. *Source*: From Ref. 9.

Placement of an arterial/venous shunt: This technique requires a catheter surgically inserted between a large artery and vein, typically the carotid artery and the jugular vein. A blood flow probe is attached to a transit time flowmeter and a beta probe is placed within this loop to simultaneously measure blood flow and count rate (Fig. 3). Using this system we have obtained flow values that were consistent with the normal blood flow values in mice as previously reported (13). Arterial/venous shunts also allow for excellent temporal resolution but are technically demanding, requiring extensive animal preparation, complex catheter manipulation, and physiological monitoring. Thus, serial studies in the same animal are not possible with this method.

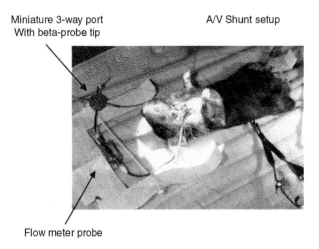

Figure 3 Simultaneous blood flow and blood time–activity curve measurement in an A/V shunt setting. A blood flow value of 0.054 ml/s was measured in mouse and 0.040 ml/s in the rat. *Source*: From Ref. 10.

Non-Invasive Methods to Measure the Blood and Myocardial Activity

Factor Analysis: In the 1980's Factor Analysis (FA) was first used for nuclear medicine studies and subsequently applied to PET to generate cardiac blood and myocardial time activity curves in canine and human studies (14–17). Using this approach both curves are derived from dynamic images without the need to place regions of interest. The FA method is based on the principle that the intensity of a given image voxel is the linear superposition of counts from several compartments which represent different anatomic structures. The method also assumes all variables within a compartment are highly correlated whereas there is a weak correlation with members of other compartments. The FA method uses four consecutive steps: *i*) data processing, *ii*) principal compartmental analysis, *iii*) determining the factor axes, and *iv*) extraction of time-activity curves. Finally, the FA blood time activity curves are normalized based on a single blood sample obtained at the end of the data collection. We and others have shown that FA can be used to extract blood and myocardial time activity curves for a variety of radiopharmaceuticals such as ^{15}O-water, ^{18}F-fluorodeoxyglucose, and ^{11}C-acetate using small animal PET (18,19). In rat heart, we can extract the three factor images (right ventricular cavity, left ventricular cavity, and left ventricular myocardium) and generate time–activity curves for each factor image using the FA method (Fig. 4). The FA blood time activity curve agreed well with that obtained

from blood sampling in rat heart (Fig. 5). However, the slope of the regression was ~25% greater than unity suggesting overestimation with the FA method. In mice, the blood time activity curve extracted from the right ventricle factor agreed well with that obtained from direct blood sampling. However, the blood time–activity curve extracted from the left ventricle factor was contaminated by myocardial activity likely owing to the small size of the mouse heart.

Use of regions of interest and ECG-gating: Deriving blood and myocardial time activity curves from region of interest placement on dynamic PET images are an attractive method approach because of its noninvasive nature. In the case of measuring blood activity the imaging protocol is easier when compared with direct blood sampling or using a beta probe. Direct measurement from the images by placing a region of interest in the left ventricle, left atrium, or the ascending aorta is commonly utilized in human studies (1,20–22). Both blood and myocardial images need to be corrected for partial volume and spillover effects. Partial volume effects reflect the underestimation of tracer activity in small (or moving) structures. Spillover represents the overestimation of activity in a structure due to high activity in an adjacent structure. In both cases, these phenomena are related to the finite spatial resolution of PET scanners and the size and movement of structures being imaged. Temporal resolution depends on the ability to create short frames during the dynamic sequence. The use of short dynamic frames will

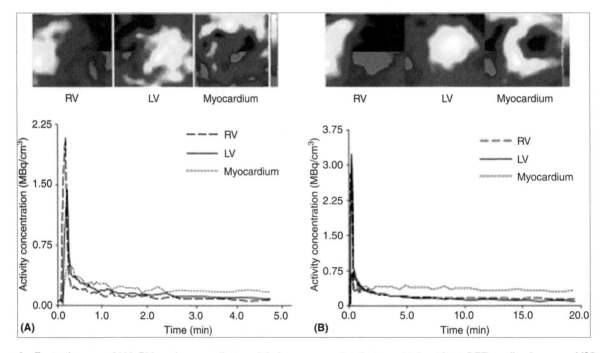

Figure 4 Factor images of LV, RV, and myocardium and their corresponding factors obtained from PET cardiac images of ^{15}O-water (**A**) or 1-^{11}C-acetate (**B**) from rats studied at rest. *Source*: From Ref. 31.

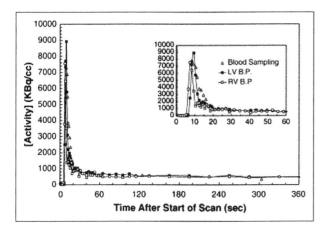

Figure 5 Blood time–activity curves extracted from rat heart images. Forty-six megabecquerel (1.26 µCi) of ^{18}F-fluorodeoxyglucose was injected. The inset shows an expanded view of the early time point illustrating the temporal resolution of the scanner, which allows the separation of the RV from the LV blood input functions. *Source*: From Ref. 10.

decrease the number of counts per frame, which leads to higher statistical uncertainty of the data. For these reasons, extracting the blood and myocardial curves using this method in rat and particularly mouse heart, is difficult. However, despite these difficulties it should be noted that myocardial ^{18}F-fluorodeoxyglucose activity calculated on static images representing 60 min post-radiopharmaceutical administration and corrected for partial volume effects using a phantom correction correlated with values obtained from well-counting (23). Moreover, myocardial blood flow (MBF) has been measured at rest and during pharmacological stress in rats using small animal PET and ^{13}N-ammonia based on region of interest analysis (24).

One approach to overcome these limitations is to perform ECG-gating during the PET acquisition. Dividing the heart cycle into several time bins (typically 8–16) in order to extract the near end-diastolic phase is an attractive technique as it allows the reconstruction of cardiac images for which the left ventricular chamber has the maximum volume.

This minimizes the spillover effect from activity in the myocardium into the left ventricular blood pool corrupting the blood time activity curve. Correction for the remaining partial volume effect of this curve is easily accomplished by normalization with a single blood sample taken at the end of the scan. ECG-gating has been used measure left ventricular function in rats and potentially in mice (25,26). Unfortunately, ECG-gating does not appear to improve the quality of the myocardial time activity curves. Further improvements in obtaining accurate blood and myocardial curves can be achieved by applying advanced image reconstruction algorithms to the ECG-gated images. An example is shown in Figure 6. This figure presents a comparison of a coronal slice through the mouse heart (0.4 mm thickness) of data reconstructed with filtered back projection (A: FBP), 2D order subset estimation maximization (B: 2D-OSEM) and maximum a posteriori (C: MAP) of the un-gated data sets (27–30). Figure 6D shows the MAP reconstruction of the end-diastolic phase. Figure 7 shows the plot of

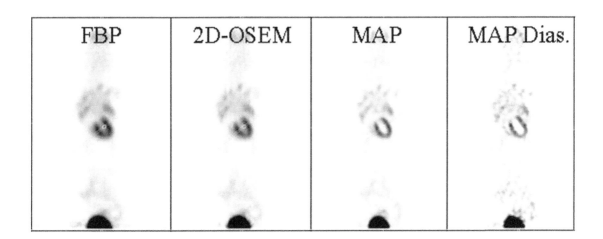

Figure 6 ^{18}F-fluorodeoxyglucose images of a BALB/C mouse obtained 3 hours post-injection of FDG imaged on the microPET FOCUS 120 scanner using different reconstructive techniques.

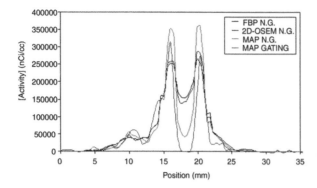

Figure 7 Activity profiles of the four images shown in Figure 6.

the profiles drawn through the myocardium of the images in Figure 6. The curves corroborate the images demonstrating the reduced spillover from myocardium to blood with appropriate reconstruction algorithms and ECG-gating.

QUANTIFICATION OF MYOCARDIAL BLOOD FLOW AND METABOLISM IN RODENTS

Once the technical aspects for quantifying blood and myocardial activity have been optimized it is then possible to apply well validated kinetic models to measure biologically relevant parameters such as MBF and myocardial metabolism. Discussed below are two such examples.

MBF: Alterations in MBF are central to normal cardiac performance and play key role in a variety of cardiac disorders such as myocardial ischemia, left ventricular hypertrophy and heart failure. Using well-validated compartmental modeling approaches MBF has been quantified accurately with PET using a variety of radiopharmaceuticals such as ^{15}O-water, ^{13}N-ammonia, and ^{11}C-acetate, in both large animals and humans. Recently, we applied some of these approaches to quantify MBF in rats. Measurements of MBF were performed with PET and ^{15}O-water using the same kinetic modeling approach as is performed in humans (31). Blood and myocardial time activity curves were generated with FA. Over wide range in flow values

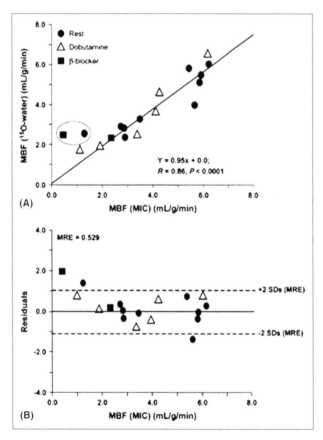

Figure 8 Correlation between MBF estimates shown in Figure 2 after forcing the intercept to zero (**A**) and the corresponding residual plot (**B**). MRE was calculated as the average of the absolute difference between measured and predicted MBF values from the regression analysis in A **Note:** •, resting perfusion; △, *Note*: perfusion during dobutamine; ■, perfusion during propranolol infusion. *Abbreviations*: MBF, myocardial blood flow; MIC, radiolabeled microspheres; MRE, mean residual error. *Source*: From Ref. 31.

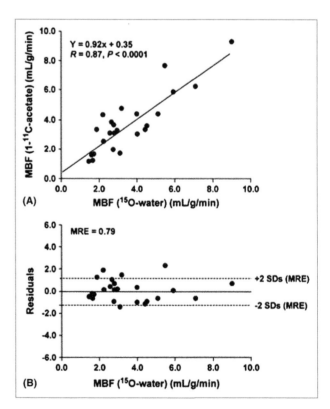

Figure 9 Correlation between MBF estimated from 1-^{11}C-acetate and MBF estimated from ^{15}O-water (**A**) and the corresponding residual plot as a function of ^{15}O-water MBF (**B**). No significant residual bias was observed. MRE was calculated as the average of the absolute difference between measured and predicted MBF values from the regression analysis in A. *Abbreviations*: MBF, myocardial blood flow; MRE, mean residual error. *Source*: From Ref. 31.

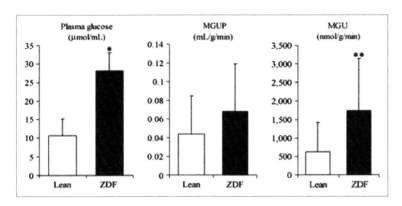

Figure 10 Glucose metabolism measurements obtained in ZDF and lean rats by compartmental modeling of 1-^{11}C-glucose PET data. MGUP 5 myocardial glucose uptake; MGU 5 myocardial glucose utilization. *Note*: *P < 0.0001; **P < 0.06. *Abbreviation*: ZDF, Zucker Diabteic Fatty rats. *Source*: From Ref. 32.

MBF values calculated by PET correlated closely with those obtained with radioactive microspheres (Fig. 8). In a separate experiment, MBF was measured with PET and ^{11}C-acetate and values were compared with those obtained with ^{15}O-water. Again FA was used to generate blood and myocardial time activity curves. The ^{11}C-acetate blood curve was corrected for $^{11}CO_2$ activity. Partial volume and spillover corrections from the ^{15}O-water study were used to correct the blood and myocardial curves. An one-compartment model was used to estimate K1, the rate constant describing the uptake of ^{11}C-acetate. This value was converted to MBF based on previously validated relationship between K1 and MBF for this tracer. There was good agreement in MBF values between the two methods (Fig. 9). Thus, it is possible to quantify MBF in rat heart using small animal PET. Translation of these findings to mouse will

require an additional stage of validation or improved resolution accomplished with either improved scanner or reconstruction algorithms.

Metabolism: Quantification of myocardial substrate metabolism has also been accomplished in rats using small animal PET. Measurements of MBF, oxygen consumption, glucose metabolism, and fatty acid metabolism were performed using ^{15}O-water, ^{11}C-acetate, ^{11}C-glucose, and ^{11}C-plamitate, respectively. The radiopharmaceutical administrations and their associated data collections were performed consecutively in Zucker Diabteic Fatty (ZDF) rats, a model of type-2 diabetes mellitus (32). Values were compared with those obtained in lean non-diabetic litter mates. The obese ZDF rats showed higher plasma hemoglobin A1c, insulin, glucose, and free fatty acid (FFA) levels than the age matched controls. Fractional myocardial glucose

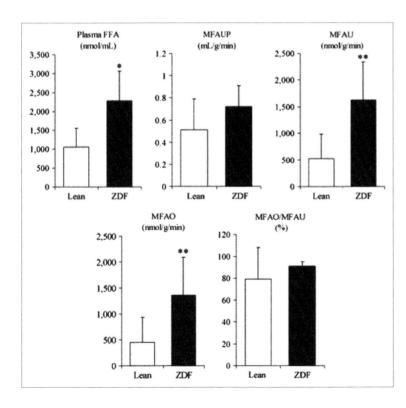

Figure 11 FFA metabolism measurements obtained in ZDF and lean rats by compartmental modeling of 1-11Cpalmitate PET data. MFAUP 5 myocardial fatty acid uptake; MFAU 5 myocardial fatty acid utilization; MFAO 5 myocardial fatty acid oxidation; MFAO/MFAU 5 myocardial FFA that was oxidized. *Note*: *P < 0.001; **P < 0.01. *Abbreviations*: FFA, free fatty acid; ZDF, Zucker diabteic fatty rats. *Source*: From Ref. 32.

uptake was not significantly different between the two groups. However, higher glucose plasma levels in the ZDF animals resulted in higher total myocardial glucose uptake quantified in nmol/g/min (Fig. 10). Similarly, fractional myocardial fatty acid uptake was not significantly different between the two groups. However, due to the higher FFA plasma levels total fatty acid uptake as well as fatty acid oxidation (again quantified in nmol/g/min) were significantly higher in the obese ZDF group (Fig. 11). The increase in myocardial fatty acid metabolism preceding the decline in myocardial glucose use is consistent with early diabetic phenotype for this model. As discussed above, further advances in instrument design and/or image reconstruction algorithms are necessary before these measurements can be obtained in mouse heart.

CONCLUSION

The convergence of rapid advances in small animal PET with respect to instrumentation design, image reconstruction algorithms, animal handling, and monitoring techniques will likely accelerate the capability quantitative measurements of a variety cardiac parameters such as myocardial perfusion, metabolism, and neuronal function in rodents. Most importantly, these measurements will become possible in mouse heart. Once this milestone is achieved the promise of seamless bi-directional translational research between bench and bedside will move closer to reality.

ACKNOWLEDGMENT

This work was supported by Grant 5PO1HL13851 from the National Institutes of Health.

REFERENCES

1. Bergmann SR, Herrero P, Matkham J, et al. Noninvasive quantitation of myocardial blood flow in human subjects with oxygen-15-labeled water and positron emission tomography. J Am Coll Cardiol 1989; 14:639.

2. Herrero P, Sharp TL, Dence C, Haraden BM, Gropler RJ. Comparison of 1-^{11}C-glucose and ^{18}F-FDG for quantifying myocardial glucose use with PET. J Nucl Med 2002; 43:1530–1541.

3. Bergmann SR, Weinheimer CJ, Markham J, Herrero P. Quantitation of myocardial fatty acid metabolism using positron emission tomography. J Nucl Med 1996; 37:1723.

4. DeGrado TR, Hutchins GD, Toorongian SA, Wieland DM, Schwaiger M. Myocardial kinetics of carbon-11-metahydroxyephedrine: Retention mechanisms and effects of norepinephrine. J Nucl Med 1993; 34: 1287–1293.

5. Kates AM, Herrero P, Dence C, et al. Impact of aging on substrate metabolism by the human heart. J Amer Col Card 2003; 41:293–299.

6. Lee HH, Davila-Roman VG, Walsh JF, et al. The dependency of contractile reserve on myocardial blood flow: Implications for the assessment of myocardial viability by dobutamine stress echocardiography. Circulation 1997; 96:1885.

7. Davila-Roman VG, Vedala G, Herrero P, et al. Altered fatty acid and glucose metabolism in idiopathic dilated cardiomyopathy. J Amer Col Card 2002; 40:271–277.

8. Herrero P, Peterson LR, McGill JB, et al. Increased myocardial fatty acid metabolism in patients with type 1 diabetes mellitus. J Am Coll Cardiol 2006; 47:598–604.

9. Sharp TL, Dence CS, Engelbach JA, Herrero P, Gropler RJ, Welch MJ. Techniques necessary for multiple tracer quantitative small-animal imaging studies. Nucl Med Biol 2005; 32:875–884.

10. Laforest R, Sharp TL, Engelbach JA, et al. Measurement of input functions in rodents: Challenges and solutions. Nucl Med Biol 2005; 32:679–685.

11. Pain F, Laniece P, Mastrippolito R, Gervais P, Hantraye P, Besret L. Arterial input function measurement without blood sampling using a β-microprobe in rats. J Nucl Med 2004; 45:1577–1582.

12. Yamamoto S, Tautani K, Suga M, Minato K, Watabe H, Iida H. Development of a phoswich detector for a continuous blood-sampling system. IEEE Trans Nucl Sci 2001; 48:1408–1411.

13. Krivitsk NM, Starostin D, Smith TL. Extracorporeal recording of the mouse hemodynamics parameters by ultrasound velocity dilution. ASAIO J 1999; 45:32–36.

14. Barber DC. The use of principal components in the quantitative analysis of gamma camera dynamic studies. Phys Med Biol 1980; 25:283–292.

15. Di Paola R, Bazin JP, Aubry F, et al. Handling of dynamic sequences in nuclear medicine. IEEE Trans Nucl Sci 1982; NS29:1310–1321.

16. Wu HM, Hoh CK, Choi Y, et al. Factor analysis for extraction of blood time–activity curves in dynamic FDG-PET studies. J Nucl Med 1995; 36:1714–1722.

17. Wu HM, Hoh CK, Buxton DB, et al. Quantification of myocardial blood flow using dynamic nitrogen-13-ammonia PET studies and factor analysis of dynamic structures. J Nucl Med 1995; 36:2087–2093.

18. Bentourkia M, Lapointe D, Selivanov V, Buvat I, Lecomte R. Determination of blood curve and tissue uptake from left ventricle using FADS in rat FDG-PET studies. Paper presented at: 1999 IEEE Nuclear Science Symposium and Medical Imaging Conference; October 24–30, 1999.

19. Kim J, Herrero P, Sharp T, et al. Minimally invasive method of determining blood input function from PET images in rodents. J Nucl Med 2006; 47:330–336.

20. van der Weerdt AP, Klein LJ, Boellaard R, Visser CA, Visser FC, Lammertsma AA. Image-derived input functions for determination of MRGlu in cardiac ^{18}F-FDG PET scans. J Nucl Med 2001; 42:1622–1629.

21. Gambhir SS, Schwaiger M, Huang SC. Simple noninvasive quantification method for measuring myocardial glucose utilization in humans employing positron emission tomography and fluorine-18 deoxyglucose. J Nucl Med 1989; 30:359–366.

22. Watabe H, Channing MA, Riddell C, et al. Non-invasive estimation of the aorta input function for measurement of tumor blood flow with [^{15}O] water. IEEE Trans Med Imaging 2001; 20:164–174.

23. Handa N, Magata Y, Tadamura E, et al. Quantitative fluorine 18 deoxyglucose uptake by myocardial positron emission tomography in rats. J Nucl Cardiol 2002; 9:616–621.

24. Croteau E, Benard F, Bentourkia M, Rousseau J, Paquette M, Lecomte R. Quantitative myocardial perfusion and coronary reserve in rats with ^{13}N-ammonia and small animal PET: Impact of anesthesia and pharmacologic stress agents. J Nucl Med 2004; 45:1924–1930.

25. Croteau E, Benard F, Cadorette J, et al. Quantitative gated PET for the assessment of left ventricular function in small animals. J Nucl Med 2003; 44:1655–1661.

26. Yang Y, Rendig S, Siegel S, Newport DF, Cherry SR. Cardiac PET imaging in mice with simultaneous cardiac and respiratory gating. Phys Med Biol 2005; 50:2979–2989.

27. Qi J, Leahy RM, Cherry SR, Chatziioannou A, Farquhar TH. High resolution 3D Bayesian image reconstruction using the microPET small-animal scanner. Phys Med Biol 1998; 43:1001–1013.

28. Chatziioannou A, Qi J, Moore A, et al. Comparison of 3-D maximum a posteriori and filtered backprojection algorithms for high-resolution animal imaging with microPET. IEEE Trans Med Imaging 2000; 19:507–512.

29. Rowland DJ, Laforest R. FDG image derived input function measurement in rats and mice from left ventricle. J Label Comp Radiopharm 2005; 48 [Abstract to the XVI International Symposium of RadioPharmaceutical Chemistry]:S312.

30. Rowland DJ, Newport DF, Laforest R, Tai YC, Welch MJ. Respiratory cardiac gating on the microPET scanner. Mol Imaging Biol 2003; 5:124.

31. Herrero P, Kim J, Sharp TL, et al. Assessment of myocardial blood flow using ^{15}O water and 1-^{11}C acetate in rats with small animal PET. J Nucl Med 2006; 47:477–485.

32. Welch MJ, Lewis JS, Kim J, et al. Assessment of myocardial metabolism in diabetic rats using small animal PET: A feasibility study. J Nucl Med 2006; 47:689–697.

16

Overview of the Potential Role of Small Animal PET

David Stout

UCLA Department of Molecular and Medical Pharmacology, Crump Institute for Molecular Imaging, Los Angeles, California, U.S.A.

INTRODUCTION

A New Breed of PET

In 1997, the small animal positron emission tomography (PET) system was built at UCLA by Dr Simon Cherry et al. (1). This giant step forward in both higher resolution and sensitivity opened up research into mice and rats, species that are readily available and well understood by the medical community. Previously, small animals were considered to be 1–9 Kg monkeys, rabbits, or small dogs, which due to their cost and handling complexities meant that relatively few experiments were possible. The small volumes and radioactivity levels needed for mice, combined with the high sensitivity of the microPET scanners, have made it possible to image over a dozen mice using the same radiochemical production run. This increase in imaging efficiency has allowed for either more animals per experiment or for multiple investigators to image their animals each day, increasing the ability to meet rising demands for imaging time.

The initial design of the small animal PET systems has been steadily improved upon, with systems now available from several different companies. These newer systems offer substantially greater sensitivity and better resolution than initial prototypes (2). These improvements are opening up the mouse and rat to research only recently possible in larger animals using clinical systems.

RADIOCHEMISTRY

With high resolution and sensitivity systems now available, the emphasis has started to shift to the radiochemist and the development of appropriately labeled compounds for the biological process under investigation. This is not a trivial task as it can take considerable time to select the right compounds, develop the right chemical precursors and synthesis routines. A new approach is emerging where the biochemical pathway under study is critically examined before a particular compound is chosen to target a specific point in the biochemistry (3).

There are considerable demands placed on the radiochemical synthesis, since the volume of injectable solution in a mouse is limited to ~0.25 mL. This requires that production runs must result in high yields and in very small volumes. For research related to molecular targets in extremely low tissue concentrations, there is the added constraint of requiring high specific activity in the final product. This is necessary so as not to occupy too many target molecules with cold compound and possibly evoke a pharmacological effect or to disturb the idea of imaging with only a trace amount of the compound.

To overcome these radiolabeling challenges, there are already companies who supply radiolabeled compounds to hospitals and research centers for PET imaging at facilities without cyclotrons. Indeed, there are currently about 85 PET radiopharmacies in the United States. In addition, there is considerable promise in the near future for developments in radiochemistry coming from the field of nanotechnology (4). It may soon be possible to quickly and efficiently produce needed compounds by simply choosing the right chip and waiting a few minutes, rather than today's typical 1–3 hours for F-18 related production run. This kind of advance coupled with the availability of off-site radiopharmaceutical delivery will likely accelerate small animal PET research by a new range of investigators who cannot afford the substantial costs of a cyclotron and radiochemical support staff.

IMAGE RECONSTRUCTION SOFTWARE

Computational methods used to create images out of the small animal PET data have progressed to the point where they can improve the image resolution and noise characteristics considerably over the old standard of filtered back projection. Routines such as ordered subset estimated maximization) and maximum a priori (MAP) image reconstruction algorithms have dramatically improved our ability to obtain detailed information from the images (5). With the increases in computational speed, routine processing of images using these methods are now possible.

ANIMAL MODELS

The ability to image mice and rats enables small animal PET research to operate using the species best understood by the biological and pharmacological research community. Moreover, the development of genetically manipulated mouse models has greatly furthered our understanding of the interrelationship between alterations in gene expression and specific biologic and disease processes. Small animal PET is now being used to accurately characterize non-invasively the phenotype of these new models or measure the response to specific interventions greatly enhancing the efficiency of such studies (6,7). There are, however, instances where certain animals models such as monkeys are best suited to research, as in the case of Parkinson's disease (8). In this case, primates are better matched for humans in terms of behavioral and biochemical markers of the disease. For this and other reasons, there will always remain a need for larger animals in PET research. There are small animal PET scanners that can be used to image larger animals up to an outer diameter of 16 cm.

DYNAMIC IMAGING

Another recent development with small animal PET systems is the ability to acquire all the data before processing into images. This sounds simple, but until recently, previous incarnations of PET systems required that the exact timing of each image be determined prior to the imaging procedure. Now, images can be tailored to the exact needs of the investigator and to match the individual characteristics of the imaging experiment. For example, the same data can be combined to make a total, or summed image, to best show all the data over time at a glance. Alternately, the data can be split into tiny durations to observe the filling of the heart chambers, lungs, and out to the kidneys. It is also possible to gate the images for signals such as respiration and heartbeat. This added flexibility further increases the informational content one can obtain from a single study, and can sometimes lead to new and unexpected findings.

SOME EXAMPLES OF CURRENT RESEARCH

Mouse Atlas
The interpretation of radiolabeled images can be a challenge, since often investigators need to relearn their anatomy as seen through the filters of the imaging systems used to obtain in-vivo data. The interpretation becomes increasingly challenging when multi-modality imaging is used requiring the fusion or registration of images with different spatial resolution and localizing capabilites. To assist investigators, mouse atlases have been created using technologies such as microCAT, microPET and referenced to mouse anatomy characterized with a cryostat (Fig. 1) (9).

Imaging Chamber
A key issue for serial multi-parametric rodent imaging is ensuring proper animal positioning and lack of movement. In this regard, systems have been developed that position and immobilize rodents while simultaneously providing barrier conditions for imaging immunocompromised animals and delivering gas anesthesia for a constant and reproducible level of anesthesia. Incorporated into these designs are heating elements that maintain the animals at a constant and normal physiological temperature, an often overlooked issue that can profoundly alter the metabolic process under investigation (Fig. 2) (10).

Antibody Imaging
A relatively new area of investigation with PET is the labeling of antibodies or antibody fragments to image

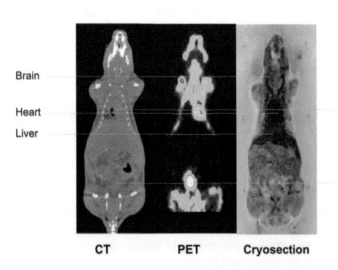

Figure 1 Examples of a mouse atlas using microCAT, microPET, and cryosections.

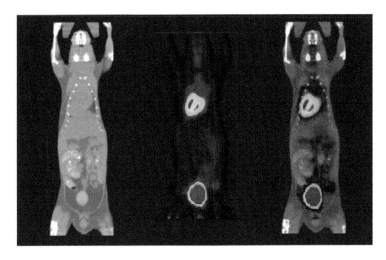

Figure 2 Combined imaging of a mouse in microCAT and microPET using a common imaging chamber.

their protein targets (Fig. 3). These experiments require isotopes with longer lives, such as I-124 or Cu-64, to account for longer clearance times of the labeled antibody from blood and tissues, some times as long as 12–48 hours (Fig. 4). Due to the nearly complete clearance of non-specific activity in these studies, anatomical imaging is essential to locate the focal uptake.

Gene Therapy

The development of new radiolabeled compounds often requires the intermediate step of confirming the specific uptake location using autoradiography, which has a resolution ~25 microns. Mice or rats can be injected with test compounds and imaged with PET, followed by sectioning using a microcryotome and imaging using light photography and autoradiography

techniques (Fig. 5). This example illustrates the insertion and imaging of a transfected viral delivered gene (reporter gene) that codes for a protein that traps the labeled PET reporter probe to image expression of the reporter gene. An alternate route of reporter gene delivery through liposomes is shown in Figure 6 in

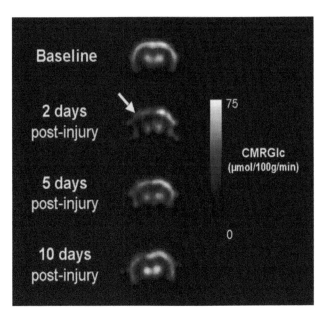

Figure 3 Brain injury and recovery in rat.

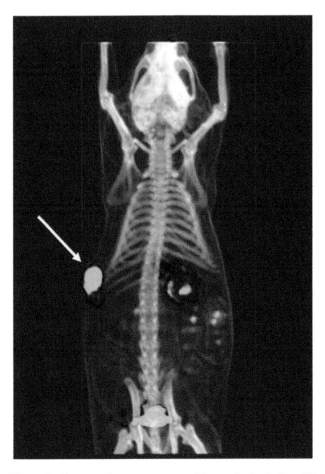

Figure 4 Human colon cancer xenograft imaged using iodine-124 labeled antibody fragment superimposed over a CT scan in a mouse.

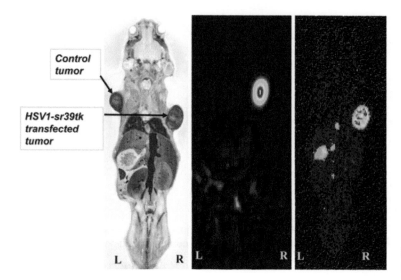

Figure 5 Cryosection light photography, microPET, and autoradiography imaging from the same mouse.

transplanted rabbit hearts. The validation of reporter gene imaging agents opens the way to non-invasive assessment of gene expression, an important step on the road to gene therapy.

Multiple Isotopes

In-vivo processes can often be investigated using multiple PET imaging probes. Indeed, the potential for measuring multiple and diverse biological parameters is one of the strengths of PET imaging. For example, Figure 7 shows the same mouse with a bone tumor imaged with both ^{18}F-fluorodeoxyglucose (FDG) and F-18 fluoride ion, showing both inflammation and bone formation, as well as bone degeneration with microCT. Figure 8 shows an example of both perfusion imaging with N-13 ammonia and metabolic imaging with FDG. Thus, the miniaturization of PET technology permits the same biological measurements that are performed in man to be obtained in rodents facilitating the bi-directional translation of research between the bench and the bedside.

Oncology

Small animal PET imaging plays a key role in furthering our understanding cancer biology providing key insights that are laying the foundation for novel

Figure 6 Transverse, coronal, and sagittal images of implanted rabbit heart showing the reporter probe, ^{18}F-fluoro-3-hydroxymet hylbutyl)guanine, localization of reporter gene expression following liposome delivered gene therapy.

diagnostic and therapy monitoring paradigms. Initially, efforts focused on imaging accelerated rates of tissue glucose metabolism, a hallmark of the neoplastic process, using the glucose analogue FDG. Small animal PET using FDG has permitted the phenotyping of new murine tumor models and to evaluate the efficacy of new therapies. More recently, novel PET probes are being developed that have improved specificity for key aspects of the neoplastic process. For example, increased cellular proliferation typlifies the cancer phenotype. 3-deoxy-3-^{18}F-fluorothymidine (FLT) is a thymidine analog whose uptake correlates with DNA replication and cellular growth (11). Thus, it is a very promising marker of tumor proliferation and reponses to therapy (Fig. 9).

Neurodegenerative Diseases

PET imaging plays a key role in the clinical management of patients with neurodegenerative disorders such as Alzheimer's and Parkinson's diseases because their imaging biomarkers are visible by PET long before symptoms appear (Fig. 10). In the case, Alzheimer's disease small animal PET is playing an important role in characterizing the binding characteristics of new PET radiopharmaceuticals such as 2-(1-(6-[2-18F-fluoro-ethyl)(methyl)amino]-2-napthyl) ethylidene) malononitrile which are designed to directly bind to amyloid-β plaques (12). Conversely, in Parkinson's disease small animal PET is being used to image endogenous effector gene expression as well as imaging of cells and vector-mediated gene expression in-vivo; all prequisites for the development and implementation of new therapies (13).

Image Reconstruction

The advances in imaging reconstruction techniques, combined with the improvements in image resolution,

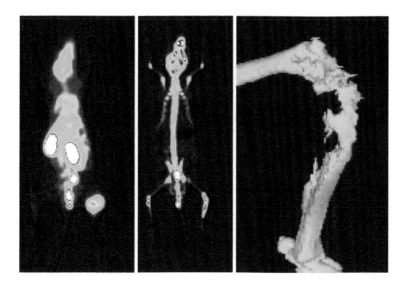

Figure 7   Examples of bone loss and formation with FDG and fluoride ion in microPET and a close up with microCT.

have lead to increasingly detailed images of mouse anatomy (Fig. 11) and are opening up new possibilities for cardiac investigations. The computational demands of these iterative reconstruction techniques are demanding, but feasible with currently available computers. For example, Figure 12 shows the improvements made in just a few short years with small animal PET systems produced by Siemens Medical Systems, starting with the prototype small animal PET in 1997, the P4 in 1999 and the Focus system in 2003. One can clearly see the improvements in image quality due to the reconstruction software. The higher resolution images enable quantification of smaller changes, thus earlier detection

of physiological changes. Perhaps the best feature is that these improvements come without making any changes or trade-offs in the acquisition hardware.

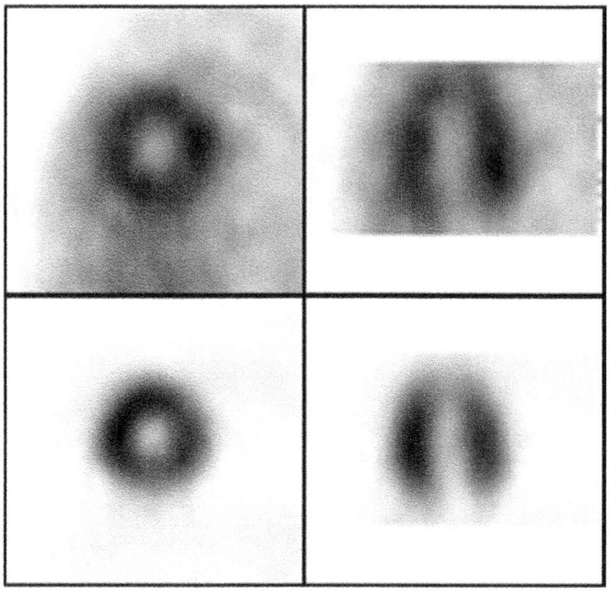

Dynamic Imaging

The ability to tailor the image sequences to match the temporal dynamics of the biology or pharmacology under investigation has rekindled the ability to study tracer kinetics or pharmacokinetics to measure rate constants of the in-vivo processes (Fig. 13). These kinds of studies can also allow short imaging sequences to look at first pass effects and parameters such as cardiac wall motion, cardiac output, and ejection fractions (Fig. 14) (14).

Gating

Due to the increasing demands of in-vivo imaging of mice, several different physiological monitoring equipment vendors now offer systems capable of measuring the high heart rates in mice (600+ bpm). Coupled with the small animal PET systems, it is possible to remove or examine the effects of cardiac and/or respiratory movements in the physiological image information (Fig. 15). Coupled with tailored dynamic imaging sequences, the use of gated images is leading to accurate measurements of blood radioactivity levels, a crucial piece of information necessary for modeling metabolic rates in-vivo.

CONCLUSIONS

The examples shown above demonstrate the wide range of biological questions under investigation with radiolabeled compounds in small animals using PET. As cardiovascular biologists and pharmacologists become more aware of the capabilities of PET, the

Figure 8   (*Top*): N-13 Ammonia perfusion images. (*Bottom*) FDG metabolism images of short axis and horizontal views in mouse heart. *Abbreviation*: FDG, ^{18}F-fluorodeoxyglucose.

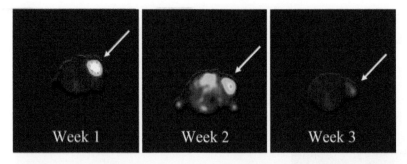

Figure 9 Decreasing FLT uptake indicating lower DNA replication following Gleevec drug therapy in a mouse tumor model. *Abbreviation*: FLT, 3-deoxy-3-^{18}F-fluorothymidine.

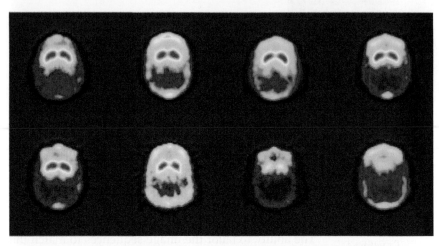

Figure 10 Squirrel monkey images of a dopamine reuptake ligand. (*Top*) Control images from 4 animals. (*Bottom*) Same subject showing increasing effects of a dopaminergic neurotoxin, replicating Parkinson's disease symptoms. Only the lower right animal showed any physical symptoms.

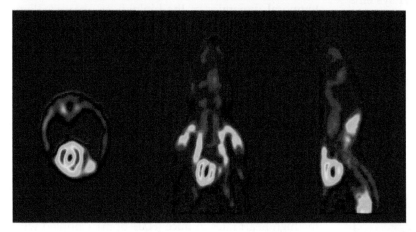

Figure 11 Transverse, coronal, and sagittal FDG images in a mouse with MAP reconstruction. *Abbreviations*: FDG, ^{18}F-fluorodeoxyglucose; MAP, maximum a priori.

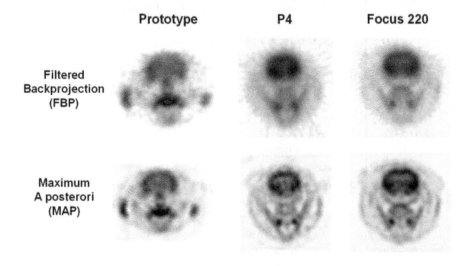

Figure 12 (*Top*) Filtered back projection images of FDG in a rat brain from prototype, P4 and Focus microPET systems. (*Bottom*) Same image data using MAP reconstruction. *Abbreviations*: FDG, ^{18}F-fluorodeoxyglucose; MAP, maximum a priori.

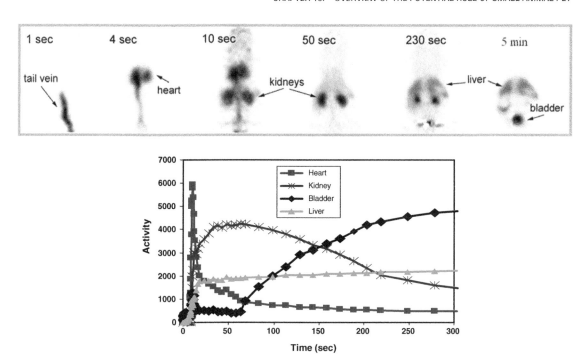

Figure 13   Dynamic imaging of a F-18 labeled drug analog, fluoropencyclovir, in a mouse with expression of the Herpes virus tk gene in the liver.

research involving this imaging technique will, with no doubt, expand greatly in the coming years. The ability of PET to show metabolic and pharmacologic processes, unlike that of most other common imaging systems that provide anatomical information, makes PET an extremely powerful tool for probing molecular processes of cellular systems in vivo. The ability to easily and frequently image animals, without invasive techniques, allows longitudinal studies in the same animals and eliminates inter-subject variability, thereby reducing the number of animals needed to reach a statistical conclusion.

Ironically, the metabolic-only information contained in PET images is actually driving the need for co-acquired anatomical images obtained from CT scans. As radiolabeling techniques and targeting to a specific biological process improve, the information about where the signal is coming from actually decreases, necessitating an overlay of anatomical information to show the exact location of the target. This need has already resulted in the first clinical PET-CT systems that have quickly taken over the vast majority of units being sold on the market today (about 70% of PET in 2003, nearly 100% in 2004). The question of whether a small animal imaging system with both PET and CT is needed still remains open but such systems are being developed.

The scope of small animal PET research is expanding rapidly. The increase in interdepartmental information sharing and collaborations between not only different departments, but also different institutions, is creating an open environment for learning and sharing of these new tools for molecular imaging. As procedures and methods become standardized and accepted, small animal imaging using radiolabeled compounds and PET will become a mainstream approach to examining biological and pharmacological processes.

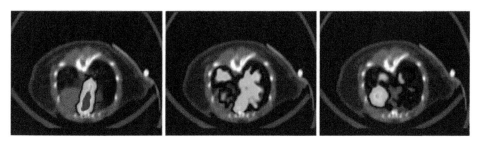

Figure 14   microPET-CT images showing FDG filling of right ventricle, lungs, and left ventricle in a mouse.

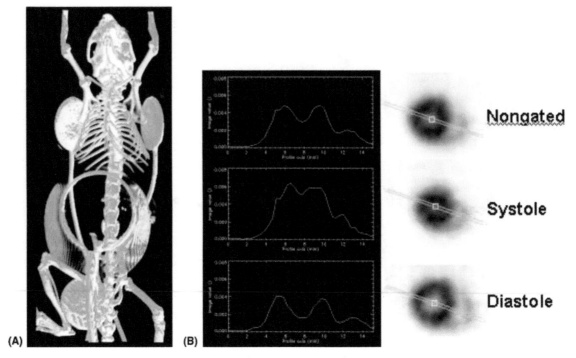

(A)   **(B)**

Figure 15   (*Left*) A microCT image of a mouse with ECG and respiration monitoring equipment. (*Right*) Line profiles through the heart images.

REFERENCES

1. Cherry SR, Shao Y, Silverman RW, et al. MicroPET: A high resolution PET scanner for imaging small animals. IEEE Trans Nucl Sci 1997; 44:1161–1166.

2. Tai YC, Chatziioannou AX, Siegel S, et al. Performance evaluation of the microPET P4: A PET system dedicated to animal imaging. Phys Med Biol 2001; 46:1845–1862.

3. Massoud TF, Gambhir SS. Molecular imaging in living subjects: Seeing fundamental biological processes in a new light. Genes Develop 2003; 17:545–580.

4. Heath JR, Phelps ME, Hood L. Nanosytems Biology. J MIBIO 2003; 5:312–325.

5. Qi J, Leahy RM, Cherry SR, Chatziioannou AX, Farquhar TH. High-Resolution 3D Bayesian image reconstruction using the microPET small-animal scanner. Phys Med Biol 1998; 43:1001–1013.

6. Chiu HC, Kovacs A, Blanton RM, et al. Transgenic expression of fatty acid transport protein 1 in the heart causes lipotoxic cardiomyopathy. Circ Res 2005; 96:225–233.

7. Simoes MV, Egert S, Ziegler S, et al. Delayed response of insulin-stimulated fluorine-18 deoxyglucose uptake in glucose transporter-4-null mice hearts. J Am Coll Cardiol 2004; 43:1690–1697.

8. Forno LS, Delaney LE, Irwin I, Langston JW. Similarities and differences between MPTP-induced Parkinsonism and Parkinson's disease. Neuropathological considerations. Adv Neurol 1993; 60:600–608.

9. Stout DB, Chow PL, Silverman R, et al. Creating a whole body digital mouse atlas with PET, CT and cryosection images [Abstract]. Mol Imaging Biol 2002; 4:27.

10. Chow PL, Stout DB, Komisopoulou E, Chatziioannou AF. A method of image registration for small animal, multi-modality imaging. Phys Med Biol 2006; 51: 379–390.

11. Leyton J, Latigo JR, Perumal M, Dhaliwal H, He Q, Aboagye EO. Quantifying the activity of adenoviral E1A CR2 deletion mutants using renilla luciferase bioluminescence and 3'-deoxy-3'-[18F]fluorothymidine positron emission tomography imaging. Cancer Res 2006; 66:9178–9185.

12. Ye L, Morgenstern JL, Lamb JR, Lockhart A. Characterisation of the binding of amyloid imaging tracers to rodent Abeta fibrils and rodent-human Abeta co-polymers. Biochem Biophys Res Commun 2006; 347:669–677.

13. Jacobs AH, Li H, Winkeler A, et al. PET-based molecular imaging in neuroscience. Eur J Nucl Med Mol Imaging 2003; 30:1051–1065.

14. Croteau E, Benard F, Cadorette J, et al. Quantitative gated PET for the assessment of left ventricular function in small animals. J Nucl Med 2003; 44:1655–1661.

17

Overview of Imaging Atherosclerosis

Vardan Amirbekian

Department of Radiology, Mount Sinai School of Medicine, New York, New York, Johns Hopkins University School of Medicine, Baltimore, Maryland, and The Sarnoff Endowment for Cardiovascular Science, Great Falls, Virginia, U.S.A.

Smbat Amirbekian

Department of Radiology, Mount Sinai School of Medicine, New York, New York, and Emory University School of Medicine, Atlanta, Georgia, U.S.A.

Juan Gilberto S. Aguinaldo

Department of Radiology, Mount Sinai School of Medicine, New York, New York, U.S.A.

Valentin Fuster

Department of Cardiology, The Zena and Michael A. Wiener Cardiovascular Institute, Mount Sinai School of Medicine, New York, New York, U.S.A.

Zahi A. Fayad

Department of Radiology, Imaging Science Laboratories, and Department of Cardiology, The Zena and Michael A. Wiener Cardiovascular Institute, Mount Sinai School of Medicine, New York, New York, U.S.A.

BACKGROUND

Atherosclerosis is the major cause of heart disease. It is also a central cause of other pathologic processes such as stroke and cerebrovascular disease, peripheral vascular disease, certain aortic aneurysms, and ischemic bowel disease. More than half of patients affected by coronary atherosclerosis experience sudden death or myocardial infarction as their first clinical manifestation of disease (1,2). Atherosclerosis accounts for roughly half of all deaths in industrial Western societies (3). It is predicted that cardiovascular diseases are going to become the main cause of death globally within the next 15 years (4). This is because of rapidly increased prevalence of cardiovascular diseases in Eastern European and developing countries as well as the increased incidence of diabetes and obesity in Western countries (4). There is significant evidence that the risk of plaque rupture is associated with plaque burden and plaque composition (5–7). The ability to detect and characterize atherosclerosis prior to adverse events will allow physicians to take aggressive measures to stabilize and reverse disease.

Molecular Imaging holds the potential to accurately and noninvasively detect atherosclerosis before dire consequences such as death, myocardial infarction or stroke arise. Furthermore, molecular imaging will provide enhanced sensitivity and specificity as well as allow us to characterize atherosclerotic plaque functionally and morphologically. We believe that molecular imaging holds the best promise of identifying plaques vulnerable to rupture and thrombosis.

IMAGING TECHNOLOGIES AND METHODS

Currently the imaging techniques that exist for atherosclerosis may be grouped as either invasive or non-invasive methods. Invasive methods include X-ray angiography (using contrast-enhanced fluoroscopic techniques), intravascular ultrasonography (IVUS), optical coherence tomography and angioscopy. Current noninvasive methods include magnetic resonance imaging (MRI), X-ray multi-detector computed tomography (MDCT) and surface B-mode ultrasonography (US). Due to superb sensitivity, radiotracer imaging technologies such as positron emission tomography (PET) will also likely have possible applications in imaging atherosclerosis (8–10), however, technical barriers related to resolution and acquisition time must be overcome before practical applications become available. Finally, new molecular ultrasonography techniques are emerging that may also theoretically provide molecular echocardiographic imaging of plaque.

The above mentioned methods have the capability to identify vessel lumen stenosis, vessel lumen diameter, and with the exception of angiography, they have the potential to assess vessel wall thickness and perhaps plaque volume. However, knowledge gained from information about plaque composition and plaque dynamics may prove to be critically important in terms of clinical decision-making (3,11–17). Currently, most of the above mentioned technologies and techniques lack the ability to fully discern atherosclerotic plaque composition (18,19). We believe that high-resolution MRI holds the potential for complete characterization of atheromatous vascular pathology. This is especially true in light of recent and emerging advances in MRI technology and techniques (20–26).

MRI can distinguish atherosclerotic plaque components based upon physical and chemical properties such as water content, molecular motion, energy and physical state, diffusion, motion as well as chemical makeup and concentration (27). The acquisition times for MRI scans have been significantly reduced thanks to dedicated pulse sequences that facilitate multi-slice imaging (20,21,23,28–30). Furthermore, black-blood MRI methods allow for accurate imaging of the arterial wall without blood-related or blood-flow artifacts (31–33).

A powerful and complimentary tool that MRI offers is magnetic resonance angiography (MRA). MRA has the capability to delineate the distribution of stenotic atheromatous lesions as well as accurately assess the severity of stenosis. It is possible to combine high-resolution black-blood techniques and MRA whereby the former will provide characterization of plaque composition and vessel wall. New phased-array coils designed for cardiac and coronary imaging have also recently been applied to carotid artery imaging allowing for improved resolution all around (34–36). In addition, there are sequences that have been developed that further improve resolution along with providing black-blood imaging (34,36).

MRI OF ATHEROSCLEROSIS

Multi-Contrast MRI of Atherosclerosis
Multi-contrast MRI entails delineating different components of atherosclerotic lesions by generating T1-weighted, T2-weighted and proton-density weighted images of plaque (18,37,38). This has largely become possible because of faster imaging methods,

the ability to perform high-resolution scans and better detection coils (RF coils). The multi-contrast MRI approach involves characterization of plaque based upon signal intensities as well as morphologic appearance of various plaque components on T1-weighted, T2-weighted and proton-density weighted images. This method has been demonstrated and validated in several independent studies (39–46).

Plaque components may be broadly grouped as lipids, fibrous tissue or thrombus/intraplaque-hemorrhage. Plaque lipids are largely comprised of unesterified cholesterol, cholesteryl esters, and some lipoproteins (44). This makes them produce a short T2 resulting in low signal intensity on T2-weighted images (Fig. 1). The short T2 of the lipid portion of atheromatous plaque is believed to be the result of: (*i*) the micelle-like structure of lipoproteins; (*ii*) exchanges between cholesteryl esters and water molecules; (*iii*) the results of oxidation and exchanges between free and bound water molecules (44,47,48). In contrast to plaque lipids, stored fat or adipose fat, such as may be found perivascularly, is mostly composed of triglycerides (49). Because of this, adipose fat found perivasculary has a different appearance on MRI compared to lipids found in atherosclerotic plaque.

The fibrous components of plaque are primarily protein and are composed of collagen and other constituents of extracellular matrix synthesized by smooth muscle cells (SMCs). This gives fibrous tissue a short T1 resulting in high signal intensity on T1-weighted images. The short T1 is the result of interactions between water and protein (50).

The appearance of a thrombus or intraplaque hemorrhage, on MRI, largely depends on the state of hemoglobin contained in the hemorrhage or retained in the thrombus. Depending on the stage of the hemorrhage/thrombus it may contain different oxidative and byproduct forms of hemoglobin such as hemosiderin, ferritin, oxyhemoglobin, deoxyhemoglobin, or methemoglobin. The different oxidative and byproduct forms of hemoglobin have particular magnetic resonance relaxation parameters that give them different signal intensities due to differences in their T1 and T2 (51–55). The progression of hemorrhage and subsequent thrombus and its appearance on MRI has been

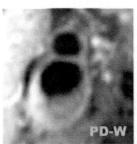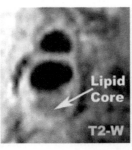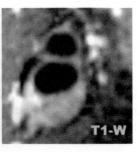

Figure 1 Multi-contrast MRI of carotid atherosclerosis. Proton-density-weighted (PD-W), T2-weighted (T2-W) and T1-weighted (T1-W) acquisitions of the same anatomical slice through a carotid atherosclerotic plaque. There is better contrast of fibrous components on the T1-weighted scans whereas there is low signal intensity and good contrast of the lipid core on the T2-weighted scan.

studied in various settings such as in the aorta and central nervous system (56–58). It is known that blood will shorten both T1 and T2 of water. It is believed that the T1 shortening is due to the generation of methemoglobin, which is paramagnetic. While the T2 shortening is largely because of magnetic susceptibility effects. Therefore, mature thrombus that is full of hemosiderin and ferritin, produces striking signal loss on T2-weighted scans.

Areas of vascular pathology may contain regions of calcification especially in older patients. Calcified regions of atherosclerotic plaque are primarily comprised of calcium hydroxyapatite thereby rendering them low signal intensity on MR images primarily because of diffusion-mediated susceptibility effects and the low proton density of calcium hydroxyapatite (59,60).

After multi-contrast images are acquired, atherosclerotic plaque can then be characterized based upon the distinctive features of components, as described previously (Fig. 1). For greater efficiency, this will theoretically be an automated process using a range of methods. An example is spatially enhanced k-mean cluster analysis, which was demonstrated by Itskovich et al. in an ex vivo study of atherosclerosis in human coronary arteries (61).

There have been recent studies examining whether MRI can detect atherosclerosis in human coronary arteries with adequate sensitivity and specificity. An ex vivo study by Nikolau et al. showed that MRI could detect 23 out of 28 human coronary plaques based upon identification using histopathology and standard pathological analysis (62). In our own ex vivo study, we looked at 22 human carotid endarterectomy samples with MRI. Afterwards we performed standard pathological and histopathological analysis. We were able to identify specific components of plaque such as lipid core, fibrous tissue, thrombus and calcification with good sensitivity and specificity (63). Another interesting finding of our study was that diffusion imaging was able to aid in detection of thrombus (63). Diffusion MR techniques images the motion of water molecules and our finding regarding thrombus detection has been demonstrated independently in the past (64).

MRI of Coronary Atherosclerosis

Artherosclerosis of the coronary arteries is the major cause of heart disease. Arguably, location of atherosclerosis within the coronary arteries leads to the most detrimental events, in terms of sheer numbers, within Western societies. There are many technical and intrinsic challenges that must be overcome before coronary atherosclerosis can be imaged reproducibly and accurately using MRI. Using cardiac and respiratory gating the myocardium can be imaged very well.

However, because of the small size, location, and tortuous course of the coronary arteries it is still not possible to image the coronary arteries even with cardiac and respiratory gating. That being said, progress is forthcoming and we believe that these technical challenges will eventually be overcome. The number of studies looking at MRI of the coronaries is gradually building with increasingly better results.

In our own studies, we looked at imaging coronary atherosclerosis using Yorkshire albino swine following induction of lesions via balloon angioplasty (65). Previous studies looking at the coronaries in pig models had been attempted with standard difficulties related to coronary arterial anatomy as well as cardiac and respiratory gating (45,65). In our study, we were able to show good correlation between MRI findings and pathological findings including histopathology (65). In the same study, we were able to visualize hemorrhage into plaques with a sensitivity of 82% and specificity of 84% based upon pathologic confirmation (65).

Based upon experience from animal studies and MRI studies of human carotid and aortic lesions, we attempted to look at human coronary atherosclerosis using black-blood MRI techniques (32). We performed high-resolution MRI of normal and atherosclerotic human coronary arteries. In this study, we used MRI performed during breath-holds in order to minimize unwanted respiratory motion interference. We found a statistically significant difference between maximal wall thickness in normal patients and patients with likely clinical atherosclerotic disease (32). In another study, Botnar et al. were able to assuage the need for breath-holds by using a black-blood fast-spin-echo technique combined with a real-time navigator for respiratory gating and real-time correction for slice position (66). There has also been other research looking at coronary arteries in healthy volunteers utilizing a respiratory navigator and a real-time motion correction gradient-echo sequence (interestingly this work was done at a 3 Tesla magnetic field) (67).

Another potential and interesting application of MRI is monitoring and examination of vessels subsequent to percutaneous intervention such as balloon angioplasty and stenting. Coulden et al. used high-resolution MRI to look at the popliteal artery before and after balloon angioplasty (68). In all of the study, patients' MRI was able to evaluate accurately the degree of atherosclerotic lesions. In fact, MRI was able to identify lesions in areas of vessels that were deemed "normal" using standard angiography (fluoroscopic contrast-enhanced X-ray angiography). In these areas that were angiographically "normal," MRI identified atheromatous plaques with cross-sectional areas ranging from 50% to 75% of lumen cross-sectional area without significant angiographic obstruction (68). This is consistent with the fact that, because of extensive outward vascular remodeling, atheromatous plaques may become rather large without obstructing the vessel lumen. In fact, it is well established that the majority of lesions involved in the pathogenesis of acute coronary syndromes are less than 50% stenosed (69,70). In the above-cited study by Coulden et al., shortly after angioplasty MRI was able to demonstrate local dissection as well as plaque fissuring (68). Subsequently, serial changes in blood flow, lumen diameter and lesion size was assessed and documented using MRI. Thus Coulden et al. showed that high-resolution MRI could be used to map out extent of disease in vessels and then monitor changes subsequent to percutaneous intervention such as angioplasty. Once challenges of imaging the coronary arteries are overcome, subsequent to percutaneous interventions (such as angioplasty and stenting) the ability to noninvasively monitor vascular remodeling and restenosis will give clinicians a powerful tool for postintervention management.

The holy grail of imaging atherosclerosis is to accurately and reproducibly image the coronary arteries and pathology thereof. In the long run, in order for coronary imaging to be valuable clinically it must entail the ability to differentiate between plaque components. Along with functional and molecular imaging, we believe that this will provide valuable information for clinical decision-making in managing disease and monitoring effectiveness of antiatherosclerosis therapy.

MRI of Aortic Atherosclerosis

Imaging the aorta with MRI may have clinical usefulness (Fig. 2). Assessing and monitoring plaque burden in the aorta will allow for monitoring response to antiatherosclerosis therapy. Aortic atherosclerosis has been previously show to be associated with a higher risk of ischemic stroke or cerebrovascular infarction (71–73). Additionally, plaque burden in the aorta may serve as a surrogate marker of systemic plaque burden. Furthermore, by imaging the aorta we may also gain screening information regarding aneurysms. The main difficulties in imaging the aorta lie in the thoracic aorta where respiratory motion and blood flow are nuisances and where it becomes a challenge to obtain submillimeter resolution for adequate sensitivity.

There has been an array of studies looking at atherosclerosis in the aorta. One such study looked at asymptomatic patients from the Framingham Heart Study. In that study, it was shown by MRI that the atherosclerotic plaque prevalence and burden increased significantly with age (74). Interestingly, they also found that in their study group there was greater burden of disease in the abdominal aorta than in the thoracic aorta (74). In a similar analysis, it was found that asymptomatic aortic atherosclerosis detected by MRI was strongly associated with the Framingham Heart Study coronary risk score and other long-term coronary risk factors (75). Taniguchi et al. recently used MRI to study the relationship between aortic atherosclerosis and risk factors for coronary artery disease (CAD) (76). They found that thoracic atherosclerosis was closely linked to traditional risk factors for CAD such as high cholesterol, age, and smoking (76).

Fayad et al. looked at atheromatous plaque in the thoracic aorta (Fig. 2). In that study, they used T1, T2 and proton-density weighted scans to evaluate plaque composition (77). Imaging was performed using rapid high-resolution fast-spin-echo sequence. Blood and blood flow artifacts were suppressed using velocity-selective flow suppression prepulses. In the same study, they compared MRI with transesophageal ultrasonography/echocardiography (TEE) in the assessment of atherosclerosis. The results of the study showed that, based on cross-sectional imaging, MRI and TEE correlated strongly in terms of mean maximum plaque thickness and plaque composition (77).

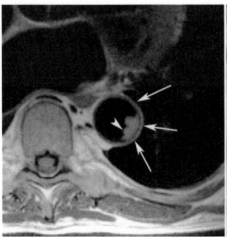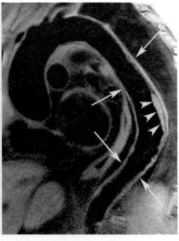

Figure 2 MRI of aortic atherosclerosis. Axial and sagittal slices through the aorta showing a large atherosclerotic lesion (*arrows* point to the wall of the aorta; *arrowheads* point to the atherosclerotic plaque).

In another study, Summers et al. used MRI to demonstrate that, when compared to a control group of patients, the wall thickness of the ascending aorta is increased in patients with homozygous mutations leading to familial hypercholesterolemia (78). In this study, the plaque components could not be analyzed accurately because only T1-weighted spin-echo images were acquired for the study.

In imaging atherosclerosis another application of MRI is the evaluation of regression subsequent to therapeutic interventions such as pharmacotherapy. Corti and Fayad et al. demonstrated that MRI could be used to evaluate the efficacy of cholesterol-lowering therapy using 3-hydroxy-3-methylglutaryl coenzyme A (HMG-CoA) reductase inhibitors (statins) in asymptomatic untreated patients with aortic and carotid atherosclerosis as well as elevated levels of low-density lipoprotein (LDL) (79). MRI was used to detect and assess aortic and carotid atherosclerosis in patients at different intervals following commencement of cholesterol-lowering therapy with simvastatin (HMG-CoA reductase inhibitor) (79,80). The results showed that MRI could be used to visualize and assess regression of atherosclerotic lesions following pharmacologic intervention (79). At 12 months, it was shown that there was a significant decrease in vessel wall area without a change in the vessel lumen area. These findings were in agreement with previous experimental animal research (81–83). Interestingly, there was a dramatic early decrease in LDL levels following initiation of treatment with simvastatin. However, despite this finding a minimum of 12 months of therapy was needed in order to observe changes in vessel walls. Even at six months, no significant changes were observed. Recently, there have been a couple of MRI studies that have compared the effect of high and low doses of statins on atherosclerosis regression (84,85). Both of these studies used MRI to demonstrate that atherosclerosis regression was probably related more to the reduction in LDL rather than the dosage of the statin used (84,85).

MRI of Carotid Atherosclerosis

Atherosclerosis of the carotid artery is the major cause of cerebrovascular/neurologic infarcts or strokes (1,3). Stroke ranks as the third leading cause of death in the United States (1,86). In contrast to dangerous or vulnerable plaque in the coronary arteries, dangerous high-risk plaques in the carotid arteries tend to be severely stenotic (1,87,88). In addition, unlike in coronary arteries, the degree of stenosis in carotid plaques is associated with increased risk of adverse events such as stroke. Similarly, the plaque content of high-risk plaques in the carotids is different when compared to high-risk plaques in the coronary arteries (89). High-risk plaque in the coronaries tend to be lipid-rich whereas high-risk plaques in the carotids

tend not to necessarily be lipid-rich but rather heterogeneous in nature with significant amount of fibrous components (although the carotids can certainly contain plaque that is predominantly lipid-rich) (87,89,90). Carotid plaques often become symptomatic due to intramural dissection, intramural hemorrhage with subsequent hematoma formation (87,90). This is thought to be a result of blood hitting up against resistant areas of stenoses, which in turn is consistent with the strong association between hypertension and risk of stroke (91). Dissection and hemorrhage involving plaque often leads to thrombus formation. Thrombus in turn can then act as major contributor of emboli leading to ischemic stroke. After formation, if carotid thrombi become chronic in nature then the risk of stroke increases further and patients often experience multiple episodes of transient ischemic attacks (TIA) that frequently precede permanent and detrimental cerebral infarcts (92).

When compared to the coronary arteries, the carotid arteries are particularly amendable to imaging. This is because they are not particularly subject to significant motion and because of their superficial location. Many studies of carotid atherosclerosis have been performed using multi-contrast MRI (Fig. 1). As discussed earlier, multi-contrast MRI involves the use of T1-weighted, T2-weighted and proton-density-weighted imaging to obtain a composite of information regarding plaque composition. Many studies have utilized high-resolution black-blood spin-echo and fast-spin-echo MRI sequences to investigate carotid atherosclerosis. In one of the first in-vivo studies, MRI was performed on patients with advanced carotid atherosclerosis who were scheduled to have carotid endarterectomy (43). In this study, the investigators were able to begin characterization of normal and abnormal areas of carotid wall. Part of the study involved characterization of short T2 components in vivo before carotid endarterectomy and correlating these values with ones obtained in vitro after carotid endarterectomy when large portions of the plaques were removed surgically for therapeutic benefit (43). In a different study, Yuan et al. examined carotid wall area and plaque size demonstrating the capabilities of MRI thereof (93). Carotid plaque size and wall area measurements may be useful in clinically following plaque progression and possible regression with appropriate treatment.

Hatsukami et al. looked at fibrous cap morphology of carotid plaques by using three-dimensional fast time-of-flight imaging (bright blood). This type of approach is a blend of T1 and proton-density weighting with enhancement of signal from flowing blood. The result is good resolution and highlighting of the fibrous cap. In that study, Hatsukami et al. were able to examine plaque fibrous cap characteristics such as thickness and to assess cap integrity in a general manner (94). Integrity of fibrous cap was studied as a possible surrogate marker/predictor of plaque rupture. In fact in a different study, it was shown that by using multi-contrast MRI on carotid atheromatous plaque it was possible to prospectively identify plaques that had unstable fibrous plaques with a fairly good sensitivity and specificity (95). Whether a plaque was unstable was determined based on comprehensive pathological analysis. The MRI prediction of instability was made blinded to the pathological results.

High-resolution black-blood MRI has been used to reasonably classify carotid atherosclerosis. Luo et al. performed a study on 37 patients with carotid artery lesions who went on to have carotid endarterectomy surgery (96). In that study, various parameters such as maximum wall-area, wall-volume, and minimum lumen-area were assessed with MRI in vivo and then ex vivo following the surgical removal of lesions. The study showed good correlation between in vivo and ex vivo MRI measurements. Subsequent further work by several groups has shown that MRI had good sensitivity and specificity in classifying carotid atherosclerotic plaques based upon the American Heart Association (AHA) classification system (97–99). In fact there has been work showing that in progressive/complex carotid atherosclerotic lesions MRI could be used to identify intraplaque hemorrhage as well as juxtaluminal hemorrhage or thrombus (100,101). Furthermore, MRI has the capability of direct thrombus imaging. This may be done via a T1-weighted turbo-field-echo sequence. This sequence optimizes detection of methemoglobin. Utilizing this method two groups have shown accurate detection of carotid lesions complicated by thrombus (102–104).

Contrast-enhanced MRI has also been used to study atherosclerosis in the carotids. By using a standard extracellular gadolinium-based contrast agent it is possible to accurately assess vessel lumen as well as create enhancement of some plaque components as demonstrated by Yuan et al. In that study, investigators compared precontrast-enhanced and postcontrast-enhanced MRI finding an 80% enhancement of fibrous components compared to low enhancement of necrotic core areas (105).

MOLECULAR IMAGING OF ATHEROSCLEROSIS

Introduction

Molecular imaging, broadly defined, encompasses imaging techniques and methods aimed at obtaining functional and dynamic molecular information about disease processes in a sensitive and specific manner. We think that molecular imaging can also help to attain functional knowledge about normal physiologic processes in addition to delineating pathology.

Molecular imaging is a multi-disciplinary field that draws on the expertise from many domains ranging from basic physics to bedside clinical practice. The National Institutes of Health (NIH) included molecular imaging and possible applications thereof as a part of the NIH Roadmap to Biomedical Research (106). Increased funding for molecular imaging research is bringing forth new advances that likely represent only a tip of greater things to come as this exciting field develops.

Within the study of atherosclerosis, a large goal of molecular imaging is to attain the capability to target specific plaque-associated molecules with agents that provide sensitive and specific contrast (107,108). This will tremendously improve detection and characterization of atherosclerotic and atherothrombotic vascular lesions (107,108). Certain plaque components such as a lipid-rich core, thin fibrous cap and macrophage infiltration are highly associated with plaque rupture, vulnerability to rupture and consequences thereof (109–111). Therefore, specific knowledge about plaque composition and morphology may prove of great clinical utility in terms of risk-stratifying patients. This specific knowledge will be gained by targeted imaging of precise constituents of atherosclerosis such as molecules related to inflammation, leukocyte and monocyte adhesion, modified lipoproteins, apoptosis, macrophages, matrix metalloproteinases (MMPs), neovascularization, thrombosis, necrosis, or apoptosis (Fig. 3). Examples of possible targets include macrophages, monocytes and foam cells. These cells represent good cellular targets for molecular imaging of atherosclerosis because they have been shown to play a critical role in the inflammation thought to be central to atherosclerosis (112–115). Furthermore, the necrotic lipid core of lesions that progress to produce acute coronary syndromes is rich in macrophage-derived foam cells (114). Inflammatory cells, mostly macrophages, produce MMPs that degrade the fibrous cap and trigger acute coronary syndromes (116,117).

Several techniques have been used to try to target specific components or molecules of atheromatous plaque with different imaging modalities and contrast mechanisms thereof. The primary and leading techniques have been nuclear imaging using radiotracers, molecular ultrasound imaging, and molecular MRI.

Molecular Imaging Using Radiotracers

Nuclear imaging can often offer superb sensitivity in detecting pathological processes. Different types of radiotracers can theoretically be loaded onto a variety of constructs that would allow for specific molecular targeting. Once specifically targeted, radiotracers have unmatched and outstanding sensitivity and specificity. However, several and diverse technical challenges will need to be overcome to make nuclear imaging more

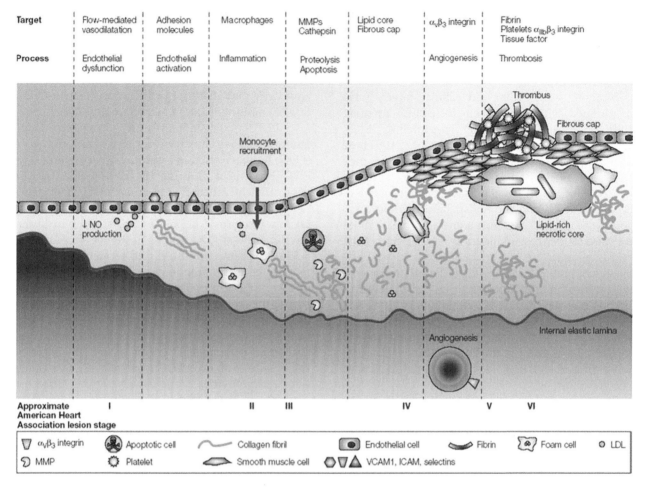

Figure 3 Molecular and cellular targets for molecular imaging of atherosclerosis. Schematic of atherosclerosis pathogenesis ranging from endothelial dysfunction (*left*) through monocyte recruitment to the development of advanced plaque complicated that can be complicated by thrombosis (*right*). This schematic is quite simplified, however, it highlights processes and components such as cell adhesion molecules, macrophages, connective tissue elements, lipid core and fibrin, apoptosis, proteolysis, angiogenesis, and thrombosis in plaques that may prove useful for imaging. *Abbreviations*: ICAM, intercellular cell adhesion molecule; LDL, low-density lipoprotein; MMP, matrix metalloproteinase; NO, nitric oxide; VCAM, vascular cell adhesion molecule. *Source*: From Ref. 107.

practical. There is much room for improvement in terms of the resolution of current nuclear techniques when compared to other imaging modalities. There is also a lack of anatomic detail on many types of scans. However, the latter problem has been somewhat overcome with the help of combined anatomical and nuclear imaging methods such as PET/CT or the newer single photon emission CT (SPECT)/CT. Another challenge is related to scan acquisition times, which can be rather long. And finally, from a practical standpoint, synthesis of certain radionuclides requires the use of a cyclotron and in general, radiotracers/radionuclides can be a challenge to work with especially if they have short half-lives. However, all things considered nuclear imaging will certainly have a prominent role in noninvasive functional assessment of atherosclerotic disease pathology.

Currently the primary nuclear imaging modalities that have been used for imaging atherosclerosis are

SPECT and PET as well as combined PET/CT. SPECT uses γ-emitting radionuclides/radiotracers and the imaging or detection of signal is setup thereof. In comparison, PET uses positron-emitting radionuclides/radiotracers and imaging is accomplished by detecting coincident 511 keV photons that arise from the decay of the positron-emitting isotope/radionuclide (118). Both SPECT and PET require acquisition of transmission maps to correct for attenuation as photons travel outside the body through different anatomical lengths and densities. With the use of PET/CT anatomical imaging is accomplished with CT while the PET scan takes place and then the two sets of scans can be viewed as an overlay or individually (119,120). Another advantage of PET/CT is that it is possible to obtain a transmission map for attenuation correction using CT X-ray photons instead of using ^{68}Germanium (^{68}Ge) segmented attenuation correction (118,121–123). At a resolution of roughly 5 mm PET is superior to SPECT, which has a resolution of approximately 15 mm. In addition, PET offers better sensitivity. Therefore, for imaging small pathological lesions, such as atherosclerotic plaques, PET is currently the more capable nuclear imaging modality.

Numerous molecular and cellular targets have been pursued with radiotracers for SPECT imaging including adhesion molecules, lipoproteins, SMCs, endothelial cells and macrophages (10,124). These studies have demonstrated feasibility, however, they have largely failed to yield strong target to background ratios. This was likely related to sluggish clearance of radiotracer from blood and nonspecific distribution in other tissues. There have been some attempts to overcome these challenges. In one study, investigators used SPECT imaging using a radionuclide-labeled antibody to the glycoprotein IIb/IIIa (GPIIb/IIIa) platelet receptors. The idea behind this was to image platelet-rich thrombus often found on atheromatous plaques that have ruptured or are vulnerable to rupture. In the study, a canine coronary arterial thrombus model was used. The results were promising in terms of identification of thrombus (10). However, newer studies have not been performed.

Tsimikas et al. have performed several studies using radiotracer-labeled monoclonal antibodies to oxidized-low–density lipoprotein (ox-LDL). These studies have been performed in both mouse and rabbit models of atherosclerosis. The methods used for the detection were autoradiography, gamma-camera scintigraphy or both. In these studies, the investigators demonstrated that radiolabeled antibodies to ox-LDL could be used for detecting atherosclerosis (125). Furthermore, they also showed that it was potentially possible to estimate plaque volume as well as follow progression and regression of atheromatous plaques (126,127).

Apoptosis is believed to play a role in the formation of necrotic debris that constitutes necrotic cores often found in unstable vulnerable plaques (5). There is data to suggest that apoptosis of macrophages contributes to the size of a necrotic core (128). Furthermore, there is evidence that apoptosis of SMCs is associated with a thin fibrous cap (129). Recently, Narula et al. used radiolabeled annexin V to target apoptotic cells in a rabbit model of atherosclerosis using gamma camera methods. Annexin V has a high affinity for phosphatidylserine (PS), which becomes exposed on the surface of apoptotic cells. In this study, the investigators measured the in vivo aortic uptake of radiolabeled annexin V. They showed that there was an almost 10-fold higher uptake in the aorta of atherosclerotic rabbits compared to control rabbits (Fig. 4) (130). Given that apoptosis is a potential determinant of plaque instability, imaging atherosclerosis using annexin V may turn out to be useful in assessing

plaque vulnerability. Quite interestingly, these methods are now being used in humans to study usefulness in assessment of carotid atherosclerosis (Fig. 5). Although preliminary, the results are suggesting that imaging carotid atherosclerosis with radiolabeled annexin V may discern carotid plaque features indicative of stability or rather lack thereof (131).

PET has shown interesting promise for imaging atherosclerosis. Thus far only 18-fluorodeoxyglucose (FDG) has been used in assessing atherosclerosis with PET. FDG is a fluorine-18 (^{18}F)-labeled modified derivative of glucose. Lederman et al. first demonstrated marked increased FDG uptake in experimental atherosclerosis (132). Several studies in humans have examined FDG uptake in the region of the aorta (133–135). However, these studies yielded little information regarding frequency, location, and intensity of uptake. Although subsequently, one study found a correlation between certain risk factors for CAD, such as age and hypercholesterolemia, and the magnitude of FDG uptake in the abdominal aorta, iliac arteries and femoral arteries (136).

Rudd et al. used FDG to perform PET on patients with symptomatic carotid atherosclerotic disease (9). Their study found that FDG accumulated to a significantly greater degree in unstable plaques compared to the stable contralateral-sided plaques (9). Furthermore, ex vivo analysis of endarterectomy samples showed that the FDG accumulated mostly in macrophages found in plaque (9). Interestingly, this suggests that FDG may be suitable for assessing plaque vulnerability since a high and active macrophage content has been linked with instability (114).

Taking advantage of combined functional and anatomical imaging Tatsumi et al. used PET/CT to study aortic FDG-uptake in patients undergoing FDG-PET/CT scans for cancer staging (137). In that study, investigators could anatomically approximate the location of FDG uptake by using the CT imaging (Fig. 6). The results showed that FDG uptake was primarily in the thoracic aorta and that it was associated with age (137). Interestingly, areas of FDG uptake were mostly distinct and separate from areas of calcification in the aorta (137).

Recently, M. Ogawa et al. performed PET/CT study correlating atherosclerosis pathology to FDG uptake in the aorta of Watanabe heritable hyperlipidemic (WHHL) rabbits. They found that there was a strong correlation between FDG uptake and the number of macrophages found in plaque (138). The findings of this study have been recently confirmed by A. Tawakol et al., who examined the relationship between FDG uptake and vascular inflammation in an induced atherosclerosis model using New Zealand white (NZW) rabbits. They also found a strong relationship between macrophage staining on immunohistopathologic slides

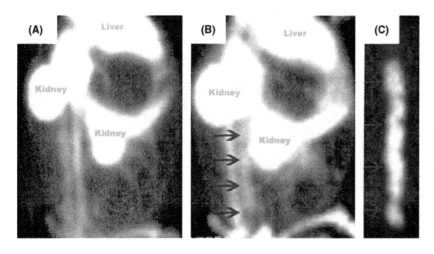

Figure 4 Gamma camera imaging of atherosclerosis using annexin V in rabbits. (**A**) Left lateral decubitus image injected with 99mTc-labeled annexin V showing blood pool activity immediately after administration. (**B**) Clear delineation of radiolabeled within the abdominal aorta at two hours. (**C**) Magnification of the abdominal aorta. *Source*: From Ref. 130.

and the amount of FDG uptake in plaques (139). However, instead of using macrophage number they assessed inflammation via RAM-11 immuno-staining, which stains for macrophages found in plaque. Both of these studies demonstrated that FDG uptake in atherosclerosis is related to macrophage content, which in turn is thought to be a marker of instability.

PET using FDG has demonstrated good promise in terms of quantifying macrophage activity and possibly inflammation. Further study is necessary to examine whether FDG-uptake correlates with future risk of atherosclerosis-related clinical events or plaque rupture. After that, investigations must be done to assess true sensitivity and specificity. However, there still remain significant challenges to the use of FDG in assessing atherosclerosis. Unfortunately, FDG is taken up into any metabolically active tissue. In fact, myocardial uptake of FDG is among the highest of all tissues, which currently excludes FDG from use in imaging coronary atherosclerosis.

Molecular Imaging Using Ultrasound

Using ultrasound for molecular imaging involves the use of micro-bubbles with a targeting entity attached such as a ligand or antibody (140). The basic principal of using micro-bubbles as an ultrasound contrast agent lies primarily in the compressibility of gases. Ultrasound micro-bubbles are gas-filled and usually smaller than the wavelength of diagnostic ultrasound. Because of this micro-bubbles undergo oscillation in an acoustic field in a way that they expand and compress at the pressure troughs and peaks, respectively (141,142). The radial oscillation of micro-bubbles leads to generation of acoustic signals that often exceed conventional ultrasound backscatter created by reflection. The signal producedp by micro-bubbles is dependent on the micro-bubble size, compressibility and density of the gas in the bubbles, the viscosity and density of the surrounding tissues, as well as the frequency and power of the ultrasound applied (143,144). The majority of micro-bubble contrast agents used for ultrasound are

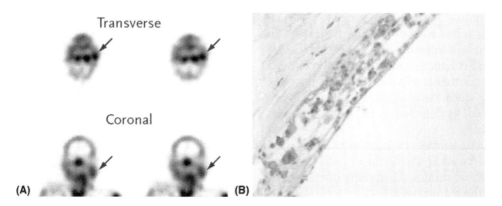

Figure 5 Human carotid atherosclerosis imaging using annexin A5. (**A**) Transverse and coronal views by single photon emission CT (SPECT) imaging in a patient with unstable atherosclerotic carotid artery lesions showing uptake of radiolabeled annexin A5 about six hours after infusion (*blue arrows*). (**B**) Immunohistochemistry of an endarterectomy specimen from the same patient showing macrophages with extensive binding of annexin A5. *Source*: From Ref. 131.

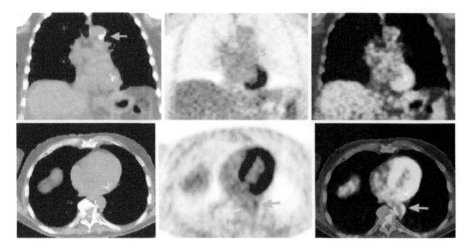

Figure 6 Fluorodeoxyglucose (FDG) uptake in the aorta. First row—coronal CT (*left*), PET (*middle*) and fused positron emission tomography/CT (*right*) images showing calcification in the aortic arch with no FDG uptake seen on the sites of calcification. Second row—transaxial images showing FDG uptake on the aortic wall corresponding to the calcification seen on the medial and lateral side of the lower descending aorta. *Source*: From Ref. 137.

encapsulated in a shell, therefore viscous and elastic damping effects of a shell contribute to properties related to signal production (143,145). The use of harmonics in ultrasound detection of micro-bubbles allows for a better contrast-to-noise ratio in differentiating bubbles from surrounding medium (143,145). The above described properties give micro-bubbles the ability to generate contrast on ultrasound imaging.

An interesting and potentially useful feature of micro-bubbles is that they can be destroyed by high acoustic powers, that may be generated at will, resulting in outward diffusion of the gas, diffusion from large shell defects, or complete fragmentation of micro-bubbles (146). Micro-bubble destruction during high-power ultrasound exposure may have potential applications in drug or gene delivery as well as perfusion imaging.

In the realm of imaging atherosclerosis there have been a few studies using molecularly targeted ultrasound contrast agents. Molecular targets that have been selected for ultrasound study include tissue-factor (TF) and endothelial adhesion molecules/integrins such as P-selectin, E-selectin, intercellular adhesion molecule-1 (ICAM-1), vascular cell adhesion molecule-1 (VCAM-1) as well as $\alpha_v\beta_3$ as a marker of neovascularization. In one of the first studies, Villanueva et al. targeted ICAM-1 with ultrasound-enhancing micro-bubbles in ex vivo preparations of endothelial cells that were either normal or activated. The study demonstrated the feasibility of imaging activated endothelial cells using targeted micro-bubbles (147).

In subsequent studies, Lindner et al. have used micro-bubbles targeted to P-selectin to image inflammation (148). In those studies, the investigators induced inflammation in the kidneys with tumor necrosis factor-α (TNF-α) (148). The study showed that it was possible to image endothelial inflammation using molecular ultrasound. Given that angiogenesis and neovascularization has been implicated in atheromatous plaque

development, micro-bubbles were used by H. Leong-Poi et al. to target alpha(v)-integrins (α_v) in order to image angiogenesis ultrasound (149). Subsequently, H Leong-Poi et al. have used ultrasound micro-bubbles to image endogenous and therapeutic angiogenesis in a hind-limb ischemia rat model (Fig. 7) (150). Ellegala et al. also used targeted micro-bubbles to image angiogenesis (151).

More recently, Hamilton et al. used micro-bubbles attached to antibodies [echogenic immunoliposomes (ELIPs)] to image a myriad of molecular targets in Yucatan mini-swine. Specifically, they used intravascular ultrasound (IVUS) to molecularly image TF, ICAM-1, VCAM-1, fibrin and fibrinogen (152). It was shown that these targeted micro-bubbles/ELIPs provided significant contrast enhancement of plaque compared to untargeted micro-bubbles and normal saline (152). Hamilton

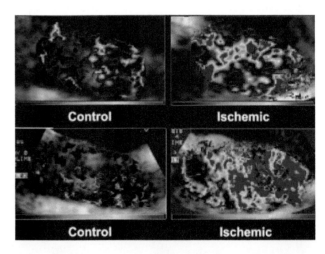

Figure 7 Contrast-enhanced ultrasound. Integrin-targeted molecular ultrasound images in a control and ischemic proximal hind-limb muscles from untreated (**A**) and fibroblast growth factor-2-treated (FGF-2 promotes angiogenesis) (**B**) rat four days after ligation showing retention fraction of integrin3-targeted micro-bubbles. *Source*: From Ref. 150.

et al. have also used immunoliposomes (targeted micro-bubbles) to detect thrombus in an experimental animal model of left ventricular thrombus (153).

Molecular Imaging Using MRI

The concept of molecular MRI is based on delivering MRI contrast agents to locations of interest using molecular targeting techniques. To accomplish this purpose various experimental targeting methods have been used. Investigators have used peptides and anti-bodies linked to MRI contrast agents. A popular approach has been using the lanthanide gadolinium to generate MRI contrast. By linking gadolinium to a molecular or cellular targeting vehicle, it is possible to generate MRI contrast at specific locations. Gadolinium is frequently used because of its paramagnetic proper-ties that generate MRI contrast and because it has been in clinical use for years in the form of nonspecific extracellular MRI contrast agents such as Gd-DTPA or Gd-DOTA (154–159).

Another approach MRI investigators have used is targeting macrophages found in atherosclerosis via iron particles in the form of superparamagnetic particles of iron oxide (SPIOs) and ultra-small superparamagnetic particles of iron oxide (USPIOs) (160–167). USPIOs and SPIOs are iron-rich and are principally taken up by mac-rophages and other cells of the reticuloendothelial system (RES). Macrophages are known to play a critical role in pathogenesis of atherosclerosis (112–114). These iron particles, that are injected, accumulate in athero-sclerotic plaque via uptake by macrophages (162,163,165–167). This causes a signal void on conventional MRI causing a marked decrease in signal intensity of the plaques where iron particles accumulate. Kooi et al. performed an interesting investigation using USPIOs on symptomatic patients (n = 11) who then underwent end-arterectomy. They showed that in areas of interest, where MRI changes were seen, there was a 24% decrease in the signal intensity on T2* (165). Furthermore, they demonstrated via pathology that 75% of ruptured or rupture-prone plaques exhibited uptake of USPIOs

whereas only one lesion representing 7% of stable lesions exhibited USPIO uptake (165).

A novel and promising molecular MR-Imaging agent for atherosclerosis is gadolinium-loaded recombi-nant high-density lipoprotein (rHDL) (168). Gadolinium is loaded into rHDL in the form of Gd-DTPA or Gd-DOTA phospholipids (168). Gadolinium-loaded rHDL is relatively easy to reconstitute and is a nanoparticle by definition because of its 7 to 12 nm size (168). Because it is endogenous to humans, it will not cause immune-mediated reactions. One of the physiologic roles of HDL is to circulate in the body and participate in cholesterol efflux from tissues including atheromatous plaques in vessels (169,170). Gadolinium-loaded rHDL (Gd-rHDL) takes advantage of HDL's physiologic role in migrating in and out of atherosclerosis to provide contrast-enhancement on MRI. Gd-rHDL was tested in vivo in apolipoprotein E (Apo-E) knockout mice and demon-strated a significant enhancement of atherosclerotic plaque at 24 hours after venous injection of contrast agent (Fig. 8) (168). Furthermore, by using confocal microscopy on tissue sections from sacrificed mice it was shown that fluorescently labeled rHDL accumu-lated in the macrophages of atheromatous plaques thus demonstrating colocalization of contrast agent and target (168).

Macrophages have been targeted, via the macro-phage scavenger receptor, with the gadolinium-containing molecular contrast agent called immuno-micelles. Gadolinium-containing immunomicelles (Gd-IM) are an interesting molecular MRI contrast agent for atherosclerosis. Immunomicelles are synthe-sized by taking specific antibodies and attaching them to micelles that have gadolinium chelates covalently linked to the polar heads of phospholipids. Immunomicelles can be precisely engineered to be any-where between 20 and 120 nm in diameter making them nanoparticles by definition (171). Lipinski et al. have used immunomicelles targeted to macrophages via the macrophages scavenger receptor to show very strong ex vivo and in vitro enhancement of plaque from

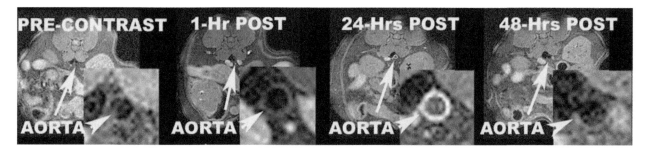

Figure 8 Atherosclerosis enhancement with gadolinium-containing recombinant high-density lipoprotein (rHDL). High density lipopro-tein—high resolution MRI in an Apolipoprotein-E knockout mouse with a baseline precontrast image and matched images at various time-points following injection of gadolinium-containing rHDL nanoparticles. (*The insets are enlargement of the aorta; the arrows point to the aorta.*)

Apo-E knockout mice (172). They were also able to demonstrate colocalization of fluorescently-labeled immunomicelles (fGd-IM) and macrophages in atherosclerotic plaque by using confocal fluorescent microscopy techniques. In the first in vivo study, V. Amirbekian et al. showed that similar immunomicelles, targeted to the macrophage scavenger receptor, produced about 50% enhancement of atheromatous plaque in the abdominal aorta of Apo-E knockout mice (Fig. 9) (173).

Another important pathologic process in atherosclerosis is neovascularization of the plaque (174–176). $\alpha_v\beta_3$ belongs to the family of integrins and is strongly associated with angiogenesis and neovascularization (177–180). Both neovascularization and $\alpha_v\beta_3$ have been targeted for molecular MRI of atherosclerosis (181–183). In a rabbit model, Winter et al. showed regions of neovascularization had a good increase in signal intensity after administration of a $\alpha_v\beta_3$-targeted nanoparticle MRI contrast agent (181).

Thrombosis often plays a detrimental role in the cascade of pathology related to atherosclerosis. Thrombus is associated with complex atherothrombotic plaques and there is evidence that it may contribute to progression of lesions (184). The capability to characterize thrombus via molecular MRI may allow for better detection of both luminal and atherothrombotic plaques. Theoretically, by targeting different components of the coagulation cascade it will be possible to differentiate late thrombus from early thrombus based upon composition. Possible key targets include P-selectin, tissue-factor, fibrin, surface markers of activated platelets and various clotting factors (185–187). There have been a few studies of thrombus imaging using conventional MRI and USPIOs (188,189). Several groups have performed investigations, in animals and on human thrombus, of MRI contrast agents linked to either antibodies or peptide ligands that bind to thrombus and components thereof (190–193). Specifically there have been a couple of studies using a fibrin-specific MRI contrast agent in a carotid crush-injury rabbit or guinea-pig models of atherosclerosis (194,195). This type animal model with crush injury is a very gross

approximation of thrombosis following plaque rupture. Sirol et al. performed a study in 12 guinea pigs using a fibrin-specific MRI contrast agent. They found that using convention MRI prior to injection of contrast agent they could only detect thrombus 42% of the time. However, after injection of the fibrin-specific contrast agent they were able to detect 100% of all thrombi (194). Furthermore, there was an almost 400% increase in signal intensity of thrombus when compared to precontrast scans (Fig. 10) (194). In a different investigation of the fibrin-specific MRI contrast agent, Botnar et al. were able to show a sensitivity and specificity of 100% in detection of arterial thrombi using a chemically induced thrombosis model in rabbits (196).

A new compound called gadofluorine-M is a possible molecular MRI contrast agent for atherosclerosis. Gadofluorine-M is a strong gadolinium chelate complex with a perfluorinated side-chain giving it unique properties. While the exact mechanisms are currently under investigation, these unique properties apparently provide for either accumulation in plaque or retention in plaque. Gadofluorine-M has been shown to enhance the aortic wall in either atherosclerosis-induced NZW rabbits or WHHL rabbits. In a study by Sirol et al., there was a 1.5-fold increase in signal intensity at one hour and two-fold increase at 24 hours after injection of the contrast agent gadofluorine-M (Fig. 11) (197). We also found that there was a correlation between signal intensity on MRI images and the plaque composition from corresponding histopathologic sections of the rabbit aortas (197).

Recently, K.A. Kelly et al. used a peptide conjugated to "magnetofluorescent" CLIO (an iron oxide based agent) particles to image VCAM-1 on endothelial cells (198). VCAM-1 plays a role in leukocyte adhesion and recruitment in the pathogenesis of atherosclerosis. The study showed that this contrast agent could be used to image VCAM-1 over-expression on endothelial cells in an Apo-E deficient mouse model of atherosclerosis (Fig. 12) (198). This approach hints at the potential of imaging inflammation, which is believed to be a central process in atherosclerosis.

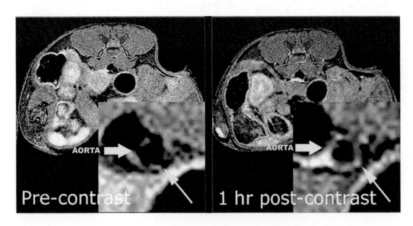

Figure 9 Enhancement of atherosclerosis using immunomicelles. On the *left* is a baseline MRI image in an Apolipoprotein-E knockout mouse. On the *right* is the same slice from the same mouse after injection of gadolinium-containing immunomicelles targeted to macrophages via the macrophage scavenger receptor (MSR). (*The insets are enlargements of the aorta; the arrows indicate the aorta.*)

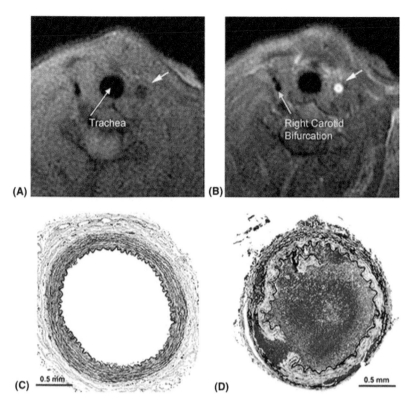

Figure 10 Detection of thrombus using a fibrin-targeted MRI contrast agent. Thrombus was induced via unilateral crush injury of the carotid artery (*short white arrows*). Panel **A** shows an MR image before injection of contrast (*the short arrow is pointing to the thrombus*). Panel **B** shows the same slice in the same animal after injection of the fibrin-targeted agent (*the short white arrow is pointing to the thrombus*). Panels **C** and **D** are normal and thrombosed immunohistopathology sections through the carotid artery.

FUTURE ROLE OF MOLECULAR IMAGING

The capability to noninvasively attain dynamic information about disease processes will give clinicians a powerful tool in managing complex chronic pathologies such as atherosclerosis. As current and emerging developments in basic science translate into clinical practice, we will come into the age of personalized medicine. As it develops further, molecular imaging will provide the power to impart clinicians with intelligence

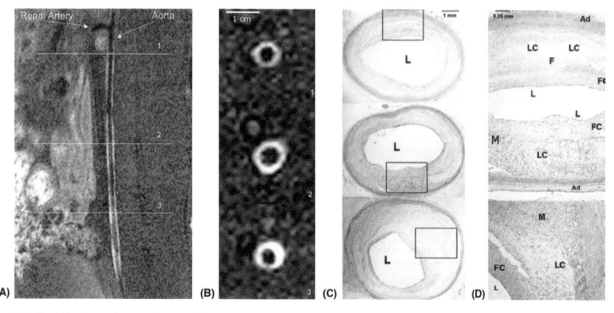

Figure 11 Gadofluorine-enhanced images of the aorta in a rabbit model of atherosclerosis. In vivo MR images [sagittal (**A**) and transverse (**B**)] of the rabbit abdominal aorta 24 hours after injection of gadofluorine showing heterogeneous enhancement in the plaque area. (**C** and **D**) Corresponding histopathological sections of atherosclerotic rabbit abdominal aorta. *Abbreviations*: Ad, adventitia; FC, fibrous cap; L, lumen; LC, lipid core; M, cells including macrophages and smooth muscle cells.

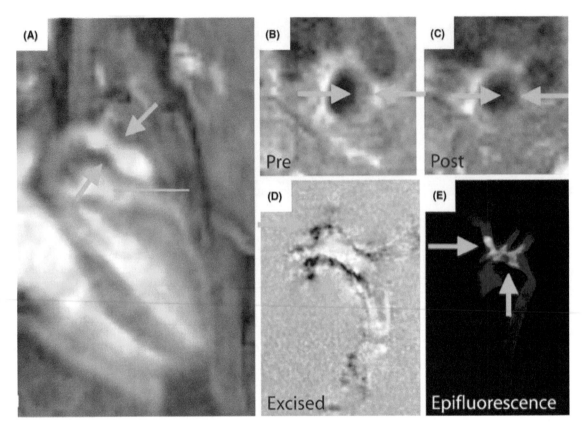

Figure 12 Targeted imaging of vascular cell adhesion molecule-1 (VCAM-1). (**A–C**) In vivo MRI of atherosclerotic lesions in apolipoprotein E -/- mice showing presumed identification of VCAM-1-expressing cells in a murine tumor necrosis factor α-induced inflammatory model. (**D and E**) Ex vivo MRI and fluorescence imaging (of fluorescently labeled VCAM-1 targeted contrast agent) confirming extensive low signal changes in the aortic wall. *Source*: From Ref. 198.

regarding pathological processes. In this way, it will play a significant role in individualized patient management. Furthermore, as technology advances and cost-of-production decreases over time, it will become possible to use molecular imaging as an effective screening tool. This will give clinicians the ability to intervene at early stages of disease before irreversible and costly consequences develop.

ACKNOWLEDGMENTS

We wish to express our gratitude to the National Institutes of Health and the National Heart, Lung, and Blood Institute for their support—grants NIH/NHLBI R01-HL-071021 and R01-HL-078667 (ZAF). We wish to thank the Stanley J. Sarnoff Endowment for Cardiovascular Science for their funding (VA). We appreciate support from The Zena and Michael A. Wiener Cardiovascular Institute, the Department of Radiology and the Department of Cardiology. The authors acknowledge the help of Drs. Michael J. Lipinksi, Juan C. Frias, Karen B. Saebo, Venkatesh Mani, Marc Sirol, Fabien Hyafil, Esad Vucic, Vitalii Itskovich, Silvia Aguiar, Hiroaki Taniguchi, Burton Drayer, John T. Fallon, Edward A. Fisher, Robin P. Choudhury and Messrs. Gabor Mizsei, Daniel D. Samber, and Frank Macaluso. Finally, we thank Lucy Feliciano and Lena Marra for their administrative help.

REFERENCES

1. AHA. Heart Disease And Stroke Statistics 2005 Update. Accessed 04 Jan 2005 http://www.americanheart.org/downloadable/heart/1103829139928HDSStats 2005Update.pdf. In 2005 ed: American Heart Association, 2005.
2. Zheng ZJ, Croft JB, Giles WH, Mensah GA. Sudden cardiac death in the United States, 1989 to 1998. Circulation 2001; 104:2158–2163.
3. Fuster V, Fayad ZA, Badimon JJ. Acute coronary syndromes: biology. Lancet 1999; 353(suppl 2):SII5–SII9.
4. Murray CJ, Lopez AD. Global mortality, disability, and the contribution of risk factors: Global Burden of Disease Study. Lancet 1997; 349:1436–1442.
5. Virmani R, Kolodgie FD, Burke AP, Farb A, Schwartz SM. Lessons from sudden coronary death: a comprehensive morphological classification scheme for atherosclerotic lesions. Arterioscler Thromb Vasc Biol 2000; 20:1262–1275.
6. Kragel AH, Reddy SG, Wittes JT, Roberts WC. Morphometric analysis of the composition of atherosclerotic

plaques in the four major epicardial coronary arteries in acute myocardial infarction and in sudden coronary death. Circulation 1989; 80:1747–1756.

7. Virmani R, Burke A, Farb A. Coronary risk factors and plaque morphology in men with coronary disease who died suddenly. Eur Heart J 1998; 19:678–680.

8. Helft G, Worthley SG, Zhang ZY, et al. Non-invasive in vivo imaging of atherosclerotic lesions using fluorine-18 deoxyglucose (18–FDG) PET correlates with Macrophage content in a rabbit model. Circulation 1999; 100:I–311.

9. Rudd JH, Warburton EA, Fryer TD, et al. Imaging atherosclerotic plaque inflammation with [18F]-fluoro-deoxyglucose positron emission tomography. Circulation 2002; 105:2708–2711.

10. Vallabhajosula S, Fuster V. Atherosclerosis: imaging techniques and the evolving role of nuclear medicine. J Nucl Med 1997; 38:1788–1796.

11. Libby P. Molecular bases of the acute coronary syndromes. Circulation 1995; 91:2844–2850.

12. Libby P, Mach F, Schonbeck U, Bourcier T, Aikawa M. Regulation of the thrombotic potential of atheroma. Thromb Haemost 1999; 82:736–741.

13. Libby P. Current concepts of the pathogenesis of the acute coronary syndromes. Circulation 2001; 104:365–372.

14. Libby P. Vascular biology of atherosclerosis: overview and state of the art. Am J Cardiol 2003; 91:3A–6A.

15. Fuster V, Badimon L, Badimon JJ, Chesebro JH. The pathogenesis of coronary artery disease and the acute coronary syndromes (2). N Engl J Med 1992; 326: 310–318.

16. Fuster V, Badimon L, Badimon JJ, Chesebro JH. The pathogenesis of coronary artery disease and the acute coronary syndromes (1). N Engl J Med 1992; 326:242–250.

17. Fuster V. Mechanisms leading to myocardial infarction: insights from studies of vascular biology. Circulation 1994; 90:2126–2146.

18. Fayad ZA, Fuster V. Clinical imaging of the high-risk or vulnerable atherosclerotic plaque. Circ Res 2001; 89:305–316.

19. Fayad ZA, Fuster V. The human high-risk plaque and its detection by magnetic resonance imaging. Am J Cardiol 2001; 88:42–45.

20. Mani V, Itskovich VV, Szimtenings M, et al. Rapid extended coverage simultaneous multisection black-blood vessel wall MR imaging. Radiology 2004; 232:281–288.

21. Mani V, Itskovich VV, Szimtenings M, et al. A new interleaved multi-slice black blood double inversion recovery technique for vessel wall imaging. Proc Intl Soc Mag Reson Med 2003; in press.

22. Vignaux OB, Augui J, Coste J, et al. Comparison of single-shot fast spin-echo and conventional spin-echo sequences for MR imaging of the heart: initial experience. Radiology 2001; 219:545–550.

23. Song HK, Wright AC, Wolf RL, Wehrli FW. Multislice double inversion pulse sequence for efficient black-blood MRI. Magn Reson Med 2002; 47:616–620.

24. Yarnykh VL, Yuan C. Multislice double inversion-recovery black-blood imaging with simultaneous slice reinversion. J Magn Reson Imaging 2003; 17:478–483.

25. Yarnykh VL, Yuan C. Multislice double inversion-recovery black-blood imaging with simultaneous slicereinversion. J Magn Reson Imaging 2003; 17(4): 478–483.

26. Griswold MA, Jakob PM, Heidemann RM, et al. Generalized autocalibrating partially parallel acquisitions (GRAPPA). Magn Reson Med 2002; 47:1202–1210.

27. Lardo AC, Fayad ZA, Chronos NAF, Fuster V. Cardiovascular Magentic Resonance—Established and Emerging Applications. New York and London: Martin Dunitz—Taylor and Francis Group Inc.; 2003.

28. Parker DL, Goodrich KC, Masiker M, Tsuruda JS, Katzman GL. Improved efficiency in double-inversion fast spin-echo imaging. Magn Reson Med 2002; 47:1017–1021.

29. Itskovich VV, Mani V, Mizsei G, et al. Parallel and non-parallel simultaneous multislice black-blood double inversion recovery techniques for vessel wall imaging. J Magn Reson Imaging 2004; 19:459–467.

30. Itskovich VV, Mani V, Aguinaldo JGS, et al. Fast inter-leaved multi-slice black blood double inversion recovery techniques for vessel wall imaging. Proc Intl Soc Mag Reson Med 2003; in press.

31. Simonetti OP, Finn JP, White RD, Laub G, Henry DA. "Black blood" T2–weighted inversion-recovery MR imaging of the heart. Radiology 1996; 199:49–57.

32. Fayad ZA, Fuster V, Fallon JT, et al. Noninvasive In vivo human coronary artery lumen and wall imaging using black-blood magnetic resonance imaging. Circulation 2000; 102:506–510.

33. Nishimura DG, Macovski A, Pauly JM. Considerations of magnetic resonance angiography by selective inversion recovery. Magn Reson Med 1988; 7:472–484.

34. Fayad ZA, Connick TJ, Axel L. An improved quadra-ture or phased-array coil for MR cardiac imaging. Magn Reson Med 1995; 34:186–193.

35. Hayes CE, Mathis CM, Yuan C. Surface coil phased arrays for high-resolution imaging of the carotid arter-ies. J Magn Reson Imaging 1996; 6:109–112.

36. Fayad ZA, Hardy CJ, Giaquinto R, Kini A, Sharma S. Improved high resolution MRI of human coronary lumen and plaque with a new cardiac coil. Circulation 2000; 102:II-399.

37. Fayad ZA. The assessment of the vulnerable athero-sclerotic plaque using MR imaging: a brief review. Int J Card Imaging 2001; 17:165–177.

38. Fayad ZA, Fuster V. Characterization of atheroscle-rotic plaques by magnetic resonance imaging. Ann N Y Acad Sci 2000; 902:173–186.

39. Merickel MB, Carman CS, Brookeman JR, Mugler JPd, Brown MF, Ayers CR. Identification and 3–D quantifi-cation of atherosclerosis using magnetic resonance imaging. Comput Biol Med 1988; 18:89–102.

40. Yuan C, Tsuruda JS, Beach KN, et al. Techniques for high-resolution MR imaging of atherosclerotic plaque. J Magn Reson Imaging 1994; 4:43–49.

41. Yuan C, Murakami JW, Hayes CE, et al. Phased-array magnetic resonance imaging of the carotid artery bifurcation: preliminary results in healthy volunteers and a patient with atherosclerotic disease. J Magn Reson Imaging 1995; 5:561–565.

42. Martin AJ, Gotlieb AI, Henkelman RM. High-resolution MR imaging of human arteries. J Magn Reson Imaging 1995; 5:93–100.

43. Toussaint JF, LaMuraglia GM, Southern JF, Fuster V, Kantor HL. Magnetic resonance images lipid, fibrous, calcified, hemorrhagic, and thrombotic components of human atherosclerosis in vivo. Circulation 1996; 94:932–938.

44. Toussaint JF, Southern JF, Fuster V, Kantor HL. T2–weighted contrast for NMR characterization of human atherosclerosis. Arterioscler Thromb Vasc Biol 1995; 15:1533–1542.

45. Worthley SG, Helft G, Fuster V, et al. High resolution ex vivo magnetic resonance imaging of in situ coronary and aortic atherosclerotic plaque in a porcine model. Atherosclerosis 2000; 150:321–329.

46. von Ingersleben G, Schmiedl UP, Hatsukami TS, et al. Characterization of atherosclerotic plaques at the carotid bifurcation: correlation of high-resolution MR imaging with histologic analysis—preliminary study. Radiographics 1997; 17:1417–1423.

47. Fisel CR, Ackerman JL, Buxton RB, et al. MR contrast due to microscopically heterogeneous magnetic susceptibility: numerical simulations and applications to cerebral physiology. Magn Reson Med 1991; 17:336–347.

48. Witztum JL, Steinberg D. Role of oxidized low density lipoprotein in atherogenesis. J Clin Invest 1991; 88:1785–1792.

49. Rapp JH, Connor WE, Lin DS, Inahara T, Porter JM. Lipids of human atherosclerotic plaques and xanthomas: clues to the mechanism of plaque progression. J Lipid Res 1983; 24:1329–1335.

50. Edzes HT, Samulski ET. Cross relaxation and spin diffusion in the proton NMR or hydrated collagen. Nature 1977; 265:521–523.

51. Rapoport S, Sostman HD, Pope C, Camputaro CM, Holcomb W, Gore JC. Venous clots: evaluation with MR imaging. Radiology 1987; 162:527–530.

52. Bass JC, Hedlund LW, Sostman HD. MR imaging of experimental and clinical thrombi at 1.5 T. Magn Reson Imaging 1990; 8:631–635.

53. Bryant RG, Marill K, Blackmore C, Francis C. Magnetic relaxation in blood and blood clots. Magn Reson Med 1990; 13:133–144.

54. Erdman WA, Jayson HT, Redman HC, Miller GL, Parkey RW, Peshock RW. Deep venous thrombosis of extremities: role of MR imaging in the diagnosis. Radiology 1990; 174:425–431.

55. Totterman S, Francis CW, Foster TH, Brenner B, Marder VJ, Bryant RG. Diagnosis of femoropopliteal venous thrombosis with MR imaging: a comparison of four MR pulse sequences [see comments]. AJR Am J Roentgenol 1990; 154:175–178.

56. Murray JG, Manisali M, Flamm SD, et al. Intramural hematoma of the thoracic aorta: MR image findings and their prognostic implications. Radiology 1997; 204:349–355.

57. Bluemke DA. Definitive diagnosis of intramural hematoma of the thoracic aorta with MR imaging [editorial; comment]. Radiology 1997; 204:319–321.

58. Bradley WG, Jr. MR appearance of hemorrhage in the brain. Radiology 1993; 189:15–26.

59. Kucharczyk W, Henkelman RM. Visibility of calcium on MR and CT: can MR show calcium that CT cannot? AJNR 1994; 15:1145–1148.

60. Kaufman L, Crooks LE, Sheldon PE, Rowan W, Miller T. Evaluation of NMR imaging for detection and quantification of obstructions in vessels. Invest Radiology 1982; 17:554–560.

61. Itskovich VV, Samber DD, Mani V, et al. Quantification of human atherosclerotic plaques using spatially enhanced cluster analysis of multicontrast-weighted magnetic resonance images. Magn Reson Med 2004; 52:515–523.

62. Nikolaou K, Becker CR, Muders M, et al. Multidetector-row computed tomography and magnetic resonance imaging of atherosclerotic lesions in human ex vivo coronary arteries. Atherosclerosis 2004; 174:243–252.

63. Shinnar M, Fallon JT, Wehrli S, et al. The diagnostic accuracy of ex vivo magnetic resonance imaging for human atherosclerotic plaque characterization. Arterioscler Thromb Vasc Biol 1999; 19:2756–2761.

64. Toussaint JF, Southern JF, Fuster V, Kantor HL. Water diffusion properties of human atherosclerosis and thrombosis measured by pulse field gradient nuclear magnetic resonance. Arterioscler Thromb Vasc Biol 1997; 17:542–546.

65. Worthley SG, Helft G, Fuster V, et al. Noninvasive in vivo magnetic resonance imaging of experimental coronary artery lesions in a porcine model. Circulation 2000; 101:2956–2961.

66. Botnar RM, Stuber M, Kissinger KV, Kim WY, Spuentrup E, Manning WJ. Noninvasive coronary vessel wall and plaque imaging with magnetic resonance imaging. Circulation 2000; 102:2582–2587.

67. Botnar RM, Stuber M, Lamerichs R, et al. Initial experiences with in vivo right coronary artery human MR vessel wall imaging at 3 tesla. J Cardiovasc Magn Reson 2003; 5:589–594.

68. Coulden RA, Moss H, Graves MJ, Lomas DJ, Appleton DS, Weissberg PL. High resolution magnetic resonance imaging of atherosclerosis and the response to balloon angioplasty. Heart 2000; 83:188–191.

69. Little WC, Constantinescu M, Applegate RJ, et al. Can coronary angiography predict the site of a subsequent myocardial infarction in patients with mild-to-moderate coronary artery disease? Circulation 1988; 78:1157–1166.

70. Ambrose JA, Tannenbaum MA, Alexopoulos D, et al. Angiographic progression of coronary artery disease and the development of myocardial infarction. J Am Coll Cardiol 1988; 12:56–62.

71. Davila-Roman VG, Barzilai B, Wareing TH, Murphy SF, Schechtman KB, Kouchoukos NT. Atherosclerosis of the ascending aorta. Prevalence and role as an independent predictor of cerebrovascular events in cardiac patients. Stroke 1994; 25:2010–2016.

72. Amarenco P, Cohen A, Tzourio C, et al. Atherosclerotic disease of the aortic arch and the risk of ischemic stroke. N Engl J Med 1994; 331:1474–1479.

73. Amarenco P, Duyckaerts C, Tzourio C, Henin D, Bousser MG, Hauw JJ. The prevalence of ulcerated plaques in the aortic arch in patients with stroke. N Engl J Med 1992; 326:221–225.

74. Jaffer FA, O'Donnell CJ, Kissinger KV, et al. MRI assessment of aortic atherosclerosis in an asymptomatic population: The Framingham Heart Study. Circulation 2000; 102:II-458.

75. O'Donnell CJ, Larson MG, Jaffer FA, Kissinger KV, Levy D, Manning WJ. Aortic atherosclerosis detected by MRI is associated with contemporaneous and longitudinal risk factors: The Framingham Heart Study (FHS). Circulation 2000; 102:II-836.

76. Taniguchi H, Momiyama Y, Fayad ZA, et al. In vivo magnetic resonance evaluation of associations between aortic atherosclerosis and both risk factors and coronary artery disease in patients referred for coronary angiography. Am Heart J 2004; 148:137–143.

77. Fayad ZA, Nahar T, Fallon JT, et al. In vivo magnetic resonance evaluation of atherosclerotic plaques in the human thoracic aorta: a comparison with transesophageal echocardiography. Circulation 2000; 101: 2503–2509.

78. Summers RM, Andrasko-Bourgeois J, Feuerstein IM, et al. Evaluation of the aortic root by MRI: insights from patients with homozygous familial hypercholesterolemia. Circulation 1998; 98:509–518.

79. Corti R, Fayad ZA, Fuster V, et al. Effects of lipid-lowering by simvastatin on human atherosclerotic lesions: a longitudinal study by high-resolution, noninvasive magnetic resonance imaging. Circulation 2001; 104:249–252.

80. Corti R, Fuster V, Fayad ZA, et al. Lipid lowering by simvastatin induces regression of human atherosclerotic lesions: two years' follow-up by high-resolution noninvasive magnetic resonance imaging. Circulation 2002; 106:2884–2887.

81. Worthley SG, Helft G, Osende JI, et al. Serial evaluation of atherosclerosis with in vivo MRI: Study of Atorvastatin and Avasimibe in WHHL Rabbits. Circulation 2000; 102:II-809.

82. McConnell MV, Aikawa M, Maier SE, Ganz P, Libby P, Lee RT. MRI of rabbit atherosclerosis in response to dietary cholesterol lowering. Arterioscler Thromb Vasc Biol 1999; 19:1956–1959.

83. Helft G, Worthley SG, Fuster V, et al. Progression and regression of atherosclerotic lesions: monitoring with serial noninvasive magnetic resonance imaging. Circulation 2002; 105:993–998.

84. Corti R, Fuster V, Fayad ZA, et al. Effects of aggressive versus conventional lipid-lowering therapy by simvastatin on human atherosclerotic lesions: a prospective, randomized, double-blind trial with high-resolution magnetic resonance imaging. J Am Coll Cardiol 2005; 46:106–112.

85. Yonemura A, Momiyama Y, Fayad ZA, et al. Effect of lipid-lowering therapy with atorvastatin on atherosclerotic aortic plaques detected by noninvasive magnetic resonance imaging. J Am Coll Cardiol 2005; 45:733–742.

86. Broderick J, Brott T, Kothari R, et al. The Greater Cincinnati/Northern Kentucky Stroke Study: preliminary first-ever and total incidence rates of stroke among blacks. Stroke 1998; 29:415–421.

87. Wasserman BA. Clinical carotid atherosclerosis. Neuroimaging Clin N Am 2002; 12:403–419.

88. Executive committee for the Asymtomatic Carotid Atherosclerosis Study. Endarterectomy for asymtomatic carotid stenosis. J Am Med Assoc 1995; 273:1421–1428.

89. Nagai Y, Kitagawa K, Matsumoto M. Implication of earlier carotid atherosclerosis for stroke and its subtypes. Prev Cardiol 2003; 6:99–103.

90. Garcia JH, Khang-Loon H. Carotid atherosclerosis. Definition, pathogenesis, and clinical significance. Neuroimaging Clin N Am 1996; 6:801–810.

91. Glagov S, Zarins C, Giddens DP, Ku DN. Hemodynamics and atherosclerosis. Insights and perspectives gained from studies of human arteries. Arch Pathol Lab Med 1988; 112:1018–1031.

92. Goldstein LB, Adams R, Becker K, et al. Primary prevention of ischemic stroke: a statement for healthcare professionals from the Stroke Council of the American Heart Association. Circulation 2001; 103:163–182.

93. Yuan C, Beach KW, Smith LH Jr., Hatsukami TS. Measurement of atherosclerotic carotid plaque size in vivo using high resolution magnetic resonance imaging. Circulation 1998; 98:2666–2671.

94. Hatsukami TS, Ross R, Polissar NL, Yuan C. Visualization of fibrous cap thickness and rupture in human atherosclerotic carotid plaque In vivo with high-resolution magnetic resonance imaging. Circulation 2000; 102:959–964.

95. Mitsumori LM, Hatsukami TS, Ferguson MS, Kerwin WS, Cai J, Yuan C. In vivo accuracy of multisequence MR imaging for identifying unstable fibrous caps in advanced human carotid plaques. J Magn Reson Imaging 2003; 17:410–420.

96. Luo Y, Polissar N, Han C, Yarnykh V, Kerwin WS, Hatsukami TS, Yuan C. Accuracy and uniqueness of three in vivo measurements of atherosclerotic carotid plaque morphology with black blood MRI. Magn Reson Med 2003; 50:75–82.

97. Chu B, Kampschulte A, Ferguson MS, et al. Hemorrhage in the atherosclerotic carotid plaque: a high-resolution MRI study. Stroke 2004; 35:1079–1084.

98. Chu B, Hatsukami TS, Polissar NL, et al. Determination of carotid artery atherosclerotic lesion type and distribution in hypercholesterolemic patients with moderate carotid stenosis using noninvasive magnetic resonance imaging. Stroke 2004; 35:2444–2448.

99. Cai JM, Hatsukami TS, Ferguson MS, Small R, Polissar NL, Yuan C. Classification of human carotid atherosclerotic lesions with in vivo multicontrast magnetic resonance imaging. Circulation 2002; 106:1368–1373.

100. Saam T, Ferguson MS, Yarnykh VL, et al. Quantitative evaluation of carotid plaque composition by in vivo MRI. Arterioscler Thromb Vasc Biol 2005; 25:234–239.

101. Zhang S, Cai J, Luo Y, et al. Measurement of carotid wall volume and maximum area with contrast-enhanced 3D MR imaging: initial observations. Radiology 2003; 228:200–205.

102. Cappendijk VC, Cleutjens KB, Heeneman S, et al. In vivo detection of hemorrhage in human atherosclerotic plaques with magnetic resonance imaging. J Magn Reson Imaging 2004; 20:105–110.

103. Moody AR. Magnetic resonance direct thrombus imaging. J Thromb Haemost 2003; 1:1403–1409.

104. Moody AR, Murphy RE, Morgan PS, et al. Characterization of complicated carotid plaque with magnetic resonance direct thrombus imaging in patients with cerebral ischemia. Circulation 2003; 107: 3047–3052.

105. Yuan C, Kerwin WS, Ferguson MS, et al. Contrast-enhanced high resolution MRI for atherosclerotic carotid artery tissue characterization. J Magn Reson Imaging 2002; 15:62–67.

106. Zerhouni E. Medicine. The NIH Roadmap. Science 2003; 302:63–72.

107. Choudhury RP, Fuster V, Fayad ZA. Molecular, cellular and functional imaging of atherothrombosis. Nat Rev Drug Discov 2004; 3:913–925.

108. Lipinski MJ, Fuster V, Fisher EA, Fayad ZA. Targeting of biological molecules for evaluation of high-risk atherosclerotic plaques with magnetic resonance imaging. Nat Clin Pract Cardiovasc Med 2004; 1:48–55.

109. Naghavi M, Libby P, Falk E, et al. From vulnerable plaque to vulnerable patient: a call for new definitions and risk assessment strategies: Part I. Circulation 2003; 108:1664–1672.

110. Naghavi M, Libby P, Falk E, et al. From vulnerable plaque to vulnerable patient: a call for new definitions and risk assessment strategies: Part II. Circulation 2003; 108:1772–1778.

111. Granada JF, Kaluza GL, Raizner AE, Moreno PR. Vulnerable plaque paradigm: prediction of future clinical events based on a morphological definition. Catheter Cardiovasc Interv 2004; 62:364–374.

112. Hansson GK. Inflammation, atherosclerosis, and coronary artery disease. N Engl J Med 2005; 352:1685–1695.

113. Smith JD, Trogan E, Ginsberg M, Grigaux C, Tian J, Miyata M. Decreased atherosclerosis in mice deficient in both macrophage colony-stimulating factor (op) and apolipoprotein E. Proc Natl Acad Sci U S A 1995; 92:8264–8268.

114. Ross R. Atherosclerosis—an inflammatory disease. N Engl J Med 1999; 340:115–126.

115. Falk E. Plaque rupture with severe pre-existing stenosis precipitating coronary thrombosis. Characteristics of coronary atherosclerotic plaques underlying fatal occlusive thrombi. Br Heart J 1983; 50:127–134.

116. Farb A, Burke AP, Tang AL, et al. Coronary plaque erosion without rupture into a lipid core. A frequent cause of coronary thrombosis in sudden coronary death. Circulation 1996; 93:1354–1363.

117. van der Wal AC, Becker AE, van der Loos CM, Das PK. Site of intimal rupture or erosion of thrombosed coronary atherosclerotic plaques is characterized by an inflammatory process irrespective of the dominant plaque morphology. Circulation 1994; 89:36–44.

118. Cohade C, Wahl RL. Applications of positron emission tomography/computed tomography image fusion in clinical positron emission tomography-clinical use, interpretation methods, diagnostic improvements. Semin Nucl Med 2003; 33:228–237.

119. Beyer T, Townsend DW, Brun T, et al. A combined PET/CT scanner for clinical oncology. J Nucl Med 2000; 41:1369–1379.

120. Townsend DW. A combined PET/CT scanner: the choices. J Nucl Med 2001; 42:533–534.

121. Chin BB, Patel PV, Nakamoto Y, et al. Quantitative evaluation of 2-deoxy-2-[18F] fluoro-D-glucose uptake in hepatic metastases with combined PET-CT: iterative reconstruction with CT attenuation correction versus filtered back projection with 68Germanium attenuation correction. Mol Imaging Biol 2002; 4:399–409.

122. Burger C, Goerres G, Schoenes S, Buck A, Lonn AH, Von Schulthess GK. PET attenuation coefficients from CT images: experimental evaluation of the transformation of CT into PET 511-keV attenuation coefficients. Eur J Nucl Med Mol Imaging 2002; 29:922–927.

123. Kamel E, Hany TF, Burger C, et al. CT vs 68Ge attenuation correction in a combined PET/CT system: evaluation of the effect of lowering the CT tube current. Eur J Nucl Med Mol Imaging 2002; 29:346–350.

124. Vallabhajosula S. Radioisotopic imaging of atheroma. In: Fuster V, ed. The Vulnerable Atherosclerotic Plaque: Understanding, Identification, and Modification. Armonk, NY: Futura Publishing; 1999:213–229.

125. Tsimikas S, Palinski W, Halpern SE, Yeung DW, Curtiss LK, Witztum JL. Radiolabeled MDA2, an oxidation-specific, monoclonal antibody, identifies native atherosclerotic lesions in vivo. J Nucl Cardiol 1999; 6:41–53.

126. Tsimikas S, Shortal BP, Witztum JL, Palinski W. In vivo uptake of radiolabeled MDA2, an oxidation-specific monoclonal antibody, provides an accurate measure of atherosclerotic lesions rich in oxidized ldl and is highly sensitive to their regression. Arterioscler Thromb Vasc Biol 2000; 20:689–697.

127. Tsimikas S. Noninvasive imaging of oxidized low-density lipoprotein in atherosclerotic plaques with tagged oxidation-specific antibodies*1. Am J Cardiol 2002; 90:L22–L27.

128. Bjorkerud S, Bjorkerud B. Apoptosis is abundant in human atherosclerotic lesions, especially in inflammatory cells (macrophages and T cells), and may contribute to the accumulation of gruel and plaque instability. Am J Pathol 1996; 149:367–380.

129. Geng YJ, Henderson LE, Levesque EB, Muszynski M, Libby P. Fas is expressed in human atherosclerotic intima and promotes apoptosis of cytokine-primed human vascular smooth muscle cells. Arterioscler Thromb Vasc Biol 1997; 17:2200–2208.

130. Kolodgie FD, Petrov A, Virmani R, et al. Targeting of apoptotic macrophages and experimental atheroma with radiolabeled annexin V: a technique with potential for noninvasive imaging of vulnerable plaque. Circulation 2003; 108:3134–3139.

131. Kietselaer BL, Reutelingsperger CP, Heidendal GA, et al. Noninvasive detection of plaque instability with use of radiolabeled annexin A5 in patients with carotid-artery atherosclerosis. N Engl J Med 2004; 350: 1472–1473.

132. Lederman RJ, Raylman RR, Fisher SJ, et al. Detection of atherosclerosis using a novel positron-sensitive probe and 18-fluorodeoxyglucose (FDG). Nucl Med Commun 2001; 22:747–753.

133. Mochizuki Y, Fujii H, Yasuda S, et al. FDG accumulation in aortic walls. Clin Nucl Med 2001; 26:68–69.

134. Yun M, Yeh D, Araujo LI, Jang S, Newberg A, Alavi A. F-18 FDG uptake in the large arteries: a new observation. Clin Nucl Med 2001; 26:314–319.

135. Machac J ZZ, Nunez R, Chen WT, Macapinlac, HA LS. The relation of F-18 FDG uptake in human thoracic aortas and risk factors for CAD (abstr). J Nucl Med 2002; 43:190P.

136. Yun M, Jang S, Cucchiara A, Newberg AB, Alavi A. 18F FDG uptake in the large arteries: a correlation study with the atherogenic risk factors. Semin Nucl Med 2002; 32:70–76.

137. Tatsumi M, Cohade C, Nakamoto Y, Wahl RL. Fluorodeoxyglucose uptake in the aortic wall at PET/CT: possible finding for active atherosclerosis. Radiology 2003; 229:831–837.

138. Ogawa M, Ishino S, Mukai T, et al. (18)F-FDG accumulation in atherosclerotic plaques: immunohistochemical and PET imaging study. J Nucl Med 2004; 45:1245–1250.

139. Tawakol A, Migrino RQ, Hoffmann U, et al. Noninvasive in vivo measurement of vascular inflammation with F-18 fluorodeoxyglucose positron emission tomography. J Nucl Cardiol 2005; 12:294–301.

140. Lindner JR. Microbubbles in medical imaging: current applications and future directions. Nat Rev Drug Discov 2004; 3:527–532.

141. Dayton PA, Morgan KE, Klibanov AL, Brandenburger GHF, KW. Optical and acoustical observations of the effects of ultrasound on contrast agents. IEEE Trans Ultrason Ferroelect Freq Contr 1999; 46:220–232.

142. Patel D, Dayton P, Gut J, Wisner E, Ferrara KW. Optical and acoustical interrogation of submicron contrast agents. IEEE Trans Ultrason Ferroelectr Freq Control 2002; 49:1641–1651.

143. DeJong N, Hoff L, Skotland T, Bom N. Absorption and scatter of encapsulated gas filled microspheres: theoretical considerations and some measurements. Ultrasonics 1992; 30:95–103.

144. Medwin H. Counting bubbles acoustically: a review. Ultrasonics 1977; 15:7–13.

145. Leong-Poi H, Song J, Rim SJ, Christiansen J, Kaul S, Lindner JR. Influence of microbubble shell properties on ultrasound signal: implications for low-power perfusion imaging. J Am Soc Echocardiogr 2002; 15:1269–1276.

146. Chomas JE, Dayton P, Allen J, Morgan K, Ferrara KW. Mechanisms of contrast agent destruction. IEEE Trans Ultrason Ferroelectr Freq Control 2001; 48:232–248.

147. Villanueva FS, Jankowski RJ, Klibanov S, et al. Microbubbles targeted to intercellular adhesion molecule-1 bind to activated coronary artery endothelial cells. Circulation 1998; 98:1–5.

148. Lindner JR, Song J, Christiansen J, Klibanov AL, Xu F, Ley K. Ultrasound assessment of inflammation and renal tissue injury with microbubbles targeted to P-selectin. Circulation 2001; 104:2107–2112.

149. Leong-Poi H, Christiansen J, Klibanov AL, Kaul S, Lindner JR. Noninvasive assessment of angiogenesis by ultrasound and microbubbles targeted to alpha(v)-integrins. Circulation 2003; 107:455–460.

150. Leong-Poi H, Christiansen J, Heppner P, et al. Assessment of endogenous and therapeutic arteriogenesis by contrast ultrasound molecular imaging of integrin expression. Circulation 2005; 111:3248–3254.

151. Ellegala DB, Leong-Poi H, Carpenter JE, et al. Imaging tumor angiogenesis with contrast ultrasound and microbubbles targeted to alpha(v)beta3. Circulation 2003; 108:336–341.

152. Hamilton AJ, Huang SL, Warnick D, et al. Intravascular ultrasound molecular imaging of atheroma components in vivo. J Am Coll Cardiol 2004; 43:453–460.

153. Hamilton A, Huang SL, Warnick D, et al. Left ventricular thrombus enhancement after intravenous injection of echogenic immunoliposomes: studies in a new experimental model. Circulation 2002; 105:2772–2778.

154. Niendorf HP, Alhassan A, Balzer T, Claub W, Geens V. Safety and Risk of Gadolinium-DTPA: Extended Clinical Experience After More Than 20 Million Applications. 3rd rev ed. Berlin: Blackwell Wissenschafts-Verlag GmbH; 1996.

155. Weinmann HJ, Laniado M, Mutzel W. Pharmacokinetics of GdDTPA/dimeglumine after intravenous injection into healthy volunteers. Physiol Chem Phys Med NMR 1984; 16:167–172.

156. Weinmann HJ, Brasch RC, Press WR, Wesbey GE. Characteristics of gadolinium-DTPA complex: a potential NMR contrast agent. AJR Am J Roentgenol 1984; 142:619–624.

157. Carr DH, Brown J, Bydder GM, et al. Gadolinium-DTPA as a contrast agent in MRI: initial clinical experience in 20 patients. AJR Am J Roentgenol 1984; 143:215–224.

158. Brugieres P, Gaston A, Degryse HR, et al. Randomised double blind trial of the safety and efficacy of two gadolinium complexes (Gd-DTPA and Gd-DOTA). Neuroradiology 1994; 36:27–30.

159. Le Mignon MM, Chambon C, Warrington S, Davies R, Bonnemain B. Gd-DOTA. Pharmacokinetics and tolerability after intravenous injection into healthy volunteers. Invest Radiol 1990; 25:933–937.

160. Schmitz SA, Coupland SE, Gust R, et al. Superparamagnetic iron oxide-enhanced MRI of atherosclerotic plaques in Watanabe hereditable hyperlipidemic rabbits. Invest Radiol 2000; 35:460–471.

161. Schmitz SA, Taupitz M, Wagner S, Wolf KJ, Beyersdorff D, Hamm B. Magnetic resonance imaging of atherosclerotic plaques using superparamagnetic iron oxide particles. J Magn Reson Imaging 2001; 14:355–361.

162. Ruehm SG, Corot C, Vogt P, Cristina H, Debatin JF. Ultrasmall superparamagnetic iron oxide-enhanced MR imaging of atherosclerotic plaque in hyperlipidemic rabbits. Acad Radiol 2002; 9(suppl 1): S143–S144.

163. Trivedi R, J UK-I, Gillard J. Accumulation of ultrasmall superparamagnetic particles of iron oxide in human atherosclerotic plaque. Circulation 2003; 108:e140; author reply e140.

164. Litovsky S, Madjid M, Zarrabi A, Casscells SW, Willerson JT, Naghavi M. Superparamagnetic iron oxide-based method for quantifying recruitment of monocytes to mouse atherosclerotic lesions in vivo: enhancement by tissue necrosis factor-alpha, interleukin-1beta, and interferon-gamma. Circulation 2003; 107:1545–1549.

165. Kooi ME, Cappendijk VC, Cleutjens KB, et al. Accumulation of ultrasmall superparamagnetic particles of iron oxide in human atherosclerotic plaques can be detected by in vivo magnetic resonance imaging. Circulation 2003; 107:2453–2458.

166. Ruehm SG, Corot C, Vogt P, Kolb S, Debatin JF. Magnetic resonance imaging of atherosclerotic plaque

with ultrasmall superparamagnetic particles of iron oxide in hyperlipidemic rabbits. Circulation 2001; 103:415–422.

167. Trivedi RA, JM UK-I, Graves MJ, et al. In vivo detection of macrophages in human carotid atheroma: temporal dependence of ultrasmall superparamagnetic particles of iron oxide-enhanced MRI. Stroke 2004; 35:1631–1635.

168. Frias JC, Williams KJ, Fisher EA, Fayad ZA. Recombinant HDL-like nanoparticles: a specific contrast agent for MRI of atherosclerotic plaques. J Am Chem Soc 2004; 126:16316–16317.

169. von Eckardstein A, Hersberger M, Rohrer L. Current understanding of the metabolism and biological actions of HDL. Curr Opin Clin Nutr Metab Care 2005; 8:147–152.

170. Assmann G, Gotto AM, Jr. HDL cholesterol and protective factors in atherosclerosis. Circulation 2004; 109: III8–14.

171. Frias JC, Aguinaldo JG, Aime S, Fallon JT, Fayad ZA. Visualizing atherosclerotic plaques with micelles: does size matter? Proc Intl Soc Mag Reson Med 2004; 12:1700.

172. Lipinski MJ, Frias JC, Aguinaldo JG, et al. In-vivo and in-vitro uptake of gadolinium-containing immunomicelles in a macrophage cell line: detection of atherosclerotic plaque using MRI. Proc Intl Soc Mag Reson Med 2004; 12:1701.

173. Amirbekian V, Lipinski MJ, Frias JC, Aguinaldo JG, Mani V, Fayad ZA. In vivo MR imaging of apolipoprotein-E knockout mice to detect atherosclerosis with gadolinium containing micelles and immunomicelles molecularly targeted to macrophages. Proc Intl Soc Mag Reson Med 2005; 13:1760.

174. Fleiner M, Kummer M, Mirlacher M, et al. Arterial neovascularization and inflammation in vulnerable patients: early and late signs of symptomatic atherosclerosis. Circulation 2004; 110:2843–2850.

175. Moulton KS, Vakili K, Zurakowski D, et al. Inhibition of plaque neovascularization reduces macrophage accumulation and progression of advanced atherosclerosis. Proc Natl Acad Sci U S A 2003; 100:4736–4741.

176. Kumamoto M, Nakashima Y, Sueishi K. Intimal neovascularization in human coronary atherosclerosis: its origin and pathophysiological significance. Hum Pathol 1995; 26:450–456.

177. Varner JA, Brooks PC, Cheresh DA. REVIEW: the integrin alpha V beta 3: angiogenesis and apoptosis. Cell Adhes Commun 1995; 3:367–374.

178. Hoshiga M, Alpers CE, Smith LL, Giachelli CM, Schwartz SM. Alpha-v beta-3 integrin expression in normal and atherosclerotic artery. Circ Res 1995; 77:1129–1135.

179. Brooks PC, Montgomery AM, Rosenfeld M, et al. Integrin alpha v beta 3 antagonists promote tumor regression by inducing apoptosis of angiogenic blood vessels. Cell 1994; 79:1157–1164.

180. Eliceiri BP, Cheresh DA. The role of alphav integrins during angiogenesis. Mol Med 1998; 4:741–750.

181. Winter PM, Morawski AM, Caruthers SD, et al. Molecular imaging of angiogenesis in early-stage atherosclerosis with alpha(v)beta3–integrin-targeted nanoparticles. Circulation 2003; 108:2270–2274.

182. Anderson SA, Rader RK, Westlin WF, et al. Magnetic resonance contrast enhancement of neovasculature with alpha(v)beta(3)-targeted nanoparticles. Magn Reson Med 2000; 44:433–439.

183. Kerwin W, Hooker A, Spilker M, et al. Quantitative magnetic resonance imaging analysis of neovasculature volume in carotid atherosclerotic plaque. Circulation 2003; 107:851–856.

184. Fuster V. [Thrombus remodeling. Key factor in the progression of coronary atherosclerosis]. Rev Esp Cardiol 2000; 53(suppl 1):2–7.

185. Corti R, Hutter R, Badimon JJ, Fuster V. Evolving concepts in the triad of atherosclerosis, inflammation and thrombosis. J Thromb Thrombolysis 2004; 17:35–44.

186. Schenone M, Furie BC, Furie B. The blood coagulation cascade. Curr Opin Hematol 2004; 11:272–277.

187. Shah PK. Insights into the molecular mechanisms of plaque rupture and thrombosis. Indian Heart J 2005; 57:21–30.

188. Johnstone MT, Botnar RM, Perez AS, et al. In vivo magnetic resonance imaging of experimental thrombosis in a rabbit model. Arterioscler Thromb Vasc Biol 2001; 21:1556–1560.

189. Schmitz SA, Winterhalter S, Schiffler S, et al. USPIO-enhanced direct MR imaging of thrombus: preclinical evaluation in rabbits. Radiology 2001; 221:237–243.

190. Flacke S, Fischer S, Scott MJ, et al. Novel MRI contrast agent for molecular imaging of fibrin: implications for detecting vulnerable plaques. Circulation 2001; 104:1280–1285.

191. Yu X, Song SK, Chen J, et al. High-resolution MRI characterization of human thrombus using a novel fibrin-targeted paramagnetic nanoparticle contrast agent. Magn Reson Med 2000; 44:867–872.

192. Winter PM, Caruthers SD, Yu X, et al. Improved molecular imaging contrast agent for detection of human thrombus. Magn Reson Med 2003; 50:411–416.

193. Johansson LO, Bjornerud A, Ahlstrom HK, Ladd DL, Fujii DK. A targeted contrast agent for magnetic resonance imaging of thrombus: implications of spatial resolution. J Magn Reson Imaging 2001; 13:615–618.

194. Sirol M, Aguinaldo JGS, Graham G, et al. Fibrin-targeted contrast agent for improvement of in vivo acute thrombus detection with magnetic resonance imaging. Atherosclerosis 2005; in press.

195. Sirol M, Fuster V, Badimon JJ, Fallon JT, Toussaint JF, Fayad ZA. Chronic thrombus detection using in-vivo magnetic resonance imaging and fibrin-targeted contrast agent. Circulation 2005; in press.

196. Botnar RM, Perez AS, Witte S, et al. In vivo molecular imaging of acute and subacute thrombosis using a fibrin-binding magnetic resonance imaging contrast agent. Circulation 2004; 109:2023–2029.

197. Sirol M, Itskovich VV, Mani V, et al. Lipid-rich atherosclerotic plaques detected by gadofluorine-enhanced in vivo magnetic resonance imaging. Circulation 2004; 109:2890–2896.

198. Kelly KA, Allport JR, Tsourkas A, Shinde-Patil VR, Josephson L, Weissleder R. Detection of vascular adhesion molecule-1 expression using a novel multimodal nanoparticle. Circ Res 2005; 96:327–336.

18

Potential Targets for Imaging Atherosclerosis

David N. Smith, Mehran M. Sadeghi, and Jeffrey R. Bender

Divisions of Cardiovascular Medicine and Immunobiology, Raymond and Beverly Sackler Foundation Cardiovascular Laboratory, Yale University School of Medicine, New Haven, Connecticut, U.S.A.

NORMAL BLOOD VESSEL

The normal blood vessel is comprised of three distinct layers or tunicae divided by internal and external elastic laminae. While mainly composed of both collagen and elastin connective tissue, each layer has particular cell types responsible for specific function. The innermost layer, the tunica intima, is separated from the vessel lumen by a thin layer of endothelial cells (EC), containing fibroblasts and myointimal cells in its subendothelial support structure. It is responsible for many regulatory mechanisms, including control of cell adhesion and transmigration as well as diffusion of fluid and macromolecules. The medial layer, or tunica media, contains vascular smooth muscle cells (VSMC), thus, contractile elements that under normal physiologic states and in response to intimal signals, determines vessel lumen diameter. The outer tunica adventitia is a supporting layer of the vessel that harbors the vascular supply (vasovasorum) as well as connective tissue and fibroblasts. Resident leukocytes, specifically T cells and monocyte derivatives [macrophages and dendritic cells (DC)] are normally present in insignificant levels within the vessel wall. This is partially due to EC-derived nitric oxide (NO), which serves as a potent inhibitor of platelet aggregation and leukocyte adhesion. Other cell types, including mesenchymal-derived pericytes, are present largely around the blood vessels in the adventitial layers.

The vascular endothelium is the largest organ in the body covering between 4000 and 7000 square meters. Ironically, study of this organ has only recently expanded, broadening our understanding of its structure, variations, and functions. The endothelium is derived embryologically from mesoderm. However, progenitor cells can also differentiate into fully functioning mature cells (1). The barrier integrity is formed and maintained through a host of intercellular adhesion molecules and constitutively expressed matrix integrins. There are three main types of EC junctions involved in forming the barrier to the blood compartment: tight junctions formed by occludens, ZO-1, and ZO-2, adherens junctions formed by cadherins linked to members of the catenin family and then to the cytoskeleton, and communicating (gap) junctions formed by connexins. A more comprehensive understanding of vascular endothelial biology is still developing as previous studies were focused on end-organ damage, the result of endothelial pathology. Yet, along the pathophysiological course of disease, many alterations of EC biophysical properties emerge, both beneficial and maladaptive, even before detection of vascular dysfunction. This review will focus on endothelial pathophysiology and altered phenotypic protein expression in two distinct vascular states—atherosclerosis and vascular remodeling. These known alterations direct the definition of novel molecular targets to enhance our understanding of vascular disease with respect to time course of events, extent of disease, and prediction of outcomes.

Under vascular pathological states, the EC responds to alterations in blood flow, tissue perfusion requirements, shear stress, inflammatory environments, and intercellular contact. Each of these states can lead to its own pattern of EC "activation" and thus modification of its normal functions in nutrient and cellular trafficking, regulation of vasomotor tone, and maintenance of an anti-inflammatory and antithrombotic milieu. This activation results in endothelial dysfunction and is one of the earliest measurable events that can result is a multitude of downstream effects. Endothelial dysfunction accompanies dysregulation of vasomotor tone, which is normally under the control of the balance between the VSMC relaxing prostacyclin (PGI2) and NO versus constricting agents (e.g., endothelin-1, angiotensin). Other effects include inefficient barrier maintenance and resultant leaky vessels, abnormal responses to flow-mediated vasodilation, establishment of a proadhesive, proinflammatory, and procoagulant local vascular environment, and even generation of a neovasculature or expanded vessel wall layers. As such, with elaboration of cytokines (e.g., interleukin-1) and chemokines (e.g., MCP-1), EC activation promotes recruitment of inflammatory cells, activation of metalloproteinases, and differentiation of VSMC to a proliferative phenotype. Vascular remodeling and neointima formation are consequences of these activation events.

ATHEROGENESIS AND ATHEROSCLEROSIS

Atherogenesis can begin early in life with EC activation (2). The pathological process is initiated by lipid deposition seen early in the subendothelial space with resultant inflammatory cell infiltration. In the presence of high circulating levels, LDL cholesterol is taken up by the endothelium. Passing through caveoli, LDL becomes trapped in the subendothelial extracellular space (3). This impairs the secretion of vasodilatory and anti-inflammatory factors, such as NO (4). The altered endothelium releases endothelin-1 (ET-1) and angiotensin II (ATII), which function as vasoconstrictors. This leads to VSMC hypertrophy and vascular remodeling (5,6). Circulating monocytes expressing type A scavenger receptors, which recognize acetylated LDL molecules, transmigrate through the endothelium to engulf the lipids within the subendothelial space. Monocytes differentiate into activated macrophages expressing higher levels of SR-A (7), CD 32 (8), SR-BI (9), CD68 (10), and the better-characterized CD36. CD36 activates the nuclear receptor peroxisome proliferator-activated receptor (PPAR)-gamma, a cell differentiating transcription factor. While PPAR-gamma is involved in induction of genes related to lipid metabolism, it also promotes the differentiation of monocytes to macrophages and foam cells upon oxidized LDL (oxLDL) exposure (11,12). Oxidized LDL is formed by active oxidases [e.g., metalloproteinases, xanthine and NAD(P)H oxidases] in the intimal layer from macrophages and EC (13). The intracellular accumulation of cholesterol overwhelms the mitochondrial metabolic capacity and endoplasmic reticulum membrane integrity leading to organelle dysfunction, cell activation and release of proinflammatory cytokines and proteases within the growing lesion. Excess reactive oxygen species generated by the metabolism of lipids are incompletely neutralized by natural antioxidants (e.g., NADPH) (14,15), leading to further EC damage.

Accumulation of oxLDL leads to further EC activation, inducing the expression of adhesion molecules, such as selectins and members of the Ig superfamily [e.g., intercellular adhesion molecule-1 (ICAM-1)]. These adhesion molecules promote tethering, firm adhesion, and endothelial transmigration of leukocytes. Vascular cell adhesion molecule (VCAM)-1 binds to VLA-4 integrin ($\alpha 4\beta 1$) expressed on monocytes and T cells. ICAM-1 binds to $\beta 2$ integrins, LFA-1 (CD11a/CD18) and Mac-1 (CD11b/CD18), expressed on most leukocytes. (16–18) The heavily sialylated homodimeric vascular adhesion glycoprotein (VAP-1) expressed on EC (and VSMC, DC as well

as most fat cells and pericytes) assists in leukocyte adhesion and trafficking (19). VAP-1 is upregulated during inflammation (20). It exhibits enzymatic properties that lead to production of oxidative substrates, promulgating EC destruction and leukocyte recruitment (21). Transmigration across the endothelium also involves junctional molecules, such as platelet-endothelial (PE)CAM-1, DNAM-1 and ZO-1 (22,23). As the inflammation ensues, EC vasomotor regulatory functions and antithrombotic properties are further lost. Specifically, activated EC over-express tissue factor, ET-1, and AT-II. This is associated with further downregulation of vasodilators, NO, and PGI2. Engagement of cell surface receptors transduces intracellular signals for nuclear gene expression via activation of transcription factors such as NF-κB. Leukocyte activation enhances secretion of proinflammatory cytokines and chemokines. Chemokine release, such as MIF, M-CSF, MCP-1, IL-8, GRO, and MIP-1, leads to increased leukocyte recruitment and differentiation. Over time, this proinflammatory milieu promotes formation of the atherosclerotic plaque (atheroma). The atheroma consists of an inflammatory lipid core with necrotic cellular debris, which is covered by a variably thickened fibrous cap composed primarily of collagen I, III, and to a lesser extent, elastin and proteoglycans (24). The balance of collagen synthesis and protease-mediated matrix degradation from activated inflammatory cells determines the density and thickness of this atheroma cap and, in part, the size of the plaque.

DEVELOPING ATHEROMA

Two distinct types of plaques, white and yellow, have been defined which culminate in two physiological fates. Angiographic studies found that the white plaques correlate with cases of stable angina (25). These plaques typically have a dense fibrous cap and lower inflammatory cell infiltration, and thus, represent a more stable plaque.

Stable Plaque

The stable plaque of chronic disease is characterized by VSMC proliferation and synthetic predominance. The VSMC proliferation is enhanced by EC and macrophage tissue factor production, which initiates the coagulation cascade to produce factor VIIa and activated factor X. Factor X binds its serine receptor on EC to elicit chemokines, adhesion molecules, proinflammatory cytokines, as well as platelet-derived growth factor (PDGF) (26). PDGF stimulates VSMC mitosis and migration into the vessel intima via protein kinase A and RhoA/Rho-associated kinase (27). PDGF specifically upregulates ICAM-1 on VSMCs but may be limited in effect by direct cell contact (28). Specific integrin subunits are increased on the VSMC in neo-intimal migration, including alpha 7 (29) and alpha V (30), which aid in migration and proliferation. Engagement of integrins on the VSMC leads to collagen synthesis and secretion of vascular endothelial growth factor (VEGF) and chemokines (31). Relevant VSMC integrin expression patterns are listed in Table 1.

Unstable Plaque

Yellow plaques, however, are similar, but unstable and comprise the vast majority of lesions correlated with acute coronary syndrome (ACS) as demonstrated in autopsy studies. They have higher red cell, cholesterol (LDL and apo-B components), and inflammatory cell content and tend to have thin fibrous caps, likely due to the levels of macrophage matrix metalloproteinase (MMP) expression. Hence, these represent highly metabolically active "hot" lesions that are more prone to rupture or erosion. Histologic studies have demonstrated that yellow plaques are filled with differentiated monocytic or foam cells, in addition to lymphocytes and apoptotic cell bodies. Inflammatory infiltration progresses, the atheroma grows and the vessel compensates by dilating (positive remodeling) to maintain blood flow. Over time, the progressive growth of the atheroma protrudes into the lumen and the vessel constricts (negative remodeling). Symptomatic disease usually occurs after 70% to 80% of the lumen is obstructed. Within the unstable plaque, the massive inflammation and release of cytokines cause weak neovasculature formation, which easily ruptures fueling macrophage accumulation in the atheroma (32). The intralesional hemorrhage deposits more cholesterol from red cell membranes (33), which contributes to intralesional thrombosis and plaque fissuring (34). EC shed membrane microparticles that both express and induce monocytic expression of tissue factor and induce intralesional thrombosis. Growing evidence supports that monocyte/macrophage expression of tissue factor is regulated both through mRNA stabilization and at the transcriptional level, adhesion-mediated phenomena that are inhibited by ICAM-1 blocking antibodies in vitro.

Generally, the greater the level of inflammation, the more prone lesions are to erosion, rupture and acute coronary events (35). Specific types and threshold levels of cytokines are markers of inflammatory activity. Among the elicited cytokines are IL-1, IL-3, IL-18, and TNF-α, which promote adhesion molecule (e.g., ICAM-1, VCAM-1, E- and P-selectin) expression on the endothelial surface to recruit more inflammatory cells. The downstream signaling cascades have been defined for particular cytokine activation events. MCP-1, G- and M-CSF are released from activated macrophages and promote cellular differentiation, proliferation, and survival within the atheroma. Epitopes on oxidized LDL (e.g., malonyldialdehyde-lysine or MDA epitopes) and endogenous heat shock protein (HSP)-60 expressed on damaged cells recruit T lymphocytes, predominantly of the proinflammatory (Th1)-type, to the atherosclerotic plaque (36). Other surface molecules (e.g., HLA and VLA) have been implicated in T cell activation (37) and recruitment.

Over time, the advanced lesion begins to calcify (38). While multiple theories exist, it is speculated that vesicles from dead macrophage/foam cells and VSMC may contain proteins that bind calcium within the plaque (38). Osteopontin, a noncollagenous glycoprotein secreted by macrophages and VSMCs, is upregulated and binds hydroxyapetite and calcium specifically in coronary artery disease (CAD) (39). Osteopontin gene expression levels correlate with severity of arterial calcification (40). Calcification increases with age (41,42). As correlated with autopsy studies, the more significant the coronary calcification, the more severe

Table 1   Vascular SMC Integrins

Receptor	Ligand	Expression pattern
$\alpha 1\beta 1$	Collagen, laminan medial cells	High in aorta, coronary medial cells, reduced in areas of intimal thickening
$\alpha 2\beta 1$	Collagen, laminan	High in ductus arteriosus and many arterial SMC
$\alpha 3\beta 1$ SMC	Collagen, laminan,	High in coronary medial fibronectin
$\alpha 4\beta 1$	VCAM-1, fibronectin	High in neovessels, atherosclerotic, osteopontin intima but not normal adult aorta
$\alpha 6\beta 1/\beta 4$	Laminan	High in small arteriole SMC, not aorta
$\alpha 7\beta 1$	Laminan	High in most medial SMC
$\alpha 8\beta 1$	Fibronectin, tenascin, vitronectin	medial SMC
$\alpha 9\beta 1$	Tenascin, thrombin-cleaved osteopontin	medial SMC
$\alpha v\beta 1$	Vitronectin, fibrinogen, agrin	fetal ductus arteriosus
$\alpha v\beta 3$	Vitronectin, fibrinogen, vWF fibronectin, laminan, osteopontin denatured collagen	High in neointimal cells of injured arteries, especially early postinjury, upregulated in atherosclerotic arteries
$\alpha v\beta 5$	Vitronectin, osteopontin	Atherosclerotic intima, upregulated early post-arterial injury

Abbreviation: SMC, smooth muscle cell.

the vessel stenosis (43). Calcification occurs early in obstructive disease and significantly increases the strength of the plaque (44). While over time, this reduces the likelihood of plaque rupture, earlier in disease, there is increased vulnerability at junctions of calcification and noncalcified vessels. These sites predispose to acute plaque rupture and acute coronary events (45). Calcification can involve the vessel elastic membrane without structural alteration or can be associated with remodeling (46). Calcium is consistently found in areas of significant vessel stenosis (47). Plaque calcification occurs significantly before the onset of clinical symptoms and has been touted as an indication for early intervention (48).

Proinflammatory signals are counter-balanced by anti-inflammatory cells. A subset of CD4-positive T lymphocytes, collectively grouped as Th2 cells, release cytokines that decrease inflammation. Some of these cytokines, such as IL-4 and IL10, have been shown to be antiatherogenic (49,50). Interferon-gamma, for example, has both pro- and anti-inflammatory effects, whereas monocyte-inhibiting factor and interleukin-10 have anti-inflammatory effects that favor LDL metabolism and maintenance of the undifferentiated monocyte. It is speculated that Th2 predominance favors plaque stabilization (51–53).

Progressive inflammation involves a complex interplay of cellular types in both the innate and adaptive immune systems. Neutrophils and mast cells localize to ruptured culprit lesions, but do not play a significant role in propagating chronic inflammation and atherogenesis. Natural killer-T (NKT) cells are more recently characterized immune cells that have functions of both natural killer and T cells and elicit robust amounts of interferon-gamma upon stimulation. NKT cells recognize glycolipid moieties on CD1d receptors of antigen presenting cells (e.g., DC, macrophages). They have also been shown to accelerate the progression and enlargement of the early atherosclerotic lesion in cholesterol fed apoE (-/-) mice via interferon-gamma signaling pathways (54–57). However, when stimulated via CD154 (CD40L) on antigen presenting cells (DC and macrophages), NKT cells also prime macrophage phagocytic activity via mechanisms independent of interferon-gamma (58,59).

The monocyte/macrophage is considered the central immune cell in atherosclerosis. In addition to the aforementioned mechanisms, monocytes express receptors that may have roles in vascular pathogenesis. For example, toll-like receptor-4 (TLR4) engagement may influence CAD. TLR4, complexed with CD14 on monocytes and activated macrophages, can be upregulated by oxLDL (60). When activated, usually by subclinical levels of endotoxin or heat-shock proteins released from damaged cells (e.g., after coronary bypass surgery), TLR4 promotes inflammation and

vascular pathology (61,62). Polymorphisms of the TLR4 gene result in reduced inflammatory responses to bacterial infections and in lesser burdens of carotid atherosclerosis (63). TLR4's response to endogenous proteins may help describe the "molecular mimicry" phenomenon in patients with infectious disease history and early acute coronary events.

A growing body of evidence illustrates monocyte-lymphocyte crosstalk in disease progression, which may be pertinent to the maintenance of chronic inflammation in atherosclerosis. CD 137, a receptor on monocytes, engages its ligand, 4-1BB, on circulating monocytes and T cells. The interaction of these proteins accounts for a monocyte proliferation above that expected by either G- or M-CSF alone (64,65). Antigen presenting cell (including monocyte)-lymphocyte interactions via CD80 (B7-1) and/or CD86 (B7-2)-CD28 engagement may also play a role. Murine knockouts of the genes encoding either B7-1 or -2 have smaller atherosclerotic lesions than wild-type mice, when crossed with cholesterol-fed, LDL-R-deficient mice (66). In this model, there is also a reduction in interferon-gamma production from murine T cells exposed to HSP60. CD40, a member of the TNF receptor family, is expressed on most leukocytes, and is critical for NKT and T cell activation. In turn, interferon-gamma from activated NKT and T cells induces the membrane expression of CD 40 via STAT-1α- mediated transcription (67). Furthermore, CD40L-CD40 interactions may stimulate adhesion molecule expression, even in the absence of cytokine stimulation (68).

In addition to direct cellular contact, activated immune cells influence progression of atherosclerosis via soluble mediators. Macrophages secrete IL-18 thereby inducing T cell interferon secretion. Interferon-gamma causes vascular intimal expansion while activating the responsive macrophage to increase the efficiency of antigen presentation to the T cell receptor (TCR). However, interferon-gamma can impair collagen synthesis in the VSMC. This could prevent thickening of the fibrous cap or disorder vessel remodeling, hence favoring unstable plaque formation. As a result of interferon-mediated activation, T cells undergo functional differentiation to a cytotoxic phenotype that releases perforin and destroys EC (69).

There is evidence that monocytes, via contact-mediated signaling, alter the phenotypic protein expression, survival and apoptotic patterns in the VSMC. The VSMC undergoes a complete differentiation from the usual contractile element to a synthetic, active inflammatory contributor. Both ICAM-1 and VCAM-1 expression are upregulated on VSMC in atherosclerosis (70,71). Local production of both AT-II and PGDF (via the EC) promotes integrin-mediated binding of monocyte to VSMC. In turn, this adhesion leads to monocytic activation of intracellular Src, phosphoinositide 3-kinase, and mitogen-activated protein kinase (MAPK) signaling intermediates that result in up-regulation of CD36 and enhancement of foam cell formation (72). Thus, the VSMC and monocyte potentiate the other's pathological effects.

The neointimal proliferation into deeper vessel layers requires specific proteases that degrade the barrier elastic membranes. Elastolysis occurs via cathepsin members of the papain family cysteine proteases. Cathepsins are expressed in T cells and activated macrophages and are known to promote angiogenesis (73). Particular cathepsins have been demonstrated in the human atheroma and activated VSMC, which may account, at least in part, for neointimal expansion (74) and, potentially, rupture of the plaque.

In combination, these events lead to destabilization, if not prohibition, of atheroma stability and perpetuation of the unstable plaque. Cells of the monocyte/macrophage lineage and, to a lesser degree, lymphocytes aggregate at either end of smaller atheromatous lesions more prone to rupture. These cells can be detected by 18-fluorodeoxyglucose (FDG) uptake and imaging as will be discussed subsequently (75,76).

The MMP in atherosclerosis is involved in plaque formation, degradation of the fibrous cap, platelet aggregation, and post-angioplasty restenosis (77). MMPs are initially secreted as proenzymes that become activated upon loss of Zn^{2+} in their catalytic site. They are categorized as collagenases (e.g., MMP-2), gelatinases (e.g., MMP-9), stromelysins (e.g., MMP-3), matrilysins (e.g., MMP-7), metalloelastases (e.g., MMP-12), or membrane-type MMPs (e.g.,MMP-14) (77–79).

Inflammatory disease states associated with elevated AT II and TNF-α induce EC and macrophages to release MMPs (80). The activated macrophage is the primary source of local MMP secretion, largely mediated through local inflammation-derived cytokine by-products (e.g., prostaglandins and TNF-α) (81). OxLDL induces MMP-14 expression in macrophages and EC. MMP-14 anchors to the cell membrane to promote pericellular degradation, as well to activate other collagenases, particularly 2 and 13 (82–86). MMPs localize in the vulnerable plaque shoulder regions, which are thought to be more susceptible to rupture and initiation of acute events (87). In these regions, cyclooxygenase and its products (e.g., PGE) induce MMPs to increase plaque instability (88).

In acute coronary events, platelets are recruited to the site and firmly adhere to fibrinogen via integrin engagement. Von Willebrand factor, fibrinogen and platelet activating factor promote platelet adhesion. Activated beta-3 integrins also bind multiple sites of fibrinogen on activated platelets under both flow and non-flow conditions (89). The glycoproteins Ib and IIb/IIIa allow for initial platelet binding and aggregation, respectively (90). Activated platelets then release ADP,

thromboxane (TXA2), and PDGF among other mediators to recruit additional platelets. Activated platelets bind to annexin V that associates with exposed phosphatidylserine on apoptotic cell bodies. The result is a huge network of fibrin, cellular debris and platelets in a proteinaceous, coagulated gel. This gelled clot can be self-limited and contained by the endogenous fibrinolytic system or can progress to partial or complete occlusion of the vessel, as in ACS or transmural myocardial infarction (MI). Released tissue-type–plasminogen activator (tPA) and thrombomodulin serve in fibrinolysis and antithrombosis, respectively. However, the remodeled, activated vessel, with tissue factor-mediated coagulation and bathed in vasoconstrictors, culminates in the acute coronary syndrome.

VASCULAR REMODELING

Vascular remodeling is defined as an enduring change in the vessel size or composition. It can be physiological or pathological, expansive (positive) or constrictive (negative). Remodeling can be focal (atherosclerosis or restenosis postangiography), segmental (aneurysm), or diffuse (graft disease). Thus, vascular remodeling encompasses either a change in vessel lumen size, a change in vessel thickness, or both. MMPs are likely responsible for vessel adaptations since degradation of the collagen-elastin architecture and extracellular matrix (ECM) allows reshaping of the vessel (91) and smooth muscle cell migration (78). Restenosis is an example of constrictive remodeling that involves MMP activation (92). The process of restenosis begins within days of the insult, as demonstrated in animal models (93). Postangiography MMP-2 levels positively correlate with restenosis within six months of the procedure (92). MMP-2 and -9 have been localized to areas of expansively remodeled atheroma and β2 integrin-mediated inflammatory retinopathy in diabetic animals (92,94).

The extent to which the cellular infiltrate, particularly of macrophages, affects constrictive remodeling is unclear (95). However, inflammation, free radicals, and cytokine production from activated leukocytes and differentiated VSMC lead to the recruitment and proliferation of adventitial myofibroblasts and adventitial microvessels (96–98). MMP expression is involved in the migration of VSMC (98) and myofibroblasts (99). Activated fibroblasts migrate, proliferate and secrete trophic factors (100). Furthermore, shear stress transmitted to VSMC through the glycocalyx activates smooth muscle progenitor cells to proliferate around injured vessels and adventitia (101–103). This restricts the expansion of the external elastic membrane, as seen in postballoon angioplasty histopathological specimen (104) and leads to a firmly enhanced matrix with altered myogenic tone and constriction of the vessel lumen (105). However, the time course of constrictive remodeling correlates better with neointimal formation than adventitial proliferation in animal studies (106).

Aneurysm formation involves the degradation of the elastic laminar membranes and has many influences, including VSMC apoptosis (107). MMP-2, -9, and caspase expression/activation, have been associated with thoracic aneurysm (108). Oxidative stress appears integrally involved in abdominal aneurysm formation (109,110). MMP expression and activity correlates with aneurysmal dilatation, possibly due to the presence of activated macrophages (111,112). MMP-3 expression increases susceptibility to aneurysm formation, while decreasing plaque size, through collagen degradation (113). Of note, the high MMP-induced matrix turnover in atherosclerosis leads to shedding of HSP-60 and other ligands for TLR4 (114). In an arterial remodeling system, a TLR4 knockout mouse exhibited blunted vascular expansion upon lipopolysaccharide (LPS) activation, demonstrating a role for TLR4 in positive remodeling (115).

The two primary manifestations of remodeling in graft arteriopathy (chronic vascular rejection) encompass both expansive remodeling and constriction (intimal hyperplasia). Interferon-gamma appears to mediate both effects in a T cell-dependent system (116). Local production of trophic factors may be largely responsible for graft arteriopathy. Receptor blockade of both VEGF and PDGF ameliorates VSMC hyperproliferation in mice (117).

It is important to note that the inflammation associated with atherosclerosis is not confined to coronary lesions (118). Aging, diabetes, and renal failure increase advance glycation end-products (AGEs) diffusely, thereby potentially stimulating EC cytokine production and promoting systemic atherosclerosis (119). Neutrophils become activated during plaque rupture, erosion, and endothelial inflammation. However, they do not remain adherent to the vessel, but pass through the coronary vasculature into the systemic circulation (120). Serological studies in atherothrombotic stroke have demonstrated a persistence of tissue factor and other markers of inflammation for months after an acute event (120,121). The presence of systemic inflammation is documented by high levels of CRP, IL-6, and serum amyloid in patients with unstable angina. CRP, the hepatic synthesis of which is induced by IL-6, tracks with LDL levels in atherosclerosis (122). However, it has recently been touted as an independent risk factor for clinical coronary artery disease. CRP induces systemic expression of adhesion molecules, IL-6, and MCP-1 from EC (123–125), thereby amplifying inflammatory events by leukocyte recruitment and activation.

In sum, atherosclerosis is largely a lipid-induced, inflammation-mediated vascular pathologic event that progresses chronically, with superimposed acute alterations resulting in clinical events. With maladapted lipid metabolism, inflammatory cell infiltration, and resultant vascular cell dysfunction, the atheroma grows and mimics tumor in its characteristic procoagulant, angiogenic, and proinflammatory properties. The atherosclerotic plaque consists of a necrotic core of cellular apoptotic bodies and debris, a dynamic atheromatous cap, vascular remodeling, and cellular proliferation. The now predictable evolution of molecular expression/activation and cellular phenotypes provides a foundation with which to develop finely targeted, molecular imaging studies that may accurately define the stage of disease and vulnerability of lesions, ideally predicting probable outcome. In our discussion above, we have described many potential target molecules. In the imaging section below, we have chosen a small, representative handful of these potential targets, based on work already performed and developed approaches. The detailed description of the pathophysiology of atherosclerosis and vascular remodeling is intended to provide a conceptual background with which to choose other promising molecular imaging targets. That is, there is an abundance of such targets, and great promise for the future.

VASCULAR MOLECULAR IMAGING

Imaging may be used to detect atherogenesis, various stages of atherosclerosis, molecular processes that predispose or lead to plaque vulnerability, subsequent thrombus formation, consequent ischemic events, as well as compensatory and repair mechanisms, whether physiologic or as a result of therapeutic interventions. The latter will be discussed elsewhere.

IMAGING ATHEROSCLEROSIS

Vascular Cell Adhesion Molecule (VCAM)-1

One of the earliest events in the course of atherogenesis is the upregulation of endothelial adhesion molecules, including VCAM-1. VCAM-1-targeted imaging is a promising approach for imaging inflammation (126). One potential limiting factor for imaging VCAM-1 expression in the vessel wall is the limited number of molecules expressed (127). Recently, a multivalent nanoparticle carrying a murine VCAM-1-specific peptide showed specific binding to TNF-activated endothelium and allowed in vivo detection of atherosclerotic lesions in apolipoprotein E$^{-/-}$ (apoE$^{-/-}$) mice via MRI (128). Whether this approach can be applied to humans, for early plaque detection, will need to be addressed in the future studies.

Lipids

Lipids are a primary component of atherosclerotic plaques and lipid content correlates with plaque vulnerability. Lipid accumulation in the plaque has been targeted with a variety of radiolabeled lipids (129–131). However, the results have been limited because of slow clearance of the tracers, oxLDL has been linked to the atherosclerosis initiation, regression and plaque instability (132). LDL oxidation is associated with the appearance of neoepitopes, which may be targeted for imaging atherosclerosis. oxLDL-specific antibodies have been obtained from immunized mice, and from patients with autoantibodies to oxLDL (132). MDA2 is a manufactured antibody that recognizes the malonyldialdehyde epitopes of oxLDL molecules in the aorta (36). MDA2 localizes to lipid rich plaques in mouse and rabbit models of atherosclerosis. Planar imaging with ^{99m}Tc-MDA2 in WHHL rabbits has been able to detect atherosclerotic plaques in the aorta, establishing the feasibility of oxLDL-targeted imaging. The use of murine antibodies may be complicated by the development of xenogeneic reactions. Therefore, efforts are focused on the development of human or humanized antibodies for imaging. Use of IK17, a human Fab monoclonal autoantibody to oxLDL selected from a phage display combinatorial library, derived from a patient with atherosclerotic cardiovascular disease and elevated oxLDL autoantibody titers (133), results in selective in vivo uptake in atherosclerotic plaques (132).

Cell Proliferation and Vascular Remodeling

Vascular cell proliferation and geometrical remodeling are key components of vascular remodeling. Although present in atherosclerosis, neointima formation is the predominant feature in postangioplasty restenosis and graft arteriosclerosis. Phenotypic changes in VSMC associated with VSMC proliferation and migration may be targeted in vivo for imaging vascular remodeling. Z2D3 is an IgM with specificity for an antigen associated with the proliferating smooth muscle cell within human atherosclerotic lesions. The exact nature of the antigen is not well defined, but the antibody can specifically recognize proliferating VSMC in vivo (134). Radiolabeled Z2D3 has been successfully used to detect rabbit experimental atherosclerotic lesions and porcine stent restenosis (134,135). Z2D3 uptake correlates with cell proliferation in both models.

Vascular cell proliferation and migration are associated with $\alpha v \beta 3$ integrin upregulation and activation. The conformational change associated with integrin activation provides an opportunity to specifically target cell proliferation in vascular diseases. RP748 is an ^{111}In-labeled quinolone antagonist of $\alpha v \beta 3$ integrin. We have demonstrated that RP748 preferentially binds with high affinity to the active conformation of the integrin, and the number of binding sites are appropriate for in vivo imaging. Using murine models of immune- and injury-mediated vascular remodeling, we have demonstrated that RP748 uptake at the site of injury tracks cell proliferation (predominantly VSMC) in vivo (Figs. 1 and 2) (136,137). Future studies will address applicability of these findings to larger animal models of vascular remodeling, and human arteriosclerosis.

Angiogenesis

The growth of atherosclerotic plaques is dependent on the formation of a neovascular network. In addition to

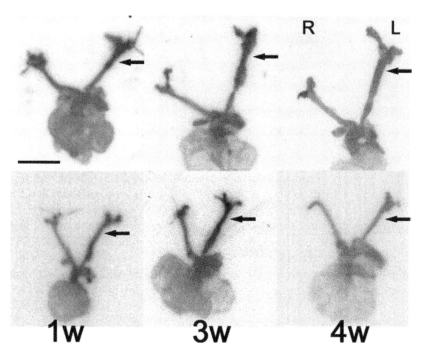

Figure 1 Autoradiographic analysis of RP748 uptake after left carotid injury. Examples of carotid autoradiographs at one, three, and four weeks (w) after left carotid injury. *Arrows* point to sites of injury. *Source*: From Ref. 136.

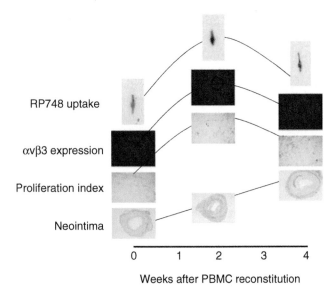

RP748 uptake

αvβ3 expression

Proliferation index

Neointima

0   1   2   3   4

Weeks after PBMC reconstitution

Figure 2   Schematic representation of the changes in neointima thickness, proliferation index αvβ3 integrin expression and RP748 uptake in transplanted human arteries at indicated times after peripheral blood mononuclear cell reconstitution, in the chimeric human/SCID mouse model. *Source*: From Ref. 137.

its role in the development of stable plaque, angiogenesis may play an important role in plaque vulnerability. Several molecular targets have been studied and validated for in vivo imaging of angiogenesis in tumors, as well as following myocardial and hindlimb ischemia (138,139). Imaging plaque angiogenesis may therefore provide important information on plaque vulnerability. The feasibility of αvβ3-targeted imaging of angiogenesis in atherosclerotic plaques was addressed in a study by Winter et al. (140). In this study, αvβ3-targeted, but not control, paramagnetic nanoparticles localized to atherosclerotic plaques in cholesterol-fed rabbits. There was prominent neointima formation with marked angiogenesis detected by immunostaining within the aortic adventitia. Nanoparticles can only target the endothelium (not VSMC) supporting the conclusion that the αvβ3 target in these experiments is expressed on angiogenic blood vessels within the plaque.

IMAGING PLAQUE VULNERABILITY

Matrix and MMP

Matrix is a primary component of the atherosclerotic plaque, and matrix degradation is a central event in plaque rupture. Imaging changes in matrix composition and enzymes responsible for such changes is a promising approach to imaging plaque vulnerability. There are a limited number of studies targeted at matrix and proteases in atherosclerosis imaging. MMP activity in the vessel wall after carotid artery ligation in

apoE$^{-/-}$ mice was targeted with ^{123}I-HO-CGS 27023A, a broad-spectrum MMP inhibitor. Although the results are important as proof of principle, the study does not address the activation state and nature of target MMPs (141). MMP-sensitive, activatable probes are an alternative and promising approach to imaging MMP-activity in vivo (142). Elastolytic cathepsins play a role in plaque rupture. A NIR cathepsin B-sensitive beacon has been successfully evaluated for detection of enzymatic activity in apoE$^{-/-}$ mice atherosclerotic plaques (143). MMP and cathepsin imaging are promising approaches for detection of vulnerable plaque.

Inflammation

Plaque vulnerability is related to the degree of intralesional inflammatory cell content. Radiolabeled leukocytes are routinely used for detection of inflammatory sites in patients. However, there is no report of successful imaging of plaque inflammation using radiolabeled leukocytes. MCP-1 is a monokine that plays an important role in monocyte-macrophage trafficking, including in atherosclerosis. MCP-1 binding sites at the sites of experimental atherosclerosis have been successfully targeted using a ^{125}I-labeled MCP-1. The uptake of ^{125}I-MCP-1 correlated with the number of plaque macrophages (144).

Leukocytes within the atheroma become activated and are probably responsible for the higher metabolic requirements in the atheromatous areas. ^{18}F-2-deoxy-D-glucose (FDG) uptake tracks glucose uptake and is used to detect metabolic activity in various tissues. FDG uptake has been demonstrated in post-mortem animal atherosclerotic plaque studies, with tracer uptake correlating with the number of plaque macrophages (75,145,146). In a study of ^{18}F-FDG uptake in human carotid arteries, tracer uptake was higher in symptomatic carotid as compared to the contralateral asymptomatic carotid artery in patients with bilateral carotid disease. Furthermore, FDG uptake localized to macrophage rich areas of the plaque (76).

Thrombus

Upon plaque rupture or erosion, the subendothelium and atheroma core are exposed to blood components, including circulating cells and plasma proteins. This leads to platelet aggregation, fibrin deposition, and thrombosis. Radiolabeled fibrinogen, antibodies with specific binding to fibrin and other components of thrombus, such as tissue plasminogen activator, have been used to target active thrombus in venous and arterial thrombosis. The use of these tracers for detection of deep venous thrombosis and pulmonary embolism has generally been limited by the lack of sensitivity and specificity, as well as the availability of other less costly, easier to use, imaging modalities with better diagnostic accuracy (147). Integrin αIIbβ3 is expressed on platelets and is activated in aggregating platelets. DMP-444 is a ^{99m}Tc-labeled cyclic peptide αIIbβ3 integrin antagonist. Pilot studies using DMP-444 in canine models of coronary injury have established the feasibility of αIIbβ3-targeted imaging of coronary thrombus (148,149). Further studies are required to establish the role of αIIbβ3 imaging for detection of acute coronary syndromes in research or clinical settings.

SUMMARY

In this chapter, we have provided a detailed description of the currently accepted pathophysiology of atherosclerosis, including features of vascular remodeling and angiogenesis. In that description, molecular concepts are discussed, laying the foundation for defining imaging targets within various pathologic stages. The molecular imaging approaches described provide examples of promising possibilities for the future, as the field advances both with regard to molecular pathophysiology, and technology.

REFERENCES

1. Brutsaert DL, De Keulenaer GW, Fransen P, et al. The cardiac endothelium: functional morphology, development, and physiology. Prog Cardiovasc Dis 1996; 39(3):239–262.
2. Strong JP, Malcom GT, McMahan CA, et al. Prevalence and extent of atherosclerosis in adolescents and young adults: implications for prevention from the Pathobiological Determinants of Atherosclerosis in Youth Study. JAMA 1999; 281(8):727–735.
3. Skalen K, Gustafsson M, Rydberg EK, et al. Subendothelial retention of atherogenic lipoproteins in early atherosclerosis. Nature 2002; 417(6890): 750–754.
4. Dancu MB, Berardi DE, Vanden Heuvel JP, Tarbell JM. Asynchronous shear stress and circumferential strain reduces endothelial NO synthase and cyclooxygenase-2 but induces endothelin-1 gene expression in endothelial cells. Arterioscler Thromb Vasc Biol 2004; 24(11): 2088–2094.
5. Amiri F, Virdis A, Neves MF, et al. Endothelium-restricted overexpression of human endothelin-1 causes vascular remodeling and endothelial dysfunction. Circulation 2004; 110(15):2233–2240.
6. Szabo C, Pacher P, Zsengeller Z, et al. Angiotensin II-mediated endothelial dysfunction: role of poly(ADP-ribose) polymerase activation. Mol Med 2004; 10(1–6):28–35.
7. Kodama T, Reddy P, Kishimoto C, Krieger M. Purification and characterization of a bovine acetyl low density lipoprotein receptor. Proc Natl Acad Sci U S A 1988; 85(23):9238–9242.
8. Stanton LW, White RT, Bryant CM, Protter AA, Endemann G. A macrophage Fc receptor for IgG is also a receptor for oxidized low density lipoprotein. J Biol Chem 1992; 267(31):22446–22451.
9. Acton SL, Scherer PE, Lodish HF, Krieger M. Expression cloning of SR-BI, a CD36–related class B scavenger receptor. J Biol Chem 1994; 269(33):21003–21009.
10. Ramprasad MP, Terpstra V, Kondratenko N, Quehenberger O, Steinberg D. Cell surface expression of mouse macrosialin and human CD68 and their role as macrophage receptors for oxidized low density lipoprotein. Proc Natl Acad Sci U S A 1996; 93(25): 14833–14838.
11. Tontonoz P, Nagy L, Alvarez JG, Thomazy VA, Evans RM. PPARgamma promotes monocyte/macrophage differentiation and uptake of oxidized LDL. Cell 1998; 93(2):241–252.
12. Steinbrecher UP, Lougheed M, Kwan WC, Dirks M. Recognition of oxidized low density lipoprotein by the scavenger receptor of macrophages results from derivatization of apolipoprotein B by products of fatty acid peroxidation. J Biol Chem 1989; 264(26): 15216–15223.
13. Stocker R, Keaney JF Jr. Role of oxidative modifications in atherosclerosis. Physiol Rev 2004; 84(4): 1381–1478.
14. Steinberg D, Parthasarathy S, Carew TE, Khoo JC, Witztum JL. Beyond cholesterol. Modifications of low-density lipoprotein that increase its atherogenicity. N Engl J Med 1989; 320(14):915–924.
15. Wentworth P Jr, Nieva J, Takeuchi C, et al. Evidence for ozone formation in human atherosclerotic arteries. Science 2003; 302(5647):1053–1056.
16. van de Stolpe A, van der Saag PT. Intercellular adhesion molecule-1. J Mol Med 1996; 74(1):13–33.
17. Plow EF, D'Souza SE. A role for intercellular adhesion molecule-1 in restenosis. Circulation 1997; 95(6):1355–1356.
18. Entwistle J, Hall CL, Turley EA. HA receptors: regulators of signalling to the cytoskeleton. J Cell Biochem 1996; 61(4):569–577.
19. Salmi M, Jalkanen S. VAP-1: an adhesin and an enzyme. Trends Immunol 2001; 22(4):211–216.
20. Jaakkola K, Nikula T, Holopainen R, et al. In vivo detection of vascular adhesion protein-1 in experimental inflammation. Am J Pathol 2000; 157(2):463–471.
21. Merinen M, Irjala H, Salmi M, Jaakkola I, Hanninen A, Jalkanen S. Vascular adhesion protein-1 is involved in both acute and chronic inflammation in the mouse. Am J Pathol 2005; 166(3):793–800.
22. Blankenberg S, Barbaux S, Tiret L. Adhesion molecules and atherosclerosis. Atherosclerosis 2003; 170(2): 191–203.
23. Libby P, Li H. Vascular cell adhesion molecule-1 and smooth muscle cell activation during atherogenesis. J Clin Invest 1993; 92(2):538–539.
24. Burleigh MC, Briggs AD, Lendon CL, Davies MJ, Born GV, Richardson PD. Collagen types I and III, collagen content, GAGs and mechanical strength of human atherosclerotic plaque caps: span-wise variations. Atherosclerosis 1992; 96(1):71–81.

25. Thieme T, Wernecke KD, Meyer R, et al. Angioscopic evaluation of atherosclerotic plaques: validation by histomorphologic analysis and association with stable and unstable coronary syndromes. J Am Coll Cardiol 1996; 28(1):1–6.

26. Gasic GP, Arenas CP, Gasic TB, Gasic GJ. Coagulation factors X, Xa, and protein S as potent mitogens of cultured aortic smooth muscle cells. Proc Natl Acad Sci U S A 1992; 89(6):2317–2320.

27. Ishikura K, Fujita H, Hida M, Awazu M. Trapidil inhibits platelet-derived growth factor-induced migration via protein kinase A and RhoA/Rho-associated kinase in rat vascular smooth muscle cells. Eur J Pharmacol 2005; 515(1–3):28–33.

28. Morisaki N, Takahashi K, Shiina R, et al. Platelet-derived growth factor is a potent stimulator of expression of intercellular adhesion molecule-1 in human arterial smooth muscle cells. Biochem Biophys Res Commun 1994; 200(1):612–618.

29. Chao JT, Meininger GA, Patterson JL, et al. Regulation of alpha7–integrin expression in vascular smooth muscle by injury-induced atherosclerosis. Am J Physiol Heart Circ Physiol 2004; 287(1):H381–H389.

30. Choi ET, Khan MF, Leidenfrost JE, et al. Beta3-integrin mediates smooth muscle cell accumulation in neointima after carotid ligation in mice. Circulation 2004; 109(12):1564–1569.

31. Peng Q, Lai D, Nguyen TT, Chan V, Matsuda T, Hirst SJ. Multiple beta 1 integrins mediate enhancement of human airway smooth muscle cytokine secretion by fibronectin and type I collagen. J Immunol 2005; 174(4):2258–2264.

32. Moulton KS, Vakili K, Zurakowski D, et al. Inhibition of plaque neovascularization reduces macrophage accumulation and progression of advanced atherosclerosis. Proc Natl Acad Sci U S A 2003; 100(8): 4736–4741.

33. Kolodgie FD, Gold HK, Burke AP, et al. Intraplaque hemorrhage and progression of coronary atheroma. N Engl J Med 2003; 349(24):2316–2325.

34. Davies MJ, Thomas AC. Plaque fissuring—the cause of acute myocardial infarction, sudden ischaemic death, and crescendo angina. Br Heart J 1985; 53(4):363–373.

35. van der Wal AC, Becker AE, Koch KT, et al. Clinically stable angina pectoris is not necessarily associated with histologically stable atherosclerotic plaques. Heart 1996; 76(4):312–316.

36. Tsimikas S, Palinski W, Halpern SE, Yeung DW, Curtiss LK, Witztum JL. Radiolabeled MDA2, an oxidation-specific, monoclonal antibody, identifies native atherosclerotic lesions in vivo. J Nucl Cardiol 1999; 6(1 Pt 1):41–53.

37. Hansson GK, Holm J, Jonasson L. Detection of activated T lymphocytes in the human atherosclerotic plaque. Am J Pathol 1989; 135(1):169–175.

38. Stary HC, Chandler AB, Dinsmore RE, et al. A definition of advanced types of atherosclerotic lesions and a histological classification of atherosclerosis. A report from the Committee on Vascular Lesions of the Council on Arteriosclerosis, American Heart Association. Circulation 1995; 92(5):1355–1374.

39. Fitzpatrick LA, Severson A, Edwards WD, Ingram RT. Diffuse calcification in human coronary arteries. Association of osteopontin with atherosclerosis. J Clin Invest 1994; 94(4):1597–1604.

40. Hirota S, Imakita M, Kohri K, et al. Expression of osteopontin messenger RNA by macrophages in atherosclerotic plaques. A possible association with calcification. Am J Pathol 1993; 143(4):1003–1008.

41. Schmermund A, Rensing BJ, Sheedy PF, Bell MR, Rumberger JA. Intravenous electron-beam computed tomographic coronary angiography for segmental analysis of coronary artery stenoses. J Am Coll Cardiol 1998; 31(7):1547–1554.

42. Sangiorgi G, Rumberger JA, Severson A, et al. Arterial calcification and not lumen stenosis is highly correlated with atherosclerotic plaque burden in humans: a histologic study of 723 coronary artery segments using nondecalcifying methodology. J Am Coll Cardiol 1998; 31(1):126–133.

43. Frink RJ, Achor RW, Brown AL Jr, Kincaid OW, Brandenburg RO. Significance of calcification of the coronary arteries. Am J Cardiol 1970; 26(3):241–247.

44. Wexler L, Brundage B, Crouse J, et al. Coronary artery calcification: pathophysiology, epidemiology, imaging methods, and clinical implications. A statement for health professionals from the American Heart Association. Writing Group. Circulation 1996; 94(5): 1175–1192.

45. Cheng GC, Loree HM, Kamm RD, Fishbein MC, Lee RT. Distribution of circumferential stress in ruptured and stable atherosclerotic lesions. A structural analysis with histopathological correlation. Circulation 1993; 87(4):1179–1187.

46. Bobryshev YV. Calcification of elastic fibers in human atherosclerotic plaque. Atherosclerosis 2005; 180(2): 293–303.

47. Simons DB, Schwartz RS, Edwards WD, Sheedy PF, Breen JF, Rumberger JA. Noninvasive definition of anatomic coronary artery disease by ultrafast computed tomographic scanning: a quantitative pathologic comparison study. J Am Coll Cardiol 1992; 20(5):1118–1126.

48. Alexopoulos D, Toulgaridis T, Davlouros P, et al. Prognostic significance of coronary artery calcium in asymptomatic subjects with usual cardiovascular risk. Am Heart J 2003; 145(3):542–548.

49. Elliott MJ, Gamble JR, Park LS, Vadas MA, Lopez AF. Inhibition of human monocyte adhesion by interleukin-4. Blood 1991; 77(12):2739–2745.

50. Yoshioka T, Okada T, Maeda Y, et al. Adeno-associated virus vector-mediated interleukin-10 gene transfer inhibits atherosclerosis in apolipoprotein E-deficient mice. Gene Ther 2004; 11(24):1772–1779.

51. Pinderski Oslund LJ, Hedrick CC, Olvera T, et al. Interleukin-10 blocks atherosclerotic events in vitro and in vivo. Arterioscler Thromb Vasc Biol 1999; 19(12):2847–2853.

52. Waehre T, Halvorsen B, Damas JK, et al. Inflammatory imbalance between IL-10 and TNFalpha in unstable angina potential plaque stabilizing effects of IL-10. Eur J Clin Invest 2002; 32(11):803–810.

53. Halvorsen B, Waehre T, Scholz H, et al. Interleukin-10 enhances the oxidized LDL-induced foam cell formation of macrophages by antiapoptotic mechanisms. J Lipid Res 2005; 46(2):211–219.

54. Malloy SI, Altenburg MK, Knouff C, Lanningham-Foster L, Parks JS, Maeda N. Harmful effects of increased LDLR expression in mice with human APOE*4 but not APOE*3. Arterioscler Thromb Vasc Biol 2004; 24(1):91–97.

55. Nakai Y, Iwabuchi K, Fujii S, et al. Natural killer T cells accelerate atherogenesis in mice. Blood 2004; 104(7): 2051–2059.

56. Formato M, Farina M, Spirito R, et al. Evidence for a proinflammatory and proteolytic environment in plaques from endarterectomy segments of human carotid arteries. Arterioscler Thromb Vasc Biol 2004; 24(1):129–135.

57. Khalil MF, Wagner WD, Goldberg IJ. Molecular interactions leading to lipoprotein retention and the initiation of atherosclerosis. Arterioscler Thromb Vasc Biol 2004; 24(12):2211–2218.

58. Scott MJ, Hoth JJ, Gardner SA, Peyton JC, Cheadle WG. Natural killer cell activation primes macrophages to clear bacterial infection. Am Surg 2003; 69(8):679–686; discussion 686–687.

59. Scott MJ, Hoth JJ, Stagner MK, Gardner SA, Peyton JC, Cheadle WG. CD40-CD154 interactions between macrophages and natural killer cells during sepsis are critical for macrophage activation and are not interferon gamma dependent. Clin Exp Immunol 2004; 137(3): 469–477.

60. Xu XH, Shah PK, Faure E, et al. Toll-like receptor-4 is expressed by macrophages in murine and human lipid-rich atherosclerotic plaques and upregulated by oxidized LDL. Circulation 2001; 104(25): 3103–3108.

61. Dybdahl B, Wahba A, Lien E, et al. Inflammatory response after open heart surgery: release of heat-shock protein 70 and signaling through toll-like receptor-4. Circulation 2002; 105(6):685–690.

62. Stoll LL, Denning GM, Weintraub NL. Potential role of endotoxin as a proinflammatory mediator of atherosclerosis. Arterioscler Thromb Vasc Biol 2004; 24(12): 2227–2236.

63. Hamann L, Kumpf O, Muller M, et al. A coding mutation within the first exon of the human MD-2 gene results in decreased lipopolysaccharide-induced signaling. Genes Immun 2004; 5(4):283–288.

64. Langstein J, Michel J, Schwarz H. CD137 induces proliferation and endomitosis in monocytes. Blood 1999; 94(9):3161–3168.

65. Langstein J, Schwarz H. Identification of CD137 as a potent monocyte survival factor. J Leukoc Biol 1999; 65(6):829–833.

66. Buono C, Pang H, Uchida Y, Libby P, Sharpe AH, Lichtman AH. B7-1/B7-2 costimulation regulates plaque antigen-specific T-cell responses and atherogenesis in low-density lipoprotein receptor-deficient mice. Circulation 2004; 109(16):2009–2015.

67. Nguyen VT, Benveniste EN. Involvement of STAT-1 and ets family members in interferon-gamma induction of CD40 transcription in microglia/macrophages. J Biol Chem 2000; 275(31):23674–23684.

68. Miller DL, Yaron R, Yellin MJ. CD40L-CD40 interactions regulate endothelial cell surface tissue factor and thrombomodulin expression. J Leukoc Biol 1998; 63(3): 373–379.

69. Liuzzo G, Baisucci LM, Gallimore JR, et al. Enhanced inflammatory response in patients with preinfarction unstable angina. J Am Coll Cardiol 1999; 34(6): 1696–1703.

70. Jang Y, Lincoff AM, Plow EF, Topol EJ. Cell adhesion molecules in coronary artery disease. J Am Coll Cardiol 1994; 24(7):1591–1601.

71. Printseva O, Peclo MM, Gown AM. Various cell types in human atherosclerotic lesions express ICAM-1. Further immunocytochemical and immunochemical studies employing monoclonal antibody 10F3. Am J Pathol 1992; 140(4):889–896.

72. Cai Q, Lanting L, Natarajan R. Growth factors induce monocyte binding to vascular smooth muscle cells: implications for monocyte retention in atherosclerosis. Am J Physiol Cell Physiol 2004; 287(3): C707–C714.

73. Joyce JA, Baruch A, Chehade K, et al. Cathepsin cysteine proteases are effectors of invasive growth and angiogenesis during multistage tumorigenesis. Cancer Cell 2004; 5(5):443–453.

74. Sukhova GK, Shi GP, Simon DI, Chapman HA, Libby P. Expression of the elastolytic cathepsins S and K in human atheroma and regulation of their production in smooth muscle cells. J Clin Invest 1998; 102(3): 576–583.

75. Lederman RJ, Raylman RR, Fisher SJ, et al. Detection of atherosclerosis using a novel positron-sensitive probe and 18-fluorodeoxyglucose (FDG). Nucl Med Commun 2001; 22(7):747–753.

76. Rudd JH, Warburton EA, Fryer TD, et al. Imaging atherosclerotic plaque inflammation with (18F)-fluorodeoxyglucose positron emission tomography. Circulation 2002; 105(23):2708–2711.

77. Jones CB, Sane DC, Herrington DM. Matrix metalloproteinases: a review of their structure and role in acute coronary syndrome. Cardiovasc Res 2003; 59(4): 812–823.

78. Henney AM, Wakeley PR, Davies MJ, et al. Localization of stromelysin gene expression in atherosclerotic plaques by in situ hybridization. Proc Natl Acad Sci U S A 1991; 88(18):8154–8158.

79. Mostafa Mtairag E, Chollet-Martin S, Oudghiri M, et al. Effects of interleukin-10 on monocyte/endothelial cell adhesion and MMP-9/TIMP-1 secretion. Cardiovasc Res 2001; 49(4):882–890.

80. Arenas IA, Xu Y, Lopez-Jaramillo P, Davidge ST. Angiotensin II-induced MMP-2 release from endothelial cells is mediated by TNF-alpha. Am J Physiol Cell Physiol 2004; 286(4):C779–C784.

81. Kim MP, Zhou M, Wahl LM. Angiotensin II increases human monocyte matrix metalloproteinase-1 through the AT2 receptor and prostaglandin E2: implications for atherosclerotic plaque rupture. J Leukoc Biol 2005; 78(1):195–201.

82. Ray BK, Shakya A, Turk JR, Apte SS, Ray A. Induction of the MMP-14 gene in macrophages of the atherosclerotic plaque: role of SAF-1 in the induction process. Circ Res 2004; 95(11):1082–1090.

83. Rajavashisth TB, Xu XP, Jovinge S, et al. Membrane type 1 matrix metalloproteinase expression in human atherosclerotic plaques: evidence for activation by proinflammatory mediators. Circulation 1999; 99(24):3103–3109.

84. Manginas A, Bei E, Chaidaroglou A, et al. Peripheral levels of matrix metalloproteinase-9, interleukin-6, and C-reactive protein are elevated in patients with acute coronary syndromes: correlations with serum troponin I. Clin Cardiol 2005; 28(4):182–186.

85. Jenkins GM, Crow MT, Bilato C, et al. Increased expression of membrane-type matrix metalloproteinase and preferential localization of matrix metalloproteinase-2 to the neointima of balloon-injured rat carotid arteries. Circulation 1998; 97(1):82–90.

86. Brown DL, Hibbs MS, Kearney M, Loushin C, Isner JM. Identification of 92–kD gelatinase in human coronary atherosclerotic lesions. Association of active enzyme synthesis with unstable angina. Circulation 1995; 91(8):2125–2131.

87. Herman MP, Sukhova GK, Libby P, et al. Expression of neutrophil collagenase (matrix metalloproteinase-8) in human atheroma: a novel collagenolytic pathway suggested by transcriptional profiling. Circulation 2001; 104(16):1899–1904.

88. Cipollone F, Prontera C, Pini B, et al. Overexpression of functionally coupled cyclooxygenase-2 and prostaglandin E synthase in symptomatic atherosclerotic plaques as a basis of prostaglandin E(2)-dependent plaque instability. Circulation 2001; 104(8):921–927.

89. Savage B, Bottini E, Ruggeri ZM. Interaction of integrin alpha IIb beta 3 with multiple fibrinogen domains during platelet adhesion. J Biol Chem 1995; 270(48): 28812–28817.

90. Lefkovits J, Plow EF, Topol EJ. Platelet glycoprotein IIb/IIIa receptors in cardiovascular medicine. N Engl J Med 1995; 332(23):1553–1559.

91. Galis ZS, Khatri JJ. Matrix metalloproteinases in vascular remodeling and atherogenesis: the good, the bad, and the ugly. Circ Res 2002; 90(3):251–262.

92. Hojo Y, Ikeda U, Katsuki T, Mizuno O, Fujikawa H, Shimada K. Matrix metalloproteinase expression in the coronary circulation induced by coronary angioplasty. Atherosclerosis 2002; 161(1):185–192.

93. Southgate KM, Fisher M, Banning AP, et al. Upregulation of basement membrane-degrading metalloproteinase secretion after balloon injury of pig carotid arteries. Circ Res 1996; 79(6):1177–1187.

94. Golubnitschaja O, Jaksche A, Moenkemann H, et al. Molecular imaging system for possible prediction of active retinopathy in patients with diabetes mellitus. Amino Acids 2005; 28(2):229–237.

95. Le Feuvre C, Tahlil O, Paterlini P, et al. Arterial response to mild balloon injury in the normal rabbit: evidence for low proliferation rate in the adventitia. Coron Artery Dis 1998; 9(12):805–814.

96. Glaser R, Lu MM, Narula N, Epstein JA. Smooth muscle cells, but not myocytes, of host origin in transplanted human hearts. Circulation 2002; 106(1):17–19.

97. Shi Y, O'Brien JE Jr, Ala-Kokko L, Chung W, Mannion JD, Zalewski A. Origin of extracellular matrix synthesis during coronary repair. Circulation 1997; 95(4): 997–1006.

98. Bendeck MP, Zempo N, Clowes AW, et al. Smooth muscle cell migration and matrix metalloproteinase expression after arterial injury in the rat. Circ Res 1994; 75(3):539–545.

99. Zalewski A, Shi Y. Vascular myofibroblasts. Lessons from coronary repair and remodeling. Arterioscler Thromb Vasc Biol 1997; 17(3):417–422.

100. Jin ZG, Melaragno MG, Liao DF, et al. Cyclophilin A is a secreted growth factor induced by oxidative stress. Circ Res 2000; 87(9):789–796.

101. Ainslie KM, Garanich JS, Dull RO, Tarbell JM. Vascular smooth muscle cell glycocalyx influences shear stress-mediated contractile response. J Appl Physiol 2005; 98(1):242–249.

102. Caplice NM, Bunch TJ, Stalboerger PG, et al. Smooth muscle cells in human coronary atherosclerosis can originate from cells administered at marrow transplantation. Proc Natl Acad Sci U S A 2003; 100(8): 4754–4759.

103. Simper D, Stalboerger PG, Panetta CJ, Wang S, Caplice NM. Smooth muscle progenitor cells in human blood. Circulation 2002; 106(10):1199–1204.

104. Sangiorgi G, Taylor AJ, Farb A, et al. Histopathology of postpercutaneous transluminal coronary angioplasty remodeling in human coronary arteries. Am Heart J 1999; 138(4 Pt 1):681–687.

105. Kim MH, Harris NR, Korzick DH, Tarbell JM. Control of the arteriolar myogenic response by transvascular fluid filtration. Microvasc Res 2004; 68(1):30–37.

106. Maeng M, Olesen PG, Emmertsen NC, et al. Time course of vascular remodeling, formation of neointima and formation of neoadventitia after angioplasty in a porcine model. Coron Artery Dis 2001; 12(4):285–293.

107. Rowe VL, Stevens SL, Reddick TT, et al. Vascular smooth muscle cell apoptosis in aneurysmal, occlusive, and normal human aortas. J Vasc Surg 2000; 31(3):567–576.

108. Taketani T, Imai Y, Morota T, et al. Altered patterns of gene expression specific to thoracic aortic aneurysms: microarray analysis of surgically resected specimens. Int Heart J 2005; 46(2):265–277.

109. Yajima N, Masuda M, Miyazaki M, Nakajima N, Chien S, Shyy JY. Oxidative stress is involved in the development of experimental abdominal aortic aneurysm: a study of the transcription profile with complementary DNA microarray. J Vasc Surg 2002; 36(2):379–385.

110. Miller FJ Jr, Sharp WJ, Fang X, Oberley LW, Oberley TD, Weintraub NL. Oxidative stress in human abdominal aortic aneurysms: a potential mediator of aneurysmal remodeling. Arterioscler Thromb Vasc Biol 2002; 22(4):560–565.

111. Kazi M, Zhu C, Roy J, et al. Difference in matrix-degrading protease expression and activity between thrombus-free and thrombus-covered wall of abdominal

aortic aneurysm. Arterioscler Thromb Vasc Biol 2005; 25(7):1341–1346.

112. Longo GM, Buda SJ, Fiotta N, et al. MMP-12 has a role in abdominal aortic aneurysms in mice. Surgery 2005; 137(4):457–462.

113. Silence J, Lupu F, Collen D, Lijnen HR. Persistence of atherosclerotic plaque but reduced aneurysm formation in mice with stromelysin-1 (MMP-3) gene inactivation. Arterioscler Thromb Vasc Biol 2001; 21(9): 1440–1445.

114. Pasterkamp G, Schoneveld AH, Hijnen DJ, et al. Atherosclerotic arterial remodeling and the localization of macrophages and matrix metalloproteases 1, 2 and 9 in the human coronary artery. Atherosclerosis 2000; 150(2):245–253.

115. Hollestelle SC, De Vries MR, Van Keulen JK, et al. Toll-like receptor 4 is involved in outward arterial remodeling. Circulation 2004; 109(3):393–398.

116. Wang Y, Burns WR, Tang PC, et al. Interferon-gamma plays a nonredundant role in mediating T cell-dependent outward vascular remodeling of allogeneic human coronary arteries. Faseb J 2004; 18(3):606–608.

117. Nykanen AI, Krebs R, Tikkanen JM, et al. Combined vascular endothelial growth factor and platelet-derived growth factor inhibition in rat cardiac allografts: beneficial effects on inflammation and smooth muscle cell proliferation. Transplantation 2005; 79(2): 182–189.

118. Haskard DO. Accelerated atherosclerosis in inflammatory rheumatic diseases. Scand J Rheumatol 2004; 33(5):281–292.

119. Rashid G, Benchetrit S, Fishman D, Bernheim J. Effect of advanced glycation end-products on gene expression and synthesis of TNF-alpha and endothelial nitric oxide synthase by endothelial cells. Kidney Int 2004; 66(3):1099–1106.

120. Naruko T, Ueda M, Haze K, et al. Neutrophil infiltration of culprit lesions in acute coronary syndromes. Circulation 2002; 106(23):2894–2900.

121. Reganon E, Vila V, Martinez-Sales V, et al. Association between inflammation and hemostatic markers in atherothrombotic stroke. Thromb Res 2003; 112(4): 217–221.

122. Zwaka TP, Hombach V, Torzewski J. C-reactive protein-mediated low density lipoprotein uptake by macrophages: implications for atherosclerosis. Circulation 2001; 103(9):1194–1197.

123. Verma S, Wang CH, Li SH, et al. A self-fulfilling prophecy: C-reactive protein attenuates nitric oxide production and inhibits angiogenesis. Circulation 2002; 106(8):913–919.

124. Pasceri V, Cheng JS, Willerson JT, Yeh ET. Modulation of C-reactive protein-mediated monocyte chemoattractant protein-1 induction in human endothelial cells by anti-atherosclerosis drugs. Circulation 2001; 103(21): 2531–2534.

125. Pasceri V, Willerson JT, Yeh ET. Direct proinflammatory effect of C-reactive protein on human endothelial cells. Circulation 2000; 102(18):2165–2168.

126. Sans M, Fuster D, Vazquez A, et al. 123Iodine-labelled anti-VCAM-1 antibody scintigraphy in the assessment of experimental colitis. Eur J Gastroenterol Hepatol 2001; 13(1):31–38.

127. Sadeghi MM, Schechner JS, Krassilnikova S, et al. Vascular cell adhesion molecule-1–targeted detection of endothelial activation in human microvasculature. Transplant Proc 2004; 36(5):1585–1591.

128. Kelly KA, Allport JR, Tsourkas A, Shinde-Patil VR, Josephson L, Weissleder R. Detection of vascular adhesion molecule-1 expression using a novel multimodal nanoparticle. Circ Res 2005; 96(3):327–336.

129. Rosen JM, Butler SP, Meinken GE, et al. Indium-111-labeled LDL: a potential agent for imaging atherosclerotic disease and lipoprotein biodistribution. J Nucl Med 1990; 31(3):343–350.

130. Virgolini I, Rauscha F, Lupattelli G, et al. Autologous low-density lipoprotein labelling allows characterization of human atherosclerotic lesions in vivo as to presence of foam cells and endothelial coverage. Eur J Nucl Med 1991; 18(12):948–951.

131. Shaish A, Keren G, Chouraqui P, Levkovitz H, Harats D. Imaging of aortic atherosclerotic lesions by (125)I-LDL, (125)I-oxidized-LDL, (125)I-HDL and (125)I-BSA. Pathobiology 2001; 69(4):225–229.

132. Tsimikas S. Noninvasive imaging of oxidized low-density lipoprotein in atherosclerotic plaques with tagged oxidation-specific antibodies. Am J Cardiol 2002; 90(10C):22L–27L.

133. Shaw PX, Horkko S, Tsimikas S, et al. Human-derived anti-oxidized LDL autoantibody blocks uptake of oxidized LDL by macrophages and localizes to atherosclerotic lesions in vivo. Arterioscler Thromb Vasc Biol 2001; 21(8):1333–1339.

134. Narula J, Petrov A, Bianchi C, et al. Noninvasive localization of experimental atherosclerotic lesions with mouse/human chimeric Z2D3 F(ab')2 specific for the proliferating smooth muscle cells of human atheroma. Imaging with conventional and negative charge-modified antibody fragments. Circulation 1995; 92(3): 474–484.

135. Johnson LL, Schofield LM, Weber DK, Kolodgie F, Virmani R, Khaw BA. Uptake of 111In-Z2D3 on SPECT imaging in a swine model of coronary stent restenosis correlated with cell proliferation. J Nucl Med 2004; 45(2):294–299.

136. Sadeghi MM, Krassilnikova S, Zhang J, et al. Detection of injury-induced vascular remodeling by targeting activated alphavbeta3 integrin in vivo. Circulation 2004; 110(1):84–90.

137. Zhang J, Krassilnikova S, Gharaei AA, et al. alphavbeta3-Targeted detection of arteriopathy in transplanted human coronary arteries: an autoradiographic study. Faseb J 2005.

138. Meoli DF, Sadeghi MM, Krassilnikova S, et al. Noninvasive imaging of myocardial angiogenesis following experimental myocardial infarction. J Clin Invest 2004; 113(12):1684–1691.

139. Hua J, Dobrucki LW, Sadeghi MM, et al. Noninvasive imaging of angiogenesis with a 99mTc-labeled peptide targeted at alphavbeta3 integrin after murine hindlimb ischemia. Circulation 2005; 111(24): 3255–3260.

140. Winter PM, Morawski AM, Caruthers SD, et al. Molecular imaging of angiogenesis in early-stage atherosclerosis with alpha(v)beta3–integrin-targeted nanoparticles. Circulation 2003; 108(18):2270–2274.

141. Schafers M, Riemann B, Kopka K, et al. Scintigraphic imaging of matrix metalloproteinase activity in the arterial wall in vivo. Circulation 2004; 109(21): 2554–2559.

142. Bremer C, Tung CH, Weissleder R. In vivo molecular target assessment of matrix metalloproteinase inhibition. Nat Med 2001; 7(6):743–748.

143. Chen J, Tung CH, Mahmood U, et al. In vivo imaging of proteolytic activity in atherosclerosis. Circulation 2002; 105(23):2766–2771.

144. Ohtsuki K, Hayase M, Akashi K, Kopiwoda S, Strauss HW. Detection of monocyte chemoattractant protein-1 receptor expression in experimental atherosclerotic lesions: an autoradiographic study. Circulation 2001; 104(2):203–208.

145. Raylman RR, Wahl RL. A fiber-optically coupled positron-sensitive surgical probe. J Nucl Med 1994; 35(5): 909–913.

146. Ogawa M, Ishino S, Mukai T, et al. (18)F-FDG accumulation in atherosclerotic plaques: immunohistochemical and PET imaging study. J Nucl Med 2004; 45(7):1245–1250.

147. Perrier A. Labeling the thrombus: the future of nuclear medicine for venous thromboembolism? Am J Respir Crit Care Med 2004; 169(9):977–978.

148. Mitchel J, Waters D, Lai T, et al. Identification of coronary thrombus with a IIb/IIIa platelet inhibitor radiopharmaceutical, technetium-99m DMP-444: A canine model. Circulation 2000; 101(14):1643–1646.

149. Sakuma T, Sklenar J, Leong-Poi H, Goodman NC, Glover DK, Kaul S. Molecular imaging identifies regions with microthromboemboli during primary angioplasty in acute coronary thrombosis. J Nucl Med 2004; 45(7):1194–1200.

19

Radiotracer Imaging of Unstable Plaque

Lynne L. Johnson
Division of Cardiology, Department of Medicine, Columbia University, New York, New York, U.S.A.

INTRODUCTION

Cardiovascular disease affects approximately 60 million people in the United States. Using available tests including coronary risk factors, coronary artery calcium scores, biomarkers, and noninvasive stress testing we can only identify asymptomatic patients with yearly mortality of 5% (1). Although myocardial perfusion imaging has proven prognostic usefulness there are patients with stable fixed obstructive lesions with large risk areas and stable courses and patients with <50% stenoses and no perfusion defects who have acute ischemic events including sudden death. Advances in molecular biology over the past 10 years has identified potential sites in atherosclerotic plaque that can be targeted with probes that produce signals that can be detected using external imaging. Experimental and clinical studies have reported the feasibility of detecting signals from atherosclerotic plaque using nuclear medicine technology and magnetic resonance imaging (MRI).

The superior resolution of MRI allows high definition anatomical imaging of plaque in the aorta and carotid arteries and by analyzing different MR signals including T1, T2 relaxation and proton density tissue composition of the components of the plaque can be identified (2–6). Characterizing the anatomy and composition of plaque in the coronary arteries using MRI has proven challenging due to small size and motion. Another approach has been the development of targeting probes that comprise nanoparticles such as fluorocarbons with attached binding site recognition arm and gadolinium to change relaxivity and produce the signal. Gadolinium (Gd) labeled probes targeting avb3 integrin on the surface of endothelial cells have been shown on in vivo MRI in experimental animal model to identify growth and proliferation of the vaso vasorum in the adventitia atherosclerotic plaque (7).

While nuclear medicine lacks the high resolution of MRI it has advantages for molecular targeting because targeting probes are in nanomolar concentrations compared to Gd enhanced targeting agents that are in millimolar concentrations (8,9). The very small size of the nuclear probes allows the probe to cross into the extravascular space and cross cell membranes. The larger MRI probes are limited to endovascular sites. Biological systems that have high affinity but relatively low abundance are potential targets for radiolabeled probes that can be administered in tracer quantities because the low administered dose has no biological effects.

A particular challenge to nuclear medicine is to image small structures such as small atherosclerotic lesions and/or small vessels. It would initially seem impossible to be able to image structures such as the coronary arteries that are below the resolution of the camera. For example, the diameter of a coronary artery is in the range of 3 mm while the spatial resolution of a conventional Anger camera is approximately 7 to 8 mm at 8 cm distance and for positron emission tomography (PET), the resolution is in the order of 4 mm (10). However, it is possible to image these small targets as beacons or hotspots if enough radiotracer can be delivered to the target and the background is low. For targets in the range of 2 to 3 mm the cts/pixel at the targeted site need to be >1000 and target to background ratio at least 2:1.

PLAQUE COMPOSITION

Advances in molecular biology and structural information provided from human autopsy studies from patients dying with vulnerable and ruptured atherosclerotic plaque have given us information on plaque evolution, plaque composition, and revealed potential targets for imaging (11,12). American Heart Association (AHA) classification of atherosclerotic lesions were recently modified and simplified by Virmani et al. based on morphology (12). Mild nonatheromatous lesions include intimal thickening due to accumulation of smooth muscle cells (SMCs) and intimal xanthoma or fatty streak. More advanced lesions are fibrous cap atheromatous lesions have a core encapsulated atheroma contains lipid-laden macrophages, extracellular lipid, and necrotic cellular debris. The fibrous cap consists of smooth muscle cells in a proteoglycan-collagen matrix with variable numbers of macrophages and lymphocytes. Plaque vulnerability to rupture with lumenal occlusion is associated with a thin cap heavily infiltrated by macrophages with rare smooth muscle cells (13). Fibroatheromata that include dense fibrous tissue and calcium are considered more stable and called fibrocalcific plaque (12). A different process called plaque erosion also leads to thrombotic lumenal occlusion at sites of endothelial denudation overlying lesions rich in SMCs and proteoglycans (plaque erosion) (14). These mural thrombi lead to either acute total vessel occlusion and myocardial infarction (MI) or sudden cardiac death, or may go undetected. In this latter case, the plaque ruptures and heals by a process similar to wound healing. Plaque erosion leading to mural thrombi without occlusion grows the plaque volume (15).

IMAGING EARLY LESIONS

Early lesions are characterized by endothelial injury and attractions of circulating monocytes via chemotactic substances such as monocyte chemoattractant protein-1 (MCP-1) that is produced by injured endothelial and SMCs as well as activated monocytes/macrophages (16). Receptors to MCP-1 are upregulated on monocytes and these receptors attract the cells to migrate to endovascular sites of injury. These activated monocytes cross into the arterial vessel wall and are transformed into macrophages as they ingest oxidized low density lipoprotein (LDL) and grow into lipid laden foam cells. This information suggested that MCP-1 is potentially a useful probe to identify early atherosclerotic lesions. In vitro and in vivo experimental studies were performed with ^{125}I labeled MCP-1 and with ^{99m}Tc labeled MCP-1 showing uptake localized to perivascular macrophages (17,18).

As an atherosclerotic lesion progresses from fatty streak to frank plaque apoptotic death of vascular smooth muscle cells increases as the neotimal lesion grows and the vessel remodels (19–21). In cell culture, plaque-derived vascular smooth muscle cells from human lesions undergo apoptosis at faster rate than nonatherosclerotic vessels (19). An early detectable change in cells undergoing programmed cell death is the externalization of phosphatidylserine (PS), a component of the cell membrane lipid bilayer (22). Normally PS is located on the inner bilayer but under internal signals the lipid components of the membrane are redistributed and PS moves from the inner to outer layer making it available for binding. Annexin V is an endogenous protein (35 kDa) that binds specifically to PS with very high affinity (22). Annexin V can be bound to ^{99m}Tc making it a probe that targets apoptosis. Because of the abundance of binding sites in tissue undergoing apoptosis and because of the commonality of programmed cell death to several prevalent diseases many experimental studies and clinical studies have been reported using ^{99m}Tc-annexin V to target apoptosis in cancer, myocardial infarction, and for vascular imaging (8,23–25).

To address the hypothesis that PS binding sites on cells undergoing apoptosis in coronary atherosclerotic lesions of a human size large animal model are abundant enough to be detectable on in vivo single photon emission CT (SPECT) imaging swine underwent coronary injury followed by high cholesterol diet. Focal uptake of ^{99m}Tc labeled annexin V was visualized in the region of the heart corresponding to the distribution of the injured vessels confirmed by phosphor screen imaging (Fig. 1) (26). Pathology showed AHA class II and III lesions with the predominant cells being smooth muscle cells. Costaining for caspase-3 revealed the cells undergoing apoptosis to be smooth muscle cells. The apoptotic index calculated as caspase-3 positive cells over total cells correlated with lesional counts and there was a cut off value of 50% that separated scans that were positive and negative. This study showed that coronary lesions appearing as hotspots can be detected on in vivo gamma imaging in a humanoid animal model. It also suggested that positive uptake seen on in vivo imaging indicates a high level of apoptosis.

IMAGING COMPLEX LESIONS

As the atherosclerotic plaque becomes more complex there are additional potential targets for imaging. The anatomical lesion that is most frequently found at autopsy in men at any age and older women dying suddenly from an acute ischemic episode is rupture of a thin cap fibroatheroma (11). The thin cap fibroatheroma comprises a large necrotic lipid core with overlying thin fibrous cap made up of SMCs. Infiltrating in the shoulder regions of the lesion are abundant macrophages (13). Rupture of the thin cap exposes the contents of the necrotic lipid core to the clotting cascade proteins in the blood to produce occlusive thrombus. Immunohistochemical staining on sections from human autopsy data with costaining for macrophages and for apoptosis with TUNEL have shown plaque macrophages undergoing apoptosis in large numbers (13). These activated macrophages secrete several matrix metalloproteinases (MMPs) that have been found in large amounts in the cap and core area. The MMPs are connective tissue-degrading enzymes that when activated by plasmin and tissue factors can degrade the connective tissue matrix. These three components of the complex and vulnerable plaque are potential targets for imaging: inflammation indicated by abundance of macrophages, apoptosis of macrophages, and MMP expression. All three of these components of the vulnerable plaque have been targeted and results reported.

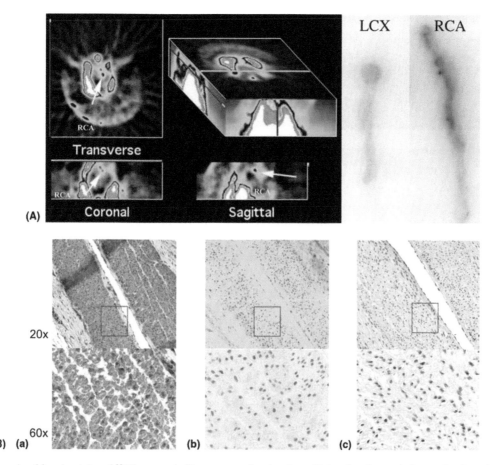

Figure 1   (**A**) Example of focal uptake of ^{99m}Tc-annexin V corresponding to sites of injury in a hypercholesterolemic swine. The in vivo single photon emission CT reconstructions are shown on the left and phosphor screen images on the *right*. The right coronary artery (RCA) was injured and the left circumflex coronary artery (LCX) was the control vessel. Phosphor screen images show tracer uptek in the proximal half of the RCA specimen. Transverse, coronal, and sagittal reconstructions of in vivo images show linear uptake of tracer in region of the RCA. (**B**) Photomicrographs of serial sections from injured vessel that showed uptake of ^{99m}Tc-annexin V on in vivo imaging. The vessel was stained for α-actin (**a**), caspase-3 (**b**), and macrophages (**c**). Lesion cells staining positively for caspase-3 (**b**) were determined to be smooth muscle cells on the basis of brown cytoplasmic staining (**a**) and negative staining for macrophages (**c**). *Source*: From Ref. 26.

Inflammation

Evidence from autopsy and from circulating markers support the premise that inflammation is an important component of atherosclerotic lesions that are prone to rupture (27,28). The vascular inflammation can be widespread was well as focal (29). The inflammatory cell type found in great abundance in vulnerable plaque is the macrophage (13). Macrophages have high basal metabolic rates, rely on external glucose source as fuel, and do not store glycogen. Glucose consumption is further when increased when macrophages are activated. Fluorodeoxyglucose (FDG) uptake by inflamed tissues is 10 to 20 times that of most other tissues. FDG uptake in tumor foci is often higher in the tumor-associated inflammatory cells than in the cancerous cells (30). Experimental studies in atherosclerotic rabbit models have shown ^{18}F-FDG uptake in aortic atheroma localized by registration with CT and by autoradiography. These studies have

shown quantitative correlation between counts and macrophage number in the lesions (31,32).

Observational studies have reported focal uptake of FDG in the aorta and carotids of patients undergoing tumor staging (33–35). One study reviewed thoracic aorta uptake of FDG and calcification on PET/ CT scans (35). These investigators found that while aortic wall calcification correlated with age, aortic uptake of FDG did not. These results support the coronary artery autopsy findings showing lipid filled thin cap fibroatheroma with macrophage infiltration more commonly in younger victims of sudden death while the fibrocalcific lesions are more commonly seen in older patients with diffuse and "stable" coronary artery disease (CAD). In one prospective study, vascular uptake of ^{18}F-FDG was localized to carotid lesions in patients with symptomatic carotid disease undergoing endarterectomy. The investigators incubated endarterectomy tissue with tritiated deoxyglucose and

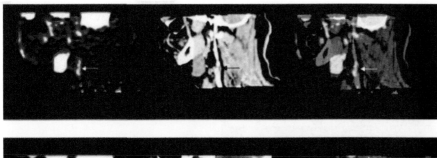

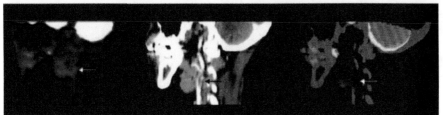

Figure 2 The *upper row* (from left to right) shows positron emission tomography (PET), contrast CT, and coregistered PET/CT images in the sagittal plane from a patient with symptomatic carotid disease. Contrast angiography and CTA showed stenosis of the proximal right internal carotid artery (*black arrow* on CT). The white arrows show [18]FDG uptake at the level of the plaque. Background [18]FDG uptake is what is normally seen in the brain, jaw muscles, and facial soft tissues. The *lower row* (from left to right) is from an asymptomatic patient with carotid stenosis on CTA (*black arrow*) with low uptake of ^{18}F-FDG at the level of the plaque on PET scan (*white arrow*). *Source*: From Ref. 36.

documented colocalization of the radioactivity to macrophages (Fig. 2) (36).

It was our hypothesis that patients with focal vascular uptake of ^{18}F-FDG have a higher cardiovascular event rate. ^{18}F-FDG scans performed for tumor staging on 500 patients from 1997 to 2002 were interpreted blinded for vascular uptake of FDG (37). A minimum of two years follow-up was obtained on all patients. All patients who died were excluded from analysis because of our inability to distinguish between death from cancer and cardiovascular (CV) death, leaving 348 patients. Of the 348 scans, 59 or 17% were positive for vascular uptake. Of these 59 patients with positive scans 17 (30%) had events, 12 of these were either stroke or myocardial infarction. Of the 289 patients with negative scans, there were only 12 CV events (4%). After risk adjustment, focal uptake of ^{18}F-FDG corresponding to the location of aorta and carotids was shown to be an independent predictor of CV endpoints including MI, transient ischemic attacks (TIA)/stroke, and symptomatic claudication requiring revascularization. These results indicate that a prospective study should be performed to identify patients with vulnerable atherosclerotic plaques who are at risk for CV events.

Imaging the coronary arteries with ^{18}F-FDG presents problems over and above limited camera resolution. ^{18}F-FDG as a glucose analogue is taken up by cardiac myocytes as a metabolic substrate. Myocardial uptake of glucose is dependent on insulin levels and on presence or absence of ischemia and uses glucose transporters such as glut-4 while uptake in inflammatory cells utilizes different mechanisms. Myocardial uptake of ^{18}F-FDG produces a high background, hiding any focal vascular uptake. It may be possible by suppressing myocardial uptake through fasting or by altering substrate, utilization to fatty acids may suppress myocardial uptake sufficiently to allow visualization of uptake in coronary plaques. This is an area for investigation.

Apoptosis of Macrophages

While apoptosis of vascular SMCs characterizes early plaque growth apoptosis of these cells has also been demonstrated in fibrotic part of advanced human atheromata (20) and SMC apoptosis may also contribute to plaque vulnerability by recruitment of inflammation via Fas-associated death domain (FADD) protein induced expression of MCP-1 and interleukin-8 (IL-8) causing migration of macrophages (21). However, apoptosis of smooth muscle cells plays a minor role in undermining plaque stability compared to apoptosis of macrophages. Using human autopsy material Kolodgie and coinvestigators defined culprit plaque as plaque rupture with intralumenal thrombosis and stable plaque as stenotic lesions with thick fibrous cap (13). Ruptured plaque showed extensive infiltration of macrophages that stained positive for caspase-1, a mammalian death protease while the stable lesions showed a dense fibrous cap with paucity of apoptotic cells (13). This same group of investigators went on to target macrophage apoptosis using radiolabeled annexin V in a rabbit model of atherosclerosis induced by aortic injury and high fat diet (Fig. 3) (38). They demonstrated in vivo uptake of ^{99m}Tc-annexin V in the aorta and showed a significant linear correlation between %ID in the lesion and macrophage death rate (38).

The group from Maastricht under the direction of Leo Hofstra reported results of a pilot study in four patients with symptomatic carotid disease (39). They demonstrated focal uptake of [99m]Tc- annexin A5 in culprit arteries from patients with recent events. Histopathology showed macrophage infiltration and intraplaque hemorrhage. Equally, stenotic lesions from patients with remote events and subsequent plaque stabilizing treatment were negative for tracer uptake and showed none of the histopathological features of instability or inflammation (39). Annexin V has a potential similar to ^{18}F-FDG to noninvasively identify

(Part I)

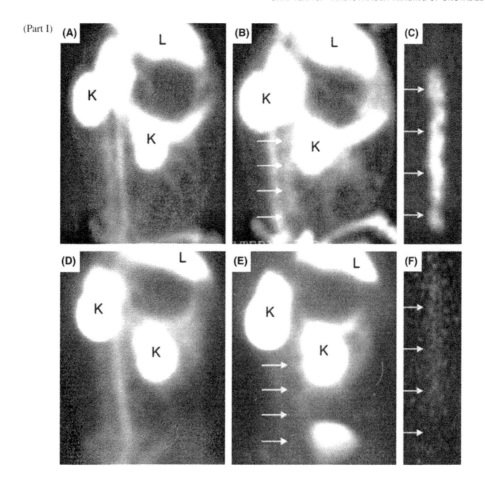

(Part II)

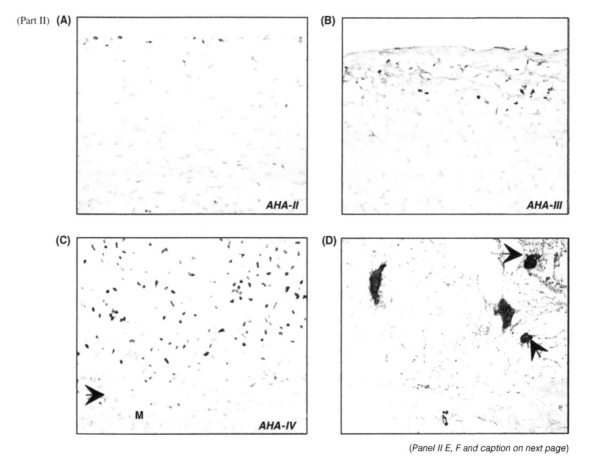

(Panel II E, F and caption on next page)

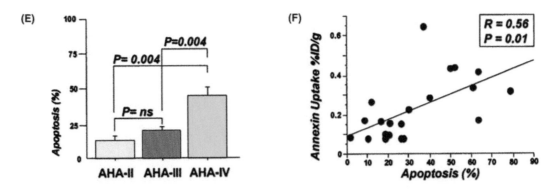

Figure 3 **Part I:** Left lateral oblique gamma images of rabbit model of atherosclerosis (**A–C**) and control rabbit (**D–F**) injected with 99m Tc-annexin V. Panels **A** and **D** represent images taken at the time of injection and panels B and D are images taken two hours after injection. Blood pool has cleared sufficiently by two hours to permit visualization of focal tracer uptake in the abdominal aorta confirmed by ex vivo images aorta (**C**). The control rabbit showed only aortic blood pool activity at the time of injection and no activity in the region of the aorta by two hours (*white arrows*) confirmed by ex vivo imaging (**F**). **Part II:** (**A–D**) show immunohistochemical stained sections from atherosclerotic rabbit aortic lesions. M indicates media and arrow heads on panel C points to internal elastic lamina. Black nuclei represent positive DNA fragment staining and methyl green is the counterstain for the nuclei. (**D**) shows ultrastructural analysis of cells in the region of the necrotic core. *Arrowheads* on this section point to nuclear condensation and cell shrinkage producing apoptotic bodies. (**E**) shows bar graph with apoptotic index related to lesion type and (**F**) shows simple regression analysis of [99mTc]- annexin V uptake and apoptotic index showing a significant correlation. *Abbreviations*: K, kidney; L, liver. *Source*: From Ref. 38.

vulnerable atherosclerotic plaque in the aorta and carotids and because of low cardiac uptake in the absence of recent myocardial infarction has greater potential than [18F]- FDG to identify vulnerable coronary artery lesions (8,40).

Metalloproteinases

Macrophages have the capacity when activated by cytokines (TNF-a, IL-1) to secrete inactive metalloproteinases (MMP) (41). The macrophages actively undergoing apoptosis at plaque rupture sites are rich in MMPs. (42) These connective tissue-degrading enzymes include interstitial collagenases (MMP1), gelatinase B (MMP9), stromolysins 1-3 (MMP3, 10, 11), and a membrane type. When activated by plasmin or by inactivation of intrinsic inhibitors in tissue, the MMP can degrade the connective tissue matrix (43). Both MMP mRNA and protein have been found in large amounts in the cap and core area (38). These enzymes are implicated in development of plaque instability by degradation of the fibrous cap. MMP activity may be neutralized by tissue inhibitors of MMP (TIMP) an approach to both therapy and to imaging. A broad based metalloproteinase inhibitor (MPI) was developed and labeled with [123I] and in vivo imaging demonstrated focal uptake in the carotids of in ApoE null mice. (44) Another broad based MPI developed by Bristol Meyer Squibb is under investigation in small and large animal models of atherosclerosis.

Oxidized Low Density Lipoprotein

Another component of the thin-capped fibroatheroma that can be targeted for imaging is lipid filled

macrophages and extracellular lipid pools. Oxidized LDL is immunogenic and comprises a large component of these lipid deposits and is found only in atherosclerotic plaques. A group of investigators has prepared murine monoclonal antibodies against malondialdehyde (MDA)-lysine epitopes and labeled these antibodies with [125I] (45–47). Uptake of the radiolabel on autoradiography was shown in the aorta of LDL-receptor deficient Watanabe heritable hyperlipidemic rabbits and corresponded to sites of Sudan staining. Using a technetium labeled antibody the investigators performed in vivo imaging in rabbits (45). Slow blood pool clearance combined with the short tracer half-life reduces scan quality that can be partially overcome by injecting cold MDA-LDA before imaging to dilute the pool. Earlier studies have directly radiolabeled LDL for in vivo imaging of atherosclerotic plaque. Any target that is present both in the blood and in the atherosclerotic plaque is challenging to detect because of the necessity to overcome or correct for blood pool activity. This is true for direct labeling of platelets and lipid particles (48,49).

PLAQUE EROSION

While thin-capped fibroatheroma are the pathological hallmark lesion for vulnerable plaque in older men and women a different lesion in found in younger women and some younger men who die suddenly from coronary thrombosis. In this lesion, the endothelium erodes exposing intima containing predominantly SMCs and proteoglycans and minimal inflammatory cells (12). Erosions are found at approximately 40% of autopsy specimens of coronary arteries from sudden death victims (14). This lesion is more commonly seen in younger patients and is seen exclusively in young women who die suddenly from an acute ischemic event. The predominant clinical association in premenopausal women is cigarette smoking. Both plaque rupture and plaque erosion may produce an acute coronary syndrome without catastrophic event or may even pass undetected. In these cases, the thrombus is not occlusive and the lesion heals and in the process leads to an incremental increase in plaque volume and reduction in luminal diameter (15). Since both plaque erosion and healed plaque erosion are characterized by SMCs that are in cycles of proliferation and quiescence, another potential target for imaging atherosclerosis is SMC proliferation.

Several radionuclide approaches to imaging SMC proliferation have been tried. Adenine analogs play a role in extracellular signaling including smooth muscle cell proliferation. Elmaleh and coinvestigators developed a radiolabeled adenine analog, ^{99m}Tc-diadenosine tetraphosphate and showed uptake in atherosclerotic plaque in a rabbit model of atherosclerosis (50). A more extensively used approach has been a radiolabeled antibody to an antigen expressed by proliferating SMCs.

Over 10 years ago in a quest to develop a method to image atherosclerosis, mice were immunized with homogenized human atherosclerotic plaques and an antibody designated as Z2D3 (IgM class, k light chain) was produced by the mice that reacted specifically with intimal proliferating SMCs in human atheroma (51). The cell line was subcloned to provide hybridoma cell lines and the antibody produced was shown to specifically cross-react with experimental atherosclerotic lesions in the rabbit aorta. The antibody colocalized with tissue staining positive for PCNA (proliferating cell nuclear antigen) and α-actin identifying the cells as proliferating SMCs. The antigen recognized by Z2D3 antibody was known to contain a complex of two or more chemically dissimilar, low molecular-weight molecules with sterol moieties. The surrogate antigens are now known to be: (i) mixture of 7-dehydrocholesterol and benzyldimethylhexadecyl-ammonium chloride and (ii) mixture of 7-dehydrocholesterol and palmitoylcholine.

Because fragments of Z2D3 IgM could not be made by enzymatic digestion the parent hybridoma line was switched to produce IgG. There was a 10-fold loss in affinity in this class switched antibody. To restore the immunoreactivity, the Z2D3 subclone IgM was genetically engineered to produce a chimera with a human IgG1 constant region. The chimeric IgG antibody showed immunoreactivity comparable to that of the parent Z2D3 IgM. This chimeric antibody was frag-

mented to F(ab')2 by pepsin digestion. These F(ab')2 fragments showed relatively high binding affinity of 3.3×10^{-7} mol/L (52). This high affinity chimeric IgG1 F(ab')2 was compared with low affinity subclones and nonspecific antibodies in rabbits with experimental atherosclerotic lesions (53,54).

A maneuver to improve the counts per pixel and target to background ratio in single photon imaging is charge modification. Torchilin et al. showed that by linking diethylenetriamine pentaacetate (DTPA) with polylysine, they could chelate a large number of trivalent metallic radiolabels (55). This process improved target to background by simultaneously increasing the amount of radiolabeled antibody delivered to the target and reducing the electrostatic attraction to nontarget cells with overall weak negative charge. Khaw et al. first demonstrated in vivo imaging of charge modified antibodies (56). Using ^{111}In labeled polylysine Z2D3 F(ab')2 in animal experiments, Narula and Khaw demonstrated that with higher antibody doses lesions could be visualized earlier (57,58). In a study in patients, Carrio and co\investigators showed focal carotid uptake of ^{111}In labeled Z2D3 F(ab')2 PL in patients with recent ischemic cerebral events (59). In a porcine model of coronary in-stent over-expansion restenosis, Johnson and coinvestigators demonstrated focal uptake of ^{111}In Z2D3 in the coronary arteries localized to sites of injury (60). The rate of SMC proliferation by histomorphometry correlated with lesional counts. Although the pathology of restenosis is different from atherosclerosis this study stands as another proof of concept that imaging uptake of radiotracer in the coronary arteries using conventional technology in humanoid model is feasible.

Despite the improved specific activity achieved at the target site using the charge-modified antibody, there are limitations due to both the ^{111}In tag and to nonspecific antibody uptake contributing to background activity. Although smaller than whole antibodies, antibody fragments are still relatively large molecules and therefore have relatively slow blood pool clearance. Transchelation of ^{111}In is a problem, especially to bone (ribs and clavicle) in SPECT imaging looking for focal uptake in coronary arteries in the thorax. There are further modifications that can improve the performance of Z2D3 imaging, which is to develop a bispecific antibody approach. In this approach, first a bispecific antibody with one functional unit that binds the antigen is injected. This molecule stays on the surface of the targeted cells but is phagocytosed into cells in the reticuloendothelial system that is a major contributing factor to high background activity. After a period of time (12 to 24 hours) to allow the bispecific antibody to clear from the blood pool and to be phagocytosed, a second small molecule with a radio tag directed against the second functional unit on the bispecific antibody is injected.

Although the bispecific antibody imaging with mono- or divalent hapten technology provide images with lower background and less nontarget organ activities, it would be of further benefit if the specific radioactivity at each antibody target site were increased to visualize even smaller targets. Khaw and Tekabe developed such an approach using bispecific antibody and radiolabeled (^{99m}Tc or ^{111}In) polymer to enhance in vivo molecular imaging for detection of specific target lesions. The bispecific antibody construct contains Z2D3 linked to anti-DTPA. The secondary binding molecule is polylysine DTPA. This secondary molecule is engineered to provide multiple DTPA binding sites for radionuclide technetium-99m. This approach for signal amplification using a target directed antibody linked to DTPA has broad application for signal amplification for molecular imaging. These investigators have demonstrated detection of femoral artery lesions comprising only several cell layers in ApoE null mice (Fig. 4) (61).

VASA VASORUM

Another potential target in the atherosclerotic plaque for imaging is the vasa vasorum. As the plaque grows the blood supply from the microvessels that normally reside in the adventitia grow and proliferate into the media occupying a larger volume of the luminal mass (62,63). Rapid growth of these thin walled leaky neovessels contributes to intraplaque hemorrhage. The accumulation of hemoglobin in the plaque probably promotes plaque instability (64). Developing radiotracers to target angiogenesis has been a lively field of research in cancer imaging and to monitor therapeutic angiogenesis. As the endothelial cells that form the walls of the neovessels grow they express integrins. Integrins are heterodimeric adhesion proteins involved in the developmental, physiological, and pathological processes. There are about 25 different integrins that mediate adhesion of cells to the extracellular matrix and other cells. They bind to RGD (arginine, glycine, aspartic acid) containing proteins.

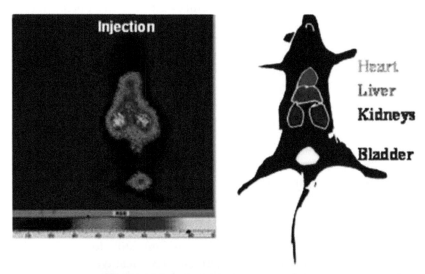

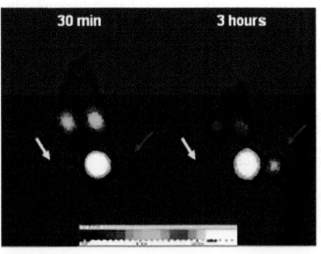

Figure 4 ApoE null mouse that had undergone left femoral artery injury was injected with ^{99m}Tc-polylysine DTPA 12 hours after receiving bispecific Z2D3/anti-DTPA antibodies. *Left upper panel* shows in vivo image obtained immediately after injection with a pin-hole camera. The adjacent mouse diagram identifies the biodistribution. The *lower panel* shows images in the same position taken at 30 minutes and 3 hours after injection of the radiolabeled tracer. As blood pool clears the focal uptake of tracer is visualized in the region of the injured femoral artery (*red arrow*) while the uninjured control right femoral artery (*yellow arrow*) shows no uptake. *Source*: From Ref. 69.

Targeted radiotracers have been developed that incorporate the RGD motif to bind to integrins for imaging vascular disease including angiogenesis in ischemic hind limb and gadolinium containing nanoparticles incorporating the RGD proteins have been developed for imaging atherosclerosis (12,65).

SUMMARY

MRI and computerized tomographic angiography (CTA) provide anatomic information on atherosclerosis including location, extent of wall thickening and with the higher resolution of MRI can provide information on plaque composition. Myocardial perfusion imaging using radionuclides provides information on regional vascular flow reserve and thereby provides information on the physiological significance of anatomic lesions. None of these approaches provides the information on plaque stability, which is the most important factor determining near future risk from acute ischemic events including stroke and MI. While MR molecular targeting approaches are being developed using specially engineered probes, nuclear medicine is the only imaging modality that routinely uses targeting probes in nanomolar concentrations having no biological effects and high affinity for relatively low abundance targets. Despite lower spatial resolution than MRI or CT small targets can be visualized as beacons on in vivo imaging when there are sufficient binding sites such as apoptotic macrophages and through radiotracer development to amplify the radioactive signal. The tracers that are presently under investigation that show considerable promise for clinical vascular imaging are ^{18}F-FDG for PET and ^{99m}Tc-annexin V for SPECT. The coronary arteries present a further challenge but from the results of experimental studies in large animals, it is probably feasible to see focal areas of radiotracer uptake in the coronary arteries in patients. Evidence from a number of sources supports the premise that plaque inflammation that characterizes vulnerability is a systemic and not just focal process. Nuclear medicine is suited for whole body imaging including imaging of the carotids, thoracic aorta, and coronaries. The advent of multi-modality imaging provides the ability to register SPECT or PET images with CT angiograms for better localization of focal uptake to atherosclerotic lesions as well as providing additional anatomic and biologic information on the lesion such as calcification.

REFERENCES

1. Pasternak RC, Abrams J, Greenland P, Smaha LA, Wilson PWF, Houston-Miller N. 34th Bethesda Conference: Task force #1–identification of coronary heart disease risk: is there a detection gap? J Am Coll Cardiol 2003; 41:1863–1874.

2. Choudhury RP, Fuster V, Badimon JJ, et al. MRI and characterization of atherosclerotic plaque emerging applications and molecular imaging. Arterioscler Thromb Vasc Biol 2002; 22:1065–1074.

3. Fayad ZA, Fuster V. Clinical imaging of the high-risk or vulnerable atherosclerotic plaque. Circ Res 2001; 89: 305–316.

4. Hatsukami TS, Ross R, Polissar NL, et al. Visualization of fibrous cap thickness and rupture in human atherosclerotic carotid plaque in vivo with high-resolution magnetic resonance imaging. Circulation 2000; 102:959–964.

5. Sirol M, Itskovich VV, Mani V, et al. Lipid-rich atherosclerotic plaques detected by gadofluorine-enhanced in vivo magnetic resonance imaging. Circulations 2004; 109:2890–2896.

6. Yuan C, Zhang S, Polissar NL, et al. Identification of fibrous cap rupture with magnetic resonance imaging is highly associated with recent transient ischemic attack or stroke. Circulation 2002; 105:181–185.

7. Winter P, Morawski AM, Caruthers SD, et al. Molecular imaging of angiogenesis in early-stage atherosclerosis with αvβ3-integrin-targeted nanoparticles. Circulation 2003; 108:2270–2274.

8. Blankenberg FG, Strauss HW. Nuclear medicine applications in molecular imaging. J Magn Reson Imaging 2002; 16:352–361.

9. Blankenberg FG, Mari C, Strauss HW. Development of radiocontrast agents for vascular imaging. Am J Cardiovasc Drugs 2002; 2:357–365.

10. Bengel FM. The atherosclerotic plaque: a healthy challenge to the limits of nuclear imaging. J Nucl Card 2005; 12:255–257.

11. Virmani R, Burke AP, Farb A. Sudden cardiac death. Cardvas Path 2001; 10:211–218.

12. Virmani R, Kolodgie FD, Burke AP, et al. Lessons from sudden coronary death. A comprehensive morphological classifacation scheme for atherosclerotic lesions. Arterioscler Thromb Vasc Biol 2000; 20: 1262–1275.

13. Kolodgie FD, Narula J, Burke AP, et al. Localization of apoptic macrophages at the site of plaque rupture in sudden coronary death. Am J Pathol 2000; 157: 1259–1268.

14. Farb A, Burke AP, Tang AL, et al. Coronary plaque erosion without rupture into a lipid core. Circulation 1996; 93:1354–1363.

15. Burke AP, Kolodgie FD, Farb A, et al. Healed plaque ruptures and sudden coronary death-evidence that subclinical rupture has a role in plaque progression. Circulation 2001; 103:934–940.

16. Abe Y, El-Masri B, Kimbal KT, et al. Soluble cell adhesion molecules in hypertriglyceridemia and potential significance on monocyte adhesion. Arterioscler Thromb Vasc Biol 1998; 18:723–731.

17. Blankenberg FG, Wen P, Dai M, et al. Detection of early atherosclerosis with radiolabeled monocyte chemoattractant protein-1 in prediabeteic Zucker rats. Pediatr Radiol 2001; 31:827–835.

18. Ohtsuki K, Hayase M, Akashi K, et al. Detection of monocyte chemoattractant protein-1 receptor expression in experimental atherosclerotic lesions. Circulation 2001; 104:203–208.

19. Bennett MR, Evan GI, Schwartz SM. Apoptosis of human vascular smooth muscle cells derived from normal vessels and coronary atherosclerotic plaques. J Clin Invest 1995; 95:2266–2274.

20. Geng YJ, Henderson LE, Levesque EB, et al. Fas is expressed in human atherosclerotic intima and promotes apoptosis of cytokine-primed human vascular smooth muscle cells. Arterioscler Thromb Vasc Biol 1997; 17:2200–2208.

21. Schaub FJ, Han DK, Liles WC, et al. Fas/FADD-mediated activation of a specific program of inflammatory gene expression in vascular smooth muscle cells. Nat Med 2000; 6:790–796.

22. England MV, Nieland LJW, Ramaekers FCS, et al. Annexin V-affinity assay: a review on an apoptosis detection system based on phosphatidylserine exposure. Cytometry 1998; 31:1–9.

23. Belhocine T, Steinmetz N, Hustinx R, et al. Increased uptake of the apoptosis-imaging agent 99mTc recombinant human annexin V in human tumors after one course of chemotherapy as a predictor of tumor response and patient prognosis. Clin Cancer Res 2002; 8: 2766–2774.

24. Strauss HW, Narula J, Blankenberg FG. Radioimaging to identify myocardial cell death and probably injury. Lancet 1992; 356:180.

25. Hofstra L, Liem IH, Dumont EA, et al. Visualisation of cell death in vivo in patients with acute myocardial infarction. Lancet 2000; 356:209–212.

26. Johnson LL, Schofield L, Donahay T, et al. 99mTc-Annexin V imaging for in vivo detection of atherosclerotic lesions in porcine coronary arteries. J Nucl Med 2005; 46:1186–1193.

27. Corti R, Hutter R, Badimon JJ, et al. Evolving concepts in the triad of atherosclerosis, inflammation and thrombosis. J Thromb Thrombolysis 2004; 17:35–44.

28. Keaney JF, Vita JA. The value of inflammation for predicting unstable angina. N Engl J Med 2002; 347:55–57.

29. Buffon A, Biasucci LM, Liuzzo G, et al. Widespread coronary inflammation in unstable angina. N Engl J Med 2002; 347:5–12.

30. Kubota R, Kubota K, Yamada S, Tada M, Ido T, Tamahashi N. Microautoradiotraphic study for the differentiation of intratumoral macrophages, granulation tissues and cancer cells by the dunamics of fluorine-18-fluorodeoxyglucse uptake. J Nucl Med 1994; 35: 104–112.

31. Ogawa M, Ishino S, Mukai T, et al. 18F-FDG Accumulation in atherosclerotic plaques: immunohistochemical and PET imaging study. J Nucl Med 2004; 45:1245–1250.

32. Tawakol A, Migrino RQ, Hoffman U, et al. Noninvasive in vivo measurement of vascular inflammation with F-18 fluorodeoxyglucose positron emission tomography. J Nucl Card 2005; 12:294–301.

33. Dunphy MPS, Freiman A, Larson SM, et al. Association of vascular 18F-FDG uptake with vascular clacification. J Nucl Med 2005; 46:1278–1284.

34. Yun M, Yeh D, Araujo LI, et al. F-18 FDG uptake in the large arteries a new observation. Clin Nucl Med 2001; 26:314–319.

35. Tatsumi M, Cohade C, Nakamoto Y, et al. Fluorodeoxyglucose uptake in the aortic wall at PET/CT: possible finding for active atherosclerosis. Radiology 2003; 229:831–837.

36. Rudd JHF, Warburton EA, Fryer TD, et al. Imaging atherosclerotic plaque inflammation with [18F]-fluorodeoxyglucose positron emission tomography. Circulation 2002; 105:2708–2711.

37. Daniels T, Bokhari S, Sciacca R, Fawwaz R, Murty R, Johnson LL. F-18 fluorodeoxyglucose uptake on PET is a marker for carotid and aortic plaque instability. Circulation 2005; 112:II-762.

38. Kolodgie FD, Petrov A, Virmani R, et al. Targeting of apoptotic macrophages and experimental atheroma with radiolabeled annexin V: A technique with potential for noninvasive imaging of vulnerable plaque. Circulation 2003; 108:3134–3139.

39. Kietselaer LJH, Reutelingsperger CPM, Heidendal AK, et al. Noninvasive detection of plaque instability with use of radiolabeld annexin A5 in patients with carotid-artery atherosclerosis. N Engl J Med 2004; 350: 1472–1473.

40. Narula J, Strauss HW PS. * I love you: implications of phosphatidyl serine (PS) reversal in acute ischemic syndromes. J Nucl Med 2003; 44:397–399.

41. Brinckerhoff CE, Matrisian LM. Matrix metalloproteinases: a tail of a frog that became a prince. Nat Rev Mol Cell Biol 2002; 3:207–214.

42. Galis G, Sukhova M, Lark M, Libby P. Increased expression of MMP degrading activity in vulnerable regions of human atherosclerotic plaques. J Clin Invest 1994; 94:2493–2503.

43. Sukhova GK, Schonbeck U, Libby P, et al. Evidence for increased collagenolysis by interstitial collagenases 1 and 3 vulnerable human atherosclerotic plaques. Circulation 1999; 99:2503–2509.

44. Schäfers M, Riemann B, Kopka K, et al. Scintigraphic imaging of matrix metalloproteinase activity in the arterial wall in vivo. Circulation 2004; 109: 2554–2559.

45. Tsimikas S, Palinski W, Halpern SE, et al. Radiolabeled MDA2, an oxidation-specific, monoclonal antibody, identifies native atherosclerotic lesions in vivo. J Nucl Cardiol 1999; 6:41–53.

46. Tsimikas S, Shortal BP, Witztum JL, et al. In vivo uptake of radiolabeled MDA2, an oxidation-specific monoclonal antibody, provides an accurate measure of atherosclerotic lesions rich in oxidized LDL and is highly sensitive to their regression. Aterioscler Thromb Vasc Biol 2000; 20:689–697.

47. Torzewski M, Shaw PX, Han KR, et al. Reduced in vivo aortic uptake of radiolabeled oxidation-specific antibodies reflects changes in plaque composition consistent with plaque stabilization. Aterioscler Thromb Vasc Biol 2004; 24:2307–2312.

48. Lam YT, Chesebro JH, Steele PM, et al. Deep arterial injury during experimental angioplasty: relation to a positive Indium-111-labeled platelet scintigram, quantitative

platelet deposition and mural thrombosis. J Am Coll Cardiol 1986; 8:1380–1386.

49. Miller DD, Boulet AJ, Tio FO, et al. In vivo Technetium-99m S12 antibody imaging of platelet-granules in rabbit endothelial neointimal proliferation after angioplasty. Circulation 1991; 83:224–236.

50. Elmaleh DR, Narula J, Babich JW, et al. Rapid noninvasive detection of experimental atherosclerotic lesions with novel 99mTc-labeled diadenosine tetraphosphates. Proc Natl Acad Sci 1998; 95:691–695.

51. Harrison DC, Calenoff E, Chen FW, Parmley WW, Khaw BA, Ross R. Plaque associated immune reactivity as a tool for the diagnosis and treatment of atherosclerosis. Trans Am Clin Climatol Assoc 1992; 103: 210–217.

52. Khaw BA, Klibanov A, O'Donnell SM, et al. Gamma imaging with negatively charge-modified monoclonal antibody: modification with synthetic polymers. J Nucl Med 1991; 32:1742–1751.

53. Narula J, Petrov A, Bianchi C, et al. Noninvasive localization of experimental atherosclerotic lesions with mouse/human chimeric Z2D3 F(ab')2 specific for the proliferating smooth muscle cells of human atheroma. Circulation 1995; 92:474–484.

54. Narula J, Petrov A, Ditlow C, Pak KY, Chen FW, Khaw BA. Maximizing radiotracer delivery to experimental atherosclerotic lesions with high-dose, negative charge-modified Z2D3 antibody for immunoscintigraphic targeting. J Nucl Cardiol 1997; 4:226–233.

55. Torchilin VP, Klibanov A, Nosiff ND, et al. Monoclonal antibody modification with chelate-linked high-molecular-weight polymers: major increases in polyvalent cation binding without loss of antigen binding. Hybridoma 1987; 6:229–240.

56. Khaw BA, Klibanov A, O'Donnell SM, et al. Gamma imaging with negatively charge-modified monoclonal antibody: modification with synthetic polymers. J Nucl Med 1991; 32:1742–1751.

57. Narula J, Petrov A, Ditlow C, et al. Gamma imaging of atherosclerotic lesions: the role of antibody affinity in in-vivo target localization. J Nucl Cardiol 1996; 3:231–241.

58. Narula J, Petrov A, Ditlow C, et al. Maximizing radiotracer delivery to experimental atherosclerotic lesions with high-dose, negative charge-modified Z2D3 antibody for immunoscintigraphic targeting. J Nucl Cardiol 1997; 4:226–233.

59. Carrió I, Pieri PL, Narula J, et al. Noninvasive localization of human atherosclerotic lesions with indium 111–labeled monoclonal Z2D3 antibody specific for proliferating smooth muscle cells. J Nucl Cardiol 1998; 5:551–557.

60. Johnson LL, Schofield LM, Weber DK, et al. Uptake of 111In-Z2D3 on SPECT imaging in a swine model of coronary stent restenosis correlated with cell proliferation. J Nucl Med 2006; 47:868–876.

61. Khaw BA, Tekabe Y, Johnson LL. Imaging experimental atherosclerotic lesions in ApoE knockout mice: enhanced targeting with Z2D3-anti-DTPA bispecific antibody and Tc-99m labeled negatively charged polymers. J Nucl Med 2006; 47:868–876.

62. Moulton KS. Plaque angiogenesis and atherosclerosis. Curr Atheroscler Rep 2001; 3:225–233.

63. Moulton KS, Heller E, Konerding MA, et al. Angiogenesis inhibitors endostatin or TNP-470 reduce intimal neovascularization and plaque growth in apolipoprotein e-deficient mice. Circulation 1999; 99:1726–1732.

64. Virmaini R, Kolodgie FD, Burke AP, et al. Atherosclerotic plaque progression and vulnerability to rupture: angiogenesis as a source of intraplaque hemorrhage. Arterioscler Thromb Vasc Biol 2005; 25:2054–2061.

65. Hua J, Dobrucki LW, Sadeghi MM, et al. Noninvasive imaging of angiogenesis with a 99mTc-labeled peptide targeted at _v_3 integrin after murine hindlimb ischemia. Circulation. 2005; 111:3255–3260.

20

Atherosclerosis Imaging with Nanoparticles

Gregory M. Lanza, Patrick M. Winter, Anne M. Neubauer, Kathy C. Crowder,
Shelton D. Caruthers, and Samuel A. Wickline
Cardiovascular Division, Department of Medicine, Washington University School of Medicine, St. Louis, Missouri, U.S.A.

INTRODUCTION

Atherosclerosis is a diffuse, progressive disease distributed heterogeneously throughout the vasculature. It is a complex pathology reflecting a multitude of interrelated processes including lipid disturbances, platelet activation, thrombosis, endothelial dysfunction, inflammation, oxidative stress, vascular smooth cell activation, altered matrix metabolism, remodeling, and genetics. Clinically, atherosclerosis may present as subtle, nonspecific symptoms or as sudden death, depending upon the circulatory territory involved, the caliber of the affected artery, local and regional flow dynamics, and the end-organ involved (1).

Simplistically, the development of plaque entails the extravasation of low density lipoprotein (LDL) particles beneath the intimal endothelium where uptake and oxidation by macrophages and smooth muscle cells occur, leading to the release of more growth factors and cytokines and the further recruitment of monocytes into the developing lesion. As this cycle propagates, foam cells accumulate in the intima, smooth muscle cells proliferate and migrate from the media, and the plaque grows (1). Angiogenic vessels, originating primarily in the adventia, expand the vasa vasorum and extend into the thickening intimal layer in response to increasing metabolic demand (2). The enlarging lesion volume is initially compensated for by remodeling the vascular wall to preserve blood flow, but eventually, encroachment on the lumen ensues (3). Often, plaque development progresses unevenly and is propelled by intimal ruptures, which lead to mural thrombus formation that may be incorporated into the lesion (4). On occasion, the rupturing process is severe, resulting in local vascular occlusion or distal embolism. Moreover, the prevalence, severity and tempo of these events are accelerated in many patients who smoke cigarettes, or suffer with diabetes, dyslipidemia, or hypertension (1). Today, risk factor assessments remain the primary means used to identify the vulnerable patient population for early therapeutic intervention in the prevention of vascular disease. Although circulating biomarkers have emerged that correlate with the risk of cardiovascular disease, the current diagnostic and therapeutic armamentarium remains focused on acute disease syndromes and secondary prevention.

This year, approximately 1,200,000 new and recurrent cases of myocardial infarction will be reported and 40% of these people will die. In the United States alone, 6,400,000 people suffer from angina, which reduces their productivity and quality of life, and 400,000 new cases are added annually (5). In short, cardiovascular atherosclerotic disease is pandemic among western societies. It threatens the longevity and quality of life for individuals and saps the productivity and resources of nations. Although emergency treatment of acute coronary syndromes and secondary prevention remain critical, we must strive to identify and limit the progression of vascular disease and unstable plaques in susceptible individuals before life-threatening vascular occlusive and embolic events occur.

FUNDAMENTALS OF MAGNETIC RESONANCE IMAGING

Traditionally, noninvasive imaging has provided anatomical information late in the disease process. Over the last decade, research efforts have begun to focus on early recognition of pathology by detection of biochemical markers that herald disease before obvious mass-effect changes manifest. Molecular imaging, that is, noninvasive, specific detection of diagnostic biosignatures, has long been associated with nuclear medicine, but over the last decade, the field has broadened to include all clinically relevant diagnostic modalities, both invasive and noninvasive. Among these modalities, magnetic resonance (MR) is emerging as an advantageous technique by virtue of its high spatial resolution and unique capability to elicit both anatomic and physiological information simultaneously. MR molecular imaging contrast agents take many forms, but nanoparticulate systems have emerged as the most successful genre to date, because they provide robust signal for noninvasive detection when targeted to tissues or cells.

To understand MR based contrast systems, a rudimentary appreciation of MR imaging and the nuclear magnetic resonance (NMR) phenomenon is required. The fundamental precept of NMR states that spins of protons (i.e., hydrogen nuclei) and electrons (6,7) when placed in a strong external magnetic field (B_0) orientate themselves either parallel (i.e., spin-up) or antiparallel (i.e., spin-down) to the magnetic field (B_0). Approximately, $1/10^6$ to $1/10^7$ protons adopt more spin-up than spin-down states per voxel. Although this is a trivial distribution imbalance, the overwhelming abundance of water in tissues magnifies this disparity and the impact is perceptible. Next, irradiating the magnetized sample with electromagnetic (radiofrequency) energy disturbs the thermal equilibrium distribution and increases the antiparallel or spin down level. The ensemble of spins exhibits a net magnetization vector tilted away from the direction of the main magnetic field (B_0). The transition from this excited state back to the ground state is known as relaxation. MR contrast is defined by the two principle NMR processes of spin relaxation, T1 (spin-lattice or longitudinal) and T2 (spin–spin or transverse) (6,7).

Most clinical MR contrast agents accelerate the rate of relaxation. While they affect both T1 and T2, paramagnetic agents principally accelerate longitudinal T1 relaxation, producing "bright" contrast in T1-weighted images. Superparamagnetic agents increase the rate of dephasing or transverse T2 relaxation, and create "dark" or negative contrast effects. T1 contrast agents directly influence proximate protons and are highly dependent on local water flux, whereas, T2 contrast agents disturb the magnetic field independent of their environment. The contrast impact of T2 agents extends well beyond their immediate surroundings, while T1 contrast agents have only very local influence. Thus, paramagnetic metals of T1 agents must ideally be exposed to water with adequate proton exchange rates within an imaging voxel. In contradistinction, superparamagnetic metals can be sequestered anywhere within a supporting matrix and still elicit T2 contrast. Regardless of the contrast platform, all targeted magnetic resonance agents must provide enough signal amplification or loss to be perceptible when the contrast is bound to the target. For sparse biosignatures, such as integrin in angiogenesis, the agent must produce enough concentration to be detected at nano to picomolar levels, which is a typical expression level for many biochemical epitopes.

NANOTECHNOLOGY IN MAGNETIC RESONANCE MOLECULAR IMAGING

Superparamagnetic Nanoparticles

Superparamagnetic nanoparticles, such as iron oxides, create image voids in a magnetic field that extend well beyond their immediate size. Detection of these agents was relatively easy on T1/T2 or highly T2-weighted images, which led to considerable early research attention. Numerous iron oxide based nanoparticles have been developed that differ in hydrodynamic particle size and surface coating material (dextran, starch, albumin, silicones) (8). In general terms, these particles are defined by nominal diameter into superparamagnetic iron oxides (SPIO, 50 nm to 500 nm) and ultrasmall superparamagnetic iron oxides (USPIO, <50 nm), which also dictates their physicochemical and pharmacokinetic properties.

USPIOs are small enough to extravasate from the systemic circulation and penetrate into plaques where they are readily phagocytosed by macrophages. The accumulated cellular iron oxide in the subendothelium is readily visualized as large regions of signal void (i.e., black) on T_2-weighted MRI images, which correlate spatially with active atherosclerosis (9,10). In this instance, the transport and uptake of iron oxide particles is nonspecific and the process is typically referred to as passive targeting.

Site-specific USPIO particles have been used to detect GP IIbIIIa receptors present on platelets in thrombi in vitro and in vivo (11). In these experiments, an RGD (arginine, glycine, aspartic acid) peptide was coupled to the surface carbohydrate of the USPIO (RGD-USPIO). Uniquely, T1-weighted imaging of targeted clots was performed at 1.5 T, using spoiled three-dimensional gradient-echo sequences at varying in-plane spatial resolutions. The contrast-enhanced clots were well visualized under ideal in vitro imaging conditions, but thrombus detection sensitivity diminished rapidly as spatial resolution decreased. These investigators insightfully noted and highlighted the confounding effects physiological motion that limits spatial resolution to the magnitude of the motion during the acquisition period. Regardless of the imaging agent, molecular imaging techniques must overcome the issues of motion artifacts, image alignments and image segmentation in order to resolve nascent pathologies expressing sparse molecular signatures in small anatomic distributions.

New T_1-weighted (i.e., bright) agents based on nanotechnology have emerged that use homing ligands to direct and concentrate particles at target sites. While some nonparticulate approaches to MR molecular imaging of vascular disease have found success against epitopes expressed at very high density, such as fibrin (12), most biosignatures of interest are naturally present at very low concentrations, again nano- or picomolar, and require marked signal amplification. In general, nanoparticulates provide enormous surface area, which can be functionalized to serve as a scaffold for targeting ligands and magnetic labels. One example is perfluoro-carbon-based nanoparticle emulsions, which are T_1-weighted molecular imaging and drug delivery agents with broad potential impact on the detection and treatment of atherosclerotic disease and its sequellae.

Targeted Perfluorocarbon Nanoparticles

Perfluorocarbon nanoparticles are ligand-targeted, lipid-encapsulated, nongaseous emulsions produced through microfluidization techniques (13). They inherently increase acoustic reflectivity (13–15) of targets when concentrated upon the surfaces or membranes (e.g., clots, endothelial cells, smooth muscle cells, synthetic membranes, etc.), but are virtually invisible to ultrasound when freely circulating . The nanoparticle platform is modified for robust MRI applications by including enormous payloads (i.e., ~100,000) paramagnetic chelates into the outer surfactant layer for T1-weighted contrast (16) or by taking advantage of the inherent high [19]fluorine content provided by the perfluorocarbon cores for imaging or spectroscopy (17,18). Active targeting to vascular biomarkers is typically accomplished by covalent cross-linking of homing ligands to the lipid surfactant, e.g., monoclonal antibodies, peptides, peptidomimetics, and others (Fig. 1). The emulsion nanoparticles have long circulatory half-lives due to their size (i.e., $t_{1/2}$ ~300 minutes) without further modification of their outer lipid surfaces with polyethylene glycol or utilization of polymerized lipids.

The relaxivity of paramagnetic nanoparticulates may be described in both ionic and macromolecular contexts (19,20). "Ionic relaxivity" is calculated with respect to absolute Gd-chelate concentration and is independent of molecular construct. "Particulate" or "molecular relaxivity" refers to the collective relaxivity

Table 1 Relaxivity of Two Paramagnetic Emulsions at Three Field Strengths

Magnetic field	Chelate	Relaxivity (mM*s)$^{-1}$	
		Ion-based (r_1)	Particle-based (r_1)
0.47 T	Gd-DTPA-BOA	21.3 ± 0.2	1,210,000 ± 10,000
0.47 T	Gd-DTPA-PE	36.9 ± 0.5	2,718,000 ± 40,000
1.5 T	Gd-DTPA-BOA	17.7 ± 0.2	1,010,000 ± 10,000
1.5 T	Gd-DTPA-BOA	33.7 ± 0.7	2,480,000 ± 50,000
4.7 T	Gd-DTPA-BOA	9.7 ± 0.2	549,000 ± 9,000
4.7 T	Gd-DTPA-BOA	15.9 ± 0.1	1,170,000 ± 10,000

Note: Gd-DTPA-BOA significantly different from Gd-DTPA-PE within field strength ($p < 0.05$).
Source: From Ref. 20.

influence of all the paramagnetic metal formulated into each nanoparticle. At 1.5 T, perfluorocarbon nanoparticles have an ionic r_1 relaxivity of ~30 (mM·s)$^{-1}$ and a molecular r_1 relaxivity of greater than 2,000,000 (mM·s)$^{-1}$, which is many fold greater than the ionic and molecular relaxivity of Gd-DTPA alone [~5 (mM·s)$^{-1}$] (Table 1) (20). This extraordinarily high relaxivity allows a voxel with as few as 100 pM of nanoparticles to be detected conspicuously, with a contrast-to-noise ratio (CNR) of 5 (21) (Fig. 2). Thus, in cardiovascular disease, the ultra high relaxivity of paramagnetic nanoparticles allow neovasculature of the vasa vasorum or microthrombi in fissuring plaques to be detected

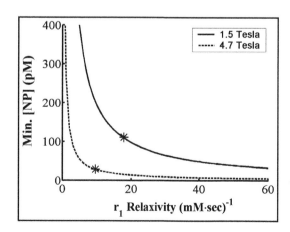

Figure 2 Effect of altering the ionic r_1 relaxivity of the nanoparticles on the minimum concentration needed for diagnostic contrast at 1.5 T and 4.7 T. Transverse relaxivity (r_2) was assumed to be 1.5 times r_1 at 1.5 T and 3 times r_1 at 4.7 T based on data obtained with the Gd-DTPA-BOA-based formulation. The asterisks indicate the ionic-based r_1 relaxivities at both field strengths from Table 1. *Source*: From Ref. 21.

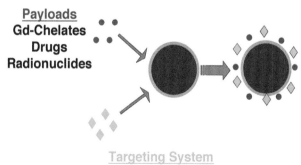

Payloads
Gd-Chelates
Drugs
Radionuclides

Targeting System
"Molecular zip codes"
Including Antibodies, Peptides, Mimetics, or Aptamers

Figure 1 Platform paradigm for ligand-targeted, perfluorocarbon nanoparticles (250 nm diameter nominal) with combination capability.

0.7 x 0.7mm   **0.1 x 0.1mm**

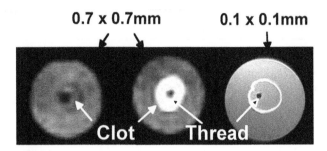

Figure 3 (*Left*) Low resolution images (3D GraSE) of fibrin-targeted clot with paramagnetic nanoparticles presenting a homogeneous, T1 weighted enhancement. (*Right*) High-resolution scan of fibrin clots on left revealing that contrast results from a thin layer of ultraparamagnetic nanoparticles along surface. *Source*: From Ref. 19.

and imaged with high resolution using standard, clinical imaging equipment.

Fibrin-Targeted Perfluorocarbon Nanoparticles

Perfluorocarbon nanoparticles have been targeted to a variety of molecular epitopes. Fibrin-targeted nanoparticles densely and specifically adhere to fibrin fibrils along the clot surface, delivering tens of thousands of gadolinium atoms with each bound particle (19). Using a typical low-resolution clinical imaging protocol, fibrin clots targeted in vitro with nanoparticles provide homogeneous T1-weighted contrast enhancement (Fig. 3). The gadolinium-rich nanoparticles overcome the partial volume dilution effect of the low-resolution voxel and appear to completely fill the clot volume with signal. However, at higher in-plane resolution, the same clot reveals that the nanoparticles were bound only at the surface and were excluded from penetration into the dense fibrin matrix (19).

In dogs, gradient echo images of thrombus targeted with antifibrin paramagnetic nanoparticles have produced high signal intensity (1780 ± 327), whereas, the contralateral control clot signal intensity (815 ± 41) was ~50% lower and similar to that of the adjacent muscle (768 ± 47) (Fig. 4) (19). The CNR between the targeted clot and blood measured with this sequence

was approximately 118 ± 21, whereas, the CNR between the targeted clot and the control clot was 131 ± 37 (19).

The concept of detecting human ruptured plaque was illustrated in vitro using carotid artery endarterectomy specimens resected from a symptomatic patient. Microscopic fibrin deposits within the "shoulders" of the ruptured plaque were readily apparent in contradistinction to control specimens (Fig. 4) (19). High-resolution, fibrin molecular imaging of ruptured plaque may someday prompt acute surgical or medical intervention, which could pre-empt stroke or myocardial infarction.

$\alpha_v\beta_3$-Targeted Perfluorocarbon Nanoparticles

One molecular signature, $\alpha_v\beta_3$-integrin, has garnered prominent early attention for angiogenic targeting applications because it is expressed on the lumenal surface of activated endothelial cells but not on mature quiescent cells. Although noninvasive and specific recognition of molecular events associated with angiogenesis in early vascular disease is not possible with traditional medical imaging techniques, $\alpha_v\beta_3$-targeted paramagnetic nanoparticles have been demonstrated to spatially localize and quantify early atherosclerotic burdens in hyperlipidemic New Zealand white rabbits (22) .

In one study, male New Zealand white rabbits were fed either 1% cholesterol or standard rabbit chow for ~80 days (22). Cholesterol-fed rabbits were divided into three groups receiving, (*i*) $\alpha_v\beta_3$-targeted paramagnetic nanoparticles, (*ii*) nontargeted paramagnetic nanoparticles, or (*iii*) pretreatment with $\alpha_v\beta_3$-targeted nonparamagnetic nanoparticles two hours prior to $\alpha_v\beta_3$-targeted paramagnetic nanoparticles. Again, the last group was designed to demonstrate the specificity of the $\alpha_v\beta_3$-integrin targeting by in vivo competitive blockade. All control-diet animals received $\alpha_v\beta_3$-targeted paramagnetic nanoparticles.

MR signal in the abdominal aorta integrated from the diaphragm to renal arteries increased after contrast injection, indicating the presence of targeted nanoparticles bound to $\alpha_v\beta_3$-integrin epitopes on the

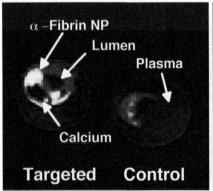
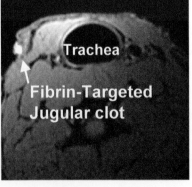

Figure 4 (*Left*) Color-enhanced MRI images of fibrin-targeted and control carotid endarterectomy specimens revealing contrast enhancement (*white*) of a small fibrin deposit on a symptomatic ruptured plaque. Calcium deposit (*black*) (3-D, fat suppressed, T1-weighted fast gradient echo). (*Right*) Thrombus in the external jugular vein targeted with fibrin-specific paramagnetic nanoparticles demonstrating T1-weighted contrast enhancement in the gradient echo image (*arrow*). *Source*: From Ref. 19.

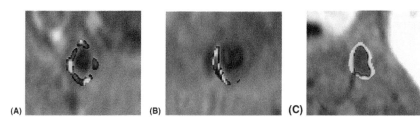

Figure 5 Top percent enhancement maps (*false-colored from blue to red*) from individual aortic segments at the renal artery (**A**), mid-aorta (**B**), and diaphragm (**C**) two hours after treatment in a cholesterol-fed rabbit given a_nb_3-targeted nanoparticles. *Source*: From Ref. 22.

neovasculature. Aortic wall contrast enhancement was variable along the circumference and length of the aorta as illustrated by the color-coded signal enhancement maps (Fig. 5) and longitudinally along the aorta for three selected rabbits (Fig. 6). However, greater signal enhancement was observed in the cholesterol-fed/targeted rabbits in practically every aortic segment. Histology of cholesterol-fed rabbit aortas revealed mild intimal thickening after 80 days, which was not appreciated in animals receiving the control diet (Fig. 7). Estimates of lumen area and wall thickness by MRI were similar for all treatment groups, reflecting early stage of vascular disease in these animals.

Quantification of the aortic signal enhancement among cholesterol-fed rabbits receiving $\alpha_v\beta_3$-targeted paramagnetic nanoparticles showed a $26 \pm 4\%$ and $47 \pm 5\%$ increase over baseline at 15 and 120 minutes, respectively, when averaged across the expanse of the aorta for each rabbit. In cholesterol-fed rabbits that received nontargeted nanoparticles, the aortic wall enhanced by $19 \pm 1\%$ within 15 minutes, but stabilized from 60 to 120 minutes at $26 \pm 1\%$, which represented about half of the signal augmentation observed for the specific $\alpha_v\beta_3$-targeted enhancement. Competitive

blockade of angiogenic $\alpha_v\beta_3$-integrins with targeted nonparamagnetic nanoparticles reduced the signal enhancement in the $\alpha_v\beta_3$-targeted paramagnetic nanoparticles group by at least 50%. In control-diet rabbits, aortic wall enhancement from $\alpha_v\beta_3$-targeted nanoparticles paralleled the contrast effect noted among the nontargeted, cholesterol-fed animals, which likely reflects delayed nanoparticle washout or slower flow in the mural vasa vasorum. Thus, the overall average signal enhancement in the aortic wall for cholesterol-fed animals was approximately twice that for control-diet animals at 120 minutes. The signal enhancement observed in the adjacent skeletal muscle due to nanoparticles in all treatment groups was negligible relative to that exhibited by the aortic wall.

In a separate study, hyperlipidemic animals received a tracer dosage of $\alpha_v\beta_3$-targeted nanoparticles (0.1 mL/kg) formulated with AlexaFluor 488 cyan dye and coadministered with a 10-fold excess of nontargeted, nonlabeled nanoparticles to minimize nonspecific entrapment of the fluorescent particles. $\alpha_v\beta_3$-targeted AlexaFluor 488 nanoparticles were localized predominantly throughout the adventitia and within intimal plaques immediately along the medial border, analogous to the typical distribution of $\alpha_v\beta_3$-integrin positive vessels. AlexaFluor 488 nanoparticles were not colocalized with cellular nuclei in either the plaque or adventitia. In particular, the distributions of $\alpha_v\beta_3$-targeted AlexaFluor 488 nanoparticles and macrophages, stained with RAM 11, had virtually no spatial overlap, which suggests that concentration of particles in the aorta was confined to the vasculature and was unrelated to phagocytic accumulation within plaque macrophages. Moreover, the lack of fluorescent label associated with intimal macrophages suggest that potential nanoparticle exchange of lipids with lipoproteins must be minimal, since fluorescently labeled lipoproteins can extravasate beyond the vasculature and could be phagocytosed.

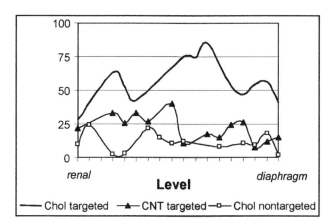

Figure 6 Spatial variation of contrast enhancement after treatment showing longitudinal signal variation from renal arteries to diaphragm for three selected animals from cholesterol-fed/targeted, control diet and cholesterol-fed/nontargeted groups. *Source*: From Ref. 22.

NANOPARTICLE-BASED DRUG DELIVERY

In contradistinction to blood-pool agents, nanoparticles provide a vehicle to transport and locally deliver

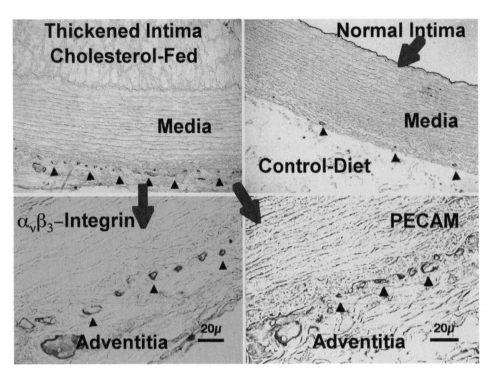

Figure 7 Histology of control (*top right*) and cholesterol-fed (*top left*) rabbit aortas revealing intimal thickening in the latter after 80 days, which was not appreciated in animals receiving the control diet. Expansion of the vasa vasorum and neovasculature in cholesterol-fed animals as indicated by PECAM (*bottom right*) and $\alpha_v\beta_3$-integrin (*bottom left*) immunochemical staining, respectively. *Source*: From Ref. 22.

drugs in a site-specific manner. We have suggested that hydrophobic compounds embedded into the outer surfactant layer of perfluorocarbon nanoparticles are delivered to targeted cells through "contact facilitated drug delivery," a mechanism in which ligand-directed binding promotes the exchange of lipids and drug from the nanoparticle to the targeted cell membrane (23). Ordinarily, this is a slow, inefficient process, but ligand binding of the nanoparticle prolongs the interaction with the target cell, minimizes the equilibrium separation of the lipid surfaces, and substantially increases the frequency and duration of lipid surface interactions (Fig. 8).

Recently, antiangiogenic agents, such as TNP-470, a water soluble form of fumagillin, were studied

by Moulton et al. (24) in Apo E$^{-/-}$ mice treated for four months (20 to 36 weeks) at a dosage of 30 mg/kg every other day, 1.68 g/kg total dose. Plaque angiogenesis and atheroma growth diminished despite persistent elevation of total cholesterol levels. We have shown the potential utility of $\alpha_v\beta_3$-targeted paramagnetic nanoparticles to deliver antiangiogenic therapy in hyperlipidemic rabbits (25).

For example, atherosclerotic rabbits were treated with $\alpha_v\beta_3$-targeted paramagnetic nanoparticles including 0 or 0.2 mole% fumagillin (25). MRI signal enhancement four hours postinjection was averaged over all imaged slices from the renal artery to the diaphragm to provide an integrated, quantitative assessment of the atherosclerotic burden. At baseline, the average MR

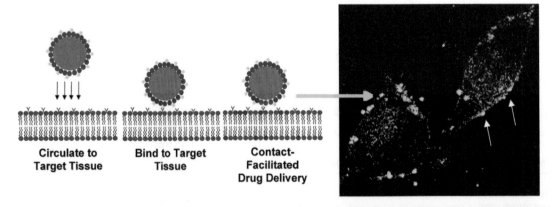

Figure 8 (*Left*) Schematic illustrating the principles of "contact-facilitated drug delivery" depicting lipid-exchange through extended membrane interactions. (*Right*) Confocal microscopic image of FITC-labeled lipid incorporated into the nanoparticle surfactant transferring onto and throughout the C-32 melanoma cell membrane. Phospholipids and drug within the nanoparticle surfactant exchange with lipids of the target membrane through a convection process rather than diffusion as is common among other targeted systems.

signal enhancement for both the fumagillin-treated (16.7 ± 1.1%) and control (16.7 ± 1.6%) rabbits were equivalent. The animals were recovered and one week later, repeat MRI studies were conducted to assess residual aortic angiogenic activity. At one week post-treatment, MR signal enhancement with $\alpha_v\beta_3$-targeted paramagnetic nanoparticles (no drug) in rabbits given targeted fumagillin therapy was markedly reduced (2.9 ± 1.6%, p < 0.05), whereas, the MR signal from the neovasculature of the control animals (i.e., targeted no drug) was unchanged (18.1 ± 2.1%). Moreover, the degree of change from week to week was proportional to the magnitude of MR signal enhancement observed at the time of treatment. Further studies have shown that the single dosage efficacy of fumagillin persists beyond two weeks before returning to baseline.

$\alpha_v\beta_3$-targeted fumagillin nanoparticles significantly reduced aortic angiogenesis with a total dose of drug more than 50,000 times lower than the oral dose used in Apo $E^{-/-}$ mice (24). We anticipate that the reduced dosage of therapeutic agent facilitated by targeted drug delivery approaches could substantially avoid the adverse effects previously reported for TNP-470 (26) or other potent agents. In clinical practice, we suggest that targeted fumagillin nanoparticles in combination with other standard care measures for hyperlipidemia (e.g., statins, dietary control, etc.) will allow prompt stabilization of plaque vulnerability and foster long-term regression.

Combination drug delivery with MR imaging spatially delineates and confirms local therapeutic delivery, it also permits the local drug concentrations to be estimated, i.e., rational drug dosing. Uniquely, perfluorocarbon paramagnetic nanoparticles offer dosimetry estimates based on ^1H, but support MR ^{19}F spectroscopy and ^{19}F imaging to quantify delivery to a specific target (18, 23).

QUANTITATIVE MAGNETIC RESONANCE ^{19}F SPECTROSCOPY AND IMAGING

In addition to the surface payload of paramagnetic chelates, perfluorocarbon (PFC) nanoparticles are 98% fluorocarbon by volume, which equates for perfluorooctylbromide (PFOB, 1.98g/mL, 498 daltons) to approximately 100 M concentration of fluorine within a nanoparticle. Perfluorocarbon nanoparticles are distinctly different from other oil-based emulsions by virtue of the physical-chemical properties of fluorine, the most electronegative of all elements (27). Fluorine has a high ionization potential and very low polarizability. Larger than hydrogen, fluorine creates bulkier, stiffer compounds that typically adopt a helical conformation. The C–F bond is chemically and thermally stable and essentially biologically inert. The dense electron cloud of fluorine atoms creates a barrier to encroachment on the perfluorinated chain

by other chemical reagents. The large surface area combined with the low polarizability presented by the fluorinated chains enhances hydrophobicity. Uniquely, perfluorinated chains are extremely hydrophobic and lipophobic simultaneously!

The biocompatibility of liquid fluorocarbons is well documented (27,28). Even at large doses, most fluorocarbons are innocuous and physiologically inactive. No toxicity, carcinogenicity, mutagenicity, or teratogenic effects have been reported for pure fluorocarbons within the 460 to 520 MW range. PFCs have tissue half-life residencies ranging from four days for perfluorooctyl-bromide up to 65 days for perfluorotripropylamine, and are not metabolized, but rather slowly reintroduced to the circulation in dissolved form by lipid carriers and expelled through the lungs. Increased pulmonary residual volumes with blood transfusion level dosages of PFC emulsions have been reported in rabbits, swine and macaque but not in mouse, dog or human (27).

Fluorine is an excellent element for MR spectroscopy and imaging because

- ^{19}F is has a gyromagnetic ratio nearly equivalent to proton (i.e., 83%)
- ^{19}F has a spin 1/2 nucleus
- ^{19}F has 100% natural abundance
- ^{19}F has essentially no detectable background concentration.

In addition, since fluorine has seven outer-shell electrons, rather than a single electron as is the case with hydrogen, the range and the sensitivity of chemical shifts to the details of the local environment are much higher for fluorine than hydrogen. As a consequence, ^{19}F MRI has been applied to study metabolism (29–31), to map physiologic pO_2 tension (32–34), and to characterize liquid ventilation (35,36). Unfortunately, most of these studies have required high magnetic field strengths, 4.7 T or greater and/or direct infusion of ^{19}F constructs to compensate for relatively low fluorine concentrations available to detect.

We have exploited the high fluorine content of the fibrin-targeted nanoparticles and the lack of background for ^{19}F imaging and spectroscopy (18). The linear relationship between ^{19}F and the gadolinium content of fibrin-targeted human plasma clots was confirmed in vitro by titrating mixtures of paramagnetic crown ether and nonparamagnetic safflower oil nanoparticles and measuring the linear decrease in signal as the amount of competing safflower oil agent increased. As expected, the number of bound paramagnetic fluorinated nanoparticles, as calculated from the normalized ^{19}F spectroscopic signal was directly proportional to the measured gadolinium content of the clots. Treatment of a fibrin clot with the crown ether emulsion provided a high number of bound nanoparticles, each composed of a large amount of perfluorocarbon, allowing acquisition of high signal-to-noise

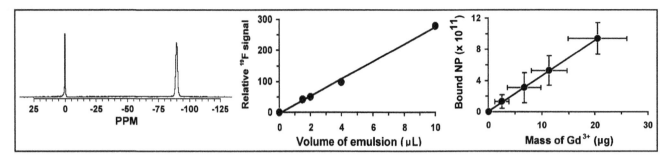

Figure 9 (*First*) Representative spectrum taken at 4.7 T of crown ether emulsion (-90 ppm) and trichlorofluormethane (0 ppm) used as a reference. (*Second*) The calibration curve for crown ether emulsion has a slope and intercept of 28.06 and −4.86, respectively, with an r^2 of 0.9968. (*Third*) The calculated number of bound nanoparticles (mean ± standard error) as calculated from ^{19}F spectroscopy versus the mass of total gadolinium (Gd^{3+}) r^2 = 0.9997. *Source*: From Ref. 18.

ratio (20.8) fluorine images at 4.7 T in less than 5 min. The corresponding ^{1}H image of the same slice showed that the ^{19}F signal from the bound nanoparticles originated from the clot surface (Fig. 9).

Human carotid endarterectomy samples have complex atherosclerotic lesions with several plaques and areas of calcification distributed throughout the vessel (18). Multi-slice ^{1}H images showed high levels of signal enhancement along the lumenal surface due to binding of targeted paramagnetic nanoparticles to fibrin deposits. A ^{19}F projection image of the artery, acquired in less than five minutes, shows an asymmetric distribution of fibrin-targeted nanoparticles around the vessel wall corroborating the signal enhancement

observed with ^{1}H MRI. Spectroscopic quantification of nanoparticle binding allowed calibration of the ^{19}F MRI signal intensity. Coregistration of the quantitative nanoparticle map with the ^{1}H image permits visualization of anatomical and pathological information in a single image (Fig. 10). In principle, estimating the exposed microthrombi surface area may predict subsequent occlusion or distal embolization. These risk data could evolve into rational guidelines for decisions to acutely intervene for plaque stabilization or to follow a more expectant course of medical therapy.

As aforementioned, the total fluorine spectrum that is quantitatively reflective of the mass of perfluorocarbon deposited within a selected voxel, can be

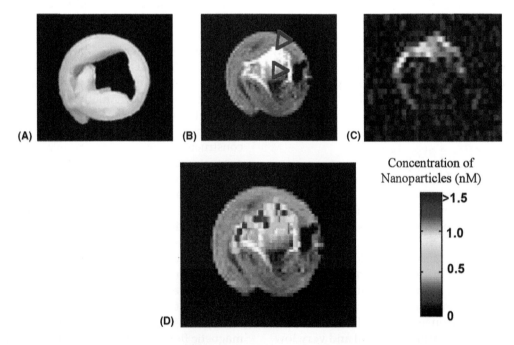

Figure 10 (A) Optical image of a 5 mm cross-section of a human carotid endarterectomy sample. This section showed moderate lumenal narrowing as well as several atherosclerotic lesions and areas of calcification. (B) A ^{1}H image acquired at 4.7 T at the same location shows signal enhancement due to the presence of gadolinium on the targeted particles (*red arrows*). (C) A ^{19}F projection image acquired at 4.7 T through the entire carotid artery sample shows high signal in the same areas due to nanoparticles bound to fibrin. (D) ^{1}H image in **B** with a false color overlay of the quantified nanoparticle concentration in the carotid as derived from the ^{19}F image. *Source*: From Ref. 18.

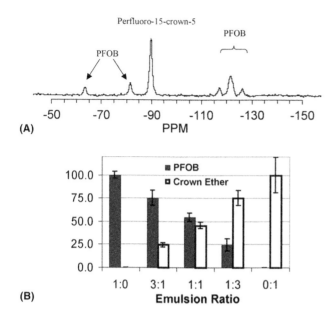

Figure 11 (**A**) ^{19}F spectrum acquired at 4.7 T of a clot treated with a mixture of fibrin-targeted crown ether and PFOB emulsions. The crown ether peak and five discernible PFOB peaks are easily detected and individually resolved. (**B**) Percentage of total ^{19}F signal attributed to crown ether or PFOB for the clots treated with different emulsion mixtures. Emulsion mixtures are listed as the ratio of PFOB to crown ether. Spectral discrimination of crown ether and PFOB allows quantification of the two nanoparticle species within a single sample. *Source*: From Ref. 18.

used to confirm and segment the proton images for easier recognition of contrast enhancement with reduced confusion induced by artifacts. [19]F imaging multi-spectral capability will permit PFC nanoparticles containing spectrally distinct fluorocarbons to be simultaneously targeted and independently measured through deconvolution of their composite spectral signal, as illustrated in Figure 11. This approach may be extended to several nanoparticle formulations to provide a multi-spectral palette with which to phenotypically characterize the biochemical nature or therapeutic sensitivity of a lesion. Such information may have prognostic utility when establishing a treatment strategy as well as retrospective value when interrogating local response to therapy. Together, ^{1}H MRI and ^{19}F imaging/spectroscopy offer many synergistic possibilities for enhancing clinical treatment strategies in atherosclerotic vascular disease.

SUMMARY

Atherosclerosis is a complex disease that continues to impart devastating losses on individuals and societies. The emergence of nanotechnology is providing new tools to detect and treat vascular disease in novel ways. In fact, the opportunities provided by targeted nanotechnology have no precedence

- never before has contrast-enhanced detection and quantification of unstable plaque noninvasively been possible;
- never before have quantitative assessments of atherosclerotic disease burden in the vascular walls been possible;
- never before has delivery of antiangiogenic or other treatments directly onto the expanding neovasculature been possible; and
- never before has the response to atherosclerotic therapy been predictable at the time of treatment or noninvasively quantifiable after therapy.

Clearly, the early opportunities that nanotechnology is creating may have profound impacts on the management of atherosclerotic disease in the future. However, the wise use of these agents will require further research into the importance, dynamics, and interplay of atherosclerotic processes.

REFERENCES

1. Faxon D, Fuster V, Libby P, et al. Atherosclerotic vascular disease conference, writing group III, pathophysiology. Circulation 2004; 109(21):2617–2625.
2. Zhang Y, Cliff WJ, Schoefl GI, et al. Immunohistochemical study of intimal microvessels in coronary atherosclerosis. Am J Pathol 1993; 143(1):164–172.
3. Glagov S, Weisenberg E, Zarins C, et al. Compensatory enlargement of human atherosclerotic coronary arteries. N Engl J Med 1987; 316:1371–1375.
4. Ojio S, Takatsu H, Tanaka T, et al. Considerable time from the onset of plaque rupture and/or thrombi until the onset of acute myocardial infarction in humans, coronary angiographic findings within 1 week before the onset of infarction. Circulation 2000; 102: 2063–2069.
5. Anonymous. Heart Disease and Stroke Statistics—2005 Update. American Heart Association. Available at, http,//www.americanheart.org/presenter.jhtml?identifier=3000090.
6. Gutierrez F, Brown J, Mirowitz S. Cardiovascular Magnetic Resonance Imaging. St. Louis, Mosby Year Book, 1992.
7. Nelson KL, Runge VM. Basic principles of MR contrast. Top Magn Reson Imaging Summer 1995; 7(3): 124–136.
8. Babes L, Denizot B, Tanguy G, et al. Synthesis of iron oxide nanoparticles used as MRI contrast agents, a parametric study. J Colloid Interface Sci 1999; 212(2): 474–482.
9. Ruehm SG, Corot C, Vogt P, et al. Ultrasmall superparamagnetic iron oxide-enhanced MR imaging of atherosclerotic plaque in hyperlipidemic rabbits. Acad Radiol 2002; 9(suppl 1):S143–S144.
10. Kooi ME, Cappendijk VC, Cleutjens KB, et al. Accumulation of ultrasmall superparamagnetic particles of iron oxide in human atherosclerotic plaques can

be detected by in vivo magnetic resonance imaging. Circulation 2003; 107(19):2453–2458.

11. Johansson LO, Bjornerud A, Ahlstrom HK, et al. A targeted contrast agent for magnetic resonance imaging of thrombus, implications of spatial resolution. J Magn Reson Imaging 2001; 13(4):615–618.

12. Botnar RM, Perez AS, Witte S, et al. In vivo molecular imaging of acute and subacute thrombosis using a fibrin-binding magnetic resonance imaging contrast agent. Circulation 2004; 109(16):2023–2029.

13. Lanza G, Wallace K, Scott M, et al. A novel site-targeted ultrasonic contrast agent with broad biomedical application. Circulation 1996; 94:3334–3340.

14. Lanza G, Wallace K, Fischer S, et al. High frequency ultrasonic detection of thrombi with a targeted contrast system. Ultrasound Med Biol 1997; 23:863–870.

15. Lanza GM, Trousil RL, Wallace KD, et al. In vitro characterization of a novel, tissue-targeted ultrasonic contrast system with acoustic microscopy. J Acoust Soc Am 1998; 104:3665–3672.

16. Lanza G, Lorenz C, Fischer S, et al. Enhanced detection of thrombi with a novel fibrin-targeted magnetic resonance imaging agent. Acad Radiol 1998; 5(suppl 1): S173–S176.

17. Yu X, Song S-K, Chen J, et al. High-resolution MRI characterization of human thrombus using a novel fibrin-targeted paramagnetic nanoparticle contrast agent. Mag Reson Med 2000; 44:867–872.

18. Morawski AM, Winter PM, Yu X, et al. Quantitative "magnetic resonance immunohistochemistry" with ligand-targeted(19)F nanoparticles. Magn Reson Med 2004; 52(6):1255–1262.

19. Flacke S, Fischer S, Scott M, et al. A novel MRI contrast agent for molecular imaging of fibrin, implications for detecting vulnerable plaques. Circulation 2001; 104: 1280–1285.

20. Winter P, Caruthers S, Yu X, et al. Improved molecular imaging contrast agent for detection of human thrombus. Mag Reson Med 2003; 50:411–416.

21. Morawski AM, Winter PM, Crowder KC, et al. Targeted nanoparticles for quantitative imaging of sparse molecular epitopes with MRI. Magn Reson Med 2004; 52:1255–1262.

22. Winter PM, Morawski AM, Caruthers SD, et al. Molecular imaging of angiogenesis in early-stage atherosclerosis with alpha(v)beta3-integrin-targeted nanoparticles. Circulation 2003; 108(18):2270–2274.

23. Lanza GM, Yu X, Winter PM, et al. Targeted antiproliferative drug delivery to vascular smooth muscle cells with a magnetic resonance imaging nanoparticle contrast agent, implications for rational therapy of restenosis. Circulation 2002; 106(22):2842–2847.

24. Moulton KS, Heller E, Konerding MA, et al. Angiogenesis inhibitors endostatin or TNP-470 reduce intimal neovascularization and plaque growth in apolipoprotein E-deficient mice. Circulation 1999; 99:1653–1655.

25. Winter PM, Morawski AM, Caruthers SD, et al. A combined drug delivery and molecular imaging agent for treatment and monitoring of plaque angiogenesis in atherosclerosis. J Am Col Cardiol 2005; (In Review).

26. Herbst RS, Madden TL, Tran HT, et al. Safety and pharmacokinetic effects of TNP-470, an angiogenesis inhibitor, combined with paclitaxel in patients with solid tumors, evidence for activity in non-small-cell lung cancer. J Clin Oncol 2002; 20(22):4440–4447.

27. Krafft M. Fluorocarbons and fluorinated amphiphiles in drug delivery and biomedical research. Adv Drug Del Rev 2001; 47:209–228.

28. Kaufman R. Clinical development of perfluorocarbon-based emulsions as red cell substitutes. In: Sjoblum J, ed. Emulsions and Emulsion Stability. New York: Marcel Dekker, 1996:343–367.

29. Ikehira H, Girard F, Obata T, et al. A preliminary study for clinical pharmacokinetics of oral fluorine anticancer medicines using the commercial MRI system 19F-MRS. Br J Radiol 1999; 72:584–589.

30. Schlemmer H, Becker M, Bachert P, et al. Alterations of intratumoral pharacokinetics of 5-fluorouracil in head and neck carcinoma during simulataneous radiochemotherapy. Cancer Res 1999; 59:2363–2369.

31. Wolf W, Presant C, Waluch V. 19F-MRS studies of fluorinated drugs in humans. Adv Drug Deliv Rev 2000; 41:55–74.

32. Noth U, Grohn P, Jork A, et al. 19F-MRI in vivo determination of the partial oxygen pressure in perfluorocarbon-loaded alginate capsules implanted into the peritoneal cavity and different tissues. Magn Reson Med 1999; 42:1039–1047.

33. Hunjan S, Zhao D, Canstandtinescu A, et al. Tumor oximetry, demonstration of an enhanced dynamic mapping procedure using fluorine-19 echo planar magnetic resonance imaging the Dunning prostate R3327-At1 rat tumor. Int J Radiat Oncol Biol Phys 2001; 49:1097–1108.

34. Fan X, River J, Zamora M, et al. Effect of carbogen on tumor oxygenation, combined fluorine-19 and proton MRI measurements. Int J Radiat Oncol Biol Phys 2002; 54:1202–1209.

35. Huang M, Ye Q, Williams D, et al. MRI of lungs using partial liquid ventilation with water-in-perfluorocarbon emulsions. Magn Reson Med 2002; 48:487–492.

36. Laukemper-Ostendorf S, Scholz A, Burger K, et al. 19F-MRI of perflubron for measurement of oxygen partial pressure in porcine lungs during partial liquid ventilation. Magn Reson Med 2002; 47:82–89.

21

Overview of Angiogenesis: Molecular and Structural Features

Arye Elfenbein and Michael Simons

Departments of Medicine and Pharmacology and Toxicology, Angiogenesis Research Center and Section of Cardiology, Dartmouth-Hitchcock Medical Center, Dartmouth Medical School, Lebanon, New Hampshire, U.S.A.

INTRODUCTION

Tissue growth and development is ideally subserved by an effective and responsive blood delivery system, one that functions to continuously adapt to the metabolic needs of its respectively dependent organs. With respect to oxygen demand, the vasculature must optimize blood flow, surface area, and structural plasticity to accommodate these varied requirements. The process of building a supportive and mature vascular network is therefore critical to establish and maintain tissue homeostasis.

Creating and preserving a structural complex that functions in these capacities requires several levels of hierarchal regulation. Excessive expansion of the vascular tree generates a wasteful allocation of resources, while inadequate development renders its dependent tissues hypoxic. Purposeful vessel expansion must also constantly preserve unidirectional blood flow and therefore distinguish the endpoints of arterial and venous growth. It follows that the modulation of vessel growth is a multi-factorial and dynamic process involving a combination of progenitor cell differentiation and vessel sprouting, countered by vascular regression. Known contributors to these events include myriad cytokines, growth factors, matrix proteins, and inducible gene transcription events.

Collectively, these intricacies and complexities of blood vessel growth lend themselves to countless potential loci of dysregulation. It is therefore not surprising that many pathological processes hinge upon the same changes in blood vessel structure seen under normal physiological conditions. As a result, angiogenic events often represent defined points of transition to cardiovascular conditions that include atherosclerosis, retinopathies, vascular malformations and cancer. In contrast, various common pathological states also serve as a primary impetus for an angiogenic response, including chronic inflammation and ischemia. The ability to effectively image and distinguish them is therefore a particularly promising tool to understand the underlying pathologies and expand the available therapeutic options in each case.

VASCULAR GROWTH: ANGIOGENESIS, ARTERIOGENESIS, AND VASCULOGENESIS

Tissue neovascularization, both under physiological conditions and in pathology, is a phenomenon that occurs via three distinct modes of vessel growth. While the entire range of vessel growth events is often collectively referred to as *angiogenesis*, this process exclusively encompasses vessel growth that originates directly from pre-existing vasculature. Angiogenesis is the primary mode of growth underlying tissue neovascularization in the adult, and it also plays an important role in the vascular development of the embryonic structures such as the neural crest.

In contrast to these angiogenic sprouts, neovascularization that involves the recruitment or development of collateral blood vessels is designated as *arteriogenesis*. This includes the functional activation of existing collateral circulation, the maturation of undeveloped secondary blood routes, or even the de novo formation of collateral vessels. Arteriogenesis often represents a localized primary response to conditions such as vascular stenosis, and represents an effective means of re-establishing adequate blood flow.

A third distinct mode of neovascularization, *vasculogenesis*, describes the process of vessel formation by differentiating endothelial or vascular progenitor cells. Vasculogenesis is responsible for the formation of the primitive tubular plexi that ultimately become the embryonic aorta, cardinal vein and other structures of the developing fetus. Originally thought to occur only during development, progenitor cell differentiation has in recent years been observed in the adult vasculature (1,2). The extent of its contribution outside a developmental context, however, still represents a field of particular controversy (3).

ANGIOGENESIS INDUCED BY HYPOXIA

Vascular responsiveness to a tissue's metabolic demands is perhaps most critically dependent upon oxygen delivery. Heightened metabolic needs and increased distances from a vascular bed therefore require an enhanced rate of blood flow. While auto-regulated vasodilation often partially diminishes an oxygen deficit, there are limits to this compensatory response. Consequently, chronic hypoxia is often only effectively relieved by the assembly of novel vascular conduits.

To this effect, several molecules are capable of transducing a hypoxic signal to ultimately initiate an appropriate angiogenic response. The most notable of these is hypoxia-inducible factor 1 (HIF1), a transcription factor that induces the expression of vascular growth factors under conditions of low oxygen concentration. HIF1 is comprised of two proteins (HIF-1α and HIF-1β), both of which directly bind to DNA.

HIF-1α is degraded under normoxic conditions by a proteosome-dependent pathway, but hypoxia confers stability that enables the protein to accumulate intracellularly. This occurs because the baseline degradation of HIF-1α depends on its post-translational modification, a process that requires oxygen as a cofactor. The Von Hippel-Lindau protein (a member of the ubiquitin ligase family) marks HIF-1α for subsequent degradation only if HIF-1α is post-translationally modified by prolyl hydroxylase-containing enzymes (4). By requiring oxygen (as well as ascorbic acid and iron), prolyl hydroxylase function ultimately results in the modulation of intracellular HIF-1α concentration. Under hypoxic conditions, hydroxylation of HIF-1α is therefore ineffective, the protein becomes less readily degraded, and it subsequently accumulates in the cell.

This increase in HIF-1α concentration also facilitates its association with HIF-1β [otherwise known as aryl hydrocarbon nuclear translocator (ARNT)] and its subsequent nuclear translocation. In the nucleus, the HIF1 complex induces an array of hypoxia responsive elements (HREs) by binding to specific DNA sequences, resulting in the activation of pathways that regulate cell migration, pH regulation, glycolysis and angiogenic growth factor expression, among others (5).

ENDOGENOUS MEDIATORS OF ANGIOGENESIS: THE VEGF FAMILY

While HIF1-mediated signaling induces a broad spectrum of cellular and extracellular events, this hypoxia-sensing system owes much of its specific angiogenic potency to the induced expression of vascular endothelial growth factor (VEGF). Initially named vascular permeability factor, VEGF has been noted for its profound ability to induce both angiogenesis and vascular leakiness (6). Members of the VEGF family [including VEGF-A, -B, -C, -D and placental growth factor (PlGF)], share certain degrees of homology and interact differentially with the three high-affinity VEGF receptor classes (VEGFRs) (7). These receptors are tyrosine kinases, dimerizing following ligand binding and signaling via pathways that involve intracellular enzymes such as MAP kinase.

VEGFR-1 (also named Flt-1) preferentially binds VEGF-A, VEGF-B and PlGF. VEGFR-2 (also named KDR/Flk-1) is a receptor for VEGF-A, -C and –D; it is predominantly involved in orchestrating the VEGF-specific effects of vascular growth and increased permeability (8). Variants of these receptors that consist only of the soluble extracellular domain exist physiologically (9). These soluble receptors serve as membrane receptor antagonists by sequestering nearby VEGF molecules and are therefore functionally angiostatic. Finally, the last member of the high-affinity VEGF receptors is VEGF-3, which binds VEGF-C and -D. This receptor has been reported to contribute to embryonic vessel maturation, yet its role in the adult is mostly limited to, and critical for, lymphatic system growth (10).

Upregulation of VEGF and its high affinity receptors is associated with hypoxia-induced angiogenesis and with tumor endothelium, rendering both potential candidates as markers of new vessel growth (11,12). However, high-affinity VEGF receptors represent only some of the membrane-bound proteins that transduce VEGF-initiated signals. The neuropilin class of membrane receptors bind both VEGF and semaphorins, with resultant effects found in neurons as well as endothelial cells (13). Neuropilins have been shown to selectively bind to particular molecular weight isoforms of VEGF; these receptors create profound downstream effects of growth and survival, in some cases amplifying VEGFR signaling (14). Further similarities between neural networks and the vasculature include observations that these molecules (along with the laminin-related family of netrins) are responsible for both axonal and vascular guidance during growth (15,16).

The range of VEGF-specific disorders reflects the importance of this growth factor in vascular growth and development. In mouse models, it has been discovered that mutations in even one allele of the VEGF gene yield disastrous vascular malformations and death (17). In humans, VEGF-associated conditions span from acute ophthalmic conditions, to progressive diseases such as rheumatoid arthritis, to congenital malformations including DiGeorge syndrome (18,19).

OTHER ANGIOGENIC MEDIATORS: FIBROBLAST GROWTH FACTORS

The VEGF family of proteins by no means represents the entire spectrum of endogenous angiogenic effectors. Other molecules that mediate vascular growth are present as matrix components, cytokines, and even metabolites of anaerobic glycolysis (20,21). Although a unified model of their in vivo interplay in the vascular system is obfuscated by the diverse effects of each in various tissues, several of them have been studied independently for the purpose of developing clinical treatments for vascular dysregulation.

Among these, the fibroblast growth factors (FGFs) represent a family of potent angiogenic effectors. Named for their original discovery in fibroblasts, FGFs are also strong inducers of growth in neurons and endothelial cells. There are 23 isolated ligands to date (FGF1–23) that bind with high affinity to 4 tyrosine kinase receptors (FGFR1–4). FGFs also mediate intracellular signaling events by binding to heparan sulfate proteoglycans, such as syndecans and perlecan. Syndecan-4, in particular, serves as a receptor for FGF-2 (also named basic FGF, or bFGF) (22,23). FGF-2 induces vascular-specific effects of increased endothelial cell migration and proliferation, rendering it a potent initiating agent of angiogenesis (24).

Other contributors to the angiogenic effects of growth factors are: platelet derived growth factor (PDGF), epidermal growth factor (EGF) and hepatocyte growth factor (HGF), each with various structural isoforms and specific receptors. The complexity of their respective interactions is still largely uncharacterized, although several discovered models of angiogenic synergy between particular growth factor combinations have been studied for their clinical relevance (25,26).

ANGIOGENIC REMODELING AND STABILIZATION: ANGIOPOIETINS

The angiogenic processes and associated growth factors described thus far primarily function to heighten cellular proliferation and enhance vascular branching. However, angiogenesis also encompasses the requisite element of structural plasticity and renovation, representing the transition between a primary vascular network and an organized system of mature vessels. In part, this involves the stabilization of budding vessels by integrating them into the surrounding matrix and recruitment of supportive pericytes.

The signaling molecules most implicated in this process of vessel remodeling are the angiopoietins, ligands that bind to two unique tyrosine kinases named Tie1 and Tie2. Although four distinct ligands have been identified (Ang-1–4), Ang-1 and Ang-2 are best understood with respect to their effects on vascular remodeling. The first insights into the involvement of angiopoietins in this role were generated by studies of Ang-1 knockout mice, which develop a normal primary vascular system that does not undergo subsequent remodeling (27).

Establishing a relatively impermeable endothelium represents a critical stage of vascular stabilization that enables a new vessel to support the flow of blood. Whereas VEGF-induced angiogenesis causes vessels to be more permeable (described above), Ang-1 opposes this effect and induces vessels to tighten their endothelial permeability barriers (28). This likely occurs by enhancing the interaction between endothelial cells, pericytes and the surrounding matrix. Another contributor to promoting maturation via matrix-cell interaction is transforming growth factor-β (TGF-β) that among many other downstream effects, activates plasminogen activator inhibitor-1 (PAI-1). This molecule prevents the degradation of matrix that surrounds forming vessels, fostering an environment that promotes maturation. However, TGF-β (like several other "angiogenic" factors) has been shown to have angiostatic effects as well, depending on which receptor it binds (29). As it becomes more apparent that angiogenic remodeling is not complete with angiopoietin signaling alone, the variable and indispensable contributions of TGF-β and related signaling molecules are likely to shape our understanding of functional development and maturation.

VESSEL DIFFERENTIATION: NOTCH-DELTA AND EPHRINS

The transition from a primary vascular network to a mature, functional system also necessitates a distinction between arterial and venous branches to maintain directional blood flow. For this reason, the differences in genetic expression between arterial and venous vessels have been intensively explored in the past decade to elucidate how the critical events of vessel differentiation occur (30). Two distinct pathways are representative of the signaling events that underlie these processes and will be described here.

The first of these is the Notch pathway, a signaling cascade that was first described in studies of fruit fly development (31). Notch signaling in human vascular physiology is best characterized by its specific association with the genetic disorder of cerebral autosomal-dominant arteriopathy with subcortical infarcts and leukoencehalopathy (CADASIL) (32). This condition is marked by a generalized arteriopathy that leads to fibrosis and eventual occlusion of small and medium-sized arteries, presenting particularly as cerebral lacunar infarcts. The implicated genetic defect is the expression of a mutated Notch 3, which presumably precludes appropriate vascular system development.

More specifically, Notch signaling is responsible for the type of intercellular signaling that is known as lateral inhibition. This mode of developmental signaling enables different fates of differentiation to be selected within a homogenous group of cells (33). The Notch receptors (Notch 1–4) are cell membrane proteins that are cleaved upon binding of their (also membrane-bound) ligands, including Delta-1, -3, -4 and Jagged-1, -2, typically from another cell. Upon protein cleavage, the intracellular domain subsequently translocates to the nucleus and affects the transcription of target genes.

Expression patterns of the Notch signaling pathways represent the first clues to their roles in arterial-venous differentiation. For example, the ligand Delta4 in the vasculature has been discovered only in the endothelium (34). Additionally, the Notch-4 receptor is limited to arterial vessels, although other Notch receptors do not share this structural specificity (35). Finally, the functional contributions of Notch signaling to arterial versus venous formation have been studied in a number of animal models. Taken together with the profound effects of mutant signaling molecules in humans, it is becoming more apparent that the role of Notch is critical for the differential development of arteries and veins from primary vascular networks.

Like the Notch signaling molecules, ephrin tyrosine kinases are a family of proteins with both membrane-bound receptors and ligands that serve to establish arterial or venous identity in developing vessels. Ephrins comprise the largest family of receptor tyrosine kinases, with 8 ligands and 14 receptors. The ephrins are subdivided into an A or B class, depending on their chemical anchors to the cell membrane. While Ephrin A ligands bind preferentially to EphA receptors (and Ephrin B ligands to EphB receptors), there is a marked degree of nonspecificity in these interactions (36). Ephrin signaling differs from that of the classical ligand-receptor model in that bidirectional signaling (from the receptor to the ligand and vice versa) is a prominent feature of its effects (37). For this reason, the localization of ephrin complexes is thought to often supersede receptor-ligand specificities in determining signaling efficacy.

It follows that ephrins mediate spatially-determined functions, including intercellular contact, cell adhesion and migration (38,39). Additionally, their importance in the vasculature is vastly extended by their contributions to vessel development. In particular, EphB4 has been shown to mark developing veins, while ephrin-B2 is specifically associated with

the arterial system (40,41). These findings, combined with the directional duality of ephrin signals, have led to an emerging model of forward signaling in one vessel type that inhibits signaling of the other (30). In this way, arterial sprouting might be enhanced while simultaneously inhibiting the formation of veins in a particular vascular bed. The ability of ephrins to function in the guidance of these forming vessels and to facilitate the formation of arterio-venous junctions consequently represents an active field of investigation.

VASCULAR REGRESSION: ANGIOPOEITIN-2

The angiogenic processes described thus far have focused on vascular expansion. However, the physiological relevance of angiogenesis becomes meaningful only within the context of the balance between vessel formation and regression. Vascular plasticity is as much a function of proliferation as degeneration, and both requisite processes ultimately dictate the efficacy of remodeling events. The significance of balanced growth and regression in vivo is underscored by the fact that endogenous initiators of both processes often act on the same pathways.

Perhaps the best example of this is Ang-2, a protein similar to Ang-1 in structure and in its affinity for the Tie2 receptor (42). However, Ang-2 prevents receptor activation, potentially opposing signals induced by Ang-1 (43). Although the physiological interaction and cross-talk between these homologues is incompletely characterized at present, the former has been implicated as a critical mediator of vessel destabilization in several models (27).

Vascular remodeling therefore depends on a dynamic equilibrium of growth, yet this balance merits even greater functional significance after vessel destabilization. This is because once a vessel has regressed to some extent, it becomes more susceptible to the growth and maturation factors described previously. However, it simultaneously becomes vulnerable to an alternate fate of complete eradication. Resolving this paradox of vascular regression relies on the context-specific activation of various parallel signaling pathways. In this way, the biological interplay between angiogenic and angiostatic factors often extends beyond a simple balance to encompass a complex model of inseparable temporal and spatial considerations.

MARKERS OF ANGIOGENESIS

The biological agents described have the capability to serve as primary mediators of angiogenic processes, yet many other critical players undergo shifts in activation level or expression pattern as a result of new blood vessel growth. Such molecules have been actively studied both for their insights into angiogenic processes and for their potential to represent specific markers of angiogenesis. While the distinction between a primary inducing agent and a secondary angiogenic byproduct is sometimes obscured by overlapping biological function, several molecules have been implicated for their specific association with new blood vessel formation. These molecules are often structural signs of angiogenesis and are of particular interest in the pursuit of effective imaging techniques for tissue neovascularization.

Despite the relative accessibility afforded by luminal markers of angiogenesis, it is important to note that many molecules associated with newly-created vasculature, particularly components of the extracellular matrix, are found in abluminal facets of these vessels. For this reason, potential markers that are not as readily distinguished by imaging tools such as antibodies or luminal bead perfusion will also be considered here.

Integrins and the Extracellular Matrix

Perhaps the most critical link between extracellular matrix proteins and intracellular signaling is the family of integrins, a diverse grouping of glycoproteins that are receptors for extracellular matrix molecules. Named for their ability to integrate signals from outside the cell, integrins form distinct noncovalent pairings of one alpha and one beta component. At least 18 alpha subunits and 8 beta subunits have been identified to date, comprising 25 known combinations of dimerization (44). Each alpha-beta combination is able to bind to particular matrix components, including fibronectin, collagen, laminin and von Willebrand factor. Although this yields a certain degree of specificity in a matrix-induced cellular response, there is also considerable overlap in the abilities of various integrin dimers to transduce responses from a particular matrix compound (45).

Adhesion of integrins to particular matrices enables the specific recruitment of cytoplasmic proteins involved with cytoskeletal remodeling. It is for this reason that integrins are well-characterized as indispensable to establish the dynamic balance between cellular adhesion and migration. In an angiogenic context, the expression of specific integrins, particularly $\alpha_V\beta_3$ and $\alpha_V\beta_5$, has been associated with blood vessel growth (46). However, the same integrins implicated in mediating angiogenic propensity have been characterized as both pro- and anti-angiogenic in different models (47,48). The involvement of $\alpha_5\beta_3$ and $\alpha_V\beta_3$ as mediators and markers of angiogenesis remains an area of active investigation, with several integrin-specific imaging studies showing varying degrees of promise toward clinical utility (49,50).

Among the other promising angiogenic markers that are involve the matrix is fibronectin, a widely prevalent glycoprotein. More specifically, a particular domain (extra-domain B, or EDB) is preferentially inserted into fibronectin molecules via splice variation at vascular remodeling sites. Targeting this domain of fibronectin by specific antibodies has been largely successful in attempts to image angiogenesis, despite the epitope's abluminal localization (18).

Luminal and Endothelial Markers

Phosphatidyl serine phospholipids are normally found on the intracellular facet of the cell membrane. However, proliferating endothelial cells have been shown to display these molecules on the outer membrane surface, making them prime targets for the detection of angiogenesis. Several preliminary studies carry promise (51), yet the value of these molecules in the imaging of angiogenic processes remains to be proven.

In contrast to the broad vascular expression pattern of phosphotidyl serine phospholipids, the membrane-bound metalloprotease known as aminopeptidase N, or CD13, has received attention on account its specific association with angiogenic endothelial cells (51). Aminopeptidase N has been shown to inactivate or activate small, extracellular, bioactive peptides by cleaving their terminal residues. However, its particular regulatory role in angiogenesis has been characterized in several models, while the involved substrates remain elusive. Furthermore, the recent discovery that CD13 inhibitors attenuate the progression of tumor angiogenesis (52) augments this molecule's candidacy as a specific marker of novel vessel growth.

Another cell surface antigen associated with new vessel growth is syndecan-4. The syndecan family of proteins consists of single-transmembrane heparan sulfate proteoglycans that transduce signals induced by heparin binding growth factors, such as FGF (23). Syndecans are ubiquitously expressed, and syndecan-4 signaling is associated with increased endothelial cell migration (53). Syndecan-4 expression is markedly upregulated in ischemic myocardium (54) and in arteries following vascular injury (55). Syndecan-4 expression is also specifically associated with the ability of satellite cells to regenerate muscle tissue following injury (56). As a result, its utility in angiogenic imaging is likely to rest in both luminal and abluminal vascular compartments.

Finally, an angiogenic marker that was discovered from cancer studies is prostate-specific membrane antigen (PSMA). Commonly used as a clinical indicator of prostatic hypertrophy, this glycoprotein has an increased expression level in angiogenic vessels supporting tumors (57). Despite this preliminary association, the functional significance of PSMA remains elusive.

Pathological Angiogenesis: Structural Features

Beyond the expression of specific angiogenic markers, various structural features serve to distinguish normally developed vasculature from that which is a product of growth dysregulation. Although experimental models of these differences often involve cancer angiogenesis, many of the same structural characteristics are present in other pathological conditions and serve as valuable guides for the imaging of angiogenesis.

To begin, nonphysiological angiogenesis is typically irregular in structural hierarchy, spatial distribution and vessel diameter, as compared with normal vasculature (58). In the case of tumors, this is often because rapidly growing parenchymal cells directly impinge upon regions of the supportive vasculature, or induce vessel growth with inappropriate subsequent maturation. The result is often increased vascular permeability, caused by inadequate structural endothelial junctions (58) and irregular patterns of blood distribution.

It is important to note that the molecular markers associated with angiogenesis are present to a certain extent in both normal and pathologic angiogenic events. For this reason, the insight derived from the greater perspective of vascular structure is invaluable in characterizing in vivo angiogenesis. Furthermore, functional measurements of a particular vascular bed are another important distinguishing parameter, as seen in the differences between immature, leaky vessels and a developed, supportive vasculature.

CONCLUSION

Although a sequential model of vessel growth, differentiation, maturation and regression is logically enticing as a representation of angiogenic processes, evidence from biological studies has made it apparent in recent years that these occurrences are necessarily parallel in nature. Individually, each process has yielded a degree of insight into the development of stable and mature vasculature, as described. However, a mechanism for how these varied events are simultaneously orchestrated and regulated still remains an elusive mystery.

The physiological roles of the described angiogenic influences are also greatly expanded by their diverse effects on nonvascular tissues. This often confounds studies of growth factors' functional specificities in eliciting or facilitating an angiogenic response. More importantly, signaling molecules often trigger pathological states via the same signaling modalities that support normal growth and development, significantly augmenting a functional definition of each in the context of neovascularization. The wide-reaching effects of these pathways have therefore in many cases caused the search for angiogenic and angiostatic therapeutics to become fraught with the challenges of specificity.

Irrespective of these difficulties, the manifold players and signaling events underlying angiogenesis lend themselves to diverse and unique markers of novel vessel growth. These can present as upregulated growth factors, altered matrix composition, macroscopic structural features, or idiosyncratically associated protein expression. The various forms of angiogenesis represent a convergence of these physiological events, with their multiplicity providing a wealth of potential for understanding, isolating and imaging specific regions of neovasculature.

REFERENCES

1. Asahara T, et al. Isolation of putative progenitor endothelial cells for angiogenesis. Science 1997; 275(5302):964–967.
2. Aicher A, et al. Essential role of endothelial nitric oxide synthase for mobilization of stem and progenitor cells. Nat Med 2003; 9(11):1370–1376.
3. Urbich C, Dimmeler S. Endothelial progenitor cells: characterization and role in vascular biology. Circ Res 2004; 95(4):343–353.
4. Ivan M, et al. HIFalpha targeted for VHL-mediated destruction by proline hydroxylation: implications for O2 sensing. Science 2001; 292(5516):464–468.
5. Harris AL. Hypoxia—a key regulatory factor in tumour growth. Nat Rev Cancer 2002; 2(1):38–47.
6. Ferrara N. Vascular endothelial growth factor: molecular and biological aspects. Curr Top Microbiol Immunol 1999; 237:1–30.
7. Li X, Eriksson U. Novel VEGF family members: VEGF-B, VEGF-C and VEGF-D. Int J Biochem Cell Biol 2001; 33(4):421–426.
8. Matsumoto T, Claesson-Welsh L. VEGF receptor signal transduction. Sci STKE 2001; 2001(112):RE21.
9. Ebos JM, et al. A naturally occurring soluble form of vascular endothelial growth factor receptor 2 detected in mouse and human plasma. Mol Cancer Res 2004; 2(6):315–326.
10. Takahashi M, Yoshimoto T, Kubo H. Molecular mechanisms of lymphangiogenesis. Int J Hematol 2004; 80(1):29–34.
11. Brekken RA, et al. Vascular endothelial growth factor as a marker of tumor endothelium. Cancer Res 1998; 58(9):1952–1959.
12. Lu E, et al. Targeted in vivo labeling of receptors for vascular endothelial growth factor: approach to identification of ischemic tissue. Circulation 2003; 108(1):97–103.
13. Neufeld G, et al. The neuropilins: multifunctional semaphorin and VEGF receptors that modulate axon guidance and angiogenesis. Trends Cardiovasc Med 2002; 12(1):13–19.

14. Bicknell R, Harris AL. Novel angiogenic signaling pathways and vascular targets. Annu Rev Pharmacol Toxicol 2004; 44:219–238.

15. Klagsbrun M, Takashima S, Mamluk R. The role of neuropilin in vascular and tumor biology. Adv Exp Med Biol 2002; 515:33–48.

16. Lu X, et al. The netrin receptor UNC5B mediates guidance events controlling morphogenesis of the vascular system. Nature 2004;

17. Ferrara N, et al. Heterozygous embryonic lethality induced by targeted inactivation of the VEGF gene. Nature 1996; 380(6573):439–442.

18. Brack SS, Dinkelborg LM, Neri D. Molecular targeting of angiogenesis for imaging and therapy. Eur J Nucl Med Mol Imaging 2004; 31(9):1327–1341.

19. Carmeliet P. Angiogenesis in health and disease. Nat Med 2003; 9(6):653–660.

20. Bellon G, Martiny L, Robinet A. Matrix metalloproteinases and matrikines in angiogenesis. Crit Rev Oncol Hematol 2004; 49(3):203–220.

21. Murray B, Wilson DJ. A study of metabolites as intermediate effectors in angiogenesis. Angiogenesis 2001; 4(1): 71–77.

22. Volk R, et al. The role of syndecan cytoplasmic domain in basic fibroblast growth factor-dependent signal transduction. J Biol Chem 1999; 274(34):24417–24424.

23. Horowitz A, Tkachenko E, Simons M. Fibroblast growth factor-specific modulation of cellular response by syndecan-4. J Cell Biol 2002; 157(4):715–725.

24. Nugent MA, Iozzo RV. Fibroblast growth factor-2. Int J Biochem Cell Biol 2000; 32(2):115–120.

25. Cao R, et al. Angiogenic synergism, vascular stability and improvement of hind-limb ischemia by a combination of PDGF-BB and FGF-2. Nat Med 2003; 9(5):604–613.

26. Carmeliet P, et al. Synergism between vascular endothelial growth factor and placental growth factor contributes to angiogenesis and plasma extravasation in pathological conditions. Nat Med 2001; 7(5):575–583.

27. Yancopoulos GD, et al. Vascular-specific growth factors and blood vessel formation. Nature 2000; 407(6801): 242–248.

28. Thurston G, et al. Angiopoietin-1 protects the adult vasculature against plasma leakage. Nat Med 2000; 6(4): 460–463.

29. Jain RK. Molecular regulation of vessel maturation. Nat Med 2003; 9(6):685–693.

30. Torres-Vazquez J, Kamei M, Weinstein BM. Molecular distinction between arteries and veins. Cell Tissue Res 2003; 314(1):43–59.

31. Portin P. General outlines of the molecular genetics of the Notch signalling pathway in Drosophila melanogaster: a review. Hereditas 2002; 136(2):89–96.

32. Kalimo H, et al. CADASIL: a common form of hereditary arteriopathy causing brain infarcts and dementia. Brain Pathol 2002; 12(3):371–384.

33. Kopan R, Turner DL. The Notch pathway: democracy and aristocracy in the selection of cell fate. Curr Opin Neurobiol 1996; 6(5):594–601.

34. Mailhos C, et al. Delta4, an endothelial specific notch ligand expressed at sites of physiological and tumor angiogenesis. Differentiation 2001; 69(2-3):135–144.

35. Uyttendaele H, et al. Vascular patterning defects associated with expression of activated Notch4 in embryonic endothelium. Proc Natl Acad Sci U S A 2001; 98(10): 5643–5648.

36. Gale NW, et al. Eph receptors and ligands comprise two major specificity subclasses and are reciprocally compartmentalized during embryogenesis. Neuron 1996; 17(1):9–19.

37. Gauthier LR, Robbins SM. Ephrin signaling: one raft to rule them all? one raft to sort them? one raft to spread their call and in signaling bind them? Life Sci 2003; 74(2-3):207–216.

38. Cheng N, Brantley DM, Chen J. The ephrins and Eph receptors in angiogenesis. Cytokine Growth Factor Rev 2002; 13(1):75–85.

39. Hamada K, et al. Distinct roles of ephrin-B2 forward and EphB4 reverse signaling in endothelial cells. Arterioscler Thromb Vasc Biol 2003; 23(2):190–197.

40. Adams RH, et al. Roles of ephrinB ligands and EphB receptors in cardiovascular development: demarcation of arterial/venous domains, vascular morphogenesis, and sprouting angiogenesis. Genes Dev 1999; 13(3): 295–306.

41. Wang HU, Chen ZF, Anderson DJ. Molecular distinction and angiogenic interaction between embryonic arteries and veins revealed by ephrin-B2 and its receptor Eph-B4. Cell 1998; 93(5):741–753.

42. Loughna S, Sato TN. Angiopoietin and Tie signaling pathways in vascular development. Matrix Biol 2001; 20(5-6):319–325.

43. Maisonpierre PC, et al. Angiopoietin-2, a natural antagonist for Tie2 that disrupts in vivo angiogenesis. Science 1997; 277(5322):55–60.

44. Hood JD, Cheresh DA. Role of integrins in cell invasion and migration. Nat Rev Cancer 2002; 2(2):91–100.

45. van der Flier A, Sonnenberg A. Function and interactions of integrins. Cell Tissue Res 2001; 305(3):285–298.

46. Friedlander M, et al. Definition of two angiogenic pathways by distinct alpha v integrins. Science 1995; 270(5241):1500–1502.

47. Brooks PC, et al. Integrin alpha v beta 3 antagonists promote tumor regression by inducing apoptosis of angiogenic blood vessels. Cell 1994; 79(7):1157–1164.

48. Reynolds LE, et al. Enhanced pathological angiogenesis in mice lacking beta3 integrin or beta3 and beta5 integrins. Nat Med 2002; 8(1):27–34.

49. Sipkins DA, et al. Detection of tumor angiogenesis in vivo by alphaVbeta3-targeted magnetic resonance imaging. Nat Med 1998; 4(5):623–626.

50. Meoli DF, et al. Noninvasive imaging of myocardial angiogenesis following experimental myocardial infarction. J Clin Invest 2004; 113(12):1684–1691.

51. Pasqualini R, et al. Aminopeptidase N is a receptor for tumor-homing peptides and a target for inhibiting angiogenesis. Cancer Res 2000; 60(3):722–727.

52. Aozuka Y, et al. Anti-tumor angiogenesis effect of aminopeptidase inhibitor bestatin against B16–BL6 melanoma cells orthotopically implanted into syngeneic mice. Cancer Lett 2004; 216(1):35–42.

53. Longley RL, et al. Control of morphology, cytoskeleton and migration by syndecan-4. J Cell Sci 1999; 112(Pt 20):3421–3431.

54. Li J, et al. Macrophage-dependent regulation of syndecan gene expression. Circ Res 1997; 81(5): 785–796.

55. Nikkari ST, et al. Smooth muscle cell expression of extracellular matrix genes after arterial injury. Am J Pathol 1994; 144(6):1348–1356.

56. Cornelison DD, et al. Essential and separable roles for Syndecan-3 and Syndecan-4 in skeletal muscle development and regeneration. Genes Dev 2004; 18(18): 2231–2236.

57. Chang SS, et al. Prostate-specific membrane antigen: much more than a prostate cancer marker. Mol Urol 1999; 3(3):313–320.

58. Jain RK, Munn LL, Fukumura D. Dissecting tumour pathophysiology using intravital microscopy. Nat Rev Cancer 2002; 2(4):266–276.

22

Imaging of Angiogenesis

Albert J. Sinusas and Lawrence W. Dobrucki
Yale University School of Medicine, New Haven, Connecticut, U.S.A.

INTRODUCTION

There are no widely available biomarkers or imaging approaches that permit the detection of angiogenesis or allow the selection of patients that are likely to respond to angiogenic therapy. Therefore, noninvasive imaging strategies will be critical for defining the temporal characteristics of angiogenesis, defining those patients that are more likely to respond to therapy, and assessing efficacy of these angiogenic therapies. Most attempts at imaging of angiogenesis have focused on imaging of the physiological consequences of the therapeutic intervention directed at stimulation of angiogenesis (1). However, there is a need for development of noninvasive approaches for direct evaluation of the molecular events associated with angiogenesis in order to more effectively predict response and track therapeutic efficacy of angiogenic therapy.

Angiogenesis is a complex biological process (see chapter X) that involves the interaction of many cell types; including monocytes/macrophages, mast cells, lymphocytes, connective tissue cells, pericytes, smooth muscle cells, endothelial cells and pluripotent progenitor cells, all of which influence the process by secreting soluble angiogenic and antiangiogenic molecules including extracellular matrix and proteolytic enzymes (2–4). Potential biological targets for imaging angiogenesis fall into three general categories: (*i*) endothelial cell markers of angiogenesis, (*ii*) nonendothelial cells involved with angiogenesis, and (*iii*) markers of the extracellular matrix (ECM) (1). The optimal application of any of the targeted imaging approaches will require registration with images reflecting the physiological changes. This chapter will focus on radiotracer and echocardiographic approaches for evaluation of angiogenesis. Chapter X focuses on magnetic resonance imaging (MRI) approaches for evaluation of angiogenesis. Both radiotracer and echocardiographic imaging approaches offers a unique opportunity to evaluate both the initial molecular signals and the physiologic consequences of the angiogenic process, and can be directly linked to targeted therapy. Radiotracer approaches offer improved sensitivity, and established methods for quantification, while echocar-diographic approaches employ targeted microbubbles that remain intravascular facilitating interrogation of molecular targets on the endothelial surface. In the future, imaging of angiogenesis will probably involve the application of multiple complimentary imaging methodologies, particularly as hybrid imaging systems become more available.

INDUCTION OF ANGIOGENESIS BY HYPOXIA OR ISCHEMIA

Arterial occlusion causes hypoxia or ischemia that are the principal natural stimuli for endogenous angiogenesis (5). Angiogenesis is an adaptation to hypoxia and/or ischemia, providing increased perfusion and oxygenation through new vessel growth. Hypoxia induces expression of several established angiogenic factors, including vascular endothelial growth factor (VEGF) (5–8), platelet-derived growth factor (PDGF), fibroblast growth factor (FGF), and tissue growth factor β-1 (TGF-β1) (9). All of these growth factors might represent potential targets for imaging. HIF-1α appears to be the most critical factor in regulation of most transcriptional hypoxic responses, although hypoxic signaling certainly involves other transcriptional control pathways. Thus, myocardial hypoxia may provide a potential marker of the initiation of angiogenesis.

THERAPEUTIC ANGIOGENESIS

The goal of therapeutic angiogenesis in management of ischemic disease is to stimulate new blood vessel growth and thereby improve perfusion, tissue oxygenation, substrate exchange, and function. The process of therapeutic angiogenesis involves growth and differentiation of new vasculature capable of restoring blood supply to the ischemic tissue. To be physiologically effective, therapeutic angiogenesis should induce capillary growth within the ischemic bed, and stimulate development of penetrating arterioles and conductance arteries. Preclinical studies have demonstrated the efficacy of angiogenic therapy in animal models, using relatively invasive measures to assess efficacy (10). However, preliminary clinical trials of stimulated angiogenesis in patients with severe ischemic disease have not demonstrated a clear benefit over placebo, when evaluated using standard clinical parameters (11–17).

In the absence of a direct angiogenesis biological marker, the evaluation of therapeutic angiogenesis has focused on a number of clinical end-points, including; symptoms, exercise tolerance, measures of quality of life, and survival (1). These studies have demonstrated a mixed and inconsistent benefit. Some studies have also employed imaging end-points, predominantly focused on the physiological consequences of the therapeutic intervention. Most of these trials have employed standard radiotracer imaging for evaluation of perfusion, and some for evaluation of function. However, this is a tremendous need for development of noninvasive approaches for direct evaluation of the molecular events associated with angiogenesis in order to more effectively track therapeutic angiogenesis.

IMAGING OF ANGIOGENESIS

Advantages of Radiotracer Imaging

Radiotracer-based imaging provides primarily images of function rather than anatomical structure, as generally produced by X-ray computed tomography (CT), MRI, or ultrasound (US) based approaches. The strength of radionuclide imaging rests on the remarkable ability to study molecular and physiological processes, however, due to the relatively poor spatial resolution, their application for imaging anatomy is somehow limited. It is evident that nuclear techniques, both single photon emission computed tomography (SPECT) and positron emission tomography (PET) possess a unique set of advantages that make them particularly suitable for evaluating processes like angiogenesis.

Both SPECT and PET imaging can provide quantitative information about the biodistribution and kinetics of a radiotracer. The accuracy of SPECT imaging is fundamentally limited by the attenuation of the low energy photons by body tissues. This introduces an error in relating the density of detected photons to the concentration of the radiopharmaceutical in an organ. Moreover, the presence of scattered radiation limits spatial resolution. The absolute quantification of SPECT and PET images is also limited by partial volume errors. While PET imaging can overcome some of the attenuation problems compared with SPECT, the PET approach is a more expensive technique. The introduction of hybrid systems (SPECT/CT and PET/CT) for imaging has greatly enhanced the performance and accuracy of nuclear imaging. The CT component

can be used for anatomical localization as well as for correction of attenuation and partial volume errors.

Principles that may apply for clinical imaging, may not be true for imaging of small animals. MicroSPECT imaging offers several advantages over microPET imaging for small animal imaging, including; availability of targeted tracers, improved physics of SPECT radiotracers, and general availability and affordability of SPECT technology. In contrast to SPECT technology, inherent resolution of PET radiotracers is fundamentally limited by physical behavior of positron decay. Moreover, both the movement of positron prior to annihilation (as much as 1–3 mm) and deviation from an exact 180° angular separation have a profound effect on PET resolution. Major advantages of targeted SPECT imaging approaches include: availability, ease of production and transport of SPECT tracers, and capability for simultaneous multiple-isotope imaging. On the other hand, microPET technology offers several unique advantages over microSPECT imaging. MicroPET imaging systems offer a significantly improved sensitivity over microSPECT imaging systems. The higher sensitivity of PET imaging permits dynamic three-dimensional image acquisitions, facilitating modeling of radiotracer kinetics and absolute quantification of radiotracer uptake for determination of tissue flow and receptor binding. Additionally, there is an established approach for attenuation correction that gives an even greater potential for image quantification. Another major advantage of microPET imaging is afforded by use of ^{11}C isotopes, which provides the ability to label a given molecule without perturbing the biological function.

Hybrid Multimodality Imaging of Angiogenesis

The addition of an X-ray CT unit to a SPECT or PET imaging system permits registration of functional radiotracer images with anatomical CT images for the purpose of radiotracer localization and improved image quantification. The CT images allow for correction of attenuation and partial volume errors associated with radiotracer imaging, permitting absolute quantification of radiotracer uptake. These corrections are critical for the development of targeted imaging of angiogenesis for quantification of associated physiological changes. The fusion of the nuclear images with CT images will enhance analysis of acquired dynamic data, by facilitating region of interest (ROI) identification in generally lower resolution radionuclide images. The utilization of hybrid imaging systems for the acquisition of registered radionuclide and CT images should also help align the nuclear images acquired at different experimental time points. Identification of the location and extent of focal uptake of a radiotracer would permit tracking of a physiological or molecular process associated with angiogenesis over time, and potentially would permit evaluation of therapies directed at altering the same physiological process. Experimental studies have demonstrated the feasibility of obtaining absolute SPECT quantification of myocardial perfusion using a combined X-ray CT and SPECT imaging system (18,19). Dedicated small animal hybrid microCT and SPECT and PET imaging systems have also been created and facilitate registration of cardiovascular anatomic structure with the more physiologically based imaging techniques like SPECT or PET (20–22).

Imaging of Physiological Consequences

Angiogenesis should result in improved perfusion and tissue oxygenation, resulting in reduced hypoxia, and diminished myocardial ischemia. Effective therapeutic angiogenesis should restore myocardial perfusion, and improve regional left ventricular function, and metabolism. Thus, angiogenesis may be indirectly evaluated noninvasively by analysis of standard physiological parameters like regional myocardial perfusion and function. However, angiogenesis may also alter vascular permeability and the effective vascular surface area for substrate exchange. In light of the changes in the microvascular structure and potentially intravascular volume, it is not clear if the effects of angiogenesis would be better evaluated by analysis with radiotracers that track changes in intravascular blood volume, or vascular and/or perfusion reserve. Evaluation of alterations in tissue oxygenation may be potentially more directly assessed by analysis of regional myocardial pH, hypoxia, or metabolism. It is important to recognize that there is tremendous interaction between all of these physiological parameters.

Angiogenesis may not alter the magnitude of stress-induced ischemia, but only delay the onset of ischemia during stress. These benefits regarding the delayed onset of stress induced ischemia would result in improvements in exercise tolerance or exercise treadmill time, or reduction in angina class. Therefore, evaluation of angiogenesis may require evaluation of these physiological parameters under conditions of stress as well as rest. Nevertheless, demonstration of physiological benefit remains a critical step in the development and evaluation of therapeutic angiogenesis. Potential imaging approaches for non-invasively evaluating therapeutic angiogenesis may include: evaluation of regional perfusion; more direct measurement of tissue oxygenation, metabolism and pH; evaluation of regional or global mechanical function; or assessment of changes in vascular permeability, vascular reserve, or intravascular blood volume. While imaging of the physiological consequences of angiogenesis can be extremely valuable, these imaging approaches may be complicated by numerous factors. As outlined previously, some of these factors include changes in

microvasculature structure, surface exchange area, vascular reactivity, and alterations in vascular permeability. These alterations in the microvascular structure and function may confound imaging approaches that utilize diffusible tracers.

Imaging of Perfusion

Either SPECT or PET imaging can be used to assess changes in myocardial perfusion. Several recent trials have used SPECT myocardial perfusion to evaluate results of newer angiogenic therapeutic approaches (23–30). The role of imaging in these clinical trials was previously reviewed (31), and summarized by a consensus panel of the American Society of Nuclear Cardiology (32). These studies used different perfusion imaging agents and protocols, different stressors (even within the same study), and generally have relied on visual or semiquantitative visual analysis of the perfusion data. Earlier nonrandomized trials have generally shown improvements in perfusion, while larger randomized placebo controlled trials have shown neutral or inconsistent results. The anticipated changes in perfusion at rest may be rather small, and therefore approaches with high spatial resolution and sensitivity may be required.

In studies of myocardial perfusion, PET tracers such as ^{13}N-ammonia, ^{15}O-water, or ^{82}Rb would theoretically be preferred due to the favorable characteristics of PET imaging in general. The PET perfusion agents generally track flow better than SPECT agents, particularly at high flows produced by pharmacological stress. PET imaging also permits absolute quantification of regional myocardial blood flow and established approaches for attenuation correction. However, quantitative PET imaging may be more difficult to apply for clinical imaging.

SPECT perfusion imaging offers a more practical approach for evaluation of changes in relative perfusion that could be easily applied in multi-center clinical trials. Several studies have addressed the ability of serial SPECT perfusion imaging to reproducibly assess changes in myocardial perfusion over a short period of time (33,34), or over a 12-month period (35). All of these studies suggest that there is both inherent biological and methodological variability to SPECT imaging of myocardial perfusion. This variability needs to be taken into account during trial design in establishing a sufficiently robust sample size, so that potentially significant changes may emerge from the variability (32). Clinical trials of angiogenesis must be carefully controlled, blinded, and centrally and uniformly processed, so as to eliminate any bias and to objectively evaluate angiogenesis techniques.

Radiotracer imaging has also been used clinically to evaluate relative skeletal muscle perfusion in patients with PAD with some success, although this technology has not been widely applied for this purpose (36). Experimental studies have employed rodent models of hindlimb ischemia to evaluate changes in tissue perfusion or flow associated with angiogenesis and arteriogenesis. Most of these studies have employed laser Doppler techniques to evaluate skeletal muscle flow. However, this widely employed approach has substantial limitations, and primarily provides an estimate of superficial flows. One study used planar ^{99m}Tc-sestamibi imaging in the rat model of hindlimb ischemia for the assessment of relative perfusion (37). In this study, planar ^{99m}Tc-sestamibi imaging significantly underestimated the relative flow deficit. Additional study used radiotracer planar imaging to quantify relative perfusion in the mouse hindlimb (38). We have been using microSPECT/CT ^{201}Tl imaging during pharmacological stress for evaluation of muscle perfusion in murine models of hindlimb ischemia (Fig. 1). This radiotracer approach may offer improved quantification of perfusion over more widely used laser Doppler techniques.

Optimal Stressor for Perfusion Imaging

Exercise is generally favored over pharmacological stress for perfusion imaging, although pharmacological stress may provide a more reproducible stress for serial imaging. In many angiogenesis trials, exercise stress is used as an indicator of functional benefit to the patient in terms of exercise time or a changing time to the onset of ischemic ST depression. Ideally, improvements in functional capacity should correspond with improvements in regional myocardial perfusion.

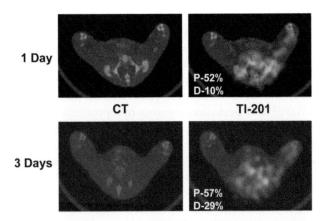

Figure 1 Transaxial microSPECT/CT ^{201}Tl perfusion images during vasodilatation with adenosine were acquired in mice at one day (*top row*) and three days (*bottom row*) following right femoral artery occlusion. Decreased adenosine ^{201}Tl perfusion (*green*) is seen in distal right leg in both mice. Relative skeletal muscle uptake (ischemic/nonischemic) was assessed quantitatively at both proximal (P) and distal (D) locations. Relative stress ^{201}Tl perfusion was significantly depressed in the ischemic leg based on quantitative analysis of the images. There was some improvement in relative perfusion by day 3.

However, if a patient demonstrates an increase in exercise time or maximal achievable workload, with a stable myocardial perfusion pattern, this may also indicate a beneficial effect, without directly demonstrating improved perfusion reserve.

While vasodilator stress may demonstrate improved perfusion or perfusion reserve, it does not necessarily link the change in perfusion to a symptomatic benefit. In addition, angiogenic therapies may lead to enhanced collateral growth in the setting of multivessel disease, which could create the potential conditions for a coronary steal (39). Under these conditions, vasodilator stress SPECT perfusion imaging may underestimate the severity of coronary disease (40). Dobutamine has been used as an alternative pharmacological stressor in some studies to overcome these problems. However, dobutamine may not create the same degree of demand stress as exercise, and dobutamine does not create the same degree of flow heterogeneity as vasodilator pharmacologic stress (41,42). Recent experimental studies suggest the value of rapid pacing as an alternative reproducible stressor for use in conjunction with perfusion imaging. The goals of the specific therapeutic trial need to be carefully analyzed when choosing a stressor for use in conjunction with radiotracer-based perfusion imaging.

Imaging of Hypoxia

Under ischemic conditions, the presence of hypoxia is a powerful angiogenic stimulus. Hypoxia imaging offers a new approach for positive imaging of myocardial ischemia, and could provide a novel noninvasive method for evaluating angiogenesis. Imaging with markers of hypoxia may provide new insights regarding the pathophysiology of natural collateral development and response to angiogenic therapies. Both, [18]F- and ^{99m}Tc-labeled nitroimidazoles have been developed for scintigraphic imaging of hypoxic tissue. The potential of imaging hypoxia with nitroimidazoles was previously reviewed (43), and the potential role of these agents in the evaluation of angiogenesis also previously discussed (31). The assessment of tissue oxygenation with nitroimidazoles may be the best indicator of the balance of flow and oxygen consumption, and hence be the most sensitive and specific predictor of myocardial ischemia. Radiotracer-based imaging of myocardial hypoxia may predict the development of collaterals and permit the future evaluation of angiogenic therapies.

Initial efforts for in vivo imaging of hypoxia in the heart employed ^{18}F-labeled nitroimidazoles (44–47). More recently, several ^{99m}Tc-labeled nitroimidazoles were employed for imaging of hypoxic tissue. Several investigators evaluated the potential of a ^{99m}Tc-labeled nitroimidazole compound (BMS-181321) for imaging of hypoxia, using both in vitro and in vivo preparations

(48–60). In an open chest, extracorporally perfused swine model, BMS-181321 retention was correlated with regional blood flow and regional metabolism under conditions of low flow ischemia (53). Rumsey et al. established the feasibility of BMS-181321 SPECT imaging for the detection of myocardial ischemia in canine models (54).

BMS-194796 (renamed BRU-59-21) is a more hydrophilic ^{99m}Tc-labeled nitroimidazole derivative of BMS-181321, which has demonstrated superior properties for imaging myocardial hypoxia in vivo relative to BMS-181321 (55,61,62). Using a canine model of transient ischemia, Rumsey et al. observed a two-fold increase in BRU-59-21 retention in postischemic regions (55). Johnson et al. demonstrated increased focal myocardial retention of BRU-59-21 in a chest swine model of partial stenosis and pacing induced demand ischemia (61). In this study, BRU-59-21 was detected on both in vivo planar and SPECT imaging in those swine in which regional myocardial lactate production was demonstrated during pacing induced ischemia. Johnson et al. also demonstrated myocardial retention of BRU-59-21 in a closed chest swine model of brief intracoronary balloon occlusion and reperfusion when the compound was injected shortly before occlusion (62). It appears that one needs to be very careful about the timing of BRU-59-21 injection in experimental models of occlusion and reperfusion.

Planar and SPECT BRU-59-21 imaging has also been used for detection of hypoxia in rat and canine studies of ischemia and infarction at early and delayed time points (63–65). Initial studies were performed in infarcted rats and demonstrated increased myocardial retention of BRU-59-21 early post-MI in the peri-infarct area (65). In a study by Sinusas et al., BRU-59-21 uptake was related to uptake of other radiolabeled compounds targeted at angiogenesis (65). Figure 2 provides an example of both in vivo and ex vivo BRU-59-21 cardiac images for a dog early postinfarction.

Dobrucki et al. demonstrated time dependent changes in regional hypoxia in a model of hindlimb ischemia, that were associated with angiogenesis, improvements in perfusion and resolution of tissue hypoxia (63). These serial in vivo imaging studies demonstrated that the hypoxic stimulus that initiates the angiogenic response, as identified with BRU-59-21 imaging peaks within days of femoral artery occlusion and declines within seven days (63). Thus, several experimental studies demonstrate retention of BRU-59-21 in ischemic tissue, and the potential for in vivo noninvasive identification of regional hypoxia. However, much additional information is needed regarding the uptake of these ^{99m}Tc-labeled nitroimidazoles under conditions of ischemia and infarction. This class of ^{99m}Tc-labeled nitroimidazoles may offer a novel approach for evaluation of angiogenesis and for assessment of the efficacy of angiogenic therapies. However, information is also needed regarding the degree or duration of hypoxia needed to stimulate angiogenesis in the heart. Clinical imaging with nitroimidazoles may be limited by extracardiac uptake, however, it is possible that intracoronary injection of such an agent may mark territories of severe resting ischemia, without confounding uptake from extracardiac organs. This technique would be potentially suitable for animal models, and may be applicable in human studies using intracoronary injections if serial catheterizations are planned per protocol. The most feasible clinical application of hypoxia imaging with the currently available radiolabeled nitroimidazoles would be for detection of skeletal muscle hypoxia associated with peripheral arteriosclerosis. This would avoid the problem of intense hepatic uptake that has complicated imaging of the heart.

Imaging of Myocardial Function

Radiotracer imaging can also be used for evaluation of regional and global left ventricular function. However, evaluation of changes in regional or global left ventricular function is likely too insensitive to predict the efficacy of therapeutic angiogenesis. The lack of improvement in function does not rule out treatment effect of angiogenic therapy, since angiogenic therapy is often applied in the setting of nontransmural myocardial injury. Under these conditions, improvement in regional myocardial perfusion would not necessarily result in improvement in regional function because of potential mechanical tethering of the revascularized

In Vivo SPECT

Ex Vivo Slices

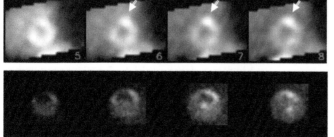

Figure 2 In vivo SPECT short axis BRU-59-21 images in a dog with acute LAD occlusion. *Arrow* identifies hypoxic region. Ex vivo BRU-59-21 hypoxia images in same dog are shown below. Early post occlusion there was a marked increase in myocardial BRU-59-21 uptake in the periinfarct anterior wall, although no uptake was seen in the central infarct area. Thus, myocardial BRU-59-21 retention is seen in early phases of coronary occlusion in the viable ischemic regions.

viable regions by adjacent necrotic regions. In contrast, an improvement in myocardial function associated with angiogenic therapy could serve as a highly specific surrogate for therapeutic efficacy.

Imaging of Regional Metabolism or pH

The potential for imaging of metabolic indices (i.e., deoxyglucose, acetate, palmitate, etc.) or indicators of tissue pH in the evaluation of myocardial angiogenesis warrants investigation, and are the focus of other chapter. For this purpose, quantitative PET or MR spectroscopic approaches may be preferred, although the potential of dynamic SPECT free fatty acid imaging with ^{123}I-labeled compounds also warrants investigation (66).

Targeted Imaging of Angiogenesis

Targeted imaging of molecular events or biological markers associated with angiogenesis will be critical for understanding the angiogenic process and tracking novel molecular or genetic therapies. Several approaches appear to be feasible for imaging of molecular events. These include the use of labeled oligonucleotides targeted to specific mRNA sequences, short peptides or peptidomimetics targeted to specific intracellular and cell surface receptors, and labeled ligand-avid imaging. Potential targets for imaging of angiogenesis would include markers of molecular events associated with the initiation of the angiogenic process, and were identified by the panel on angiogenesis at the Lake Tahoe Invitation Meeting of the American Society of Nuclear Cardiology (32), and are discussed in Chapter X. Potential biological targets for imaging angiogenesis fall into three principal categories: (*i*) endothelial cell markers of angiogenesis, (*ii*) nonendothelial cells involved with angiogenesis, and (*iii*) markers of the extracellular matrix (Table 1).

The molecular targets include evaluation of the altered expression or activation of integrins ($\alpha v \beta 3$, $\alpha v \beta 5$), VEGF and FGF receptors. In the following sections, the discussion will focus on the potential for targeted imaging of the $VEGF_{121}$ receptors and $\alpha v \beta 3$ integrin in models of angiogenesis.

Table 1 Potential Targets for Imaging Angiogenesis

Non-endothelial cell targets	Endothelial cell targets	Extracellular matrix
Monocytes	CD13	Selective matrix metalloproteinases
Stem cells	$\alpha v \beta 3$	Perlecan
	Syndecan-4	Del-1
	Vascular endothelial growth factor -R	

Source: From Ref. 32.

Imaging of VEGF Receptors

VEGF is a fundamental mediator of angiogenesis, affecting many cellular functions including, release of other growth factors, cell proliferation, migration, survival and angiogenesis (67). VEGF has at least five isoforms and two principal receptors; VEGFR-1 (Flt-1) and VEGFR-2 [also known as kinase domain region (KDR) and Flk-1]. Among the VEGF isoforms, $VEGF_{121}$ is unique in that it lacks affinity for heparin-like molecules, and binds only to its high-affinity endothelial cell-specific tyrosine kinase receptors, flt-1 and KDR (68). Studies have shown that the genes for VEGF, flt-1 and KDR are all responsive to hypoxia (6,68,69). Thus, VEGF receptor over-expression is a specific marker of hypoxic stress within tissue, and may provide a target for imaging of ischemic mediated angiogenesis.

Early efforts for targeted radiotracer imaging of tumor angiogenesis employed compounds targeted at VEGF. Collingridge et al. developed a novel positron-emitting radiotracer based upon a human monoclonal anti-VEGF antibody. This radiotracer comprises a monoclonal antibody (VG76e) that binds to human VEGF, labeled with [124]I, a positron-emitting radionuclide. These investigators demonstrated specific binds to VEGF, and feasibility for in vivo imaging of tumor angiogenesis. Li et al. subsequently developed a ^{123}I-labeled tracer for SPECT imaging of the $VEGF_{165}$ receptor for in vivo visualization of tumor angiogenesis. More recently, Blankenberg et al. described a novel imaging construct comprised of a standard ^{99m}Tc-labeled protein noncovalently bound to a "docking tag" fused to a "targeting protein" (70). The assembly of this complex was based on interactions between "adapter protein" (human 109-amino acid, HuS), "docking tag" (15-amino acid fragments of ribonuclease I, Hu-tag), and $VEGF_{121}$. Planar and SPECT images performed in mice implanted with mammary adenocarcinoma cells demonstrated significant uptake of ^{99m}Tc-HuS/Hu-$VEGF_{121}$ within subcutaneous tumor. This study suggested that it was possible to identify tumor neovasculature in lesions as small as several millimeters in soft tissue. Moreover, this approach can be adapted for in vivo delivery of other targeting proteins of interest without affecting their bioactivity (70).

VEGF receptors also appear to be reasonable targets for imaging of ischemia-induced angiogenesis. Lu et al. recently demonstrated the feasibility of imaging angiogenesis associated with hindlimb ischemia using radiolabeled-$VEGF_{121}$ in a rabbit model (71). Planar imaging of ^{111}In-labeled recombinant human $VEGF_{121}$ demonstrated focal uptake in the ischemic hindlimb. Immunohistochemistry confirmed increased expression of KDR and Flt-1 receptors within the ischemic hindlimbs, although the skeletal muscle was not evaluated for capillary density to confirm angiogenesis. The key findings of this study are summarized in

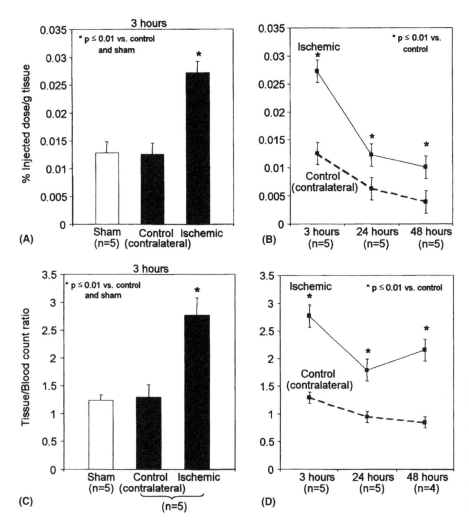

Figure 3   ^{111}In-VEGF retention in model of rabbit hindlimb ischemia. ^{111}In-VEGF activity was measured by gamma well counting in deep muscles of hindlimb (**A, B**), and hindlimb-to-blood ratios (**C, D**) at three hours (**A, C**) and up to 48 hours (**B, D**) after radiotracer injection. Control indicates contralateral nonischemic hindlimb in rabbits with unilateral hindlimb ischemia. *Solid bars*, group with unilateral ischemia. *Open bars*, group with sham operated hindlimb. *Source*: From Ref. 71.

Figure 3. This preclinical study suggests that it is possible to identify ischemic tissue by radiolabeling angiogenic receptors using a naturally occurring ligand as the imaging probe. The use of radiolabeled VEGF[121] as an imaging agent takes advantage of the specificity of VEGF[121] for hypoxia-inducible endothelial cell VEGF receptors. The availability of recombinant human form of VEGF[121] avoids the potential problem of immunogenicity associated with the use of antibodies as targeting ligands (72). Despite favorable blood clearance, ^{111}In-VEGF[121] was strongly retained in the liver and kidneys, which limits its use in myocardial imaging. Moreover, this approach depends strongly on total VEGF[121] receptor density.

VEGF receptor imaging could complement routine clinical perfusion imaging by providing additional information relevant to hypoxic stress. Dual isotope imaging with [99mTc] or ^{111}In-labeled VEGF-targeted probe and ^{201}Tl chloride or [99mTc]-sestamibi perfusion tracers could be useful for identifying sites of ongoing angiogenesis and regions at risk of ischemic injury (31). Targeted approaches for VEGF imaging could improve the evaluation of therapeutic angiogenesis, and the selection of sites for local delivery of proangiogenic agents. Despite successful applications of VEGF-targeted imaging approaches in models of cancer and hindlimb ischemia, further studies in experimental and clinical models of myocardial ischemia will be required to validate this imaging concept for application in patients with ischemic heart disease.

Imaging of Integrins

The angiogenic response is modulated by the composition of the ECM and intercellular adhesions, including integrins (73,74). Integrins are a family of $\alpha\beta$ heterodimeric cell surface receptors that mediate divalent cation-dependent cell–cell and cell–matrix adhesion through tightly regulated interactions with ligands (75). Integrin-ligand binding is dependent on conformational changes in the integrin structure (76). Members of the integrin family are capable of mediating an array of cellular processes, including cell adhesion, migration, proliferation, differentiation, and survival (77,78). The $\alpha v\beta3$ integrin is expressed by endothelial cells, as well as smooth muscle cells (79), platelets, growth factor-stimulated monocytes and T lymphocytes (80,81), and osteoclasts (82), permitting the interaction of these cells with a wide variety of ECM components (74). The effects of the $\alpha v\beta 3$ integrin may be modulated by changes in expression of the $\alpha v\beta 3$ integrin or conformational changes resulting in activation (83). Fortunately, $\alpha v\beta 3$ expression in quiescent endothelial cells is very low, while "angiogenic" endothelial cells demonstrate marked activation and upregulation of $\alpha v\beta 3$ expression. This biological behavior offers a tremendous advantage for targeted imaging, by providing a favorable target to background ratio. Thus, the $\alpha v\beta 3$ integrin is expressed specifically in angiogenic vessels and is known to modulate angiogenesis, and therefore represents a potential novel target for imaging angiogenesis (84,85).

SPECT Imaging of Integrins

Haubner et al. reported the synthesis and characterization of a series of radiolabeled $\alpha v\beta 3$ antagonists, reporting kinetics in both in vitro and in vivo preparations (85–87). Their work has focused on the use of cyclic Arg-Gly-Asp (RGD) peptides, known to bind to the $\alpha v\beta 3$ integrin. These radiolabeled RGD peptides exhibited high affinity for the $\alpha v\beta 3$ integrin, and specific binding in several tumor cell lines expressing $\alpha v\beta 3$. While these compounds demonstrated rapid clearance from blood, they are cleared predominantly through the hepatobiliary system, which may complicate imaging of myocardial angiogenesis.

Harris et al. recently reported the high affinity and selectivity of an ^{111}In-labeled quinolone (^{111}In-RP748) for the $\alpha v\beta 3$ integrin using assays of integrin-mediated adhesion (88). These investigators also demonstrated a rapid blood clearance and favorable biodistribution of ^{111}In-RP748, as well as the feasibility for tumor imaging. This preliminary work in imaging tumor angiogenesis supports the potential of ^{111}In-RP748 for targeted $\alpha v\beta 3$ imaging of myocardial angiogenesis. Sadeghi et al. evaluated a cy3-labeled homologue (TA145) of ^{111}In-RP748 using cultured endothelial cell preparations incubated in the presence and absence of established integrin activators (89,90). TA145 localized to $\alpha v\beta 3$ at focal cell–cell contact points, (89,90) and colocalized with LM609 an established $\alpha v\beta 3$ antibody (91).Under these in vitro experimental conditions, TA145 appears to exhibit preferential binding to the activated form of $\alpha v\beta 3$ integrin. This targeted peptidomimetic demonstrated preferentially binding to Mn^{2+}-activated $\alpha v\beta 3$ integrin on endothelial cells (ECs) (89). Saturation binding assays demonstrated a higher apparent affinity and number of binding sites per EC for RP748 in the presence of Mn^{2+}. The in vitro binding parameters suggested preferential binding to the activated form of the $\alpha v\beta 3$ integrin, and favorable conditions for in vivo imaging of activated

$\alpha v\beta 3$ integrin by ^{111}In-RP748. This suggests that ^{111}In-RP748 may also exhibit selective binding to activated $\alpha v\beta 3$ integrin.

Meoli et al. were the first to report the potential of ^{111}In-RP748 for in vivo imaging of myocardial angiogenesis (92–94). ^{111}In-RP748 demonstrated favorable kinetics for imaging of ischemia induced angiogenesis in the heart. These investigators used established canine model of myocardial infarction, which are known to produce nontransmural infarction and peri-infarct ischemia resulting in myocardial angiogenesis. In vivo SPECT imaging in these dogs demonstrated focal uptake of ^{111}In-RP748 in the infarct region associated with activation of the $\alpha v\beta 3$ integrin. Relative ^{111}In-RP748 activity was increased >3.5 fold in the infarcted region at three weeks postreperfusion (91). However, reconstruction and interpretation of the ^{111}In-RP748 "hot spot" images of the $\alpha v\beta 3$ integrin required careful coregistration of the targeted images with perfusion images (95). In vivo and ex vivo dual isotope SPECT short axis ^{111}In-RP748 and ^{99m}Tc-sestamibi images are shown for representative dogs at different time points following myocardial infarction (Fig. 4). The findings derived from ^{111}In-RP748 and ^{99m}Tc-sestamibi images were confirmed by gamma well counting of myocardial tissue (Fig. 5). The presence of ischemia-induced angiogenesis in this model was confirmed within the infarct region by staining with an endothelial specific lectin. Immunohistochemistry also confirmed increased expression of the $\alpha v\beta 3$ integrin in the microvasculature. These histological findings are also summarized in Figure 5.

The specificity of ^{111}In-RP748 for targeted imaging of the $\alpha v\beta 3$ integrin was confirmed by nonimaging studies employing a rat model of injury-induced myocardial angiogenesis, and a nonspecific isomeric negative control compound (^{111}In-RP790) (91). These rat studies of nontransmural infarction demonstrated that only ^{111}In-RP748 was selectively retained in regions of injury-induced angiogenesis where ^{201}Tl perfusion was reduced (Fig. 6). At two weeks after infarction, a two-fold increase in retention of ^{111}In-RP748 was observed in the most ischemic regions.

These investigators subsequently evaluated the uptake and clearance of ^{111}In-RP748 at early and late time points following ischemic injury, in relationship to myocardial perfusion, tissue hypoxia, and immunohistochemical markers of the angiogenic process. The myocardial retention of ^{111}In-RP748 was increased within 3 to 10 hours of reperfusion, in both canine and rat models of ischemic injury, suggesting increased expression or possibly simply the activation of the $\alpha v\beta 3$ integrin within hours of ischemic injury (92–94). An additional series of rat studies demonstrated that the regional myocardial retention of ^{111}In-RP748 in the reperfused infarcted region correlated with the uptake of a radiolabeled nitroimidazole (BRU-59-21), which

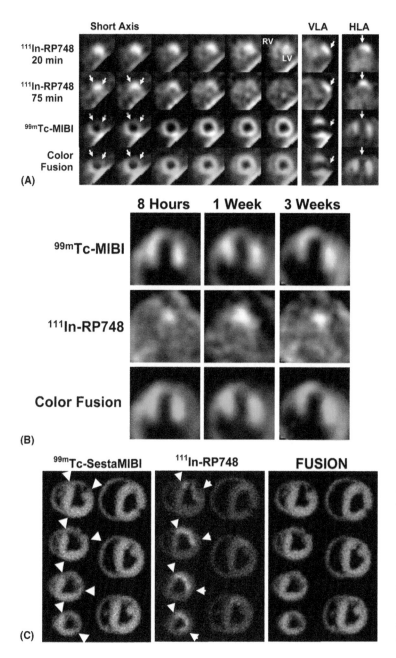

Figure 4 In vivo and ex vivo ^{111}In-RP748 and ^{99m}Tc-sestamibi (^{99m}Tc-MIBI) images from dogs with chronic infarction. Serial in vivo ^{111}In-RP748 SPECT short axis, vertical long axis (VLA), and horizontal long axis (HLA) images in a dog three weeks post-LAD infarction at 20 minutes, and 75 minutes postinjection in standard format (**A**). ^{111}In-RP748 SPECT images were registered with ^{99m}Tc-MIBI perfusion images (*third row*). The 75 minutes ^{111}In-RP748 SPECT images were colored red and fused with MIBI images (*green*) to better demonstrate localization of ^{111}In-RP748 activity within the heart (*color fusion, bottom row*). Right ventricular (RV) and left ventricular (LV) blood pool activity are seen at 20 minutes. *Filled arrows* indicate region of increased ^{111}In-RP748 uptake in anterior wall. This corresponds to the anteroapical ^{99m}Tc-sestamibi perfusion defect (*open arrow*). Sequential ^{99m}Tc-sestamibi (*top row*) and ^{111}In-RP748 in vivo SPECT HLA images at 90 minutes postinjection (*middle row*) from a dog at eight hours (acute), and one and three weeks post-LAD infarction (**B**). Increased myocardial ^{111}In-RP748 uptake is seen in anteroapical wall at all three time points, although appears to be maximal at one week postinfarction. Color fusion ^{99m}Tc-MIBI (*green*) and ^{111}In-RP748 (*red*) images (*bottom row*) demonstrate ^{111}In-RP748 uptake within ^{99m}Tc-MIBI perfusion defect. Ex vivo ^{99m}Tc-sestamibi (*left*) and ^{111}In-RP748 (*center*) images of myocardial slices from a dog three weeks post-LAD occlusion, with color fusion image on right (**C**). Short axis slices are oriented with anterior wall on top, RV on left. *Open arrows* indicate anterior location of nontransmural perfusion defect region, and *filled arrows* indicate corresponds area of increased ^{111}In-RP748 uptake. *Source*: From Ref. 91.

has been shown to be retained in hypoxic myocardium (64). This supports the role of ^{111}In-RP748 as a targeted marker of angiogenesis, which is stimulated in regions of myocardial hypoxia.

Experimental studies have also demonstrated the value of a ^{99m}Tc-labeled peptide (NC100692, GE Healthcare, U.K.) for targeted imaging of the $\alpha v \beta 3$ integrin in rodent models of hindlimb ischemia using high resolution pinhole planar imaging (96) and models of myocardial infarction using microSPECT imaging (21,94). NC100692 is a chelate-peptide conjugate containing an RGD motif in a configuration that allows high affinity (Ki ~1nM) and specific binding to the $\alpha v \beta 3$ integrin (97). The potential of NC100692 for

targeted in vivo imaging of angiogenesis was confirmed in murine models of hindlimb ischemia, in which angiogenesis was induced downstream from the femoral artery occlusion. In these validation studies, mice were sacrificed after NC100692 imaging at different time points postischemia for gamma well counting and immunohistological analysis of muscle tissue distal to the vascular occlusion (96). A significant increase in NC100692 activity was observed in the ischemic limb at three and seven days after the occlusion which normalized by 14 days postocclusion (Fig. 7). Immunohistochemical staining for lectin, an endothelial cell marker, confirmed a progressive increase in capillary density in the ischemic hindlimb

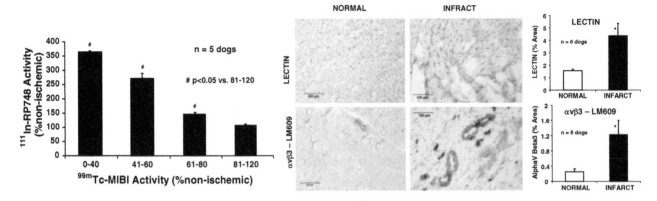

Figure 5 Postmortem analyses of canine studies with ^{111}In-RP748. Results from gamma well counting of myocardial ^{111}In-RP748 activity in relationship to ^{201}Tl activity are shown for dogs three weeks following myocardial infarction (*left*). Myocardial segments were segregated based on ^{201}Tl perfusion. Relative myocardial RP748 activity was 350% increased in the most ischemic regions. Quantitative histological analysis (*right*) confirmed increased capillary density based on endothelial specific lectin staining. Increased expression of the $\alpha v \beta 3$ integrin on these vessels was confirmed by staining with LM609, a known antibody for the $\alpha v \beta 3$ integrin. *Source*: From Ref. 91.

at these time points. The observed changes in regional NC100692 uptake in the ischemic tissue derived from quantification of the noninvasive imaging were confirmed by gamma well counting of tissue. However, the imaging approach tended to underestimate the relative increases of tissue NC100692 uptake. These differences probably reflect errors due to attenuation and partial volume effects associated with imaging. Representative in vivo NC100692 images from this study are shown in Figure 7 (96). Subsequent studies by these same investigators confirmed the time dependent changes in NC100692 uptake in relationship to

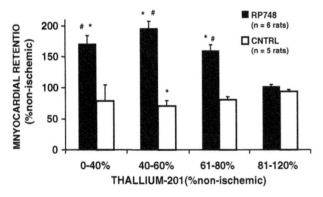

Figure 6 Gamma well counting of myocardial radiotracer activity in relationship to ^{201}Tl uptake in chronic rat model. Data are shown for rats injected with either ^{111}In-RP748 or control compound. Uptake of ^{111}In-RP748 was highest in infarcted regions with reduced ^{201}Tl retention. In contrast, myocardial uptake of the control compound was not associated with ^{201}Tl perfusion. On average, the relative myocardial retention of ^{111}In-RP748 in the postischemic and infarcted regions was nearly twice that in regions with normal ^{201}Tl perfusion, however, no selective retention of the nonspecific control compound was observed. *Source*: From Ref. 91.

tissue hypoxia within the ischemic hindlimb using serial in vivo imaging (98). As illustrated in Figure 8, analysis of angiogenesis in the setting of hindlimb ischemia can be improved by application of hybrid microSPECT/CT imaging. Another group of investigators has also demonstrated the value of a ^{123}I-labeled RGD peptide for in vivo imaging of angiogenesis using a similar murine model of hindlimb ischemia (99). Using an analogous ^{125}I-labeled RGD peptide, they also observed maximal uptake of the αv targeted compound at three days postischemia.

Su et al. evaluated the myocardial uptake of NC100692 within the early period following myocardial infarction in a chronic rat model of reperfused infarction, and at three days, and one and two weeks postinfarction. As with the ^{111}In-labeled peptidomimetic (RP748), an approximately two-fold increase in NC100692 retention was observed in the infarct area relative to the nonischemic regions, which persisted for at least two weeks postinfarction. NC100692 imaging has also been used to evaluate quantitative changes in myocardial angiogenesis in mice following myocardial infarction. A microSPECT/CT image from a mouse one week following myocardial infarction is shown in Figure 9. In murine studies, NC100692 microSPECT imaging demonstrated increased angiogenesis in the infarct region of transgenic mice with a deletion of matrix metalloproteinase-9 (MMP-9) relative to wild-type mice (21). The observed differences in myocardial NC100692 uptake were associated with histological differences in angiogenesis. The value of the $\alpha v \beta 3$ targeted imaging approach for assessment of myocardial angiogenesis was recently confirmed by another group of investigators that injected an ^{123}I-labeled RGD peptide in pigs with chronic ischemia treated with direct intramyocardial injection of phVEGF$_{165}$ (100).

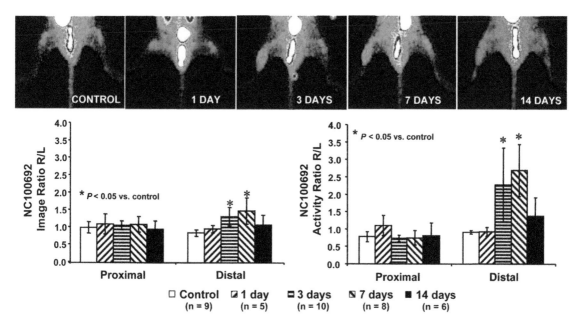

Figure 7 In vivo planar pinhole images of control mouse and mice at variable time points after right femoral occlusion are shown following intravenous injection with ^{99m}Tc-NC100692. "Hot spots" were seen in the ischemic limb distal to the occlusion on day 3 and 7, and decreased on day 14. Imaging analysis (*left graph*) showed a significant ($P < 0.05$) increase in radiotracer ischemic-to-nonischemic retention ratio on day 3 and day 7 versus the control group. Gamma well counting muscle yielded a higher initial ratio of radiotracer activity in ischemic to nonischemic contralateral hindlimb (*right graph*) compared with image analysis. A significant ($P < 0.05$) increase in ischemic-to-nonischemic tissue activity ratio was observed on day 3 and day 7. The regions proximal to the occlusion showed no difference in the radiotracer retention. *Source*: From Ref. 96.

These experimental studies suggest that the radio-labeled $\alpha v \beta 3$ targeted agents may be valuable noninvasive marker of angiogenesis following ischemic injury. Additional experimental studies will be required to define the duration of $\alpha v \beta 3$ integrin expression/activation following ischemic injury or following stimulated angiogenesis. The changes in expression/activation of $\alpha v \beta 3$ integrin will also need to be related to changes in more functional parameters like myocardial perfusion, regional mechanical function, permeability, and regional hypoxia. The potential for targeted imaging of other integrins, like $\alpha v \beta 5$, must also be considered.

PET Imaging of Integrins

There are a limited number of studies utilizing RGD-based PET tracers for targeted imaging of integrins,

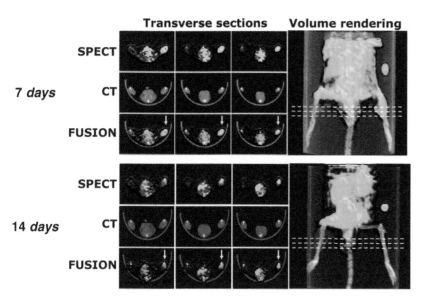

Figure 8 Application of hybrid microSPECT/CT imaging of mice with hindlimb ischemia following intravenous injection with ^{99m}Tc-NC100692 permits more accurate assessment of regional radiotracer uptake. Shown are microSPECT, microCT and fused images in mice at 7 and 14 days following left femoral artery occlusion. The fusion of the microSPECT images with microCT images facilitates anatomic localization of the radiotracer uptake. Increased activity is seen in the ischemic limb distal to the occlusion on day 7, and decreased on day 14.

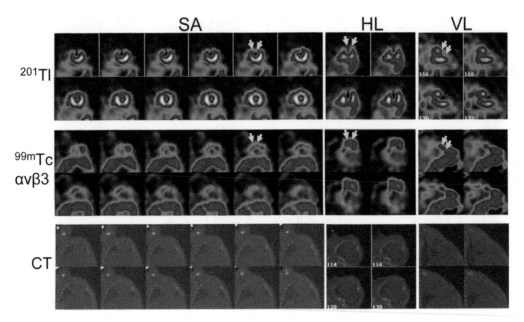

Figure 9 MicroSPECT/CT images in a mouse one week following surgical infarction of the anterior wall. Shown are microSPECT short axis (SA), horizontal long axis (HLA), and vertical long axis (VLA) imaging in standard format obtained following injection of ²⁰¹Tl (*top row*), injection of ⁹⁹ᵐTc-NC100692 (*middle row*), and corresponding microCT images. The arrows indicate and area of decrease ²⁰¹Tl perfusion in the area of the anterior wall infarction. Focal uptake of ⁹⁹ᵐTc-NC100692 is seen in the same area.

however, these studies have been restricted to the assessment of tumor angiogenesis. With the increasing availability of PET scanners, there is also a considerable interest in developing novel positron-emitting tracers targeted at $\alpha v \beta 3$ to noninvasively track angiogenesis in the clinical setting (101).

Chen et al. synthesized and applied a positron emitter ⁶⁴Cu-labeled PEGylated dimeric RGD peptide radiotracer for lung cancer imaging in mice (102). The radiotracer revealed rapid blood clearance via renal system and minimum nonspecific activity accumulation in normal tissue. PEGylation improved tumor-targeting efficacy and reduced biliary excretion, however, the tumor uptake was lowered due to lower receptor binding affinity than the dimeric RGD peptide. Although, this agent is an excellent PET tracer for integrin-positive tumor imaging, biodistribution data strongly suggest that peptide ligands of this class may be promising for imaging integrin expression in the myocardium.

More recently, the general advantage of the glycosylation approach in designing peptide-based tracers with favorable imaging properties for clinical applications has been confirmed. ¹⁸F-galacto-RGD, a glycosylated cyclic pentapeptide (Arg-Gly-Asp-DPhe-Val) was synthesized and characterized by Haubner et al. This tracer also showed high affinity and selectivity for the $\alpha v \beta 3$ integrin and receptor-specific accumulation in $\alpha v \beta 3$-positive tumors as well as rapid predominantly renal elimination (103). The radiotracer uptake was correlated with the measurement of $\alpha v \beta 3$ integrin expression assessed by Western blotting and immunohistochemistry of human $\alpha v \beta 3$ (LM609) and murine $\beta 3$. In the addition to the microPET imaging in mice, these investigators performed initial evaluation of tumor angiogenesis in humans. In all patients, rapid, renal excretion was observed, resulting in fast tracer elimination from blood, low background activity, and, on average, a nine-fold higher activity accumulation was found in the tumor than in muscle.

To further improve the retention of $\alpha v \beta 3$ radioligands, multimeric (composed of several identical subunits) RGD peptides were recently introduced. Although all work was done in tumor imaging, these multimeric RGD peptides showed increased binding affinities in vitro and improved tracer accumulation compared with the monomeric compounds. These compounds may also have potential for use in imaging of myocardial angiogenesis (103).

Echocardiographic Imaging of Integrin
Contrast enhanced ultrasound with targeted microbubble contrast agents have recently been used to evaluate integrin expression in vivo in models of tumor angiogenesis (104,105). As echocardiographic contrast microbubbles are pure intravascular tracers, the disease processes assessed must be characterized by antigens that are expressed on within the vascular compartment. Ellegala et al. proposed the use of targeted contrast ultrasound agents to evaluate increased expression of the $\alpha v \beta 3$ integrin in tumor angiogenesis in combination with the evaluation of regional tissue perfusion and

intravascular blood volume. This represents a novel approach for potentially evaluating both targeted biological markers in combination with important indicators of the physiological consequences of angiogenesis. These investigators used a targeted microbubble conjugated to echistatin, which unfortunately is not selective for the $\alpha v \beta 3$ integrin. While these early studies have this potential limitation they show tremendous promise for the contrast ultrasound approach for sequential targeted molecular and physiological imaging.

Other Markers of Angiogenesis

Since the natural ligand for several integrins are collagen, radiolabeled fragments of collagen have the potential for imaging of angiogenesis (106). Several PET radiotracers have already been developed for tumor imaging, which are modeled after selective MMP inhibitors (107,108). Targeting of MMPs may also prove useful for imaging of myocardial angiogenesis. Other potential avenues for noninvasive evaluation of angiogenesis involve the detection of HIF-1α activation, monitoring of the influx of blood-derived macrophages or circulating endothelial precursor cells, as well as expression of other markers specific for developing vasculature.

Linking Targeted Imaging with Targeted Therapeutics

Investigators are trying to develop targeted therapies for treatment of tumor angiogenesis that employ long half-life beta-emitting radiotracers like yttrium-90, copper-67, which are linked to antibody fragments, peptides, or peptidomimetics that target molecular markers of the angiogenic process. By switching the radiolabel from a short half-life gamma-emitting tracer (^{99m}Tc) to a longer half-life beta-emitting radiotracer (^{90}Y), a targeted ^{99m}Tc-labeled radiotracer like ^{99m}Tc-NC100692, optimized for in vivo imaging, can be modified for therapeutic benefit.

SUMMARY

With the development of novel gene therapies for treatment of ischemic heart disease directed at stimulation of angiogenesis, noninvasive imaging strategies will be critical for defining the pathophysiology of angiogenesis, defining those patients that will likely respond to angiogenic therapy, and assessing efficacy of these therapies. The demonstration of physiological benefit remains a critical step in the development and evaluation of these new therapies. Unfortunately, no large controlled clinical trial to date has clearly demonstrated an improvement in regional perfusion or function associated with angiogenesis. It may be that evaluation of changes in relative perfusion and function may be insensitive to detect subtle physiological changes. This

may represent an inherent limitation of SPECT imaging for this purpose. The use of PET perfusion imaging may be useful to establish initial proof of principle of angiogenic therapy, based on the potential quantification of regional flow and flow reserve. Evaluation of other PET metabolic indices may also be useful for selection of patients for clinical therapeutic angiogenesis trials, as well as for evaluation of therapeutic effect.

Hypoxia imaging with radiolabeled nitroimidazoles could provide an alternative noninvasive method for evaluation of the physiological consequences of angiogenesis. However, initial studies may require the use of intracoronary delivery to improve target to background ratios. Imaging of hypoxia may be best suited for evaluation of angiogenesis associated with limb ischemic in patients with peripheral vascular disease.

The future for noninvasive imaging of angiogenesis may rest on the development of targeted biological markers of angiogenesis. This targeted imaging may be accomplished with radiolabeled tracers, as well as targeted microbubbles. VEGF receptors could serve as targets for imaging of angiogenesis, by imparting physiologic information on hypoxic stress within viable tissue. VEGF receptor imaging could also be potentially useful in the evaluation of therapeutic angiogenic strategies. The $\alpha v \beta 3$ integrin appears to be another important target for imaging of angiogenesis. Several radiolabeled ligands targeted at the $\alpha v \beta 3$ integrin have already proved useful for tracking angiogenesis in both experimental models of myocardial ischemia and hindlimb ischemia. This targeted molecular imaging should include imaging of both endothelial cell and nonendothelial cell markers, along with markers of the extracellular matrix and intercellular adhesions, including integrins, involved with the angiogenic process. The optimal application of any of the targeted imaging approaches will require registration with physiological images, and anatomical structure. Eventually we may be able to link targeted molecular diagnostic imaging with targeted therapy.

ACKNOWLEDGMENTS

I gratefully acknowledge the input of Dr. Mehran Sadeghi, Dr. Michael Simons, and Dr. Flordeliza Villanueva in the preparation of this chapter.

REFERENCES

1. Simons M, Bonow RO, Chronos NA, et al. Clinical trials in coronary angiogenesis: issues, problems, consensus: An expert panel summary. Circulation (Online) 2000; 102(11):E73–E86.
2. Carmeliet P. Mechanisms of angiogenesis and arteriogenesis. Nat Med 2000; 6(4):389–395.

3. Giordano FJ. Angiogenesis: mechanisms, modulation, and targeted imaging. J Nucl Cardiol 1999; 6(6): 664–671.

4. Haas TL, Madri JA. Extracellular matrix-driven matrix metalloproteinase production in endothelial cells: implications for angiogenesis. Trends Cardiovasc Med 1999; 9(3–4):70–77.

5. Shweiki D, Itin A, Soffer D, et al. Vascular endothelial growth factor induced by hypoxia may mediate hypoxia-initiated angiogenesis. Nature 1992; 359(6398): 843–845.

6. Brogi E, Schatteman G, Wu T, et al. Hypoxia-induced paracrine regulation of vascular endothelial growth factor receptor expression. J Clin Invest 1996; 97(2):469–476.

7. Banai S, Jaklitsch MT, Shou M, et al. Angiogenic-induced enhancement of collateral blood flow to ischemic myocardium by vascular endothelial growth factor in dogs. Circulation 1994; 89(5):2183–2189.

8. Li J, Brown L, Hibberd M, et al. VEGF, flk-1, and flt-1 expression in a rat myocardial infarction model of angiogenesis. Am J Physiol 1996; 270(5 Pt 2):H1803–H1811.

9. Kourembanas S, Hannan R, Faller D. Oxygen tension regulates the expression of platelet-derived growth factor-B chain gene in human endothelial cells. J Clin Invest 1990; 86:670–674.

10. Giordano FJ, Ping P, McKirnan MD, et al. Intracoronary gene transfer of fibroblast growth factor-5 increases blood flow and contractile function in an ischemic region of the heart. Nat Med 1996; 2(5):534–539.

11. Kleiman NS, Califf RM. Results from late-breaking clinical trials sessions at ACCIS 2000 and ACC 2000. American College of Cardiology. J Am Coll Cardiol 2000; 36(1):310–325.

12. Henry TD, Annex BH, McKendall GR, et al. The VIVA Trial: vascular endothelial growth factor in ischemia for vascular angiogenesis. Circulation 2003; 107(10): 1359–1365.

13. Grines CL, Watkins MW, Helmer G, et al. Angiogenic Gene Therapy (AGENT) Trial in patients with stable angina pectoris. Circulation 2002; 105(11):1291–1297.

14. Stewart D. Late-Breaking Clinical Trial Abstracts. Circulation 2002; 106(23):2986–a-.

15. Hedman M, Hartikainen J, Syvanne M, et al. Safety and feasibility of catheter-based local intracoronary vascular endothelial growth factor gene transfer in the prevention of postangioplasty and in-stent restenosis and in the treatment of chronic myocardial ischemia: phase II results of the Kuopio Angiogenesis Trial (KAT). Circulation 2003; 107(21):2677–2683.

16. Lederman R, Mendelsohnb F, Andersonc R, et al. Therapeutic angiogenesis with recombinant fibroblast growth factor-2 for intermittent claudication (the TRAFFIC study): a randomised trial. Lancet 2002; 359:2053–2058.

17. Rajagopalan S, Mohler ER, III, Lederman RJ, et al. Regional angiogenesis with vascular endothelial growth factor in peripheral arterial disease: a phase II randomized, double-blind, controlled study of adeno-viral delivery of vascular endothelial growth factor 121 in patients with disabling intermittent claudication. Circulation 2003; 108(16):1933–1938.

18. Blankespoor S, Wu X, Kalki K, et al. Attenuation correction of SPECT using x-ray CT on an emission-transmission CT system: myocardial pefusion assessment. IEEE Trans Nucl Sci 1996; 43:2263–2274.

19. Kalki K, Brown J, Blankespoor S, et al. Myocardial pefusion imaging with correlated X-ray CT and SPECT system: an animal study. IEEE Trans Nucl Sci 1996; 43:2000–2007.

20. Bentley MD, Ortiz MC, Ritman EL, et al. The use of microcomputed tomography to study microvasculature in small rodents. Am J Physiol Regul Integr Comp Physiol 2002; 282(5):R1267–R1279.

21. Lindsey ML, Escobar GP, Dobrucki LW, et al. Matrix metalloproteinase-9 gene deletion facilitates angiogenesis after myocardial infarction. Am J Physiol Heart Circ Physiol 2006; 290(1):H232–H239.

22. Su H, Spinale FG, Dobrucki LW, et al. Noninvasive targeted imaging of matrix metalloproteinase activation in a murine model of postinfarction remodeling. Circulation 2005; 112(20):3157–3167.

23. Losordo D, Vale P, Symes J, et al. Gene therapy for myocardial angiogenesis: initial clinical results with direct myocardial injection of phVEGF165 as sole therapy for myocardial ischemia. Circulation 1998; 98(25): 2800–2804.

24. Rosengart T, Lee L, Patel S, et al. Angiogenesis gene therapy: phase I assessment of direct intramyocardial administration of an adenovirus vector expressing VEGF121 cDNA to individuals with clinically significant severe coronary artery disease. Circulation 1999; 100:468–474.

25. Hendel R, Henry T, Rocha-Singh K, et al. Effect of intracoronary recombinant human vascular endothelial growth factor on myocardial perfusion: evidence for a dose-dependent effect. Circulation 2000; 101(2): 118–121.

26. Laham R, Sellke F, Edelman E, et al. Local perivascular delivery of basic fibroblast growth factor in patients undergoing coronary bypass surgery: results of a phase I randomized, double-blind, placebo-controlled trial. Circulation 1999; 100:1865–1871.

27. Vale P, Losordo D, Milliken C, et al. Left ventricular electromechanical mapping to assess efficacy of phVEGF(165) gene transfer for therapeutic angiogenesis in chronic myocardial ischemia. Circulation 2000; 102(9):965–974.

28. Udelson J, Dilsizian V, Laham R, et al. Therapeutic angiogenesis with recombinant fibroblast growth factor-2 improves stress and rest myocardial perfusion abnormalities in patients with severe symptomatic chronic coronary artery disease. Circulation 2000; 102(14):1605–1610.

29. Simons M, Annex B, Laham R, et al. Pharmacological treatment of coronary artery disease with recombinant fibroblast growth factor-2: double-blind, randomized, controlled clinical trial. Circulation 2002; 105(7):788–793.

30. Losordo D, Vale P, Hendel R, et al. Phase 1/2 placebo-controlled, double-blind, dose-escalating trial of myocardial vascular endothelial growth factor 2 gene transfer by catheter delivery in patients with chronic

myocardial ischemia. Circulation 2002; 105(17): 2012–2018.

31. Sinusas AJ. Imaging of angiogenesis. J Nucl Cardiol 2004; 11(5):617–633.

32. Cerqueira M, Udelson J. Lake Tahoe Invitation Meeting 2002. J Nucl Cardiol 2003; 10(2):223–256.

33. MacDonald L, Elliott M, Leonard S, et al. Variability of myocardial perfusion SPECT: contribution or repetitive processing, acquisition, and testing. J Nucl Med 1999; 40:126P.

34. Mahmarian J, Moye L, Verani M, et al. High reproducibility of myocardial perfusion defects in patients undergoing serial exercise thallium-201 tomography. Am J Cardiol 1995; 75(16):1116–1119.

35. Burkhoff D, Jones J, Becker L. Variability of myocardial perfusion defects assessed by thallium-201 scintigraphy in patients with coronary artery disease not amenable to angioplasty or bypass surgery. J Am Coll Cardiol 2001; 38(4):1033–1039.

36. Wolfram RM, Budinsky AC, Sinzinger H. Assessment of peripheral arterial vascular disease with radionuclide techniques. Semin Nucl Med 2001; 31(2): 129–142.

37. Mack C, Magovern C, Budenbender K, et al. Salvage of angiogenesis induced by adenovirus-mediated gene transfer of vacular endothelial growth factor protects against ischemic vascular occlusion. J Vasc Surg 1998; 27:699–709.

38. Babiak A, Schumm AM, Wangler C, et al. Coordinated activation of VEGFR-1 and VEGFR-2 is a potent arteriogenic stimulus leading to enhancement of regional perfusion. Cardiovasc Res 2004; 61(4):789–795.

39. Becker L. Conditions for vasodilator-induced coronary steal in experimental myocardial ischemia. Circulation 1978; 57:1103–1110.

40. Arrighi J, Dione D, Condos S, et al. Adenosine Tc99m-sestamibi SPECT underestimates ischemia compared with N-13 ammonia PET in a chronic canine model of ischemia. J Nucl Med 1999; 40:6P.

41. Fung A, Gallagher K, Buda A. The physiologic basis of dobutamine as compared with dipyridamole stress interventions in the assessment of critical coronary stenosis. Circulation 1987; 76:943–951.

42. Lafitte S, Matsugata H, Peters B, et al. Comparative value of dobutamine and adenosine stress in the detection of coronary stenosis with myocardial contrast echocardiography. Circulation 2001 2001; 103(22): 2724–2730.

43. Sinusas A. The potential for myocardial imaging with hypoxia markers. Semin Nucl Med 1999; 29(4): 330–338.

44. Martin G, Caldwell J, Graham M, et al. Noninvasive detection of hypoxic myocardium using fluorine-18-fluoromisonidazole and positron emission tomography. J Nucl Med 1992; 33(12):2202–2208.

45. Martin G, Caldwell J, Rasey J, et al. Enhanced binding of the hypoxia cell marker [3H]fluoromisomidazole in ischemic myocardium. J Nucl Med 1989; 30(2): 194–201.

46. Shelton M, Dence C, Hwang D, et al. In vivo delineation of myocardial hypoxia during coronary occlusion using fluorine-18 fluoromisonidazole and positron emission tomography: a potential approach for identification of jeopardized myocardium. J Am Coll Cardiol 1990; 16:477–485.

47. Shelton M, Dence C, Hwang D, et al. Myocardial kinetics of fluorine-18 misonidazole: a marker. J Nucl Med 1989; 30:351–358.

48. Kusuoka H, Hashimoto K, Fukuchi K, et al. Kinetics of a putative hypoxic tissue marker, technetium-99m-nitroimidazole (BMS181321), in normoxic, hypoxic, ischemic and stunned myocardium. J Nucl Med 1994; 35(8):1371–1376.

49. Ng CK, Sinusas AJ, Zaret BL, et al. Kinetic Analysis of technetium-99m-labeled nitroimidazole (BMS-181321) as a tracer of myocardial hypoxia. Circulation 1995; 92(5):1261–1268.

50. Rumsey W, Patel B, Linder K. Effect of graded hypoxia on retention of technetium-99m-nitroheterocycle in perfused rat heart. J Nucl Med 1995; 36:632–636.

51. Fukuchi K, Kusuoka H, Watanabe Y, et al. Ischemic and reperfused myocardium detected with technetium-99m-nitroimidazole. J Nucl Med 1996; 37(5): 761–766.

52. Weinstein H, Reinhardt CP, Leppo JA. Direct detection of regional myocardial ischemia with technetium-99m nitroimidazole in rabbits. J Nucl Med 1998; 39(4):598–607.

53. Stone CK, Mulnix T, Nickles RJ, et al. Myocardial kinetics of a putative hypoxic tissue marker, 99mTc-labeled nitroimidazole (BMS-181321), after regional ischemia and reperfusion. Circulation 1995; 92(5): 1246–1253.

54. Rumsey WL, Kuczynski B, Patel B, et al. SPECT imaging of ischemic myocardium using a technetium-99m-nitroimidazole ligand. J Nucl Med 1995; 36(8): 1445–1450.

55. Rumsey W, Patel B, Kuczynski B, et al. Comparison of two novel technetium agents for imaging ischemic myocardium. Circulation 1995; 92(suppl I):I-181.

56. Archer C, Edwards B, Kelly J, et al. Technetium labeled agents for imaging tissue hypoxia in vivo. In: Nicolini M, Bandoli G, Mazzi U, eds. Technetium and Rhenium in Chemistry and Nuclear Medicine. Padova, Italy: SGE Ditoriali Publishers, 1995:535–539.

57. Okada RD, Johnson Gr, Nguyen KN, et al. 99mTc-HL91. Effects of low flow and hypoxia on a new ischemia-avid myocardial imaging agent. Circulation 1997; 95(7):1892–1899.

58. Okada RD, Johnson G, 3rd, Nguyen KN, et al. 99mTc-HL91: "hot spot" detection of ischemic myocardium in vivo by gamma camera imaging. Circulation 1998; 97(25):2557–2566.

59. Johnson G, 3rd, Nguyen KN, Liu Z, et al. Technetium 99m-HL-91: a potential new marker of myocardial viability assessed by nuclear imaging early after reperfusion. J Nucl Cardiol 1998; 5(3):285–294.

60. Melo T, Duncan J, Ballinger JR, et al. BRU59-21, a second-generation 99mTc-labeled 2-nitroimidazole for imaging hypoxia in tumors. J Nucl Med 2000; 41(1): 169–176.

61. Johnson LL, Schofield L, Mastrofrancesco P, et al. Technetium-99m-nitroimadazole uptake in a swine

model of demand ischemia. J Nucl Med 1998; 39(8): 1468–1475.

62. Johnson LL, Schofield L, Donahay T, et al. Myocardial uptake of a (99m)Tc-nitroheterocycle in a swine model of occlusion and reperfusion. J Nucl Med 2000; 41(7): 1237–1243.

63. Dobrucki L, Hua J, Bourke B, et al. Non-invasive imaging of hypoxia induced angiogenesis following hindlimb ischemia in mice. J Nucl Cardiol 2004; 11:372.

64. Meoli D, Bourke B, Hu L, et al. Regional hypoxia correlates with radiolabeled targeted markers of myocardial angiogenesis in ischemic rat model. J Nucl Med 2002.

65. Sinusas A, Meoli D, Sadeghi M, et al. Serial evaluation of myocardial hypoxia post myocardial infarction with technetium-99m labeled nitroimidazole. Circulation 2002; 106:II-581.

66. Shi C, Young L, Daher E, et al. Correlation of myocardial para-123I-Iodophenyl-pentadecanoic acid retention with 18F-Fluorodeoxyglucose accumulation during experimental low flow ischemia. J Nucl Med 2002; 43:421–431.

67. Ferrara N, Gerber H, LeCouter J. The biology of VEGF and its receptors. Nat Med 2003; 9(6):669–676.

68. Ferrara N, Davis-Smyth T. The biology of vascular endothelial growth factor. Endocr Rev 1997; 18(1):4–25.

69. Tuder R, Flook B, Voelkel N. Increased gene expression for VEGF and the VEGF receptors KDR/Flk and Flt in lungs exposed to acute or to chronic hypoxia. Modulation of gene expression by nitric oxide. J Clin Invest 1995; 95(4):1798–1807.

70. Blankenberg FG, Mandl S, Cao YA, et al. Tumor imaging using a standardized radiolabeled adapter protein docked to vascular endothelial growth factor. J Nucl Med 2004; 45(8):1373–1380.

71. Lu E, Wagner WR, Schellenberger U, et al. Targeted in vivo labeling of receptors for vascular endothelial growth factor: approach to identification of ischemic tissue. Circulation 2003; 108(1):97–103.

72. Goldman S. Receptor imaging: competitive or complementary to antibody imaging. Semin Nucl Med 1997; 27(2):85–93.

73. Brooks P, Clark R, Cheresh D. Requirement of vascular integrin alpha v beta 3 for angiogenesis. Science 1994; 264(5158):569–571.

74. Brooks P, Montgomery A, Rosenfeld M, et al. Integrin alpha v beta 3 antagonists promote tumor regression by inducing apoptosis of angiogenic blood vessels. Cell 1994; 79(7):1157–1164.

75. Xiong JP, Stehle T, Diefenbach B, et al. Crystal structure of the extracellular segment of integrin alpha Vbeta3. Science 2001; 294(5541):339–345.

76. Humphries M. Integrin activation: the link between ligand binding and signal transduction. Curr Opin Cell Biol 1996; 8(5):632–640.

77. Schwartz M, Schaller M, Ginsberg M. Integrins: emerging paradigms of signal transduction. I. Annu Rev Cell Dev Biol 1995; 11:549–599.

78. Hynes R. Integrins: bidirectional, allosteric signaling machines. Cell 2002; 110(6):673–687.

79. Shattil S. Function and regulation of the beta 3 integrins in hemostasis and vascular biology. Thromb Haemost 1995; 74(1):149–155.

80. Murphy J, Bordet J, Wyler B, et al. The vitronectin receptor (alpha v beta 3) is implicated in cooperation with P-selectin and platelet-activating factor, in the adhesion of monocytes to activated endothelial cells. Biochem J 1994; 304:537–542.

81. Huang S, Endo R, Nemerow G. Upregulation of integrins alpha v beta 3 and alpha v beta 5 on human monocytes and T lymphocytes facilitates adenovirus-mediated gene delivery. J Virol 1995; 69:2257–2263.

82. Horton MA. The alpha v beta 3 integrin "vitronectin receptor". Int J Biochem Cell Biol 1997; 29(5): 721–725.

83. Legler D, Wiedle G, Ross F, et al. Superactivation of integrin (&agr)v(&bgr)3 by low antagonist concentrations. J Cell Sci 2001; 114(8):1545–1553.

84. Sipkins D, Cheresh D, Kazemi M, et al. Detection of tumor angiogenesis in vivo by alphaVbeta3-targeted magnetic resonance imaging. Nat Med 1998; 4(5): 623–626.

85. Haubner R, Wester H, Reuning U, et al. Radiolabeled alpha(v)beta3 integrin antagonists: a new class of tracers for tumor targeting. J Nucl Med 1999; 40(6): 1061–1071.

86. Haubner R, Wester H, Weber W, et al. Noninvasive imaging of alpha(v)beta3 integrin expression using 18F- labeled RGD-containing glycopeptide and positron emission tomography. Cancer Res 2001; 61(5): 1781–1785.

87. Haubner R, Wester H, Burkhart F, et al. Glycosylated RGD-containing peptides: tracer for tumor targeting and angiogenesis imaging with improved biokinetics. J Nucl Med 2001; 42(2):326–336.

88. Harris T, et al. Design, Synthesis, and evaluation of radiolabeled integrin v 3 receptor antagonists for tumor imaging and radiotherapy. Cancer Biother Radiopharm 2003; 18(4):631–645.

89. Sadeghi M, Krassilnikova S, Zhang J, et al. Detection of injury-induced vascular remodeling by targeting activated avß3 integrin in vivo. Circulation 2004; (in press).

90. Zhang J, Krassilnikova S, Gharaei AA, et al. $\alpha v \beta 3$-Targeted detection of arteriopathy in transplanted human coronary arteries: an autoradiographic study. FASEB J 2005; 05–4130fje.

91. Meoli D, Sadeghi M, Krassilnikova S, et al. Noninvasive imaging of myocardial angiogenesis following experimental myocardial infarction. J Clin Invest 2004; 113(12):1684–1691.

92. Meoli D, Sadeghi M, Giordano F, et al. Pilot study of targeted imaging of angiogenesis. J Nucl Cardiol 2001; 4:S133.

93. Meoli D, Sadeghi M, Krassilnikova S, et al. Noninvasive imaging of myocardial angiogenesis post myocardial infarction. Paper presented at: Molecular, Integrative, and Clinical Approaches to Myocardial Ischemia, 2001; Seattle, WA.

94. Su H, Hu X, Bourke B, et al. Detection of myocardial angiogenesis in chronic infarction with a novel

technetium-99m labeled peptide targeted at avß3 integrin. Circulation 2003; 108:SIV-278–279.

95. Liu Y, Fernando G, Hall D, et al. A new method for quantification of SPECT "hot-spot" cardiac imaging: A phantom validation. J Nucl Cardiol 2003; 10:S8.

96. Hua J, Dobrucki LW, Sadeghi MM, et al. Noninvasive imaging of angiogenesis with a 99mtc-labeled peptide targeted at {alpha}v{beta}3 integrin after murine hind-limb ischemia. Circulation 2005; 111(24):3255–3260.

97. Morrison M, Davis J, Ricketts S-A, et al. Monitoring of tumour response to therapy with a novel angiogenesis imaging agent. Mol Imaging Biol 2003; 2:272.

98. Hua J, Dobrucki L, Bourke B, et al. Serial non-invasive imaging of angiogenesis using radiotracer targeted at v 3 integrin in model of murine hindlimb ischemia. Mol Imaging Biol 2004; 6:96.

99. Lee KH, Jung KH, Song SH, et al. Radiolabeled RGD uptake and alphav integrin expression is enhanced in ischemic murine hindlimbs. J Nucl Med 2005; 46(3): 472–478.

100. Johnson L, Haubner R, Schofield L, et al. Radiolabeled RGD peptide to image angiogenesis in swine model of hibernating myocardium. Circulation 2003; 108: IV-405.

101. Lewis MR. Radiolabeled RGD peptides move beyond cancer: PET imaging of delayed-type hypersensitivity reaction. J Nucl Med 2005; 46(1):2–4.

102. Chen X, Sievers E, Hou Y, et al. Integrin alpha v beta 3-targeted imaging of lung cancer. Neoplasia 2005; 7(3):271–279.

103. Haubner R, Weber WA, Beer AJ, et al. Noninvasive visualization of the activated alphavbeta3 integrin in cancer patients by positron emission tomography and [18F]Galacto-RGD. PLoS Med 2005; 2(3):e70.

104. Leong-Poi H, Christiansen J, Klibanov AL, et al. Noninvasive assessment of angiogenesis by ultrasound and microbubbles targeted to {alpha}v-integrins. Circulation 2003; 107(3):455–460.

105. Ellegala DB, Leong-Poi H, Carpenter JE, et al. Imaging tumor angiogenesis with contrast ultrasound and microbubbles targeted to {alpha}v{beta}3. Circulation 2003; 108(3):336–341.

106. Edwards W, Anderson C, Fields G, et al. Evaluation of radiolabeled type IV collagen fragments as potential tumor imaging agents. Bioconjug Chem 2001; 12: 1057–1065.

107. Furumoto S, Takashima K, Kubota K, et al. Tumor detection using 18F-labeled matrix metalloproteinase-2 inhibitor. Nucl Med Biol 2003; 30(2):119–125.

108. Zheng Q-H, Fei X, Liu X, et al. Synthesis and preliminary biological evaluation of MMP inhibitor radiotracers [11C]methyl-halo-CGS 27023A analogs, new potential PET breast cancer imaging agents. Nucl Med Biol 2002; 29(7):761–770.

23

Targeted Magnetic Resonance Imaging of Angiogenesis in the Vascular System

Ebo D. de Muinck

Department of Medicine and Department of Physiology, Angiogenesis Research Center and Section of Cardiology, Dartmouth-Hitchcock Medical Center, Dartmouth Medical School, Lebanon, New Hampshire, U.S.A.

Justin D. Pearlman

Department of Medicine and Department of Radiology, Angiogenesis Research Center and Section of Cardiology, Dartmouth-Hitchcock Medical Center, Dartmouth Medical School, Lebanon, New Hampshire, U.S.A.

INTRODUCTION

Targeted imaging of neovascularization in the cardiovascular system and in tumors is a rapidly developing field that promises to enhance our ability to understand how new blood vessels grow in patients and to monitor this growth in response to treatment. A requirement that needs to be met to realize this promise is to understand how new blood vessels impact organ perfusion, metabolism and function (1). To this end, a combination of targeted vascular imaging with readouts of perfusion, metabolism and function may be combined in a multi-modal imaging approach. Magnetic resonance imaging (MRI) seems ideally suited for this purpose because of its ability to acquire multiple parameters during a single imaging session. Combining these imaging capabilities with targeted MRI of neovessels therefore is an attractive proposition, especially when the impact of neovessels on cardiac (patho)physiology is to be studied. The current chapter will summarize nontargeted MRI approaches to neovascularization, as well as targeted strategies. Considerations regarding choice of MR contrast material will be addressed, together with approaches that improve signal:noise, new developments such as "smart" contrast agents, and direct imaging of superparamagnetic particles.

NEOVASCULARIZATION: ANGIOGENESIS AND ARTERIOGENESIS

Three patterns of new blood vessel formation have been identified to date. Embryonic vascular development is initiated by a process called vasculogenesis (2), and subsequent maturation and expansion of the embryonic vascular bed rely on angiogenesis and arteriogenesis (3). The extent to which vasculogenesis

contributes to new vessel formation in adult organisms has not been established, but both angiogenesis and arteriogenesis are well-recognized modes of new vessel growth in the adult. New vessels grow in response to hypoxia, and inflammation in tumors, atherosclerotic plaques, and virtually all organs (4). Angiogenesis describes a process characterized by the formation of endothelial tubes that are stabilized by mural cells, i.e., pericytes, and arteriogenesis denotes the formation of larger diameter vessels with one or more layers of vascular smooth muscle cells (5). The cascade of events involved in angiogenesis is being studied by many groups and an ever increasing body of data contributes to our understanding of this process, arteriogenesis, however, is far less well studied and understood (4). Likewise, MRI strategies that aim to show new blood vessel growth have focused on imaging of angiogenic blood vessels, primarily in tumors (6). This chapter focuses primarily on angiogenesis imaging in the vascular system and will review the endothelial cells surface markers that have been targeted to date.

PRINCIPLE OF MAGNETIC RESONANCE IMAGING

Contrary to the acquisition of electric signals from the body, the recording of magnetic signal is less easily accomplished because of the low strength of these signals, requiring strong magnets and extremely sensitive sensors. The MR signal is created by altering the intrinsic angular momentum or "spin" of protons (i.e., hydrogen nuclei) and electrons. To this end, a strong external magnetic field is applied and this causes the spins to orient themselves either parallel (spin-up) or antiparallel (spin-down) to the field. The total impact, which is a function of the strength of the external magnetic field, is minimal, about 0.01 to 0.1 eV or around 10^{-6} to 10^{-7} more spin-up than spin-down states per voxel, but because tissues are predominantly water, this small distribution imbalance is perceptible. Importantly, imbalance between spin-up and spin-down orientation results in a net magnetization that is tilted away from the main magnetic field.

The transition from the magnetized state to the ground state is called relaxation phase. Relaxation is characterized by the time (T in sec.) it takes to return to the ground state and by the rate (R = 1/T in sec^{-1}) at which this occurs. Also, relaxation occurs in two directions, longitudinally along the main magnetic field, and transversely perpendicular to the main magnetic field. Longitudinal relaxation is described by T_1 and R_1, and transverse relaxation is described by T_2 and R_2. Electrons and protons do not only spin around their longitudinal axis, but they also exhibit precession, i.e., an axis of rotation that is outside the longitudinal axis and at an angle to it, comparable to a spinning top, a gyroscope or the earth. The dispersion of precession frequencies within an inhomogeneous magnetic field, such as a patients' body in an MR magnet, is described by T_2^* and R_2^*. Contrast occurs because of the varying density of protons and electrons in tissues and because relaxation rates differ between tissues due to their different chemical and physical composition. Naturally occurring contrast can be enhanced with exogenous magnetic particles. These typically are the rare-earth element gadolinium (Gd) or iron oxides. The strong magnetization of these particles is used to alter the weaker signal from the intrinsic protons and electrons, thus creating stronger magnetization fields. It has not been possible to image magnetic particles directly in vivo, but a method that could accomplish this has recently been published after validation in vitro (7), and will be summarized later in this chapter. Gadolinium is a paramagnetic agent and it accelerates longitudinal relaxation and therefore is called a T_1 agent, it enhances MR contrast and therefore brightens the image in T_1-weighted images. Iron oxides are superparamagnetic agents that increase the rate of transverse or dephasing relaxation, and thus are called T_2 agents. The increased rate of transverse relaxation causes a negative contrast effect, it darkens the image.

NONTARGETED IMAGING OF ANGIOGENESIS

The primary steps of the angiogenic cascade involve vasodilatation mediated to a large extent by nitric oxide, followed by an increase in vessel permeability in response to increased levels of vascular endothelial growth factor (VEGF) (3). This increased permeability is caused by the formation of vesiculo-vacuolar organelles and trans-endothelial openings that have been viewed by different groups as trans-endothelial cell

pores or inter-endothelial cell gaps (3,8,9). Increased permeability allows for the extravasation of plasma proteins that provide a scaffold for migrating endothelial cells. Through the same principle leaky angiogenic vessels can be visualized after the injection of low molecular weight (0.5-3.0 kDa) contrast media such as gadodiamide (10) or high molecular weight (12-15 kDa) contrast agents, typically gadolinium bound to larger molecules such as albumin or dendrimers, for example, biotin-bovine serum albumin-Gd-diethylenetriamine pentaacetic acid (biotin-BSA-Gd-DTPA) (11). Upon further vessel maturation, the vascular wall is stabilized by angiopoietin-1 (12) and other factors (3), and leakiness disappears. Indeed, vessel leakiness seems to characterize a phase in vessel development that is VEGF dependent (11). Vascular leakiness disappeared in tumor bearing mice in response to treatment with an anti-VEGF antibody, and the same effect was seen after administration of an agent that suppressed hypoxia inducible factor 1-α (HIF1-α), presumably because of reduced VEGF levels in response to HIF1-α suppression (13). Acute administration of VEGF in normal tissue lead to immediate extravasation of macromolecular biotin-BSA-Gd-DTPA, which attenuated over a time course that was compatible with inactivation of the VEGF protein, and which recurred after a second injection of VEGF (14).

Vessel permeability is most often studied with dynamic contrast enhanced T1-weighted MRI. It measures both vessel permeability and blood volume in the voxels of a selected volume of tissue (15,16). Dynamic contrast enhanced MRI can measure the transfer of contrast material to the extravascular space and it can measure the rate of transfer back to the blood (6,17). Low molecular contrast agents such as gadodiamide and gadopentetate have short circulation times and fast diffusion rates, and in a comparison of low versus high molecular weight agents, the longer circulating more slowly diffusing high molecular agents showed more favorable kinetics for permeability imaging. The assessment of the rate at which a low molecular weight contrast transfers from the blood to the extravascular space was found to be confounded by the fact that this measurement is flow dependent. Due to their fast diffusion rates, increased flow will result in increased transfer to the extravascular space, in addition flow in angiogenic areas is highly heterogeneous and thus further confounds the assessment of permeability with low molecular weight agents (17). Thus the low molecular agents transit back and forth between the circulation and the extravascular space, and are cleared primarily by diffusion, whereas macromolecular contrast agents such as biotin-BSA-Gd-DTPA have been shown to be cleared mainly by interstitial convection and lymphatic uptake (14,18,19). The longer circulation time of the high molecular weight agents and the fact that they appear to be cleared from the tissue by lymphatic drainage resulting in longer residence time of the agents in the tissue, allows for saturation of the extravascular space and thus an assessment of permeability that is not confounded by other phenomena. The size of the agents is not a limiting factor in the transit from blood to extravasular space because the pore diameter of angiogenic endothelium has been shown to range from 380 to 780 nm in tumors (20), and the diameter of contrast agents is <1 nm for low molecular agents to 200 nm for lipisome based nanoparticles that can carry approximately 50,000 Gd atoms (21). Dynamic contrast enhanced MRI has been used in the heart to assess microvascular damage in the setting of myocardial infarction and ischemia-reperfusion injury (22), but it has not been applied to image angiogenesis.

A limitation of dynamic contrast enhanced MRI for measuring vascular permeability is the fact that each pixel reflects the signal from a relatively large volume (23,24), approximately $1 \times 1 \times 3$ mm, thus micro-environmental differences within angiogenic tissue will not be picked up. In fact, the signal from dynamic contrast enhanced MRI reflects an amalgam of vascular permeability, blood flow, vascular surface area and interstitial pressure (6,24), and these different influences need to be taken into account when interpreting these images. Despite these caveats, dynamic contrast enhanced MRI has been shown to correlate well with microvessel density upon histology (6,25,26), and several groups are pursuing permeability imaging as a noninvasive strategy to monitor tumor angiogenesis and its therapeutic modulation (13,14).

Another approach to image neovessel formation in a nontargeted manner is perfusion sensitive imaging. Paramagnetic contrast arrival in the heart is marked by bright signal in the right atrium, followed by increased signal from the right ventricle, the left atrium, and finally, the myocardium. This progression of brightness through the heart can be summarized in graphs of intensity versus time, and in ischemic myocardium signal enhancement will be delayed compared to normally perfused myocardial segments (Figs. 1 and 2). Based on this principle, we have developed a first pass or bolus transit imaging method to quantify myocardial perfusion, and we have shown that in ischemic hearts treated with VEGF, there was significant improvement of perfusion in ischemic territories compared to controls and that this improvement was accompanied by smaller infarcts and better ventricular function than in controls (27).

In summary, permeability imaging is the resultant of multiple characteristics of angiogenic vasculature, but nevertheless appears to identify an early, VEGF dependent hyperpermeable phase of angiogenesis. High molecular weight agents seem to be most suited for dynamic contrast enhanced MRI to assess vascular permeability in angiogenic areas. Perfusion sensitive, first pass imaging of ischemic myocardium represents an alternative to permeability imaging that seems well suited to assessment of the effects of new vessel growth in the heart.

TARGETED IMAGING OF ANGIOGENESIS IN THE VASCULAR SYSTEM

Targeted imaging strategies rely on the availability of an antibody or nonantibody ligand that recognizes a target that is selectively or preferentially expressed on or inside cells in the area of interest. Because this approach recognizes substrates at the molecular level it has also been called molecular imaging (28). When MRI was first introduced, the concept of molecular imaging had already been established with radiolabeled antibodies in nuclear medicine (29), and it was a natural extension of this expertise to apply it to the field of MR using magnetically labeled monoclonal antibodies. As with permeability imaging, most of the work has been performed in tumors, but this section will focus on targeted imaging of angiogenesis in the vascular system.

Before discussing the approaches that have been used to date in more detail, a few remarks that relate to the field in general can be made. The target needs to be available in sufficient quantity to allow for successful binding by the ligand, it needs to be expressed in a manner that is specific in time and space for the process that is to be tracked, and the biological process that is characterized by expression of the target must be thoroughly understood. Furthermore, the affinity of the ligand for its target and its other pharmacokinetic characteristics needs to be known, in the case of MR with its high background noise levels, there should be significant signal:noise enhancement, toxicity profiles should be acceptable, the targeting agents should be easy to produce and to facilitate their acceptance in the clinic it should be possible to image them with standard clinical magnets. Not all of these criteria have been met by the approaches that will be discussed below, but the field is evolving at an increasing pace, with multiple groups developing approaches that may reach the clinic at some point.

Magnetic Resonance Imaging of Endothelial Targets in the Cardiovascular System

A summary of the endothelial targets that have been visualized with targeted MRI, is presented in Table 1, and will be discussed in more detail in the subsequent section.

Magnetic Resonance Imaging of the $\alpha_v\beta_3$ Integrin

Integrins are a family of noncovalently associated heterodimeric cell surface receptors that are composed of a α- and β-subunit, and they regulate adhesion of

Table 1 Targeted MRI of Endothelial Cell Surface Markers in the Vascular System

Endothelial address	Targeting moiety	In vivo model	Evidence for homing to target
$\alpha_v\beta_3$ (39)	Lipid monolayer around perfluorocarbon, decorated with approx. 94,200 Gd and vitronectin peptidomimetic	Rabbit atherosclerosis model	Stronger signal in atherosclerotic animals Signal reduction after injection of targeted non-paramagnetic particles
VCAM-1 (55)	Fluorescently labeled CLIO particle decorated with peptide homing sequence (VHSPNKK)	TNF-α induced inflammation in mouse ear Mouse atherosclerosis model (ApoE-/-)	Particles localize to inflamed endothelium in vivo Inflamed vessels show VCAM-1 and particles upon histology Colocalization of particle and VCAM-1 throughout entire atherosclerotic lesion
ICAM-1 (56)	Fluorescently labeled liposomes with 29.5% gadolinium chelator lipid	Mouse autoimmune encephalitis model	Localization of liposomes to brain microvasculature in encephalitis Liposome distribution matches previously seen pattern of ICAM-1 distribution in this model
E-selectin (61)	Gadolinium-DTPA labeled with two molecules of sialyl-Lewis^x, a mimetic of an E-Selectin ligand	Mouse hepatitis model	Liver enhancement in mice with hepatitis, no enhancement after injection of no-targeted gadolinium, no liver enhancement in normal mice

Abbreviations: CLIO, cross linked iron oxide; DTPA, diethylenetriamine pentaacetic acid; ICAM-1, intercellular adhesion molecule-1; VCAM-1, vascular cell adhesion molecule-1.

cells with the extracellular matrix and adhesion between cells (30). Sixteen members of the integrin family have been shown to be relevant in vascular biology, and seven of these are expressed on endothelial cells ($\alpha_1\beta_1$, $\alpha_2\beta_1$, $\alpha_3\beta_1$, $\alpha_5\beta_1$, $\alpha_6\beta_1$, $\alpha_v\beta_3$, $\alpha_v\beta_5$) at varying times (30). The $\alpha_v\beta_3$ integrin is the most extensively studied vascular address in targeted imaging and treatment of angiogenesis and different imaging modalities have been used, both in tumors (31–38), and in the cardiovascular system (39–42). Multiple targeting moieties have been developed to direct imaging molecules and therapeutics to this endothelial cell surface marker, and these include the tripeptide RGD (arginine-glycine-aspartate) (36), monoclonal antibodies (33), peptidomimetics (39), and quinolones (40). The fact that it is one of the earliest endothelial addresses used for endothelial targeting (43), and the availability of multiple ligands in all likelihood have contributed to its popularity as a vascular target. In phage display experiments, the tripeptide RGD was identified as a homing sequence that targets intergrins on vascular endothelium (43,44). The importance of the RGD sequence in integrin binding is underlined by its identification as the integrin binding sequence in a number of integrin ligands, such as fibronectin, laminin, vitronectin and collagens (45–47), and contact regions for the RGD sequence have been identified in integrin subunits (48). The $\alpha_v\beta_3$ integrin has also been identified as an important regulator of angiogenesis (47), thus making it an attractive vascular address for targeted therapy (36).

Because the RGD binding site has been identified in a number of integrin subunits, the tripeptide is not specific for the $\alpha_v\beta_3$ integrin, and other molecules have been used to home MRI contrast agents to this vascular address. In tumors $\alpha_v\beta_3$ integrin has been imaged with gadolinium polymerized liposomes decorated with antibodies against $\alpha_v\beta_3$ (LM 609, Chemicon International, Inc.) (33), and in atherosclerotic plaque $\alpha_v\beta_3$ has been visualized using perfluorocarbon (perfluorooctylbromide) encapsulated by a lipid-surfactant monolayer, decorated with a peptidomimetic vitronectin antagonist, and containing approximately 94,200 Gd atoms per particle (39). Targeted imaging was performed in rabbits with mild, intramural aortic atherosclerosis (no luminal narrowing), that showed plaque neovessels positive for $\alpha_v\beta_3$ upon histology. Control groups were atherosclerotic animals injected with nontargeted nanoparticles and nonatherosclerotic animals injected with targeted nanoparticles. In the atherosclerotic animals, there was greater signal enhancement in practically all segments of the aorta after injection of the targeted nanoparticle compared to controls, and signal enhancement was reduced by at least 50% after competitive blockade with nonparamagnetic targeted nanoparticles.

Three surface molecules that promote cell adhesion and that are upregulated as part of endothelial cell activation in response to inflammatory stimuli, have been targeted with paramagnetic or superparamagnetic agents. These molecules are vascular cell adhesion molecule-1 (VCAM-1), intercellular adhesion molecule-1 (ICAM-1) and E-selectin, they overlap partially in their expression during inflammation, both in time and place within the vasculature (49), and all three molecules play a role in angiogenesis (50–52).

Magnetic Resonance Imaging of Vascular Cell Adhesion Molecule 1

Endothelial upregulation of VCAM-1 occurs under various inflammatory conditions including atherosclerosis and it has been shown that it is a critical regulatory molecule in atherosclerotic lesion development (49,53,54). Using the phage display technique, Kelly et al. identified a peptide sequence containing the VHSPNKK motif (valine-histidine-serine-proline-asparagine-lysine-lysine) that has homology to the α-chain of very late antigen, a known ligand for VCAM-1 (55). Using FITC labeled mono-valent VHSPNKK peptides and flow cytometry, they showed that exposure of murine cardiac endothelial cells to peptide resulted in a 12-fold greater rise of fluorescence intensity above background than exposure of fluorescently labeled monoclonal antibody against VCAM-1 (anti-VCAM-1-FITC). To generate multivalent fluorescent, paramagnetic nanoparticles they used a CVHSPNKKC motif that was extended at the C-terminus by the sequence GGSKGK (glycine-glycine-serine-lysine-glycine-lysine) and attached it to a fluorescently labeled cross-linked iron oxide nanoparticle (CLIO-Cy5.5). Using intravital confocal microscopy in a model of TNF-α induced vascular inflammation in the mouse ear, they showed that the nanoparticles localized to the endothelium of the inflamed ear but not the normal ear. The particles were cleared from the circulation in four hours but persisted in the inflamed ear up to four hours after injection. Upon histology, VCAM-1 expression was demonstrated on the inflamed blood vessels, and by using an anti-FITC antibody localization of the peptide to inflamed vascular endothelium was confirmed. The nanoparticles were also injected in a mouse atherosclerosis model (ApoE$^{-/-}$ mice), and extensive signal attenuation was observed, associated with iron oxide nanoparticle accumulation in atherosclerotic lesions. The localization of signal attenuation corresponded with aortic narrowing upon MRI, accumulation of Cy5.5 fluorescence in the aortic wall upon en face examination of the aorta ex vivo, and there was colocalization of Cy5.5 fluorescence and VCAM-1 antibody binding throughout the atherosclerotic lesions upon immunohistochemistry. Thus, the authors present multiple lines of evidence for the binding of the peptide to endotehial cells that express VCAM-1. This study presents a significant advancement in the field of targeted vascular MRI, but unequivocal demonstration of the ligand binding its vascular address after injection in vivo is still lacking in this study, as in other work performed in the field thus far. The localization of the nanoparticle throughout the entire atherosclerotic lesion is compatible with binding of the particle to other cells that express VCAM-1. In advanced atherosclerotic lesions, VCAM-1 is also expressed on intimal cells including vascular smooth muscle cells (49). The precise delineation of the address that is sought out by targeting agents after injection in vivo remains an incompletely resolved issue. The degree of specificity of endothelial targeting agents for their respective vascular addresses needs to be defined for each agent if we are to gain a full understanding of the biological significance of the images, because colocalization with the same target on other cells will obscure the message emanating from the targeted tissue.

Magnetic Resonance Imaging of Intercellular Adhesion Molecule-1

ICAM-1 is another angiogenic cellular adhesion molecule that is upregulated in vascular inflammation, and its expression in atherosclerosis is spatially and temporally different from VCAM-1 expression (49). Using antibody-conjugated paramagnetic liposomes, Sipkins et al. imaged vascular inflammation in a mouse model of experimental autoimmune encephalitis (56). The paramagnetic liposomes were composed of 60% pentacosadiynoc acid, 29.5% Gd chelator lipid, 10% amine-terminated lipid and 0.5% biotinylated lipid, and after the addition of avidin, targeted nanoparticles were formed by exposing the avidin coated particles to biotinylated antibody against ICAM-1. To confirm that the nanoparticles homed to blood vessels in mice with encephalitis, nanoparticles were labeled Texas Red, and brains were harvested 24 hours after injection of the probe. Upon histology, the particles were shown to localize to the brain microvasculature, and the location of nanoparticle binding appeared to correlate with a pattern of ICAM-1 expression previously described by immunohistochemistry. Subsequently, MRI of half brains was performed ex vivo and the encephalitic brains showed substantial increases in MR signal intensity in the cerebellar and cerebral cortex, which the authors found consistent with the localization of the fluorescent signal.

Magnetic Resonance Imaging of E-Selectin

E-Selectin (CD62E) is a well established marker of inflammatory endothelial cell activation, it has been shown to be upregulated on proliferating endothelial cells in vitro (57), and in tumor tissue (58,59). In hemangioma, placenta and neonatal foreskin, it has been shown to colocalize with dividing endothelial cells in the microvasculature (59), and soluble E-selectin promotes angiogenesis (50). Thus, like VCAM-1 and ICAM-1, E-selectin is a marker of angiogenic endothelial cells and it plays a functional role in neovascularization. Kang et al. decorated CLIO nanoparticles with F(ab')$_2$ fragments of a monoclonal antibody against human E-selectin (H18/7, Vascular Research Division, Department of Pathology, Brigham and Women's Hospital, Boston, Massachusetts, U.S.A.), and they showed specific binding of the targeted CLIO to human umbilical vein endothelial cells (HUVEC) that were induced to express E-selectin by exposing them to interleukin-1β (60). Another approach was taken by Boutry et al. (61) who targeted E-selectin with a mimetic of an E-selectin ligand, sialyl-Lewis[x]. Two molecules of the sialyl-Lewis[x] mimetic were linked to Gd-DTPA, and in a mouse hepaptitis model, significant enhancement of the liver was shown after intravenous injection of targeted Gd-DTPA, whereas no enhancement was seen in the hepatitis model after injection of nontargeted Gd-DTPA, and injection of targeted Gd-DTPA in healthy mice.

In summary, these studies demonstrate that targeted MRI of markers of endothelial cell activation during neovascularization is feasible. More precise demonstration that the targeting agents colocalize with their vascular address after injection in vivo will enable the field to move ahead with more confidence. Targeted imaging of cardiac neovascularization and imaging of arteriogenesis are obvious next steps in the development of targeted neovascular imaging in the cardiovascular system.

IMAGING OF MOLECULAR VASCULAR TARGETS WITH MAGNETIC RESONANCE IMAGING: THE NEED FOR SIGNAL ENHANCEMENT

The capability to depict endothelial cell surface markers is constrained by the fact that these molecular targets are engulfed by a "sea" of protons and electrons in the body, resulting in high back ground and low signal: noise ratios. Thus, together with a judicious choice of contrast agent, strategies to enhance the signal emanating from the molecular target are a key element of a targeted MRI approach.

As outlined previously, the choice is between agents that enhance MR contrast, i.e., paramagnetic T1 agents based on Gd that "brighten" the MR image, and agents that have a negative contrast effect, superparamagnetic T2 agents based on iron that darken the image. The effect of a single gadolinium atom on longitudinal relaxation is far less pronounced than the effect of one iron atom on transverse relaxation, and detectable concentrations of Gd in vivo are in the millimolar range, whereas iron atoms can be detected at nanomolar levels in vivo (21). Because of their greater relaxivity, superparamagnetic metals exert effects well beyond their size, and this means that these agents create an extensive imaging void that may obscure structural aspects such as micro-vessel architecture of surrounding tissue (62). Because T1 agents do not obscure their surroundings, they allow for an imaging strategy that could identify a region of interest at low resolution, which could then be examined in more detail at high resolution. The low relaxivity thus may

constitute an advantage if surrounding tissue is to be imaged. A disadvantage of Gd is that it is not biocompatible and therefore needs to be administered in the form of chelates such as Gd-DTPA. Little is know about the potential toxicity following cellular dechelation after the agents have reached the cytoplasm. Iron based agents on the other hand are composed of biodegradable iron, which is biocompatible and can be processed using normal cellular pathways for iron metabolism.

Signal enhancement for both agents is achieved through methods that compact a large number of the atoms on the cell surface or inside the cell, with the compaction requirements obviously being less stringent for the iron based agents. The concentration of atoms at the cellular level depends on receptor density and percent ligand occupation on the one hand, and on the number of atoms that can be directed to the cell on the other hand. An example where signal to noise ratios were at detectable levels because of high target density is an experiment in which fibrin in thrombi was visualized with a targeted peptide that carried a relatively small "payload" of 4 Gd atoms (63). Here, thrombus in ruptured plaque could be detected successfully in a rabbit model, because of the shear abundance of the target, i.e., fibrin in the clot. However, most molecular targets are expressed at nanomolar or picomolar concentrations and must be detected at clinical field strengths (1.5 T or 3 T) with voxels optimistically sized between 2.0×10^7 μm^3 and 5.0×10^9 μm^3. Therefore, many applications use nanoparticles or macromolecules that can carry a high "payload" of magnetic atoms, as described in the examples reviewed in the previous section. Another, complementary strategy relies on the compaction of the magnetic particles inside the cell where further concentration of the atoms can occur.

Gadolinium has been linked to liposomes, dendrimers, emulsions, polymer based nanoparticles and fullerenes, increasing the number of Gd atoms per particle up to approximately 50,000 for liposome-bound Gd in a 200 nm particle (21). Liposomes are vesicles with an aqueous volume entirely enclosed by a lipid bilayer membrane, and the encapsulated Gd atoms are sequestered from extracellular water fluxes, thus minimizing their effect on R1 (62). To overcome this limitation, amphipathic paramagnetic chelates were developed that allowed for incorporation of Gd into lipid membranes (64). However, considerable amounts (up to 50%) of the Gd remained within the internal leaflet of the amphipathic chelates, resulting in suboptimal relaxivity. An approach was developed that decorated cross-linked lipid bilayer nanoparticles on the membrane surface (31) exposing all Gd atoms to water, and these particles were used successfully to image $\alpha_v\beta_3$ on breast carninoma vasculature (33) and to target therapeutic genes to cancer cells in tumor bearing mice (34). The lipid monolayers mentioned in the previous section are a further modification of this concept, they no longer have an aqueous core, but surround perfluorocarbon instead and they have been decorated with approx. 94,200 gadolinium atoms (39).

Iron oxide for MRI was introduced shortly after the introduction of Gd chelates in the form of superparamagnetic iron oxide particles (SPIO, 50–500 nm) (28), later followed by ultra small paramagnetic iron oxide nanoparticles (USPIO, <50 nm) (65), monocrystaline iron oxide nanoparticles (MION, around 20 nm) (66) and cross linked iron oxide (CLIO, around 25 nm) (67). To prevent aggregation of the particles in circulation, they typically are coated with (aminated) dextran, which presents a suitable platform for further surface modification with targeting moieties. The pharmacokinetics of the nanoparticles influences their targeting behavior in circulation. For example, the larger SPIO are phagocytized when injected into the circulation and accumulate in normal liver and spleen parenchyma (62), whereas the smaller USPIO have a longer half-life in circulation and because of their smaller size can pass through fenestrated or hyperpermeable immature endothelium as reviewed earlier. These factors complicate issues of dose and target specificity, and need to be taken into account when the selectivity of the targeting agent for its vascular address is assessed.

Intracellular compaction of imaging agents is another approach to increase signal:noise, however, because Gd atoms rely on exposure to water to exert their effect on R1, this strategy will be more effective for iron based agents than for agents based on Gd. For example, internalization and intracellular compaction has been shown to increase T2 and T2* of MION by approximately 400% (68). Transfer to the cytoplasm can be achieved in theory when this translocation is a specific result of the ligand receptor interaction of the vascular target that is being imaged. However, it is likely that the artificial ligand, i.e., the targeted nanoparticle or macromolecule will not always initiate receptor mediated internalization because it lacks many characteristics of the natural ligand. An alternative approach is to equip the targeting particle with a membrane translocation signal, which makes internalization receptor independent. Several translocation signals have been identified, for example, the third helix domain of Antennapedia (69), peptide derived from anti-DNA monoclonal antibodies (70), and VP22 herpesvirus protein (71). One frequently used approach to move MR contrast into cells relies on the membrane-translocating sequence of the tat protein of HIV (72) or tat protein derived peptide sequences (73). Both MION and CLIO decorated with tat protein or peptide have been successfully targeted to cells and translocated into the cytoplasm (67,74,75).

In summary, in practically all cases of molecular MRI, a strategy to improve signal:noise is warranted. The use of targeted nanoparticles or macromolecules

to deliver a higher "payload" to the molecular address has the added advantage that the particles can be decorated with therapeutics so live feedback could be obtained on the arrival of the targeted therapeutic at its molecular address.

FUTURE DEVELOPMENTS

"Smart" Contrast Agents

Several agents have been developed that respond to the presence of certain molecules in their vicinity by a change in signal, and these have been called "smart" contrast agents or probes (67). Examples of this imaging approach are intracellular calcium indicators (76,77), molecules that announce the presence of β-galactosidase activity (78) and fluorescent substrates for various hydrolytic enzymes (79). In a recent development, Kircher et al. (67) developed a dual fluorochome probe for imaging protease activity in cells that they linked to CLIO particles, thus creating a particle that would allow for location of a lesion by MRI, and subsequent interrogation at closer range with optical techniques. One fluorophore (Cy5.5) was attached to the N-terminus of a proteolytically cleavable L-polyarginine peptide that was linked to the CLIO particle, whereas the other probe (Cy7) was directly attached to the CLIO carrier and could not be cleaved. Fluorescence of CY5.5 remained quenched until the linker was cleaved while Cy7 fluorescence served as an internal standard to correct for variations in Cy5.5 fluorescence intensity due to substrate concentration and differences in lesion size and depth. The feasibility of combined MRI and

Angiogenesis MRI
Monitors Treatment Effect

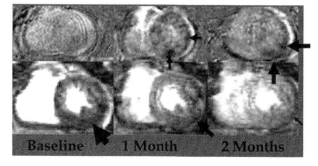

Figure 1   The top row shows hybrid T1,T2* imaging which mutes susceptibility effects from contrast arrival to the ventricles but displays a blackout in zones of angiogenesis, with no angiogenesis at baseline (*left*), intense angiogenesis one month after stimulation with FGF2 (*center*), and inward extension covering zone of posterolateral ischemia at two months (*right*). The second row shows peak tissue enhancement in perfusion-sensitive series, with initially marked hypoperfusion of the posterolateral wall (*left*), improvement one month after therapy (*center*), and resolution of perfusion delay at two months (*right*).

Perfusion-Sensitive MRI

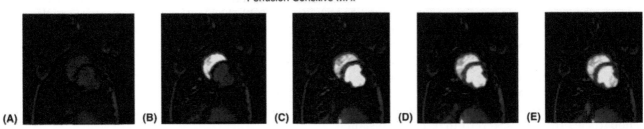

(A)   (B)   (C)   (D)   (E)

Figure 2 A series of images are obtained of each target area of interest, one image per heartbeat, under changing conditions. Typically T1 contrast agent (Gd-DPTA) is injected as a bolus (e.g., 10 cc Magnevist at 3 cc/s into a 75 kg subject). T1-sensitive imaging is adjusted so the target is dark at steady-state (4+ images into test series). Contrast agent-labeled blood arrival to target tissue is marked by increased signal due to shortened T1 (faster magnetization recovery). Select frames from a temporal series showing the mid short axis view of the heart are shown left to right: (**A**) steady state baseline (4th image into series); (**B**) arrival of contrast agent to the right ventricle; (**C**) arrival of contrast agent to the left ventricle; (**D**) arrival of contrast agent to the inter-ventricular septum; (**E**) late and markedly reduced arrival to the lateral wall due to left circumflex artery occlusion.

optical imaging in vivo has been demonstrated before in mice (80), but if this application is to move into the clinical vascular field optical imaging would have to be achieved by an intravascular method because of the low penetration depth of the fluorescent signal.

Imaging of Superparamagnetic Particles

As mentioned before, iron based superparamagnetic particles generate a signal void well beyond their size and this precludes closer inspection of the tissue in proximity to the particles but also the exact localization of the particle within the tissue. An imaging technique that looks directly at superparamagnetic particles was published by Gleich and Weizenecker (7). They applied an inhomogeneous magnetic field to a plastic plate with holes filled with undiluted (0.5 mol Fe/L) commercially available superparamagnetic contrast agent (Resovist, Schering AG, Berlin, Germany), and the contrast was further irradiated by a weak radio-frequency field. Because the magnetic field was inhomogeneous, there were saturated and unsaturated magnetic particles. Gleich and Weizman showed that the unsaturated particles responded to the radio-frequency signal with an oscillating magnetization, and by changing the location of the low intensity spot in the magnetic field (either mechanically or by auxiliary magnetic fields), the sample could be scanned bit by bit, thus generating a map of the spatial distribution of the particles. Thus far, this approach has not been transferred to in vivo experiments, but it could hold the key to the resolution of the complex 3D structure of angiogenic vessels visualized by targeted superparamagnetic particles.

SUMMARY AND CONCLUSIONS

The aggregate of the seminal work of multiple groups shows that targeted MRI of four angiogenic molecules on activated endothelium is feasible, and one of the

next steps is to take this work to the heart and image cardiac angiogenesis. Molecular imaging of cardiac angiogenesis has been achieved with radionuclides targeted to $\alpha_v\beta_3$ (40). Bringing targeted MRI of angiogenesis to the heart and combining it with MRI of cardiac function, perfusion and metabolism will move us a significant step forward in our understanding of the impact of neovessels on these parameters. The recently published strategy for direct imaging of iron based contrast may facilitate the 3D resolution of (micro)vascular networks visualized through targeted MRI.

Understanding the stage of vessel development that is marked by the expression of the endothelial target that is being visualized, more precise demonstration of colocalization of targeting moiety and endothelial address, and a delineation of the functional impact of neovessels on the heart will not only advance the field of targeted vascular MRI, but the entire field of targeted imaging.

Finally, the application of "smart" contrast agents to image intracellular processes in vivo and the use of nanoparticles for targeted delivery of imaging agents combined with therapeutics are opportunities waiting to be realized by the groups who are active in the rapidly evolving field of molecular vascular targeting.

REFERENCES

1. Villanueva FS. Molecular images of neovascularization: art for art's sake or form with a function? Circulation 2005; 111:3188–3191.
2. Risau W, Flamme I. Vasculogenesis. Annu Rev Cell Dev Biol 1995; 11:73–91.
3. Carmeliet P. Mechanisms of angiogenesis and arteriogenesis. Nat Med 2000; 6:389–395.
4. de Muinck E, Simons M. Re-evaluating therapeutic neovascularization. J Moll Cell Cardiol 2004; 36:25–32.
5. Carmeliet P. Angiogenesis in health and disease. Nat Med 2003; 9:653–660.

6. Neeman M, Dafni H. Structural, functional, and molecular MR imaging of the microvasculature. Annu Rev Biomed Eng 2003; 5:29–56.

7. Gleich B, Weizenecker J. Tomographic imaging using the nonlinear response of magnetic particles. Nature 2005; 435:1214–1217.

8. Dvorak HF, Nagy JA, Feng D, Brown LF, Dvorak AM. Vascular permeability factor/vascular endothelial growth factor and the significance of microvascular hyperpermeability in angiogenesis. Curr Top Microbiol Immunol 1999; 237:97–132.

9. Neeman M. Functional and molecular MR imaging of angiogenesis: seeing the target, seeing it work. J Cell Biochem Suppl 2002; 39:11–7.

10. Rissanen TT, Markkanen JE, Arve K, et al. Fibroblast growth factor 4 induces vascular permeability, angiogenesis and arteriogenesis in a rabbit hindlimb ischemia model. Faseb J 2003; 17:100–102.

11. Ziv K, Nevo N, Dafni H, et al. Longitudinal MRI tracking of the angiogenic response to hind limb ischemic injury in the mouse. Magn Reson Med 2004; 51:304–311.

12. Thurston G, Rudge JS, Ioffe E, et al. Angiopoietin-1 protects the adult vasculature against plasma leakage. Nat Med 2000; 6:460–463.

13. Jordan BF, Runquist M, Raghunand N, et al. Dynamic contrast-enhanced and diffusion MRI show rapid and dramatic changes in tumor microenvironment in response to inhibition of HIF-1alpha using PX-478. Neoplasia 2005; 7:475–485.

14. Dafni H, Landsman L, Schechter B, Kohen F, Neeman M. MRI and fluorescence microscopy of the acute vascular response to VEGF165: vasodilation, hyper-permeability and lymphatic uptake, followed by rapid inactivation of the growth factor. NMR Biomed 2002; 15:120–131.

15. de Lussanet QG, Backes WH, Griffioen AW, van Engelshoven JM, Beets-Tan RG. Gadopentetate dimeglumine versus ultrasmall superparamagnetic iron oxide for dynamic contrast-enhanced MR imaging of tumor angiogenesis in human colon carcinoma in mice. Radiology 2003; 229:429–438.

16. Rudisch A, Kremser C, Judmaier W, Zunterer H, DeVries AF. Dynamic contrast-enhanced magnetic resonance imaging: a non-invasive method to evaluate significant differences between malignant and normal tissue. Eur J Radiol 2005; 53:514–519.

17. de Lussanet QG, Langereis S, Beets-Tan RG, et al. Dynamic contrast-enhanced MR imaging kinetic parameters and molecular weight of dendritic contrast agents in tumor angiogenesis in mice. Radiology 2005; 235:65–72.

18. Dafni H, Gilead A, Nevo N, Eilam R, Harmelin A, Neeman M. Modulation of the pharmacokinetics of macromolecular contrast material by avidin chase: MRI, optical, and inductively coupled plasma mass spectrometry tracking of triply labeled albumin. Magn Reson Med 2003; 50:904–914.

19. Dafni H, Israely T, Bhujwalla ZM, Benjamin LE, Neeman M. Overexpression of vascular endothelial growth factor 165 drives peritumor interstitial convection

and induces lymphatic drain: magnetic resonance imaging, confocal microscopy, and histological tracking of triple-labeled albumin. Cancer Res 2002; 62:6731–6739.

20. Hobbs SK, Monsky WL, Yuan F, et al. Regulation of transport pathways in tumor vessels: role of tumor type and microenvironment. Proc Natl Acad Sci U S A 1998; 95:4607–4612.

21. Artemov D. Molecular magnetic resonance imaging with targeted contrast agents. J Cell Biochem 2003; 90:518–524.

22. Saeed M, Wendland MF, Watzinger N, Akbari H, Higgins CB. MR contrast media for myocardial viability, microvascular integrity and perfusion. Eur J Radiol 2000; 34:179–195.

23. Pearlman JD, Laham RJ, Post M, Leiner T, Simons M. Medical imaging techniques in the evaluation of strategies for therapeutic angiogenesis. Curr Pharm Des 2002; 8:1467–1496.

24. McDonald DM, Choyke PL. Imaging of angio-genesis: from microscope to clinic. Nat Med 2003; 9:713–725.

25. Tuncbilek N, Karakas HM, Altaner S. Dynamic MRI in indirect estimation of microvessel density, histologic grade, and prognosis in colorectal adenocarcinomas. Abdom Imaging 2004; 29:166–172.

26. Cheng HL, Chen J, Babyn PS, Farhat WA. Dynamic Gd-DTPA enhanced MRI as a surrogate marker of angiogenesis in tissue-engineered bladder constructs: a feasibility study in rabbits. J Magn Reson Imaging 2005; 21:415–423.

27. Pearlman JD, Hibberd MG, Chuang ML, et al. Magnetic resonance mapping demonstrates benefits of VEGF-induced myocardial angiogenesis. Nat Med 1995; 1:1085–1089.

28. Bulte JW, Kraitchman DL. Iron oxide MR contrast agents for molecular and cellular imaging. NMR Biomed 2004; 17:484–499.

29. Dobrucki LW, Sinusas AJ. Cardiovascular molecular imaging. Semin Nucl Med 2005; 35:73–81.

30. Rupp PA, Little CD. Integrins in vascular development. Circ Res 2001; 89:566–572.

31. Guccione S, Li KC, Bednarski MD. Vascular-targeted nanoparticles for molecular imaging and therapy. Methods Enzymol 2004; 386:219–236.

32. Li KC, Bednarski MD. Vascular-targeted molecular imaging using functionalized polymerized vesicles. J Magn Reson Imaging 2002; 16:388–393.

33. Sipkins DA, Cheresh DA, Kazemi MR, Nevin LM, Bednarski MD, Li KC. Detection of tumor angiogenesis in vivo by alphaVbeta3-targeted magnetic resonance imaging. Nat Med 1998; 4:623–626.

34. Hood JD, Bednarski M, Frausto R, et al. Tumor regression by targeted gene delivery to the neovasculature. Science 2002; 296:2404–2407.

35. Curnis F, Gasparri A, Sacchi A, Longhi R, Corti A. Coupling tumor necrosis factor-alpha with alphaV integrin ligands improves its antineoplastic activity. Cancer Res 2004; 64:565–571.

36. Chen X, Plasencia C, Hou Y, Neamati N. Synthesis and biological evaluation of dimeric RGD peptide-paclitaxel conjugate as a model for integrin-targeted drug delivery. J Med Chem 2005; 48:1098–1106.

37. Schmieder AH, Winter PM, Caruthers SD, et al. Molecular MR imaging of melanoma angiogenesis with alphanubeta3-targeted paramagnetic nanoparticles. Magn Reson Med 2005; 53:621–627.

38. Winter PM, Caruthers SD, Kassner A, et al. Molecular imaging of angiogenesis in nascent Vx-2 rabbit tumors using a novel alpha(nu)beta3-targeted nanoparticle and 1.5 tesla magnetic resonance imaging. Cancer Res 2003; 63:5838–5843.

39. Winter PM, Morawski AM, Caruthers SD, et al. Molecular imaging of angiogenesis in early-stage atherosclerosis with alpha(v)beta3–integrin-targeted nanoparticles. Circulation 2003; 108:2270–2274.

40. Meoli DF, Sadeghi MM, Krassilnikova S, et al. Noninvasive imaging of myocardial angiogenesis following experimental myocardial infarction. J Clin Invest 2004; 113:1684–1691.

41. Sadeghi MM, Krassilnikova S, Zhang J, et al. Detection of injury-induced vascular remodeling by targeting activated alphavbeta3 integrin in vivo. Circulation 2004; 110:84–90.

42. Hua J, Dobrucki LW, Sadeghi MM, et al. Noninvasive imaging of angiogenesis with a 99mTc-labeled peptide targeted at alphavbeta3 integrin after murine hindlimb ischemia. Circulation 2005; 111: 3255–3260.

43. Arap W, Pasqualini R, Ruoslahti E. Cancer treatment by targeted drug delivery to tumor vasculature in a mouse model. Science 1998; 279:377–380.

44. Pasqualini R, Koivunen E, Ruoslahti E. A peptide isolated from phage display libraries is a structural and functional mimic of an RGD-binding site on integrins. J Cell Biol 1995; 130:1189–1196.

45. Pierschbacher MD, Hayman EG, Ruoslahti E. Location of the cell-attachment site in fibronectin with monoclonal antibodies and proteolytic fragments of the molecule. Cell 1981; 26:259–267.

46. Pierschbacher MD, Ruoslahti E. Cell attachment activity of fibronectin can be duplicated by small synthetic fragments of the molecule. Nature 1984; 309:30–33.

47. Ruoslahti E, Engvall E. Integrins and vascular extracellular matrix assembly. J Clin Invest 1997; 99: 1149–1152.

48. Haas TA, Plow EF. Integrin-ligand interactions: a year in review. Curr Opin Cell Biol 1994; 6:656–662.

49. Iiyama K, Hajra L, Iiyama M, et al. Patterns of vascular cell adhesion molecule-1 and intercellular adhesion molecule-1 expression in rabbit and mouse atherosclerotic lesions and at sites predisposed to lesion formation. Circ Res 1999; 85:199–207.

50. Koch AE, Halloran MM, Haskell CJ, Shah MR, Polverini PJ. Angiogenesis mediated by soluble forms of E-selectin and vascular cell adhesion molecule-1. Nature 1995; 376:517–519.

51. Joussen AM, Poulaki V, Le ML, et al. A central role for inflammation in the pathogenesis of diabetic retinopathy. Faseb J 2004; 18:1450–1452.

52. Kevil CG, Orr AW, Langston W, et al. Intercellular adhesion molecule-1 (ICAM-1) regulates endothelial cell motility through a nitric oxide-dependent pathway. J Biol Chem 2004; 279:19230–19238.

53. Cybulsky MI, Gimbrone MA Jr. Endothelial expression of a mononuclear leukocyte adhesion molecule during atherogenesis. Science 1991; 251:788–791.

54. Cybulsky MI, Iiyama K, Li H, et al. A major role for VCAM-1, but not ICAM-1, in early atherosclerosis. J Clin Invest 2001; 107:1255–1262.

55. Kelly KA, Allport JR, Tsourkas A, Shinde-Patil VR, Josephson L, Weissleder R. Detection of vascular adhesion molecule-1 expression using a novel multimodal nanoparticle. Circ Res 2005; 96:327–336.

56. Sipkins DA, Gijbels K, Tropper FD, Bednarski M, Li KC, Steinman L. ICAM-1 expression in autoimmune encephalitis visualized using magnetic resonance imaging. J Neuroimmunol 2000; 104:1–9.

57. Bischoff J, Brasel C, Kraling B, Vranovska K. E-selectin is upregulated in proliferating endothelial cells in vitro. Microcirculation 1997; 4:279–287.

58. Mayer B, Spatz H, Funke I, Johnson JP, Schildberg FW. De novo expression of the cell adhesion molecule E-selectin on gastric cancer endothelium. Langenbecks Arch Surg 1998; 383:81–86.

59. Kraling BM, Razon MJ, Boon LM, et al. E-selectin is present in proliferating endothelial cells in human hemangiomas. Am J Pathol 1996; 148:1181–1191.

60. Kang HW, Josephson L, Petrovsky A, Weissleder R, Bogdanov A Jr. Magnetic resonance imaging of inducible E-selectin expression in human endothelial cell culture. Bioconjug Chem 2002; 13:122–127.

61. Boutry S, Burtea C, Laurent S, Toubeau G, Vander Elst L, Muller RN. Magnetic resonance imaging of inflammation with a specific selectin-targeted contrast agent. Magn Reson Med 2005; 53:800–807.

62. Lanza GM, Winter PM, Caruthers SD, et al. Magnetic resonance molecular imaging with nanoparticles. J Nucl Cardiol 2004; 11:733–743.

63. Botnar RM, Perez AS, Witte S, et al. In vivo molecular imaging of acute and subacute thrombosis using a fibrin-binding magnetic resonance imaging contrast agent. Circulation 2004; 109:2023–2029.

64. Kabalka GW, Davis MA, Holmberg E, Maruyama K, Huang L. Gadolinium-labeled liposomes containing amphiphilic Gd-DTPA derivatives of varying chain length: targeted MRI contrast enhancement agents for the liver. Magn Reson Imaging 1991; 9:373–377.

65. Kooi ME, Cappendijk VC, Cleutjens KB, et al. Accumulation of ultrasmall superparamagnetic particles of iron oxide in human atherosclerotic plaques can be detected by in vivo magnetic resonance imaging. Circulation 2003; 107:2453–2458.

66. Shen T, Weissleder R, Papisov M, Bogdanov A Jr, Brady TJ. Monocrystalline iron oxide nanocompounds (MION): physicochemical properties. Magn Reson Med 1993; 29:599–604.

67. Kircher MF, Weissleder R, Josephson L. A dual fluorochrome probe for imaging proteases. Bioconjug Chem 2004; 15:242–248.

68. Weissleder R, Moore A, Mahmood U, et al. In vivo magnetic resonance imaging of transgene expression. Nat Med 2000; 6:351–355.

69. Derossi D, Calvet S, Trembleau A, Brunissen A, Chassaing G, Prochiantz A. Cell internalization of the

third helix of the Antennapedia homeodomain is receptor-independent. J Biol Chem 1996; 271:18188–18193.

70. Avrameas A, Ternynck T, Nato F, Buttin G, Avrameas S. Polyreactive anti-DNA monoclonal antibodies and a derived peptide as vectors for the intracytoplasmic and intranuclear translocation of macromolecules. Proc Natl Acad Sci U S A 1998; 95:5601–5606.

71. Phelan A, Elliott G, O'Hare P. Intercellular delivery of functional p53 by the herpesvirus protein VP22. Nat Biotechnol 1998; 16:440–443.

72. Fawell S, Seery J, Daikh Y, et al. Tat-mediated delivery of heterologous proteins into cells. Proc Natl Acad Sci U S A 1994; 91:664–668.

73. Schwarze SR, Ho A, Vocero-Akbani A, Dowdy SF. In vivo protein transduction: delivery of a biologically active protein into the mouse. Science 1999; 285:1569–1572.

74. Koch AM, Reynolds F, Kircher MF, Merkle HP, Weissleder R, Josephson L. Uptake and metabolism of a dual fluorochrome Tat-nanoparticle in HeLa cells. Bioconjug Chem 2003; 14:1115–1121.

75. Josephson L, Tung CH, Moore A, Weissleder R. High-efficiency intracellular magnetic labeling with novel superparamagnetic-Tat peptide conjugates. Bioconjug Chem 1999; 10:186–191.

76. Takahashi A, Camacho P, Lechleiter JD, Herman B. Measurement of intracellular calcium. Physiol Rev 1999; 79:1089–1125.

77. Li WH, Parigi G, Fragai M, Luchinat C, Meade TJ. Mechanistic studies of a calcium-dependent MRI contrast agent. Inorg Chem 2002; 41:4018–4024.

78. Louie AY, Huber MM, Ahrens ET, et al. In vivo visualization of gene expression using magnetic resonance imaging. Nat Biotechnol 2000; 18:321–325.

79. Haugland RP. Detecting enzymatic activity in cells using fluorogenic substrates. Biotech Histochem 1995; 70:243–251.

80. Josephson L, Kircher MF, Mahmood U, Tang Y, Weissleder R. Near-infrared fluorescent nanoparticles as combined MR/optical imaging probes. Bioconjug Chem 2002; 13:554–560.

24

Cardiac Sympathetic Neuroimaging

David S. Goldstein

Clinical Neurocardiology Section, National Institute of Neurological Disorders and Stroke, National Institutes of Health, Bethesda, Maryland, U.S.A.

BACKGROUND

Symptoms or signs of abnormal autonomic nervous system function occur commonly in many diseases. Clinical evaluations have depended on physiologic, pharmacologic, and neurochemical approaches. Recently, imaging of sympathetic noradrenergic innervation has been introduced and applied especially in the heart.

METHODS AND RESULTS

Most studies have used ^{123}I-metaiodobenzylguanidine (^{123}I-MIBG). Parkinson's disease (PD) with orthostatic hypotension features cardiac sympathetic denervation, whereas multiple system atrophy, often difficult to distinguish clinically from PD, features intact cardiac sympathetic innervation. Decreased cardiac uptake or increased "washout" of ^{123}I-MIBG-derived radioactivity is associated with worse prognosis or more severe disease in a variety of conditions.

Dysautonomias are conditions where altered activity of the autonomic nervous system adversely affects health (1).

The autonomic nervous system has at least five components—enteric, parasympathetic cholinergic, sympathetic cholinergic, sympathetic noradrenergic, and adrenomedullary hormonal. Langley introduced the term, "autonomic nervous system," about a century ago, referring to neurons in ganglia outside the brain and spinal cord that seemed to function independently, or autonomously, of the central nervous system. He identified enteric, sympathetic, and parasympathetic components. In the early 20th century, Cannon added an adrenal hormonal component. According to Cannon, the sympathetic nervous system and adrenal gland would act together to maintain "homeostasis" (a word he coined) in emergencies. The notion of a unitary "sympathoadrenal" system remains generally accepted, despite evidence for differential regulation and dysregulation of sympathetic neuronal and adrenomedullary hormonal effectors (2). The

autonomic nervous system therefore is not just neuronal, and its components do not function autonomously from the central nervous system. One might instead refer to them as "automatic" neuroendocrine systems, regulated by the brain in concert with behaviors and emotions.

Failure of a particular component of the autonomic nervous system produces characteristic clinical manifestations. In particular, due to the absolute requirement of intact sympathetic neurocirculatory function for humans to tolerate simply standing up, sympathetic noradrenergic failure presents as orthostatic intolerance and orthostatic hypotension (OH).

CLINICAL METHODS TO ASSESS CARDIOVASCULAR AUTONOMIC FUNCTION

In 1628, William Harvey noted the effects of emotions on the heart: "For every affection of the mind that is attended with either pain or pleasure, hope or fear, is the cause of an agitation whose influence extends to the heart ..." In the second half of the 19th century, Pavlov and others istudied effects of cardiac sympathetic nerve stimulation, and in 1895, Oliver and Schäfer reported for the first time the potent cardiovascular effects of injected adrenal extract, the active principle of which was discovered to be epinephrine (adrenaline) several years later. Thus, probably the first method to assess cardiovascular autonomic function was physiologic. Physiologic measures continue to play a key role in clinical laboratory assessment of sympathetic noradrenergic function (3–6).

Neuropharmacologic measures were introduced early in the 20th century, by Dale, Loewi, and others who studied drugs affecting ganglionic neurotransmission. Use of this approach increased greatly in the mid-20th century, after Ahlquist's description of effects of catecholamines mediated by specific receptors (7). Modern autonomic function testing includes measurements of cardiovascular responses to any of a variety of neuropharmacologic agents, including tyramine, trimethaphan, edrophonium, desipramine, yohimbine, clonidine, atropine, phenylephrine, nitroglycerine, and isoproterenol.

Also beginning early in the 20th century, researchers began to develop chemical means to assess activity of what Cannon called the "sympathico-adrenal" system. Bioassays used by Cannon were the first to detect successfully epinephrine release into the circulation. Cannon later developed and exploited a preparation based on the magnitude of the increase in heart rate in animals with denervated hearts; abolition of the increase by adrenalectomy confirmed the hormone's adrenal source (8). Subsequent chemical methods depended on fluorescence detection, radioenzymatic assays, or liquid chromatography with electrochemical detection.

CARDIAC SYMPATHETIC NEUROIMAGING

A fourth clinical method to assess the cardiovascular regulation by the autonomic nervous system is by imaging. To date, autonomic neuroimaging has focused almost entirely on noradrenergic innervation of the heart, which possesses dense sympathetic postganglionic nerves.

Imaging agents to visualize cardiac sympathetic innervation share two characteristics. First, they are substrates for the cell membrane norepinephrine transporter. Second, they are substrates for the vesicular monoamine transporter. Because of these characteristics, sympathoneural imaging agents are taken up into sympathetic nerves by the Uptake-1 process; once inside the nerves they are sequestered by storage vesicles. The basis for sympathetic neuroimaging therefore is radiolabeling of the vesicles in sympathetic nerves. Ligands for adrenoceptors or other receptors on sympathetic nerves or for the cell membrane norepinephrine transporter have so far not attained the success of ligands taken up by and stored in sympathetic nerves.

Two types of sympathoneural agents can visualize cardiac sympathetic innervation successfully. The first type, exemplified by ^{123}I-MIBG, are substrates for the cell membrane and vesicular monoamine transporters but not for key enzymes that degrade norepinephrine, monoamine oxidase and catechol-O-methyltransferase. These agents are not catecholamines but are sympathomimetic amines. They provide excellent anatomic depiction of sympathetic innervation; however, the kinetics of radioactivity concentrations in the tissue cannot be assumed to indicate the fate of endogenous norepinephrine. Those of the second type, exemplified by 6-[^{18}F]fluorodopamine, are catecholamines.

Analysis of time-activity curves relating radioactivity concentrations with time could provide not only anatomic but also functional information about specific aspects of cardiac sympathetic function; however, the metabolic fate of these agents is complex, and it is by no means established that alterations in the time-activity curves can actually provide functional information such as about the rate of sympathetically mediated exocytosis. Either type of agent should theoretically be able to detect altered turnover of the vesicular contents, which should be related to postganglionic sympathetic nerve traffic and exocytosis.

By far the most commonly used method for cardiac sympathetic neuroimaging worldwide is scintigraphy or single photon emission computed tomography after i.v. injection of ^{123}I-MIBG. Utilization of this agent in the United States has lagged behind.

PARKINSON'S DISEASE WITH ORTHOSTATIC HYPOTENSION

Perhaps the clearest evidence for dysautonomia in PD has come from assessments of sympathetic noradrenergic innervation of the heart. At NIH, the OH is defined as a fall in systolic pressure of 20 mmHg or more and in diastolic pressure of 10 mm or more between lying supine for 15 minutes and then standing for five minutes. Depending on the definition used and the referral pattern to the evaluation center, OH occurs in one-fifth to up to one-half of patients with PD.

Primary chronic autonomic failure occurs in other diseases besides PD. Pure autonomic failure (PAF) features severe neurogenic OH due to generalized sympathetic denervation, without symptoms or signs of central neurodegeneration. In multiple system atrophy (MSA), OH occurs with evidence of central neurodegeneration, subclassified into parkinsonian, cerebellar, or mixed forms.

It is generally well accepted that in PAF, the lesion is postganglionic, whereas in MSA the lesion is preganglionic. Until recently, the site of the lesion in PD + OH was unknown. OH in patients with parkinsonism has been thought to be a side-effect of treatment with levodopa, to develop only late in the disease, or, if prominent and early with respect to disordered movement, to indicate a different disease, such as MSA. Instead, regardless of levodopa treatment and duration of disease, patients with PD + OH have clear evidence for baroreflex-cardiovagal failure, baroreflex-sympathoneural failure, and as highlighted below, loss of sympathetic innervation diffusely in the left ventricular myocardium. These findings indicate that in PD, OH results from the disease process, not the treatment, although administration of drugs that directly or indirectly produce vasodilation can worsen orthostatic tolerance and decrease blood pressure when the patient stands up.

A score of studies have by now agreed remarkably on the finding that most patients with PD have at least partial loss of sympathetic innervation of the heart, as indicated by low myocardial concentrations of radioactivity after injection of the sympathoneural imaging agents, ^{123}I-MIBG and 6-[^{18}F]fluorodopamine, or by neurochemical assessments during right heart catheterization. Recently the notion of cardiac sympathetic denervation in PD received important pathologic confirmation, with the demonstration of abnormal absence of tyrosine hydroxylase staining in post-mortem cardiac tissue from patients with PD (9,10). These findings imply that, analogously to PAF, PD involves a postganglionic lesion, in turn implying that PD is not only a disease of the central nervous system but also a dysautonomia.

Among patients with PD lacking OH, about half have a loss of 6-[^{18}F]fluorodopamine-derived radioactivity diffusely in the left ventricular myocardium, and most of the remaining half have loss localized to the lateral or inferior walls, with relative preservation in the septum or anterior wall (11) The loss of 6-[^{18}F]fluorodopamine-derived radioactivity progresses over time (12). The extent of loss of sympathetic innervation in PD seems to vary remarkably among organs and is especially prominent in the heart (13–15). The bases for this organ selectivity remain unknown.

Perhaps as remarkable as the finding that all patients with PD + OH have cardiac sympathetic denervation is the finding that all patients with MSA, with or without OH, have intact cardiac sympathetic innervation (9,11,16,17). Patients with normal cardiac ^{123}I-MIBG-derived radioactivity have had an absence of nigrostriatal Lewy bodies, have glial cytoplasmic inclusions, thought to be characteristic of MSA, and have normal myocardial tyrosine hydroxylase staining (9,10). According to a proposed pathophysiologic classification of primary chronic autonomic failure, PD + OH features a postganglionic, sympathetic, noradrenergic lesion, whereas the parkinsonian form of MSA does not.

CARDIAC SYMPATHONEURAL DYSFUNCTION

Many studies using ^{123}I-MIBG imaging have noted decreased cardiac "uptake" of the agent, detected at about 15 minutes after its i.v. injection, or increased "washout" of the agent, from the extent of decline in radioactivity by three to four hours after the injection. This pattern has been shown to have prognostic significance, with a worse outlook or more severe disease in patients with essential hypertension, coronary artery disease, congestive heart failure, arrhythmias, myocarditis, cardiomyopathy, syndrome X, or myocardial infarction. Moreover, successful drug treatment results in recession of this pattern (18–24).

The impressive variety of conditions where abnormal cardiac uptake or washout of ^{123}I-MIBG-derived radioactivity relates to disease severity or prognosis suggest common mechanisms relatively independent of disease-specific pathophysiology. One such mechanism may be a high rate of postganglionic cardiac sympathetic traffic.

From kinetic modeling one can predict that, all other things being the same, translocation of vesicles to the membrane surface and an increased rate of exocytosis will produce a pattern of decreased peak ^{123}I-MIBG-derived radioactivity and more rapid loss of the radioactivity. Another mechanism may be high enough circulating catecholamine concentrations to compete with ^{123}I-MIBG for neuronal uptake via the cell membrane norepinephrine transporter (25). High local norepinephrine concentrations at uptake-1 sites in the hearts of patients with high cardiac sympathetic nerve traffic might also compete for neuronal uptake of ^{123}I-MIBG and promote accelerated loss of radioactivity by attenuating reuptake. Moreover, general intraneuronal pathophysiologic changes, such as cellular hypoxia or acidosis, would be expected to decrease the efficiency of the vesicular monoamine transporter or cell membrane norepinephrine transporter, activities of which depend on energy-requiring processes.

CONCLUSIONS

PD entails a postganglionic sympathetic noradrenergic lesion, implying that the disease is not only a movement disorder, from dopamine loss in the nigrostriatal system of the brain, but also a dysautonomia, from norepinephrine loss in the sympathetic nervous system of the heart. Abnormal time-activity curves for myocardial ^{123}I-MIBG-derived radioactivity may reflect local sympathetic nervous activation.

In conclusion, cardiac sympathetic neuroimaging is an important new technique, for assessing local sympathetic innervation and possibly function in a large number of neurocardiologic disorders.

REFERENCES

1. Goldstein D, Eisenhofer G, Robertson D, Straus R, Esler M. Dysautonomias: clinical disorders of the autonomic nervous system. Ann Intern Med 2002; 137:753–763.
2. Pacak K, Palkovits M, Yadid G, Kvetnansky R, Kopin IJ, Goldstein DS. Heterogeneous neurochemical responses to different stressors: a test of Selye's doctrine of nonspecificity. Am J Physiol 1998; 275:R1247–R1255.
3. Goldstein DS, Tack C. Non-invasive detection of sympathetic neurocirculatory failure. Clin Auton Res 2000; 10:285–291.

4. Grassi G, Esler M. How to assess sympathetic activity in humans. J Hypertens 1999; 17(6):719–734.

5. Hagbarth KE, Burke D. Microneurography in man. Acta Neurol (Napoli) 1977; 32(1):30–34.

6. Wallin BG, Elam M. Microneurography and autonomic dysfunction. In: Low PA, ed. Clinical Autonomic Disorders. Boston: Little, Brown and Company, 1993:243–252.

7. Ahlquist RP. A study of adrenotropic receptors. Am J Physiol 1948; 153:586–600.

8. Cannon WB, Rapport D. Further observations on the denervated heart in relation to adrenal secretion. Am J Physiol 1921; 58:308–337.

9. Orimo S, Ozawa E, Oka T, et al. Different histopathology accounting for a decrease in myocardial MIBG uptake in PD and MSA. Neurology 2001; 57:1140–1141.

10. Orimo S, Oka T, Miura H, et al. Sympathetic cardiac denervation in Parkinson's disease and pure autonomic failure but not in multiple system atrophy. J Neurol Neurosurg Psychiatry 2002; 73:776–777.

11. Goldstein DS, Holmes C, Dendi R, Bruce S, Li S-T. Orthostatic hypotension from sympathetic denervation in Parkinson's disease. Neurology 2002; 58:1247–1255.

12. Li ST, Dendi R, Holmes C, Goldstein DS. Progressive loss of cardiac sympathetic innervation in Parkinson's disease. Ann Neurol 2002; 52(2):220–223.

13. Takatsu H, Nishida H, Matsuo H, et al. Cardiac sympathetic denervation from the early stage of Parkinson's disease: clinical and experimental studies with radiolabeled MIBG. J Nucl Med 2000; 41(1):71–77.

14. Reinhardt MJ, Jungling FD, Krause TM, Braune S. Scintigraphic differentiation between two forms of primary dysautonomia early after onset of autonomic dysfunction: value of cardiac and pulmonary iodine-123 MIBG uptake. Eur J Nucl Med 2000; 27:595–600.

15. Taki J, Nakajima K, Hwang EH, et al. Peripheral sympathetic dysfunction in patients with Parkinson's disease without autonomic failure is heart selective and disease specific. Eur J Nucl Med 2000; 27:566–573.

16. Braune S, Reinhardt M, Schnitzer R, Riedel A, Lucking CH. Cardiac uptake of [123I]MIBG separates Parkinson's disease from multiple system atrophy. Neurology 1999; 53:1020–1025.

17. Senard JM, Brefel-Courbon C, Rascol O, Montastruc JL. Orthostatic hypotension in patients with Parkinson's disease: pathophysiology and management. Drugs Aging 2001; 18:495–505.

18. Soeki T, Tamura Y, Bandou K, et al. Long-term effects of the angiotensin-converting enzyme inhibitor enalapril on chronic heart failure. Examination by 123I-MIBG imaging. Jpn Heart J 1998; 39(6):743–751.

19. Sakata K, Shirotani M, Yoshida H, et al. Effects of amlodipine and cilnidipine on cardiac sympathetic nervous system and neurohormonal status in essential hypertension. Hypertension 1999; 33(6):1447–1452.

20. Agostini D, Belin A, Amar MH, et al. Improvement of cardiac neuronal function after carvedilol treatment in dilated cardiomyopathy: a 123I-MIBG scintigraphic study. J Nucl Med 2000; 41(5):845–851.

21. Lotze U, Kaepplinger S, Kober A, Richartz BM, Gottschild D, Figulla HR. Recovery of the cardiac adrenergic nervous system after long-term beta-blocker therapy in idiopathic dilated cardiomyopathy: assessment by increase in myocardial 123I-metaiodobenzylguanidine uptake. J Nucl Med 2001; 42(1):49–54.

22. Matsui T, Tsutamoto T, Maeda K, Kusukawa J, Kinoshita M. Prognostic value of repeated 123I-metaiodobenzylguanidine imaging in patients with dilated cardiomyopathy with congestive heart failure before and after optimized treatments—comparison with neurohumoral factors. Circ J 2002; 66(6):537–543.

23. Kasama S, Toyama T, Hoshizaki H, et al. Dobutamine gated blood pool scintigraphy predicts the improvement of cardiac sympathetic nerve activity, cardiac function, and symptoms after treatment in patients with dilated cardiomyopathy. Chest 2002; 122(2):542–548.

24. de Milliano PA, Tijssen JG, van Eck-Smit BL, Lie KI. Cardiac 123 I-MIBG imaging and clinical variables in risk stratification in patients with heart failure treated with beta blockers. Nucl Med Commun 2002; 23(6):513–519.

25. Shouda S, Kurata C, Mikami T, Wakabayashi Y. Effects of extrinsically elevated plasma norepinephrine concentration on myocardial 123I-MIBG kinetics in rats. J Nucl Med 1999; 40(12):2088–2093.

25

Imaging of Beta-Receptors in the Heart

Jeanne M. Link
Division of Nuclear Medicine, University of Washington, Seattle, Washington, U.S.A.

John R. Stratton
Division of Cardiology, Veterans Administration Medical Center, Seattle, Washington, U.S.A.

Wayne C. Levy and Jeanne Poole
Division of Cardiology, University of Washington, Seattle, Washington, U.S.A.

James H. Caldwell
Divisions of Cardiology and Nuclear Medicine, University of Washington, Seattle, Washington, U.S.A.

OVERVIEW

The adrenergic system plays the major role in the regulation of heart rate and contractility and is abnormal in most of the major cardiac diseases. Beta-adrenergic receptors (β-AR) are an important imaging target because they modulate the contractile response. PET imaging of β-ARs is a noninvasive procedure that helps us to understand this receptor in cardiac disease. An image of β-AR alone is unlikely to provide a useful clinical diagnostic tool for most cardiac diseases due to insufficient sensitivity. However, imaging both pre- and postsynaptic function may provide a more clinically relevant assessment of adrenergic function in various cardiac disease states.

ADRENERGIC SYSTEM AND BETA-ADRENERGIC RECEPTORS

The adrenergic system plays the major role in modulation of heart rate and contractility (1). Adrenergic signaling is initiated by the binding of the adrenergic agonists, norepinephrine, and epinephrine, to adrenergic receptors. Three types of adrenergic receptors, alpha 1, alpha 2, and beta receptors, are known, and each of these has three subtypes. (β-ARs also have a fourth "putative" subtype. All the types of adrenergic receptors are found in the heart, but the ratio of β to α receptors in the heart is 10:1 (1,2) and approximately 75% to 80% of the cardiac β-AR are the $\beta1$ subtype. Binding of adrenergic agonists to the active β-AR ($\beta1$ and $\beta2$) on the outer surface of the sarcolemma causes a conformational change in the transmembrane

receptor G-coupled protein that initiates a series of secondary intracellular molecular interactions. Specifically, cyclic AMP production and a number of cellular events including phosphorylation of the calcium channel and the release of calcium by the sarcoplasmic reticulum (2–4). In the healthy heart, the β-AR are part of an exquisitely balanced and complex regulatory system. While we are far from understanding all the details, in addition to contraction, the system includes myocyte growth and apoptosis (2). The β-AR are the receptors with the greatest effect on augmenting or maintaining contractility; despite the need for a maximal response from all of the available β-AR to induce the required inotropic response (1–3).

Importance of β-AR in Disease
While it remains unclear how many cardiac diseases result from derangements in cardiac β-AR (5,6), β-AR concentrations are abnormal in many cardiac diseases, including congestive heart failure, myocardial infarction, ischemia, and cardiomyopathy, as well as in diabetes and in thyroid-induced muscle disease (7–9). β-ARs exhibit decreased response to agonists with aging (10) and changes in β-AR responsiveness are associated with the formation of ventricular arrhythmias (11). Not only are β-AR concentrations abnormal, but also in some cardiac diseases the receptors are dysfunctional. For example, $\beta1$ receptors are down-regulated, while $\beta2$ receptors are uncoupled (deactivated) in congestive heart failure (CHF) (1). Clinically the β-AR antagonists, beta-blockers, have provided some of the most effective therapies to date for reducing mortality in patients with various cardiovascular diseases (12,13).

These associations of β-AR with disease provide a compelling reason for further elucidation of the pathways signaled by β-AR. Our understanding of the adrenergic system has benefited from in vitro tests, from animal models, and from analysis of human biopsies. One can measure β-AR in humans by tissue sampling of the myocardium, but the procedure is invasive and multiple tissue samples are likely to be needed due to regional variation in β-AR density. Noninvasive imaging has potential for increasing our understanding of in vivo function. However, in vivo human cardiac β-AR function remains poorly measured and understood. There has been substantial progress in the development of radiotracers and techniques for nuclear imaging of β-AR in vivo.

IMAGING OF β-AR

To understand the strengths and limitations of imaging studies of β-AR in vivo, one must first understand how the receptor functions. The structural requirements for molecules to bind to β-AR are well known (2,3,14). The binding of agonists, epinephrine and norepinephrine, to β-AR creates a conformational change in the transmembrane receptor protein to release a sub-protein inside the cell. This release initiates the second-messenger signaling cascade. Beta blockers are antagonists that bind at a different site of the trans-membrane loop of the beta receptor and prevent agonists from binding to the receptor and initiating the intracellular signaling (2,14).

The rate of binding of an agonist or antagonist (ligand binding) to the receptor follows a second order rate equation (15). Agonists tend to bind quickly (large association constant k_{on}) but to not remain bound for very long (large dissociation constant k_{off}). The ratio of k_{off}/k_{on} is termed the equilibrium dissociation constant, K_D. The K_D is a measure of the stability of the ligand bound to the receptor and has a significant impact on the potential of a radioligand to be a useful imaging agent. Antagonists have a range of kinetic constants, and thus a range of K_Ds. In general, an agonist is not used for β-AR imaging because it will also be taken up in the pre-synaptic system by the norepinephrine transporter (NET1, also known as uptake-1). Consequently, an agonist becomes a marker of both pre- and postsynaptic binding, primarily presynaptic, so the image cannot be uniquely interpreted for β-AR.

Interpretable β-AR imaging requires a radiolabeled tracer that binds rapidly and tightly (slow k_{off}) to the receptor. However, the binding cannot be so tight that the tracer is irreversibly bound to the receptor nor so rapid that it is first-pass extracted and thus is a measure of delivery by blood flow. A radiotracer antagonist with a K_D in the nanomolar range usually meets

these requirements (16). As mentioned, the radiotracer must be specific for the β-AR and not bind to other receptors or be taken up nonspecifically in the heart. The image of a high specific activity bound radiopharmaceutical may be sufficient to give an indication of the regional distribution of the receptor within the heart but, by itself, does not provide individual values for the receptor density or affinity. The measured product of receptor density and K_D is referred to as the *binding potential*. It is more useful to know the density of receptor concentration because that is the number of available receptors and is a measure of the potential responsiveness of the system. K_D may or may not be an important parameter to determine from imaging, depending on whether or not it changes as part of the disease process. For mixed agonist/antagonist ligands, the starting assumption must be that it does change with disease, but there is little information in humans as yet.

In order to quantify β-AR in various myocardial regions, activity per volume of tissue is determined from the image. This number can be converted to pmol per volume if one accurately knows the specific activity, i.e., activity that was injected per mass. However, that analysis only measures the pmol of receptor ligand (antagonist) bound in a given volume of tissue at the time of imaging, not the total amount of receptors in a volume or B'_{max}. To determine β-AR density and affinity, the binding kinetics need to be modeled and corrected for any radioactive metabolites and for any nonspecific uptake. This can be done by imaging after multiple injections of varying specific activity and by applying mathematical modeling of the kinetic data from the imaging time activity curves to solve the reversible binding equation. B'_{max} is the total potentially available beta receptor density and is related to total, bound and free receptor concentrations. The general principles using ligand binding to assess receptor concentration and function are outlined for in vitro testing in text books (15) and have been used for several receptor systems with PET data (17,18).

Because assessing receptor density and affinity requires quantification, the only imaging modality that has been used successfully, thus far, is PET. The radioligand that has been most thoroughly validated for assessing β-AR density in vivo is [11C]-CGP12177. [11C]-CGP12177 is a selective β-AR antagonist, but it is not selective for the β1 over the β2 subtype; it also has a slight agonist activity and a measured K_D in vitro of ~0.18 nM (19,20). Labeled CGP12177 has proved especially useful for imaging because it does not diffuse across membranes because it is sufficiently hydrophilic. The result is that CGP12177 only binds to cell surface β-AR that are thought to be the active receptors, and it has low nonspecific binding. This property simplifies the analysis of its binding and uptake in vivo. [^{11}C]-CGP12388 is a closely related tracer and hydrophilic beta adrenergic antagonist (21). The synthesis of high specific activity [^{11}C]-CGP12388 has been easier than [^{11}C]-CGP12177 for some laboratories.

Development of mathematical modeling and validation of [^{11}C]-CGP12177 for assessing β-AR in vivo has been done principally by researchers from Orsay, France (22–27), with additional modeling development by researchers at the Hammersmith Hospital in London, England (28). The initial experimental approach used a three-injection protocol. The first was a high specific activity injection that resulted in a maximal uptake of ligand by receptor, while only a low percentage of the β-AR is occupied by radioactive ligand. The second lower specific activity injection occupies ideally ~50% of the receptors. The final injection was nonradioactive CGP12177 of known concentration to displace the radioactive CGP12177. A *three compartment* model was applied to the time activity curves from the PET images from this protocol and tissue samples from human heart were used to confirm the accuracy of the model (22). These same investigators then simplified the mathematical model to a graphical analysis (24,27) that reduced the PET study to a two-injection protocol. This method uses an initial injection of high specific activity [^{11}C]-CGP12177, resulting in low receptor occupancy, and then a lower specific activity injection resulting in approximately 50% of the available β-AR receptors being occupied by CGP12177. The graphical analysis can be used to determine B'_{max} but not K_D. This work was extended to account for binding of the radioligand to blood cells (25). Similar modeling procedures have been applied to [^{11}C]-CGP12388 (29).

Other tracers have been developed to image the alpha and the β-AR receptors in vivo and these have been reviewed by Rieman and Elsinga et al. (30,31). Of these agents, [^{11}C]-carazolol and [^{18}F]-fluorocarazolol have been the most investigated. However, carazolol is lipophilic and will freely diffuse across cellular membranes. The development of mathematical models that can accurately separate the small amount of specific binding at the cell membrane from cytosolic and nonspecific binding for any of the lipophilic β-AR radiotracers has been challenging.

RESULTS OF PET IMAGING OF [^{11}C]-CGP12177 AND [^{11}C]-CGP12388 IN VIVO

Despite more than 15 years of a quantitative PET methodology for measuring β-AR, there are few reports of imaging human β-AR density in disease. Merlet et al. found that B'_{max} in the left ventricle was decreased in patients with idiopathic cardiomyopathy; 3.12 ± 0.51 pmol/mL in patients vs 6.60 ± 1.18 pmol/mL in normals (23). Schafers et al. found that in patients with

hypertrophic cardiomyopathy B'_{max} was decreased from 10.2 ± 2.9 (n = 19) pmol/g in normals to 7.3 ± 2.6 (n = 13) pmol/g in patients (32). The same group observed down regulation of B'_{max} to 5.9 ± 1.3 pmol/g of tissue in patients with arrhythmogenic right ventricular cardiomyopathy (33). In ischemic patients with CHF we have found 9.8 ± 7.3 pmol/g for LV B'_{max} of patients versus 12.8 ± 3.9 pmol/g for LV B'_{max} in normals. Qing et al. found that for untreated asthma patients B'_{max} was not significantly different, 9.1 ± 3.3 pmol/g versus 8.8 ± 2.3 pmol/g for normals, but the measured B'_{max} decreased 19% from 8.4 ± 2.03 pmol/g in asthmatics treated with Albuterol (34). The only report using [^{11}C]-CGP12388 for imaging human cardiac disease reported a 36% decrease in B'_{max} in idiopathic cardiomyopathy, but no experimental details were provided by the authors (31). These imaging results are consistent with each other, and are also consistent with the decrease of 25% in β-AR density measured in vitro for cardiac tissue biopsies from heart failure patients (35). In these patients, the regional distribution of β-AR can be quite heterogeneous. The downregulation of β-AR in cardiomyopathies measured in vivo by imaging has provided interesting physiological confirmation of animal studies and human biopsy measures. The system is so tightly regulated that a 20% density difference is often all that is seen in many cardiac diseases from in vitro binding assays of tissue samples. Despite the small change, some clinical utility may come from imaging β-AR. For example, B'_{max} is reported to predict left ventricular volume six months postacute myocardial infarction (36). It may be that this difference is even greater in other diseases, for example, in diabetes (37).

In CHF, more β1 receptors are lost than β2 receptors, so a selective β1 imaging agent might uncover more clinically useful information than is available with the nonselective CGP12177, which images both β1 and β2 receptors. Several groups have worked to develop a β1 selective radiotracer. It is, however, likely that the new information obtained from imaging a selective β1 tracer in the heart will be marginal, because β1 receptors already compose 75% to 80% of the ventricular receptors. A selective agent that measured solely β1 could probably be only 20% more sensitive to change in β-AR. For example, a β1 selective agent might measure a 25% downregulation while a mixed β1 + β2 radiotracer might measure only a 20% downregulation. β2-AR are abundant in the lung and thus not a good cardiac imaging target due to the overwhelming lung background.

β-AR imaging may have more clinical utility in diseases where there are likely to be regional differences, such as in patients with coronary artery disease (CAD) that have some degree of scarring and contractile differences. Pathological examination of tissue sections from CAD patients has shown that innervation is spotty and heterogeneous: such adrenergic heterogeneity is potentially arrhythmogenic. In addition to the report of predicting volume, in ischemic CHF, heterogeneity of β-AR densities might be predictive of a poorer outcome. Also β-AR measured before and after treatment might prove useful to evaluate response. These are only speculations as the work has not yet been done.

It is our opinion, based upon our preliminary work and that of others, that imaging to measure global and regional β-AR densities alone is not likely to have a major clinical diagnostic role, although it is clearly useful in understanding disease processes. This does not imply that imaging β-AR is unimportant. Our hypothesis is that examination of just the pre- or postsynaptic system using imaging is not sufficient for clinical interpretation. Examination of the presynaptic system can be done with [^{11}C]-meta-hydroxyephedrine (mHED) or MIBG or several PET agents including fluorodopamine and phenylephrine. The balance between the two parts of the signaling system, pre- and postsynaptic, should be imaged to understand clinically relevant physiology and may also be useful for assessing prognosis in individuals. For example, heterogeneous mismatch of the sympathetic system in ischemic disease likely has more prognostic and mechanistic value in understanding the etiology of sudden cardiac death than do any global measures of sympathetic function.

Few studies have used imaging to evaluate both pre- and postsynaptic function in the same subjects. Nozawa et al. used salt hypertensive rats as a model to show in compensated hypertrophy that pre- and postsynaptic function was maintained, whereas neuronal loss and β-AR downregulation were significant by the time heart failure had developed (38). Mardon et al. used a similar rat model and methods to test the long standing hypothesis that increased circulating catecholamines will lead to downregulation of β-AR (measured using titrated CGP12177) and NET-1 of the presynaptic tracer, in this case [^{123}I]-MIBG (39). The hypothesis was supported: NET-1 decreased 29% to 33 % and β-AR decreased 31% to 36%. Ungerer et al. imaged presynaptic function and performed tissue assays of β-AR from biopsy samples of cardiomyopathic human heart prior to transplantation (40). The investigators verified that presynaptic imaging with mHED correlated with the density of the NET-1 sites. They also found that the β-AR and presynaptic sites were heterogeneously distributed, the β-AR less so than the NET-1 density, and most importantly that mHED and β-AR densities were poorly correlated with each other, i.e., mismatched. Schafers et al. used imaging to study both pre- and postsynaptic function in six patients with hypertrophic cardiomyopathy (41). These authors found that NET-1 volume of distribution decreased to 33.4 ± 4.3 mL/g (n = 9) from 71.0 ± 18.8 mL/g of tissue (n = 13) in normals and β-AR decreased

to 7.3 ± 2.6 pmol/g (n = 10) from 10.2 ± 2.9 pmol/g tissue (n = 19) in normals. In the six patients who had both pre- and postsynaptic studies, there was no correlation between their pre- and postsynaptic measures.

MEASURING PRE- AND POSTSYNAPTIC FUNCTION AT THE UNIVERSITY OF WASHINGTON

We have imaged both pre- and postsynaptic function in 20 healthy sedentary older volunteers (65 to 79 years) and 10 patients with CAD and a left ventricular ejection fraction <35% (60 to 75 years, p = ns vs. normals), including some subjects whose results were previously reported (42). Five of the patients were on the beta blocker metoprolol, but had not taken the drug for more than 24 hours. No patient was taking medications known to directly affect presynaptic sympathetic function. Postsynaptic function (B'$_{max}$) was measured by imaging [^{11}C]-CGP12177 and analyzing the data using the graphical method and in-house software (24,25,27,43). Presynaptic function was measured using imaging of [^{11}C]-mHED, a NE analog, and the metabolite-corrected mHED left atrial blood curve as input for either the calculation of RF (44) or our distributed model to estimate neuronal uptake of mHED into the nerve terminal (PS$_{nt}$, mL/min/g) (45).

As a control, myocardial blood flow (MBF) was measured by imaging [^{15}O]-water (46,47). The radiopharmaceuticals were synthesized and purified, and the metabolites of mHED were measured as described previously (42,48,49). A GE Advance tomograph was used with rapid dynamic imaging in a ~2.5 hour protocol. For analysis, each dynamic PET image was reconstructed with a 10 mm Hanning filter and reoriented into short-axis cardiac projections and the heart images were divided into 12 regions that excluded the most apical regions as well as regions of myocardial infarction [MBF < 0.16 (mL/min/g)] in the CHF subjects. The mean global MBF at rest was 0.76 ± 0.23 mL/min/g in the CHF patients versus 0.78 ± 0.16 mL/min/g in the normals (p = ns). As illustrated in Figure 1A, MBF was not significantly different in any regions between normals and CHF patients, verifying that there were no large regions of infarcted tissue.

Global PS$_{nt}$ for mHED was significantly lower in CHF (0.22 ± 0.09 mL/min/g) than in normals (0.77 ± 0.29, p < 0.001, Fig. 1B) and regional PS$_{nt}$ was lower in 6 of 12 regions. RF showed similar results. In contrast, global β-AR density (B'$_{max}$) (Fig. 1C), tended to be lower, although not significantly different, in the CHF patients vs. normals, 9.8 ± 7.3 vs. 12.8 ± 3.9 pmol/mL, respectively. Regionally, the same trend was seen. There was no difference in mean B'$_{max}$ between patients who had previously taken beta-blockers and those who had not. The global mismatch between pre- and

Average function

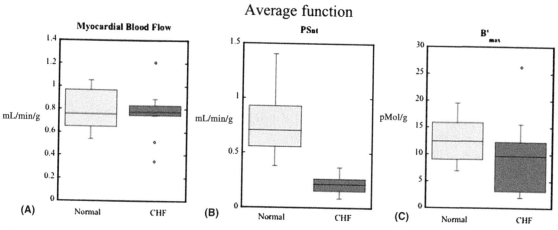

(A) Normal CHF   **(B)** Normal CHF   **(C)** Normal CHF

Figure 1 Box plots for the average values for the patient groups for the measured function for the entire left ventricle. The horizontal line is the median value. The boxes encompass 25% to 75% of the values and the error bars show values within 1.5 times the box limit. Outliers are shown as *open circles*. (**A**) No significant difference in myocardial blood flow between the normal and CHF subjects. (**B**) The significant difference between normal and CHF subjects. (**C**) The trend toward downregulation of the β-AR density in the CHF compared with the normal subjects.

postsynaptic function was expressed as B'_{max}:PS_{nt} ratio. Compared to normals, patients with CHF had both a greater B'_{max}:PS_{nt} ratio (86 ± 88 vs. 20 ± 8) and a much more variable ratio as evidenced by the standard deviation. Figure 2 shows mismatch in a graph of the 12 regions of the left ventricle for each subject. In the normals, B'_{max}:RF was homogenous over all 12 regions. In contrast, B'_{max}:RF mismatch was greater and much more variable in the majority of regions in most of the CHF patients compared to the normals, This difference reached statistical significance in 6 of the 12 regions (3 lateral and 3 inferior). The difference between normal

and CHF subjects is dominated by the presynaptic component and the extent of mismatch between pre- and postsynaptic function is highly variable between individual patients with CHF and normal subjects. The variability in mismatch shown in Figure 2 may be a visual indicator of potential arrhythmogenicity and should be further evaluated.

There are important clinical implications for these findings. First, the study demonstrates clinically relevant differences in pre- and postsynaptic function, when measured sequentially and within a short time period, between patients with ischemic CHF and

Mismatch between post- and pre-synaptic function

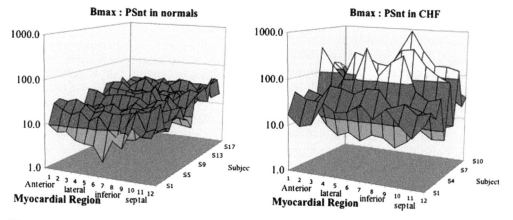

Figure 2 The difference in mismatch, presented as the ratio of postsynaptic (B'_{max}) to presynaptic (PS_{nt}) function, between the normal subjects and those with CHF. The 12 left ventricular regions for an individual subject are plotted on the X axis. The individual subjects are plotted on the Y axis and denoted as S followed by a #. The Z vertical axis shows the extent of mismatch and is a logarithmic scale. The normal subjects are on the *left* and the CHF patients are on the *right*. The figures show that the postsynaptic to presynaptic ratio is relatively uniform between regions and across normal subjects (*left panel*). In contrast, the ratio of post- to presynaptic function is greater in almost all of the CHF patients with much more heterogeneity between regions and between subjects than is seen in the normal subjects.

age-matched normals. Decreased presynaptic function measured by mHED imaging has been previously reported in patients with ischemic heart disease (49–52). Secondly, in our patients with CHF, global B'$_{max}$ was 25% lower than in the normals and similarly decreased for all regions. Although this was not statistically significant, it is directionally consistent with results from in vitro assays of biopsy samples (7,23,53).

The mechanism for the observed mismatch is unclear. The presynaptic component of the cardiac sympathetic nervous system has been shown to be very sensitive to ischemic insult (54–56). Impaired reuptake because of partial sympathetic nerve dysfunction could lead to excess local catecholamine concentrations in the myo-neuronal junction with a compensatory decrease in β-AR density. The small decrease is consistent with several previous reports (7,55). The extent and/or location of reinnervation after an ischemic event could also affect mismatch and the potential for generation of arrhythmias.

SUMMARY

Marked regional mismatch between pre- and postsynaptic left ventricular sympathetic function can be detected by imaging and is present in patients with severe CHF. This could serve as a mechanism for inducing lethal ventricular arrhythmias. We have demonstrated, using PET imaging, that global and regional presynaptic function is decreased in patients with ischemic CHF as compared to age-matched normals, whereas postsynaptic β-AR density is slightly but not significantly decreased. This results in a mismatch between pre- and postsynaptic function that varies regionally and could potentially produce local myocardial conditions that are arrhythmogenic and should be studied further. We have also demonstrated that this PET protocol can be done in a relatively short time that minimizes potential for changes in sympathetic function and is clinically acceptable. The radiotracers do not cause clinically significant hemodynamic changes, even in patients with moderately severe CHF.

In addition to the need for more studies of both pre- and postsynaptic β-AR in the same patients with heart disease to evaluate clinical utility, other sympathetic studies may prove useful that have yet to be done but are clearly needed (57,58). The matching or mismatching of presynaptic function to postsynaptic α-AR may also have value for assessing prognosis. The challenging experiment of evaluating the second messenger system induced by β-AR using imaging has potential value in cardiac disease, especially CHF, but has not yet been attempted. We are at a stage where, as more of the cardiac molecular pathways are determined, we have the opportunity to evaluate them in humans to understand the functional balance of these pathways in health and disease, not just their test tube concentrations. These are important goals and a challenge for the future that will require considerable work but will be worth the effort.

ACKNOWLEDGMENTS

We gratefully acknowledge the numerous discussions that led to this work from Ken Krohn, James Bassingthwaighte, Itamar Abrass, Michael Bristow and Christopher Rhodes. We also thank the NIH for support for our work: Grants R01 HL50239 and AG15462. We acknowledge the many contributions of Mr. Erik Butterworth, Ms. Marilou Gronka and Ms. Katherine Seymour in data collection and image analysis, Ms. Barbara Lewellen and the other PET suite personnel, and Steve Shoner and other radiochemists for radiochemical synthesis and quality control.

REFERENCES

1. Brodde OE. Beta 1– and beta 2–adrenoceptors in the human heart: properties, function, and alterations in chronic heart failure. Pharmacol Rev 1991; 43(2): 203–242.
2. Rockman HA, Koch WJ, Lefkowitz RJ. Seven-transmembrane-spanning receptors and heart function. Nature 2002; 415(6868):206–212.
3. Dzimiri N. Regulation of β-adrenoceptor signaling in cardiac function and disease. Pharmacol Rev 1999; 51(3): 465–502.
4. Bers DM. Cardiac excitation-contraction coupling. Nature 2002; 415:198–205.
5. Liggett SB. β-Adrenergic receptors in the failing heart: the good, the bad, and the unknown. J Clin Invest 2001; 107:947–948.
6. Engelhardt S, et al. Progressive hypertrophy and heart failure in beta1-adrenergic receptor transgenic mice. Proc Natl Acad Sci U S A 1999; 96(12):7059–7064.
7. Bristow MR, et al. Differences in beta-adrenergic neuroeffector mechanisms in ischemic versus idiopathic dilated cardiomyopathy. Circulation 1991; 84(3): 1024–1039.
8. Bristow MR. Pathophysiologic and pharmacologic rationales for clinical management of chronic heart failure with beta-blocking agents. Am J Cardiol 1993; 71(9):12C–22C.
9. Fowler MB, et al. Assessment of the β-adrenergic receptor pathway in the intact failing human heart: progressive receptor down-regulation and subsensitivity to agonist response. Circulation 1986; 74(6): 1290–1302.
10. Lakatta EG, Sollott SJ. Perspective on mammalian cardiovascular aging: humans to molecules. Comp Biochem Physiol; A Molecular Integrative Physiology 2002; 132(Part A):699–721.

11. Sosunov EA, et al. β1 and β2–adrenergic receptor subtype effects in German shepherd dogs with inherited lethal ventricular arrhythmias. Cardiovas Res 2000; 48:211–219.

12. Capricorn I. Effect of carvedilol on outcome after myocardial infarction in patients with left-ventricular dysfunction: the CAPRICORN randomised trial. Lancet 2001; 357:1385–1390.

13. Ellis K, et al. Mortality benefit of beta blockade in patients with acute coronary syndromes undergoing coronary intervention: pooled results from the EPIC, EPILOG, EPISTENT, CAPTURE and RAPPORT trials. J Interv Cardiol 2004; 16:299–305.

14. Kobilka B. Adrenergic and muscarinic receptors of the heart. In: Roberts R, ed. Molecular Basis of Cardiology. Blackwell Publishing, 1993.

15. Cooper JR. Receptors. In: Cooper JR, Bloom FE, Roth RH, eds. The Biochemical Basis of Neuropharmacology, 6 ed. Oxford, England: Oxford University Press, 1991:88–110.

16. Link JM, Krohn KA. Interpreting enzyme and receptor kinetics: keeping it simple, but not too simple. Nucl Med Biol 2003; 30(8):819–826.

17. Syrota A. In vivo investigation of myocardial perfusion, metabolism and receptors by positron emission tomography. Int J Microcirc Clin Exp 1988; 8:411–422.

18. Lammertsma AA. Radioligand studies: imaging and quantitative analysis. Eur J Neuropsychopharmacol 2002; 12(6):513–516.

19. van Waarde A, et al. Uptake of radioligands by rat heart and lung in vivo: CGP 12177 does and CGP 26505 does not reflect binding to beta-adrenoceptors. Eur J Pharmacol 1992; 222(1):107–112.

20. Contreras ML, Wolfe BB, Molinoff PB. Kinetic analysis of the interactions of agonists and antagonists with beta adrenergic receptors. J Pharmacol Exp Therap 1986; 239:136–143.

21. Elsinga PH, et al. Synthesis and evaluation of (S)-4-(3-(2'-[^{11}C]isopropylamino)-2-hydroxypropoxy)-2H-benzimidazol-2-one ((S)-[^{11}C]CGP 12388) and (S)-4-(3-((1'-[^{18}F]-fluoroisopropyl)amino)-2-hydroxypropoxy)-2H- benzimidazol-2-one ((S)-[^{18}F]fluoro-CGP 12388) for visualization of beta-adrenoceptors with positron emission tomography. J Med Chem 1997; 40(23):3829–3835.

22. Delforge J, Syrota A, Mazoyer BM. Identifiability analysis and parameter identification of an in vivo ligand-receptor model from PET data. IEEE Trans Biomed Eng 1990; 37:653–661.

23. Merlet P, et al. Positron emission tomography with ^{11}C CGP-12177 to assess β-adrenergic receptor concentration in idiopathic dilated cardiomyopathy. Circulation 1993; 87:1169–1178.

24. Delforge J, et al. Cardiac beta-adrenergic receptor density measured in vivo using PET, CGP 12177, and a new graphical method. J Nucl Med 1991; 32:739–748.

25. Delforge J, et al. In vivo quantification and parametric images of the cardiac β-adrenergic receptor density. J Nucl Med 2002; 43:215–226.

26. Boullais C, Crouzel C, Syrota A. Synthesis of 4-(3-t-butylamino-2-hydroxypropoxy)-benzimidazol-2[C-11]-1 (CGP 12177). J Label Comp Radiopharm 1986; 26:490–491.

27. Delforge J. Correction of a relationship that assesses beta-adrenergic receptor concentration with PET and carbon-11-CGP 12177. J Nucl Med 1994; 35(5):921.

28. Lefroy DC, et al. Diffuse reduction of myocardial beta-adrenoceptors in hypertrophic cardiomyopathy: a study with positron emission tomography. J Am Coll Cardiol 1993; 22(6):1653–1660.

29. Doze P, et al. Quantification of beta-adrenoceptor density in the human heart with (S)-[^{11}C]CGP 12388 and a tracer kinetic model. Eur J Nucl Med 2002; 29(3):295–304.

30. Riemann B, et al. High non-specific binding of the β1-selective radioligand 2-^{125}I-ICI-H. Nuklearmedizin 2003; 42:173–180.

31. Elsinga PH, van Waarde A, Vaalburg W. Receptor imaging in the thorax with PET. Eur J Pharmacol 2004; 499:1–13.

32. Schafers M, et al. Myocardial presynaptic and post-synaptic autonomic dysfunction in hypertrophic cardiomyopathy. Circ Res 1998; 82(1):57–62.

33. Wichter T, et al. Abnormalities of cardiac sympathetic innervation in arrhythmogenic right ventricular cardio-myopathy: quantitative assessment of presynaptic nor-epinephrine reuptake and postsynaptic beta-adrenergic receptor density with positron emission tomography. Circulation 2000; 101(13):1552–1558.

34. Qing F, et al. Effect of long-term beta2-agonist dosing on human cardiac beta-adrenoceptor expression in vivo: comparison with changes in lung and mononuclear leukocyte beta-receptors. J Nucl Cardiol 1997; 4(6):532–538.

35. Bristow MR, et al. β1- and β2-adrenergic receptor subpopulations in nonfailing and failing human ventricular myocardium: coupling of both receptor subtypes to muscle contraction and selective β1-adrenoceptor down-regulation in heart failure. Circ Res 1986; 59:297–309.

36. Spyrou N, et al. Myocardial beta-adrenoceptor density one month after acute myocardial infarction predicts left ventricular volumes at six months. J Am Coll Cardiol 2002; 40(7):1216–1224.

37. Matsuda N, et al. Diabetes-induced down-regulation of beta1-adrenoceptor mRNA expression in rat heart. Biochem Pharmacol 1999; 58(5):881–885.

38. Nozawa T, et al. Dual-tracer assessment of coupling between cardiac sympathetic neuronal function and downregulation of β-receptors during development of hypertensive heart failure of rats. Circulation 1998; 97:2359–2367.

39. Mardon K, et al. Uptake-1 carrier downregulates in parallel with the beta-adrenergic receptor desensitization in rat hearts chronically exposed to high levels of circulating norepinephrine: implications for cardiac neuroimaging in human cardiomyopathies. J Nucl Med 2003; 44(9):1459–1466.

40. Ungerer M, et al. Regional in vivo and in vitro characterization of autonomic innervation in cardiomyopathic human heart. Circulation 1998; 97(2):174–180.

41. Schafers M, et al. Myocardial pre- and post-synaptic autonomic dysfunction in patients with hypertrophic cardiomyopathy assessed by PET with C-11–CGP 12177 and C-11–hydroxyephedrine. J Nucl Med 1996; 37:71.

42. Link JM, et al. PET measures of pre- and post-synaptic cardiac beta adrenergic function. Nucl Med Biol 2003; 30(8):795–803.

43. Butterworth EA, Link JM, Caldwell JH. Automated left ventricular region of interest: definition on cardiac PET images reduces operator time. J Nucl Med 1999; 40:239.

44. Schwaiger M, et al. Noninvasive evaluation of sympathetic nervous system in human heart by positron emission tomography. Circulation 1990; 82:457–464.

45. Caldwell JH, et al. [C-11]-Meta-hydroxyphedrine (MHED) transport capacities and volumes of distribution by PET during catecholamine stimulation. J Nucl Med 1996; 37:70.

46. Bergmann SR, et al. Noninvasive quantitation of myocardial blood flow in human subjects with oxygen-15-labeled water and positron emission tomography. J Am Coll Cardiol 1989; 14:639–652.

47. Herrero P, Markham J, Bergmann SR. Quantitation of myocardial blood flow with $H2[^{15}O]$ and positron emission tomography: assessment and error analysis of a mathematical approach. J Comput Assist Tomogr 1989; 13:862–873.

48. Link JM, Caldwell JH, Krohn KA. Catalysts and carrier carbon in production of ^{11}C-phosgene. J Nucl Med 2001; 42(suppl 5):20.

49. Link JM, et al. High speed liquid chromatography of phenylethanolamines for the kinetic analysis of [^{11}C]-meta-hydroxyephedrine and metabolites in plasma. J Chromatog B 1997; 693:31–41.

50. Allman KC, et al. Carbon-11 hydroxyephedrine with positron emission tomography for serial assessment of cardiac adrenergic neuronal function after acute myocardial infarction in humans. J Am Coll Cardiol 1993; 22:368–375.

51. Caldwell JH, et al. Evaluation of pre- and post-synaptic sympathetic function in normals and subjects with CAD and LV dysfunction. J Nucl Med 2002; 43(5):141.

52. Caldwell JH, et al. Global and regional pre- and post-synaptic function in sudden cardiac death survivors. J Nucl Cardiol 2001; 8:S82.

53. White DC, et al. Preservation of myocardial beta-adrenergic receptor signaling delays the development of heart failure after myocardial infarction. Proc Natl Acad Sci U S A 2000; 97(10):5428–5433.

54. Inoue H, Zipes DP. Time course of denervation of efferent sympathetic and vagal nerves after occlusion of the coronary artery in the canine heart. Circ Res 1988; 62:1111–1120.

55. Kozlovskis PL, et al. Regional beta-adrenergic receptors and adenylate cyclase activity after healing of myocardial infarction in cats. J Mol Cell Cardiol 1990; 22(3):311–322.

56. Matsunari I, et al. Extent of cardiac sympathetic neuronal damage is determined by the area of ischemia in patients with acute coronary syndromes. Circulation 2000; 101(22):2579–2585.

57. Gordeladze JO, et al. Enhanced responsiveness of the myocardial beta-adrenoceptor-adenylate cyclase system in the perfused rat heart (I). Biosci Rep 1998; 18(5): 229–250.

58. Morita K, Kuge Y, Tamaki N. What is the clinical role of neuronal imaging? J Nucl Med 2003; 44(9):1457–1468.

26

SPECT Imaging of Cardiac Adrenergic Receptors

Ignasi Carrió
Nuclear Medicine Department, Autonomous University of Barcelona, Hospital Sant Pau, Barcelona, Spain

SUMMARY

Cardiac neurotransmission imaging with single photon emission computed tomography (SPECT) allows in vivo assessment of presynaptic reuptake and storage of neurotransmitters (Figs. 1 and 2). SPECT of cardiac neurotransmission offers characterization of the myocardial neuronal function in primary cardioneuropathies, in which there is no significant structural abnormality of the heart, and in secondary cardioneuropathies caused by the metabolic and functional changes that take place in different diseases of the heart. In patients with heart failure, the assessment of sympathetic activity has important prognostic implications and will result in better therapy and outcome. In diabetic patients, scintigraphic techniques allow the detection of autonomic neuropathy in early stages of the disease. In conditions with risk of sudden death such as idiopathic ventricular tachycardia and arrhythmogenic right ventricular cardiomyopathy, SPECT demonstrates altered neuronal function when no other structural abnormality is seen. In patients with ischemic heart disease, heart transplantation, drug induced cardiotoxicity, and dysautonomias, the assessment of neuronal function can be helpful in the characterization of the disease and improved prognostic stratification.

Future directions in cardiac neurotransmission SPECT include the development of tracers for new types of receptors, targeting of second messenger molecules, and the early assessment of cardiac neurotransmission in genetically predisposed subjects for prevention and early treatment of heart failure.

CARDIAC NEUROTRANSMISSION SPECT IN PRIMARY CARDIONEUROPATHIES

Dysautonomias

Cardiac ^{123}I-MIBG uptake may be impaired in some patients with neurologic disorders of the central and peripheral nervous system with autonomic dysfunction. In clinical practice, it is difficult to identify the cause of a parkinsonian syndrome in many patients, in particular when symptoms have begun recently and are confined to the extrapyramidal system. Even with longer duration of symptoms, about 24% of patients with parkinsonian syndrome are misdiagnosed as Parkinson's disease (PD) and subsequently discovered to have alternative diseases such as multiple system atrophy (MSA), striatonigral degeneration (SND) and progressive supranuclear palsy (PSP). The differentiation is particularly difficult in patients with PD also showing symptoms of autonomic failure, which is not infrequent. Selective investigation of postganglionic cardiac neurons possibly enables a safe differentiation between both entities (1) as patients with PD show reduced ^{123}I-MIBG uptake, independently of duration and severity of autonomic and parkinsonian symptoms. Furthermore, the mean value of heart to mediastinum ratio (HMR) in patients with PD is significantly lower than those with SND, PSP, or no disease, regardless of disease severity or intensity of anti-parkinsonian treatment (2). Therefore, cardiac scintigraphy with ^{123}I-MIBG may be used in the diagnosis and characterization of akinetic-rigid syndromes, especially PD (Fig. 3).

Heart Transplantation

During orthotopic heart transplantation, the entire recipient heart is excised except for the posterior atrial walls, to which the donor atria are anastomosed. During the process, the allograft becomes completely denervated. Lack of autonomic nerve supply is associated with major physiologic limitations. The inability to perceive pain does not allow symptomatic recognition of accelerated allograft vasculopathy, and heart transplant patients often develop acute ischemic events or left ventricular dysfuction or die suddenly. In addition, denervation of the sinus node does not allow adequate acceleration of heart rate during stress and efficient increase in cardiac output. Furthermore, loss of vasomotor tone may adversely affect the physiologic alterations in blood flow, produce altered hemodynamic performance at rest and during exercise, and decrease exercise capacity. Scintigraphic uptake of ^{123}I-MIBG supports the concept of spontaneous reinnervation

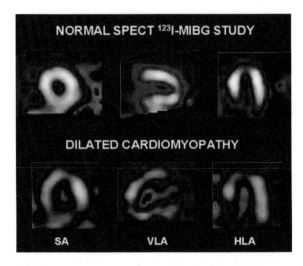

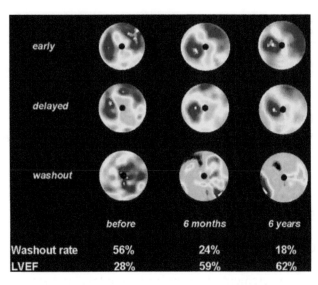

Figure 1 Normal MIBG Study: SPECT images of the heart show homogeneous distribution of ¹²³I-MIBG over the myocardial wall. Heart to mediastinum ratio was 1.96. Dilated cardiomyopathy: SPECT images of the heart show heterogeneous distribution of ¹²³I-MIBG over the myocardial wall in a dilated heart. Heart to mediastinum ratio was 1.40.

Figure 2 ¹²³I-MIBG SPECT polar maps of patients used to demonstrate regional distribution and wash-out rates.

taking place after transplantation (3–5). All studies performed up to five years after heart transplantation suggest that reinnervation is likely to be a slow process and occurs only after one year post-transplantation (6).

Idiopathic Ventricular Tachycardia and Fibrillation
In patients presenting with idiopathic ventricular tachycardia and fibrillation, no structural or functional abnormalities of the myocardium can be demonstrated by conventional imaging and clinical testing. Early diagnosis and treatment are of clinical importance because ventricular fibrillation is the most common arrhythmia at the time of sudden death (7). Typical arrhythmias in these patients can be provoked by physical or mental stress or by catecholamine application. Schaffers et al. (8), using ¹²³I-MIBG, ¹¹C-hydroxyephedrine and

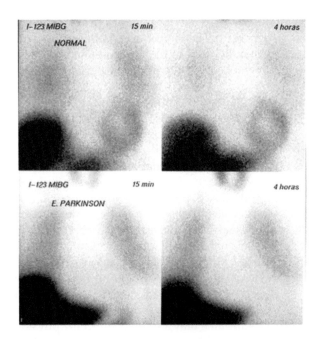

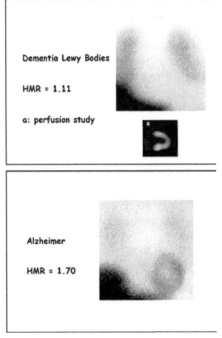

Figure 3 Cardiac MIBG in Parkinson's disease: MIBG image of the heart in a patient with Parkinson's disease show severely reduced MIBG uptake in the heart, as compared to a normal control, indicating damage of cardiac postganglionic sympathetic nerves. Cardiac MIBG image of a patient with Lewy Bodies dementia showing absent myocardial uptake, note the normal perfusion scan of the same patient. Cardiac MIBG image of a patient with Alzheimer's disease showing normal MIBG uptake and distribution.

[11]C-CGP, have demonstrated that in patients with idiopathic right ventricular outflow tract tachycardia, both the presynaptic myocardial catecholamine reuptake and the postsynaptic myocardial β-adrenoceptors density are reduced despite normal blood catecholamine levels. These scintigraphic findings represent the only demonstrable myocardial abnormality in patients with idiopathic tachycardia and fibrillation, and suggest that myocardial β-adrenoceptor downregulation in these patients occurs subsequently to increased local synaptic catecholamine levels caused by impaired catecholamine reuptake (9).

CARDIAC NEUROTRANSMISSION SPECT IN SECONDARY CARDIOMYOPATHIES

Dilated Cardiomyopathies

After the onset of myocardial failure, enhanced sympathetic nervous system activity plays an important role in supporting the cardiovascular system by increasing heart rate, contractility, and venous return. Blood pressure to preserve organ perfusion is supported by systemic arterial constriction. But on the other hand, increased sympathetic activity has deleterious effects on the cardiovascular system. Altered sympathetic cardiac adrenergic function may also cause arrhythmias, desensetization of postsynaptic beta-adrenoceptors, and activation of other neurohumoral systems such as the renin-angiotensin system, which may themselves exert adverse effects and contribute to progression of myocardial dysfunction. In addition, prolonged exposure to norepinephrine may contribute to disease progression by acting directly on the myocardium to modify cellular phenotype and result in myocyte death (10,11).

In patients with dilated cardiomyopathies (DCM), because of the increased concentration of circulating catecholamines resulting from heart failure, the myocardial responsiveness to beta-adrenoceptor agonists is blunted. Frequently, alterations of cardiac sympathetic innervation contribute to fatal outcomes in patients with heart failure. Merlet et al. (12–14) studying patients with functional classes II-IV and left ventricular ejection fraction less than 40%, and using as endpoints transplantation and cardiac or noncardiac death, have demonstrated that the only independent predictors of mortality are low metaiodobenzylguanidine (MIBG) uptake and left ventricular ejection fraction. In addition, MIBG uptake and circulating norepinephrine concentration were the only predictors for life duration when using multivariate life table analysis. Therefore, it seems that impaired cardiac adrenergic innervation as assessed by MIBG imaging is strongly related to mortality in patients with heart failure. Along the same line, Maunory et al. (15) have described that cardiac adrenergic neuronal function is impaired in children with idiopathic dilated cardiomyopathy. Patients with congestive cardiomyopathy typically have accelerated wash-out rates (>25% from 15 to 85 minutes) as compared to controls (<10%). Wakabayashi et al. (16) reported ^{123}I-MIBG imaging as the most powerful independent long-term prognostic value for both ischemic and idiopathic cardiomyopathy patients, which hints at cardiac autonomic dysfunction as a common endpoint leading to cardiac death, regardless of the underlying etiology of the cardiac disease.

Gerson et al. (17) studied the effect of chronic carvedilol treatment in patients with heart failure and cardiac sympathetic nerve dysfunction of varying severity due to idiopathic cardiomyopathy. Most patients showed a favorable response in LV function to carvedilol treatment, regardless of the baseline level of cardiac sympathetic nervous system function, as assessed by cardiac ^{123}I-MIBG imaging. Patients with relatively advanced cardiac sympathetic dysfunction (baseline HMR <1.40 in MIBG studies) were the most likely to show evidence of improved cardiac sympathetic nervous system function in response to carvedilol therapy, indicating that impairment of cardiac sympathetic nerve function is reversible in these patients. Other studies have shown improvement of cardiac MIBG parameters in response to treatment performed in patients treated with various β-adrenergic blockers (18–21).

Coronary Artery Disease

The sympathetic nervous tissue may be more sensitive to the effects of ischemia than the myocardial tissue. It has been shown that the uptake of ^{123}I-MIBG is significantly reduced in the areas of myocardial infarction, and acute and chronic ischemia (22–24). A decrease in MIBG uptake in ischemic tissue represents the loss of integrity of postganglionic, presynaptic neurones (Fig. 4). It is likely that ischemia induces damage to sympathetic neurons, which may take a long time to regenerate and that repetitive episodes of ischemia should result in a relatively permanent loss of MIBG uptake. Early after infarction, sympathetic denervation in adjacent noninfarcted regions is frequently observed (25).

Several studies have shown (26) reduced MIBG uptake and a more extensive area of reduced uptake than the thallium perfusion defect. These regions may be associated with spontaneous ventricular tachyarrhythmias after myocardial infarction (27). The eventual concordance between the extent of MIBG defect at rest and perfusion defect at exercise in patients with coronary artery disease, suggests that mild degrees of ischemia may be injurious to the integrity of myocardial sympathetic neurones (28–30). Correlation between

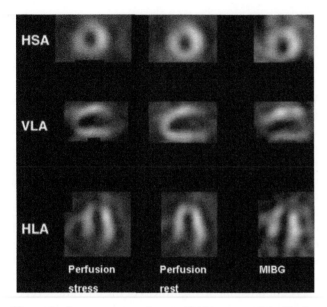

Figure 4 Coronary artery disease (CAD): myocardial perfusion (⁹⁹ᵐTc-Tetrofosmin) and ¹²³I-MIBG SPECT images of a patient with CAD show stress inducible ischemia over the infero-apical wall. Note that the MIBG defect clearly matches the region with stress inducible ischemia indicating neuronal damage.

the MIBG defect and the area at risk and the presence of angina further supports the concept of neuronal damage in the ischemic territory.

Gaudino et al. (31) have provided evidence of good correspondence between ¹²³I-MIBG imaging and the presence or absence of sympathetic cardiac nerves by direct immunohistochemical staining in patients with LV aneurysms due to long-lasting anterior myocardial infarction. The authors indicate that most of the original nerves are destroyed by the ischemic insult and that aneurysm-associated nerves are the result of an anomalous reinnervation process. Furthermore, they observed a ¹²³I-MIBG/²⁰¹Tl mismatch (¹²³I-MIBG defect where ²⁰¹Tl uptake is normal) in areas distant from the previous myocardial infarction (inferior wall), hypothesizing that at least some of the sympathetic nerves innervating the inferior wall have a course through the anterior wall, following the annular and/or oblique shape of the muscular structures of the myocardium rather than the longitudinal course of the epicardial vessels.

Sudden death accounts for 50% of all cardiovascular deaths in developed countries. Most of sudden cardiac deaths result from ventricular tachyarrhythmias. While prevention of arrhythmic deaths is ineffective with pharmacologic treatment, implantable cardioverter defibrillators (ICD) reduce the mortality rate in subgroups of patients thought to be at risk. However, identification of patients who most benefit from these devices remains difficult and implantation of the device in patients who will not benefit, leads to unnecessary morbidity with increased medical costs. Unfortunately, current available methods for precise identification of candidates at risk for sudden death, such as conventional risk factors and standard cardiovascular testing (e.g., assessment of ventricular function, Holter monitoring, signal-averaged electrocardiography, and electrophysiologic testing) are inadequate. The autonomic nervous system plays an important role in triggering and sustaining malignant ventricular arrhythmias in patients with susceptible substrate. Along this line, Arora et al. (32) recently evaluated the use of ¹²³I-MIBG myocardial imaging (as means of local myocardial sympathetic innervation) and spectral analysis of heart rate variability (HRV) (as means of central autonomic tone) in patients with implantable defibrillators. The combined use of ¹²³I-MIBG scintigraphy and HRV analysis correlated with the occurrence of an appropriate discharge. ICD discharges were associated with lower HMR of early ¹²³I-MIBG imaging, more extensive ¹²³I-MIBG defects and more ¹²³I-MIBG/⁹⁹ᵐTc-sestamibi mismatch segments (denervation in areas of myocardial viability), as compared with patients without previous ICD discharge. Furthermore, all patients with ¹²³I-MIBG uptake and an HRV value below the mean had an ICD discharge, and conversely, no patients with ¹²³I-MIBG uptake and an HRV value above the mean had an ICD discharge. These results indicate that the combined noninvasive evaluation of local cardiac autonomic innervation and systemic autonomic function by means of ¹²³I-MIBG and HRV allows identification of patients at risk for potentially fatal arrhythmias and sudden cardiac death, who are most likely to benefit from an ICD.

REFERENCES

1. Braune S, Reinhardt M, Schnitzer R, Riedel A, Lücking CH. Cardiac uptake of [123I]MIBG separates Parkinson's disease from multiple system atrophy. Neurology 1999; 53:1020–1025.
2. Yoshita M. Differentiation of idiopathic Parkinson's disease from striatonigral degeneration and progressive supranuclear palsy using iodine-123 meta-iodobenzylguanidine myocardial scintigraphy. J Neurol Sci 1998; 155:60–67.
3. De Marco T, Dae M, Yuen MS, et al. Iodine-123 MIBG scintigraphic assessment of the transplanted human heart: evidence for late reinnervation. J Am Coll Cardiol 1995; 25:927–931.
4. Schwaiger M, Hutchins GB, Kalff V. Evidence for regional caatecholamine uptake and storage sites in the transplanted human heart by positron emission tomography. J Clin Invest 1991; 87:1681–1690.
5. Dae M, DeMarco T, Botvinick E, et al. Scintigraphic assessment of MIBG uptake in globally denervated

human and canine hearts: implications and clinical studies. J Nucl Med 1992; 33:1444–1450.

6. Estorch M, Campreciós M, Flotats A, et al. Sympathetic reinnervation of cardiac allografts evaluated by 123I-MIBG imaging. J Nucl Med 1999; 40:911–916.

7. Lerch H, Wichter T, Schamberger R, et al. Sympathetic myocardial innervation in idiopathic ventricular tachycardia and fibrillation. Eur J Nucl Med 1995; 22:805.

8. Wichter T, Lerch H, Schafers M, et al. reduction of postsynaptic beta-receptor density in arrhythmogenic right ventricular dysplasia: assessment with positron emission tomography. Circulation 1996; 94(suppl I):I-543.

9. Schafers M, Lerch H, Wichter T, et al. cardiac sympathetic innervation in patients with idiopathic right ventricular outflow tract tachycardia. J Am Coll Cardiol 1998; 32:181–186.

10. Ungerer M, Bohm M, Elce J, et al. Altered expression of beta-adrenergic receptor kinase and beta-adrenergic receptors in the failing human heart. Circulation 1993; 87:454–463.

11. Henderson EB, Kahn JK, Corbet J, et al. Abnormal I-123-MIBG myocardial wash-out and distribution may reflect myocardial adrenergic derangement in patients with congestive cardiomyopathy. Circulation 1988; 78:1192–1199.

12. Merlet P, Dubois JL, Adnot S, et al. Myocardial beta-adrenergic desensitization and neuronal norepinephrine uptake function in idiopathic dilated cardiomyopathy. J Cardiovasc Pharmacol 1992; 19:10–16.

13. Merlet P, Valette H, Dubois JL, et al. Prognostic value of cardiac metaiodobenzylguanidine imaging in patients with heart failure. J Nucl Med 1992; 33:471–477.

14. Merlet P, Benvenuti C, Moyse D, et al. Prognostic value of MIBG imaging in idiopathic dilated cardiomyopathy. J Nucl Med 1999; 40:917–923.

15. Maunoury C, Agostini D, Acar Ph, et al. Impairment of cardiac neuronal function in childhood dilated cardiomyopathy: an I123–MIBG scintigraphic study. J Nucl Med 2000; 41:400–404.

16. Wakabayashi T, Nakata T, Hashimoto A, et al. Assessment of underlying etiology and cardiac sympathetic innervation to identify patients at high risk of cardiac death. J Nucl Med 2001; 42:1757–1767.

17. Gerson MC, Craft LL, McGuire N, et al. Carvedilol improves left ventricular function in heart failure patients with idiopathic dilated cardiomyopathy and a wide range of sympathetic nervous system function as measured by iodine123 metaiodobenzylguanidine. J Nucl Cardiol 2002; 9:608–615.

18. Agostini D, Belin A, Amar MH, et al. Improvement of cardiac neuronal function after carvedilol treatment in dilated cardiomyopathy: a 123I-MIBG scintigraphic study. J Nucl Med 2000; 41:845–851.

19. Lotze U, Kaepplinger S, Kober A, Richartz BM, Gottschild D, Figulla HR. Recovery of the cardiac adrenergic nervous system after long-term beta blocker therapy in idiopathic dilated cardiomyopathy: assessment by increase in myocardial 123I-metaiodobenzylguanidine uptake. J Nucl Med 2001; 42:49–54.

20. Kasama S, Toyama T, Kumakura H, et al. Effect of spironolactone on cardiac sympathetic nerve activity and left ventricular remodeling in patients with dilated cardiomyopathy. J Am Coll Cardiol 2003; 41:574–581.

21. Yamada T, Shimonagata T, Fukunami M, et al. Comparison of the prognostic value of cardiac Iodine-123 metaiodobenzylguanidine imaging and heart rate variability in patients with chronic heart failure. A prospective study. J Am Coll Cardiol 2003; 41:231–238.

22. Fagret D, Wolf JE, Comet M. Myocardial uptake of meta-[123I]-iodobenzylguanidine ([123I]-MIBG) in patients with myocardial infarct. Eur J Nucl Med 1989; 15:624–628.

23. McGhie AI, Corbett JR, Akers MS, et al. Regional cardiac adrenergic function using I-123 meta-iodobenzylguanidine tomographic imaging after acute myocardial infarction. Am J Cardiol 1991; 67:236–242.

24. Nishimura T, Oka H, Sago M, et al. Serial assessment of denervated but viable myocardium following acute myocardial infarction in dogs using iodine-123 MIBG and thallium-201 chloride myocardial single photon emission tomography. Eur J Nucl Med 1992; 19:25–29.

25. Minardo JD, Tuli MM, Mock BH, et al. Scintigraphic and electrophysiological evidence of canine myocardial sympathetic denervation and reinnervation produced by myoacardial infarction or phenol application. Circulation 1988; 78:1008–1019.

26. Hartikainen J, Mäntysaari M, Kuikka J, Länsimies E, Pyörälä K. Extent of cardiac autonomic denervation in relation to angina on exercise test in patients with recent acute myocardial infarction. Am J Cardiol 1994; 74:760–763.

27. Hartikainen J, Kuikka J, Mantsayaari M, et al. Sympathetic reinnervation after acute myocardial infarction. Am J Cardiol 1996; 77:5–9.

28. Kramer CM, Nicol PD, Rogers WJ, et al. Reduced sympathetic innervation underlies adjacent noninfarcted region dysfunction during left ventricular remodelling. J Am Coll Cardiol 1997; 30:1079–1085.

29. Podio V, Spinnler MT, Spandonari T, et al. Regional sympathetic denervation after myocardial infarction: a follow-up study using [123I]MIBG. Q J Nucl Med 1995; 39:40–43.

30. Matsunari I, Schricke U, Bengel FM, et al. Extent of cardiac sympathetic neuronal damage is determined by the area of ischemia in patients with acute coronary syndromes. Circulation. 2000;22:2579–2585.

31. Gaudino M, Giordano A, Santarelli P, et al. Immunohistochemical-scintigraphic correlation of sympathetic cardiac innervation in postischemic left ventricular aneurysms. J Nucl Cardiol 2002; 9:601–607.

32. Arora R, Ferrick KJ, Nakata T, et al. 123I-MIBG imaging and heart rate variability analysis to predict the need for an implantable cardioverter defibrillator. J Nucl Cardiol 2003; 10:121–131.

27

Molecular and Cellular Imaging of Myocardial Inflammation

David K. Glover

Cardiovascular Division, Department of Medicine, University of Virginia School of Medicine, Charlottesville, Virginia, U.S.A.

INTRODUCTION

Inflammation plays an injurious role in a number of diseases or pathologic conditions affecting the myocardium including myocarditis (1) cardiac allograft rejection (2) or in the setting of reperfusion injury following coronary revascularization (3) or post-transplantation. Accordingly, there has been considerable interest in the development of noninvasive imaging techniques for detecting the presence and severity of myocardial inflammation as well as for serial assessment of the efficacy of treatment of these conditions.

[99m]Technetium or [111]Indium-labeled leukocytes and ^{67}Ga-citrate have been widely used in nuclear medicine as imaging agents for the detection of inflammatory sites (4–6). Although these techniques have proven to be useful in certain settings, they are not ideal for several reasons. Radiolabeling of leukocytes involves withdrawal of whole blood from the patient, isolation of leukocytes, followed by incubation of the white cells with either Tc99m-hexamethylpropylene amine oxime (HMPAO, Ceretec) or In-111 oxine and reinjection of the labeled cells back into the patient. One major drawback to this technique is that the procedure for isolating the leukocytes from whole blood is known to activate them (6). Premature activation may cause the radiolabeled leukocytes to bind at sites other than the target of interest potentially reducing the sensitivity of the test. In addition, the isolation and ex vivo labeling procedure is lengthy and requires handling of potentially infectious whole blood by the nuclear medicine technologist. Finally, such a procedure would not be recommended or perhaps even possible in patients who may be neutropenic due to disease or chemotherapy. ^{67}Ga-citrate is also not ideal due to its relatively nonspecific localization and poor imaging characteristics (7,8). ^{67}Ga-citrate does not bind to specific inflammatory cells per se. Instead, following i.v. injection, ^{67}Ga-citrate binds to transferin in the blood and is released into the interstitial space at inflammatory sites due to the leakiness of capillary walls made permeable to intravascular protein by the inflammatory process.

MYOCARDITIS

Myocarditis is defined as inflammation and injury of the myocardium in the absence of ischemia (1). Myocarditis is most commonly caused by viruses but can also be caused by other infectious agents such as bacteria, fungi, and various parasites, as well as by numerous noninfectious agents such as certain drugs or chemical toxins. Spontaneous recovery usually occurs. However, sudden death or progression to dilated cardiomyopathy can occur in up to 10% of cases (9). The "gold standard" for diagnosis in patients with suspected myocarditis is percutaneous right ventricular endomyocardial biopsy (10). However, biopsy techniques are laborious and carry significant risk, particularly since the disease is often benign. Furthermore, sampling errors limit the sensitivity and specificity of biopsy techniques (10,11). It was estimated that it would require up to 17 biopsy specimens to achieve an 80% sensitivity for myocarditis (12). In addition, it is not possible to predict which of the acute cases will progress to chronic myocarditis and dilated cardiomyopathy based on the analysis of biopsy samples. For these reasons, the diagnosis of myocarditis is usually made through the exclusion of coronary artery disease on the basis of clinical and laboratory findings. A noninvasive imaging approach for accurately and specifically diagnosing and monitoring myocarditis is needed.

RADIONUCLIDE IMAGING

[111]In Antimyosin Antibodies

[111]In antimyosin scintigraphy has been used to visualize active myocardial damage associated with myocardial infarction (13,14) as well as inflammation (15–18). Antimyosin antibody uptake occurs in irreversibly damaged myocytes in which the sarcolemma is disrupted and heavy chain myosin is exposed (19–21). Clinical studies demonstrated that myocardial In-111 antimyosin antibody imaging exhibits favorable

diagnostic accuracy for the detection of acute myocarditis (18,22,23). Antimyosin uptake is also seen in a large subset of patients with dilated cardiomyopathy, suggesting ongoing myocyte injury in this population (24–26). Matsumori et al. observed that 70% of patients with dilated cardiomyopathy showed positive antimyosin scans and their left ventricular ejection fraction was significantly lower than that of the patients who showed negative antimyosin uptake, suggesting that ^{111}In-antimyosin antibody imaging might be useful for evaluating patient prognosis and for selecting patients for cardiac transplantation (25). Although the sensitivity of antimyosin imaging is relatively high (82–100%) (15,24,27), the specificity is quite low (25–53%) (16,17,22). The low specificity is mainly due to the fact that the tracer binds to necrotic myocytes regardless of the exact cause of the necrosis.

Antitenascin-C Monoclonal Antibodies

Tenascin-C is an extracellular matrix protein, which appears in various pathological states, such as wound healing, cancer invasion, or inflammation (28–30). In a murine model of myocarditis, tenascin-C was reported to be expressed in the initial stage before necrosis or inflammatory cell infiltration was histologically apparent (31). Recently, a radiolabeled antitenascin-C monoclonal antibody has been developed for imaging inflammation related to myocarditis (32). In a rat model of autoimmune myocarditis produced by immunization with porcine myosin twice over a seven day interval, Sato et al. demonstrated high accumulation of ^{111}In-labeled antitenascin-C antibody in the inflamed myocardial region on ex vivo autoradiographs of heart slices (Fig. 1) (32). Furthermore, using a dual isotope approach

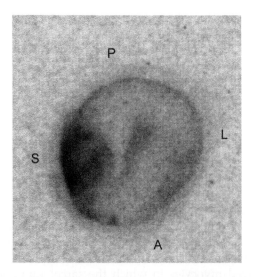

Figure 1 Autoradiographic image of a heart slice from a rat with autoimmune myocarditis showing ^{111}In-antitenascin-C Fab' uptake. *Abbreviations*: A, anterior; S, septal; P, posterior; L, lateral. *Source*: From Ref. 32.

with ^{111}In-antitenascin-C and ^{99m}Tc-sestamibi and SPECT imaging, these investigators demonstrated that the focal uptake of ^{111}In antitenascin-C antibody in the septal wall could easily be visualized in vivo (Fig. 2) (32).

CARDIOVASCULAR MAGNETIC RESONANCE IMAGING

Over the past 15 years, there have been a number of studies examining the usefulness of cardiovascular magnetic resonance (CMR) imaging for diagnosing acute myocarditis using both T2- and T1-weighted sequences, with and without gadolinium enhancement and the results have been promising with very high sensitivities and specificities reported (33–38). In one study, Friedrich et al. used gadolinium-enhanced T1-weighted images to follow the progression and regression of acute myocarditis (35). Early focal enhancement was observed on day 2 but then became more diffuse over time. The enhancement persisted for several weeks before gradually returning to the same level as in control patients by day 84. In another recent study, Mahrholdt et al. used contrast CMR with segmented inversion recovery gradient-echo (IR-GRE) pulse sequences in 32 patients with suspected acute myocarditis and directly compared the visualized regions of contrast enhancement with histopathology of biopsy samples taken from the same areas using the CMR as a guide (38). The advantage of the IR-GRE pulse sequence is that it provides markedly higher contrast between damaged and normal tissues (39). Using this technique, Mahrholdt et al. reported that contrast enhancement was present in 28 of the 32 patients (88%) and was usually found in one or several foci within the epicardial quartile of the wall, most often the lateral free wall (Fig. 3) (38). The pattern of contrast enhancement in the setting of myocarditis inflammation is different from that observed with myocardial infarction which typically originates in the subendocardium (40). However, the epicardial localization of the inflammatory foci is consistent with postmortem findings in other studies of myocarditis (12,41). Also shown in Figure 3 are the corresponding immunohistochemical staining sections from the biopsies taken from the contrast enhanced areas. In the patients with active myocarditis (patients 6, 7, and 14), there was evidence of myocyte damage and macrophage infiltration (38). Importantly, these investigators also followed the progression of the disease by performing follow-up (FU) CMR imaging at three months in 20 of the patients that had shown contrast enhancement in the acute stage (38). Figure 4 shows typical examples of acute and FU CMR images. The average contrast enhanced areas were diminished or disappeared altogether and there was a corresponding improvement in

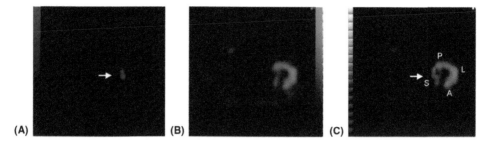

Figure 2   Transverse dual-isotope tomographic images of ^{111}In-anti-TNC Fab' and ^{99m}Tc-MIBI in a rat with experimentally induced autoimmune myocarditis. (**A**) *Arrow* shows focal uptake of ^{111}In-anti-TNC Fab'. (**B**) ^{99m}Tc-MIBI myocardial perfusion image. (**C**) Fused image demonstrates that the focal uptake of ^{111}In-anti-TNC Fab' was located on the interventricular septum. *Source*: From Ref. 32.

mean ejection fraction and a decrease in mean maximal end diastolic volume (EDV) and end systolic volume (ESV) (38). Results such as these demonstrate that contrast CMR has the potential to be an extremely valuable tool for diagnosing myocardial inflammation in the setting of active myocarditis as well as for following its regression during healing.

Cardiac Transplantation Rejection

Acute rejection is the leading cause of mortality in heart transplant recipients. Specific alloimmune responses in cardiac rejection are mediated by lymphocytes, which represent the only cell type capable of specifically recognizing and distinguishing among different antigenic determinants. Among several functionally distinct subtypes of lymphocytes, T lymphocytes (T-cells) play the

most important role in the initiation and regulation of immune responses to protein antigens and the elimination of foreign cellular material, such as microbes, virally infected cells, and tissue allografts. Acute cellular rejection occurs most frequently in the first year post-operatively, often as early as two to three weeks after transplantation, even with modern immunosuppressive therapy. Chronic rejection manifests itself in a pathologic process known as graft arteriosclerosis, accelerated arteriosclerosis, or allograft vascular disease, which differ in many respects from atherosclerosis and is recognized as transplant vasculopathy (42). These two manifestations after transplantation are both mediated by T-cells and remain important causes of death after successful cardiac transplantation.

Routine detection of acute cardiac rejection is currently performed with periodical endomyocardial biopsy (43,44). However, for the same reasons as with myocarditis, the sensitivity of detection of transplant rejection using biopsy techniques is generally poor.

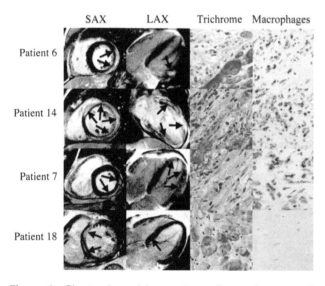

Figure 3   Short axis and long axis cardiovascular magnetic resonance images and histopathologic staining of biopsy samples taken from the areas of contrast enhancement (*black arrows*) in patients with suspected myocarditis. The top three rows are from patients with active myocarditis showing myocyte damage (trichome staining) and infiltration of macrophages (PGM1 mAb) while the bottom row is from a patient without active myocarditis but with hypertrophic cardiomyopathy. *Source*: From Ref. 38.

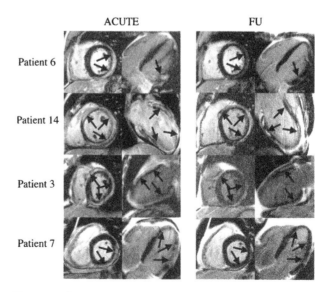

Figure 4   Cardiovascular magnetic resonance images acquired from the same patients in Fig. 3 in the acute and three-month follow-up periods. *Source*: From Ref. 38.

In addition, transplant patients must undergo these procedures repeatedly.

RADIONUCLIDE IMAGING

Detection of transplantation rejection-related myocardial damage by evaluating myocardial uptake of ^{111}In-labeled monoclonal antimyosin antibodies has been established by several groups (45–52). Quantitative assessment of the presence and intensity of radiolabeled antibody uptake can be made by calculating a simple ratio between the activities in regions-of-interest drawn on the heart and lungs on images (53). The antimyosin antibody technique has been shown to have high sensitivity for detecting acute rejection, even in the presence of a normal endomyocardial biopsy (51,54,55). A prospective analysis by Ballester's group demonstrated that patients with decreased antimyosin antibody uptake during the first three months after transplantation appear to be free from severe rejection-related complications during the first year, whereas persistent antimyosin uptake during the first three months is predictive of rejection-related complications during that interval (51). It has also been reported that 80% of heart transplant recipients had a positive antimyosin scan during the first year (51,55). This may be due to the fact that a positive scan may reflect not only the acute rejection episode but also myocardial infection occurring early after operation or from prolonged exposure of myosin from damaged myocytes even after the episode rejection has improved as a result of treatment (56). For this reason, patient management decisions during the first year post-op, such as the need for additional immunosuppressive therapy, should probably be based on the biopsy results for detecting acute rejection because overimmunosuppression might result if antimyosin scintigraphy was employed as the gold standard.

After the first year postoperatively, invasive endocardial biopsies can be avoided and annual antimyosin scintigraphy has been proposed to manage patients and treat for rejection (48,57). The impact of suppressing biopsies beyond the first year by using noninvasive antimyosin imaging is favorable in terms of cost, patient comfort, and allocation of resources for other nontransplant needs. It also provides a rationale for long-term noninvasive risk stratification and patient management (57).

CARDIOVASCULAR MAGNETIC RESONANCE IMAGING

Cellular imaging by CMR usually involves the use of iron oxide nanoparticles. Inflammatory processes can be imaged following an i.v. injection of the particles due to the fact that they are readily engulfed by macrophages and their relaxivity properties results in a hypointensity observed on T2*-weighted images. Several groups of investigators have reported on using ultra-small superparamagnetic iron oxide (USPIO) nanoparticles for imaging cardiac transplant rejection with CMR (58,59). While the results of these studies are promising, the small size of the USPIO nanoparticles (20–30 nm) limits the sensitivity and ultimate contrast obtained. In a recent study, Wu et al. reported on an in vivo labeling approach that used much larger micrometer-sized paramagnetic iron oxide (MPIO) spheres to label immune cells in a rat model of heart transplant rejection (60). As shown in Figure 5, there was punctuate, high contrast hypoenhancement in the rejecting allograft hearts (and lungs) compared with no hypoenhancement in the isograft transplant. The highly punctuate pattern suggests that the MPIO technique has the ability to image single immune cells compared with the USPIO technique that has a more diffuse pattern of localization. Furthermore, Wu et al. also showed that it was possible to follow the temporal progression of rejection over a period of three days after a single injection of MPIO particles (Fig. 6) (60). They showed that there was a pericardium-to-endocardium progression pattern of macrophage infiltration into the left ventricle

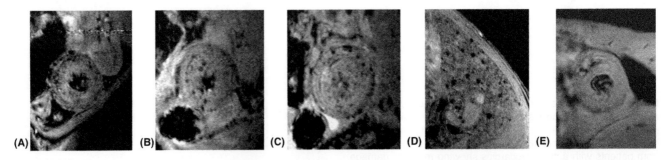

Figure 5   In vivo MRI of allograft hearts and lungs 24 hours after i.v. injection of micrometer-sized paramagnetic iron oxide particles. (**A**) Allograft heart five days post-transplant; (**B, C**) allograft heart six days post-transplant; (**D**) allograft lung six days post-transplant; (**E**) isograft heart six days post-transplant. *Source*: Ref. 60.

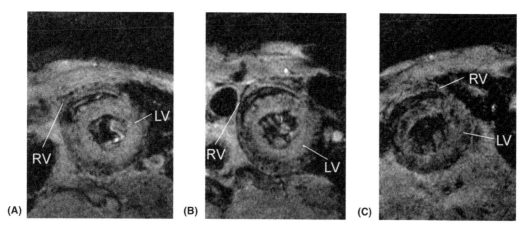

Figure 6 Serial magnetic resonance images of a rat allograft over time. Micrometer-sized paramagnetic iron oxide particles were administered only once, on post-transplant day 3.5, and imaging performed on (**A**) 3.5, (**B**) 4.5, and (**C**) 5.5 days post-transplant, respectively. *Source*: From Ref. 60.

as the rejection progressed in the allograft hearts, with no loss of signal contrast over time.

MYOCARDIAL CONTRAST ECHOCARDIOGRAPHY

As described in detail in Chapter 6, inflammation can be imaged using myocardial contrast echocardiography (MCE) due to the fact that lipid-shelled microbubbles are retained by activated leukocytes that are adherent to the vascular endothelium in local areas of inflammation and can also be phagocytosed intact by tissue resident macrophages (61,62). Furthermore, the interaction between microbubbles and activated leukocytes or endothelium can be enhanced by lipid shell surface modifications, including the addition of different monoclonal antibodies targeted to various adhesion molecules such as intercellular adhesion molecule-1 (ICAM-1) (63–65).

Although T-cells play the most important role in the immune responses leading to allograft rejection, monocytes and macrophages are also recruited to the

area through local chemokine release from these T-cells. Using a rat model of allograft rejection, Weller et al. were the first group to show that MCE with ICAM-1 targeted microbubbles was capable of identifying acute rejection of the transplanted heart (64). In a later study, Kondo et al. showed that the video intensity after microbubble injection was highest in the untreated allograft heart and that the signal was inhibited in a dose dependent manner with cyclosporine-A (CsA), an agent that inhibits T-cells as well as reduces monocyte and macrophage infiltration (Fig. 7) (65). Furthermore, these investigators showed that the video intensity of the retained microbubbles was related to the degree of acute rejection and could be used to grade the severity of rejection (Fig. 8) (65).

POSTISCHEMIC INFLAMMATION

Radionuclide Imaging
It has long been known from necropsy studies that polymorphonuclear leukocytes (PMNs) infiltrate

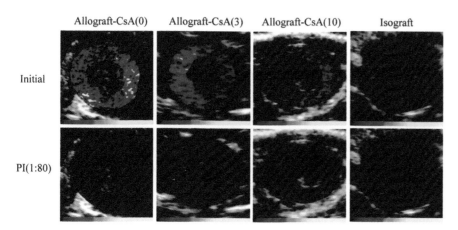

Figure 7 Background-substracted color-coded ultrasound images at post-transplantation day 3. The initial frames were acquired 10 minutes after injection of microbubbles. The first three panels shown are of an allograft heart in the absence (*left image*) or presence of either a low (*middle image*) or high (*right image*) concentration of cyclosporin-A. The image on the far right is of an isograft heart at the same 10 minutes time point. *Source*: From Ref. 65.

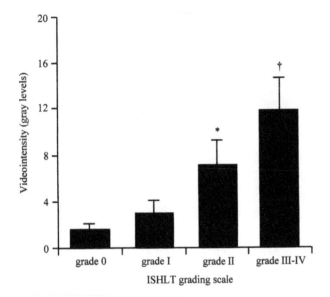

Figure 8 Relation of video intensity from retained microbubbles with International Society for Heart and Lung Transplantation rejection criteria. *Note:* *p < 0.001; †p < 0.01 versus isograft. *Source*: From Ref. 65.

regions of acute myocardial infarction (66). In 1979, Thakur et al. were the first group to use radiolabeled autologous leukocytes to image the time course and pattern of neutrophil infiltration in a closed-chest canine model of acute myocardial infarction produced by a catheter plug embolization technique (67). Radiolabeling of the neutrophils in this study was accomplished by withdrawing blood from the dog, isolating the white blood cells, incubating them with ^{111}In-oxine, followed by reinjection of the labeled cells back into the dog (67). As mentioned earlier, there are a number of drawbacks to ex vivo PMN labeling techniques, particularly for routine clinical use. Previous attempts at in vivo leukocyte labeling have been successful using monoclonal antibodies to neutrophils. However, the clinical use of these tracers has been limited by their poor specificity due to the fact that they target a large pool of antigens in myeloid bone marrow.

More recently, there has been considerable effort to develop new, receptor-targeted inflammation imaging agents with high specificity (68–72). For example, ^{99m}Tc-RP128, an antagonist of the tuftsin receptor on phagocytic cells (68) and ^{99m}Tc-EPI-HNE-2, a radiolabeled neutrophil elastase inhibitor (69) have been shown to detect inflammatory sites either in humans (^{99m}Tc-RP128) or in monkeys (^{99m}Tc-EPI-HNE-2) following intravenous injection.

LTB$_4$ is a lipid mediator biosynthesized from arachidonic acid through the 5-lipoxygenase pathway (70). LTB$_4$ is known to be synthesized by polymorphonuclear neutrophils, monocytes, macrophages,

endothelial cells and tracheal epithelial cells (71,72) and is a potent chemotactic and chemokinetic agent. LTB$_4$ stimulates neutrophil aggregation, lysosomal enzyme release, superoxide production and endothelial adhesion through binding to leukotriene B$_4$ receptors and is known to be an important mediator in both acute and chronic inflammation (73).

^{99m}Tc-RP517 is a new technetium-labeled leukotriene B$_4$ (LTB$_4$) receptor antagonist that was developed for imaging acute inflammation or infection (Fig. 9) (74,75). After intravenous injection, ^{99m}Tc-RP517 has been shown to localize at sites of inflammation in a guinea pig model of peritonitis, and in rabbit models of *E. coli*-induced infection, phorbol-ester-induced inflammatory bowel, and *S. aureus*-induced infection, making this tracer a potential candidate for imaging inflammation (74).

Recently, Riou et al. investigated the specific human leukocyte subtypes to which ^{99m}Tc-RP517 is bound (75). Fluorescence activated cell sorter (FACS) analysis was performed on whole human blood as well as on isolated neutrophils using a fluorescent analog of ^{99m}Tc-RP517, [F]-RP517. The results of the FACS analysis are shown in Figure 10. The left panel of Figure 10 depicts the three distinct subpopulations of leukocytes—neutrophils, monocytes, and lymphocytes—separated from the whole human blood by FACS in the absence of [F]-RP517. The right panel of Figure 10 shows the leukocyte subpopulations after incubation of whole human blood with [F]-RP517. As shown in red, there was preferential staining of the neutrophil pool by [F]-RP517. Mean fluorescent intensity (MFI) values were 61.7 ± 0.1, 14.6 ± 0.1, and 5.5 ± 0.1 for neutrophils, monocytes, and lymphocytes, respectively. Riou et al. also examined the specificity of binding of [F]-RP517 (500 nmol/L) on isolated neutrophils with a competition binding assay using both the receptor agonist LTB$_4$ as well as nonfluorescent RP517 (75).

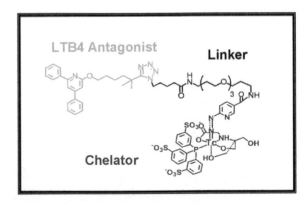

Figure 9 ^{99m}Tc-RP517 structure. The structure consists of an LTB$_4$ receptor antagonist moiety linked with a ^{99m}Tc-chelating domain. *Source*: From Ref. 75.

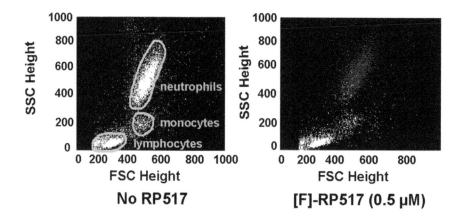

Figure 10 Identification of white blood cell subtypes by flow cytometry in whole human blood (*left panel*), and preferential [F]-RP517 neutrophil binding to neutrophils after incubation with whole blood (*right panel, red dots*). *Source*: From Ref. 75.

[F]-RP517 binding was reduced by 44% from 19.3 ± 1.4 to 10.9 ± 0.9 ($p < 0.001$) when LTB_4 (400 nmol/L) was present. Likewise, nonfluorescent RP517 inhibited the binding of [F]-RP517 on isolated neutrophils in a dose-dependent manner, with a 50% effective concentration (EC_{50}) = 26 ± 1 nmol/L. These in vitro results demonstrated that [F]-RP517 preferentially labeled the neutrophil pool in whole human blood.

In addition to these in vitro studies, Riou et al. investigated the ^{99m}Tc-RP517 uptake pattern in a canine model of postischemic myocardial inflammation (75). ^{99m}Tc-RP517 was injected i.v. in 9 anesthetized, open-chest dogs before coronary occlusion (90 min) and reperfusion (120 min). There was an inverse exponential relationship between ^{99m}Tc-RP517 uptake and occlusion flow (r = 0.73). In the same 15 segments, ^{99m}Tc-RP517 uptake was highly correlated with the neutrophil enzyme myeloperoxidase (r = 0.91). As

shown in Figure 11, when the myocardial segments were grouped according to the severity of flow reduction during the occlusion, a graded distribution of ^{99m}Tc-RP517 was observed, with segments in the lowest flow ranges showing the highest ^{99m}Tc-RP517 uptake (75). Figure 12A shows a photograph of a blue dye and TTC-stained heart slice from a representative occlusion/reperfusion dog. Figure 12B shows an ex vivo image of ^{99m}Tc-RP517 in the same slice (75). Note that ^{99m}Tc-RP517 was localized to the risk area with a higher degree of uptake in the central ischemic zone. Also note that the uptake of the tracer in the normal

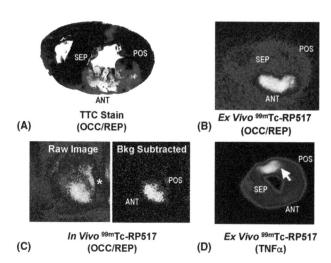

Figure 12 Triphenyltetrazolium chloride-stained heart slice (**A**) and ex vivo ^{99m}Tc-RP517 image (**B**) of the same heart slice. (**C**) Raw (*left*) and background-subtracted (*right*) in vivo ^{99m}Tc-RP517 images acquired from a dog 60 minutes after reperfusion. Background subtraction was performed to eliminate the surgically related tracer uptake in the field of view. The shadow on the raw image denoted by an asterisk is the metal rib spreader. Note that focal ^{99m}Tc-RP517 uptake was readily observed in the inflamed anteroseptal region of the heart on both ex vivo and in vivo images. Tracer uptake was negligible in the normal, posterior wall. (**D**) Typical ex vivo ^{99m}Tc-RP517 image after intramyocardial TNFα injection. *Abbreviations*: ANT, anterior; POS, posterior; SEP, septum; TTC, triphenyltetrazolium chloride. *Source*: From Ref. 75.

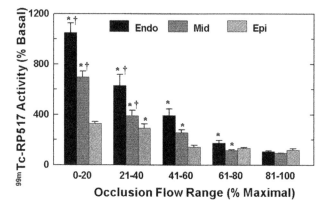

Figure 11 Myocardial localization of ^{99m}Tc-RP517. The endocardial (endo), midwall (mid), and epicardial (epi) segments from eight dogs (n = 573) were grouped according to the flow reduction during occlusion. Note the strong, inverse relationship between occlusion flow and ^{99m}Tc-RP517 myocardial uptake in the endocardial and midwall layers. Also note the graded increase in ^{99m}Tc-RP517 from the epicardial to the endocardial layer in flow ranges, 60% of normal. *Note*: *P, 0.05 versus adjacent flow range; †P, 0.05 versus adjacent layer. *Source*: From Ref. 75.

posterior wall was negligible. The ischemic to normal zone count ratio of ^{99m}Tc-RP517 obtained from quantification of ex vivo images was not significantly different from the MPO tissue concentration ratio from the same regions (2.7 ± 0.2 and 2.6 ± 0.7, respectively, P = NS). Figure 12C depicts a raw and background-subtracted in vivo ^{99m}Tc-RP517 image that was acquired 60 minutes after reperfusion in one dog. Note that there was focal ^{99m}Tc-RP517 uptake that delineated the area of inflammation in the occluded-reperfused zone of the anteroseptal wall.

In these experimental studies, ^{99m}Tc-RP517 was injected during the baseline period, prior to coronary occlusion. More recently, our laboratory has investigated whether this tracer can also be given early after reperfusion to assess postischemic myocardial inflammation. We found that ^{99m}Tc-RP517 produces a strong signal in the reperfused zone whether it is administered at baseline or at the time of reperfusion, and that the signal intensity increases over a period of several hours after injection (unpublished data).

In three additional dogs, Riou et al. produced myocardial inflammation by direct intramyocardial injection of the inflammatory cytokine TNFα rather than by ischemia (75). Figure 12D depicts a ^{99m}Tc-RP517 image of a heart slice from one dog that received an intramyocardial injection of TNFα. Note the prominent uptake of the tracer observed at the injection site (*white arrow*). In this slice, there was a 75% increase in tracer uptake at the site of TNFα injection corresponding to a 51% increase in MPO in the same region. For all three dogs, the mean ^{99m}Tc-RP517 uptake and MPO ratios (injection site/contralateral wall) were 1.55 ± 0.13 and 1.45 ± 0.06, respectively (P = NS) (75).

Although ^{99m}Tc-RP517 was shown to localize at sites of inflammation or infection following i.v. injection, one disadvantage of this tracer for general nuclear medicine applications is that, due to its lipophilic nature, it exhibits high hepatobiliary clearance resulting in a significant amount of the tracer appearing in the gut. Recently, van Eerd et al. reported on a new more hydrophilic ^{111}In-labeled LTB$_4$ receptor antagonist (^{111}In-DPC11870) that was also developed for scintigraphic imaging of infection or inflammation (76). This new compound consists of the same LTB$_4$ receptor antagonist as the RP517 compound, however, it is conjugated with a diethylenetriaminepentaacetic acid (DTPA) moiety to allow radiolabeling with [111]In. These investigators reported that ^{111}In-DPC11870 bound to its receptor with relatively high affinity (IC$_{50}$ = 10 nmol/L) with the majority of uptake occurring on granulocytes (76). In a rabbit model of infection induced by *E. coli* injection into the thigh muscle, van Eerd et al. demonstrated that the abscess could be visualized as early as two hours after tracer injection with no uptake in the non-infected contralateral thigh and the target:background

ratio improved over time (76). The tracer was not observed in the digestive tract. Further studies are necessary to validate ^{111}In-DPC11870 in various models of myocardial inflammation and to determine its usefulness for assessing inflammation.

Myocardial Contrast Echocardiography

Christiansen et al. demonstrated that myocardial inflammation after ischemia/reperfusion can be imaged noninvasively using MCE with leukocyte-targeted microbubbles (see Chapter 6, Fig. 5 for details) (77). The video intensity pattern obtained with targeted microbubbles in the reperfused myocardial zone is different from that observed with 99mTc-RP517 (77). The reason for this difference is that the microbubbles are confined to the intravascular compartment and represent active leukocyte recruitment whereas 99mTc-RP517 labeled neutrophils can transmigrate into the myocardial tissue representing total tissue burden. Advantages of the MCE technique in this setting is the ability to image inflammation early after reperfusion combined with a relatively short protocol duration.

REFERENCES

1. Aretz HT, Billingham, ME, Edwards WD, et al. Myocarditis: a histopathologic definition and classification. Am J Cardiovasc Pathol 1986; 1:3–14.
2. Ascher NL, Hoffman RA, Hanto DW, Simmons RL. Cellular basis of allograft rejection. Immunol Rev 1984; 77:217–232.
3. Frangogiannis NG, Smith CW, Entman ML. The inflammatory response in myocardial infarction. Cardiovasc Res 2002; 53:31–47.
4. Lavander JP, Lowe J, Baker JR, et al. Gallium-67 citrate scanning in neoplastic and inflammatory lesions. Br J Radiol 1971; 44:361–366.
5. Peters AM, Danpure HJ, Osman S, et al. Clinical experience with 99mTc-hexamethylpropylene-amineoxime for labeling leucocytes and imaging inflammation. Lancet 1986; 2:946–949.
6. Peters AM, Saverymuttu SH. The value of indium-labeled leucocytes in clinical practice. Blood Rev 1987; 1:65–76.
7. Tsan MF. Mechanism of gallium-67 accumulation in inflammory lesions. J Nucl Med 1985; 26:88–92.
8. Peters AM. The use of nuclear medicine in infections. Br J Radiol 1998; 71:252–261.
9. Fuster V, Gersh BJ, Giuliani ER, et al. The natural history of idiopathic dilated cardiomyopathy. Am J Cardiol 1981; 47:525–531.
10. Mason JW, O'Connell JB, Herskowitz A, et al. A clinical trial of immunosuppressive therapy for myocarditis. The Myocarditis Treatment Trial Investigators. N Eng J Med 1995; 333:269–275.
11. Shanes JG, Ghali J, Billingham ME, et al. Interobserver variability in the pathologic interpretation of endomyocardial biopsy results. Circulation 1987; 75:401–405.

12. Hauck AJ, Kearney DL, Edwards WD. Evaluation of postmortem endomyocardial biopsy specimens from 38 patients with lymphocytic myocarditis: implications for role of sampling error. Mayo Clin Proc 1989; 64:1235–1245.

13. Khaw BA, Strauss HW, Moore R, et al. Myocardial damage delineated by indium-111 antimyosin Fab and technetium-99m pyrophosphate. JNM 1987; 28: 76–82.

14. Johnson LL, Seldin DW, Becker LC, et al. Antimyosin imaging in acute transmural myocardial infarctions: results of a multicenter clinical trial. JACC 1989; 13:27–35.

15. Yasuda T, Palacios IF, Dec GW, et al. Indium 111-monoclonal antimyosin antibody imaging in the diagnosis of acute myocarditis. Circulation 1987; 76:306–311.

16. Dec GW, Palacios I, Yasuda T, et al. Antimyosin antibody cardiac imaging: its role in the diagnosis of myocarditis. JACC 1990; 16:97–104.

17. Obrador D, Ballester M, Carrio I, et al. Active myocardial damage without attending inflammatory response in dilated cardiomyopathy. JACC 1993; 21: 1667–1671.

18. Kuhl U, Lauer B, Souvatzoglu M, Vosberg H, Schultheiss HP. Antimyosin scintigraphy and immunohistologic analysis of endomyocardial biopsy in patients with clinically suspected myocarditis— evidence of myocardial cell damage and inflammation in the absence of histologic signs of myocarditis. JACC 1998; 32:1371–1376.

19. Khaw BA, Fallon JT, Beller GA, Haber E. Specificity of localization of myosin-specific antibody fragments in experimental myocardial infarction. Histologic, histochemical, autoradiographic and scintigraphic studies. Circulation 1979; 60:1527–1531.

20. Khaw BA, Mattis JA, Melincoff G, Strauss HW, Gold HK, Haber E. Monoclonal antibody to cardiac myosin: imaging of experimental myocardial infarction. Hybridoma 1984; 3:11–23.

21. Khaw BA, Gold HK, Yasuda T, et al. Scintigraphic quantification of myocardial necrosis in patients after intravenous injection of myosin-specific antibody. Circulation 1986; 74:501–508.

22. Narula J, Khaw BA, Dec GW, et al. Diagnostic accuracy of antimyosin scintigraphy in suspected myocarditis. J Nucl Cardiol 1996; 3:371–381.

23. Margari ZJ, Anastasiou-Nana MI, Terrovitis J, et al. Indium-111 monoclonal antimyosin cardiac scintigraphy in suspected acute myocarditis: evolution and diagnostic impact. Int J Cardiol 2003; 90:239–245.

24. Obrador D, Ballester M, Carrio I, Berna L, Pons-Llado G. High prevalence of myocardial monoclonal antimyosin antibody uptake in patients with chronic idiopathic dilated cardiomyopathy. J Am Coll Cardiol 1989; 13:1289–1293.

25. Matsumori A, Yamada T, Sasayama S. Antimyosin antibody imaging in clinical myocarditis and cardiomyopathy: principle and application. Int J Cardiol 1996; 54:183–190.

26. Nanas JN, Margari ZJ, Lekakis JP, et al. Indium-111 monoclonal antimyosin cardiac scintigraphy in men

with idiopathic dilated cardiomyopathy. Am J Cardiol 2000; 85:214–220.

27. Lekakis J, Nanas J, Moustafellou A, Kostamis P, Moulopoulos S. Antimyosin scintigraphy for detection of myocarditis. Scintigraphic follow-up. Chest 1993; 104:1427–1430.

28. Jones FS, Jones PL. The tenascin family of ECM glycoproteins: structure, function, and regulation during embryonic development and tissue remodeling. Dev Dyn 2000; 218:235–259.

29. Erickson HP. Tenascin-C, tenascin-R and tenascin-X: a family of talented proteins in search of functions. Curr Opin Cell Biol 1993; 5:869–876.

30. Chiquet-Ehrismann R, Mackie EJ, Pearson CA, Sakakura T. Tenascin: an extracellular matrix protein involved in tissue interactions during fetal development and oncogenesis. Cell 1986; 47:131–139.

31. Imanaka-Yoshida K, Hiroe M, Yasutomi Y, et al. Tenascin-C is a useful marker for disease activity in myocarditis. J Pathol 2002; 197:388–394.

32. Sato M, Toyozaki T, Odaka K, et al. Detection of experimental autoimmune myocarditis in rats by 111In monoclonal antibody specific for tenascin-C. Circulation 2002; 106:1397–1402.

33. Gagliardi MG, Bevilacqua M, DiRenzi P, et al. Usefulness of magnetic resonance imaging for diagnosis of acute myocarditis in infants and children, and comparison with endomyocardial biopsy. Am J Cardiol 1991; 99: 1089–1091.

34. Gagliardi MG, Pollett B, DiRenzi P. MRI for the diagnosis and follow-up of myocarditis. Circulation 1999; 99:458–459.

35. Friedrich MG, Strohm O, Schulz-Menger J, et al. Contrast media-enhanced magnetic resonance imaging visualizes myocardial changes in the course of viral myocarditis. Circulation 1998; 97:1802–1809.

36. Roditi GH, Hartnell GG, Cohen MC. MRI changes in myocarditis—evaluation with spin echo, cine MR angiography, and contrast enhanced spin echo imaging. Clin Radiol 2000; 55:752–758.

37. Laissy JP, Messin B, Varenne O, et al. MRI of acute myocarditis: a comprehensive approach based on various imaging sequences. Chest 2002; 122:1638–1648.

38. Mahrholdt H, Goedecke C, Wagner A, et al. Cardiovascular magnetic resonance assessment of human myocarditis. A comparison to histology and molecular pathology. Circulation 2004; 109:1250–1258.

39. Simonetti OP, Kim RJ, Fieno DS, et al. An improved MR-imaging technique for the visualization of myocardial infarction. Radiology 2001; 218:215–223.

40. Mahrholdt H, Wagner A, Judd RM, et al. Assessment of myocardial viability by cardiovascular magnetic resonance imaging. Eur Heart J 2002; 23:602–619.

41. Shirani J, Freant LJ, Roberts WC. Gross and semiquantitative histologic findings in mononuclear cell myocarditis causing sudden death, and implications for endomyocardial biopsy. Am J Cardiol 1993; 72:952–957.

42. Dengler TJ and Pober JS. Cellular and molecular biology of cardiac transplant rejection. JNC 2000; 7:669–685.

43. Hosenpud JD, Novick RJ, Breen TJ, Keck B, Daily P. The Registry of the International Society for Heart and Lung

Transplantation: twelfth official report—1995. J heart Lung Transplant 1995; 14:805–815.

44. Arizón JM. Registro Nacional de Transplante Cardíaco. 6° Informe (1984/1994). Rev Esp Cardiol 1995; 48:792–797.

45. Frist W, Yasuda T, Segall G, et al. Noninvasive detection of human cardiac transplant rejection with indium-111 antimyosin (Fab) imaging. Circulation 1987; 76(5 Pt 2): V81–V85.

46. Ballester-Rodes M, Carrio-Gasset I, Abadal-Berini L, Obrador-Mayol D, Berna-Roqueta L, Caralps-Riera JM. Patterns of evolution of myocyte damage after human heart transplantation detected by indium-111 monoclonal antimyosin. Am J Cardiol 1988; 62:623–627.

47. De Nardo D, Scibilia G, Macchiarelli AG, et al. The role of indium-111 antimyosin (Fab) imaging as a noninvasive surveillance method of human heart transplant rejection. J Heart Transplant 1989; 8:407–412.

48. Ballester M, Obrador D, Carrio I, et al. Indium-111-monoclonal antimyosin antibody studies after the first year of heart transplantation. Identification of risk groups for developing rejection during long-term follow-up and clinical implications. Circulation 1990; 82:2100–2108.

49. Schutz A, Fritsch S, Kugler C, et al. Indium-111 monoclonal antimyosin for diagnosis of cardiac rejection. Transplant Proc 1990; 22:1464–1465.

50. Crespo MG, Pulpon LA, Dominguez P, et al. Detection of human cardiac transplant rejection with indium-111 monoclonal antimyosin antibody imaging. Transplant Proc 1990; 22:1463.

51. Ballester M, Obrador D, Carrio I, et al. Early postoperative reduction of monoclonal antimyosin antibody uptake is associated with absent rejection-related complications after heart transplantation. Circulation 1992; 85:61–68.

52. Hesse B, Mortensen SA, Folke M, Brodersen AK, Aldershvile J, Pettersson G. Ability of antimyosin scintigraphy monitoring to exclude acute rejection during the first year after heart transplantation. J Heart Lung Transplant 1995; 14:23–31.

53. Carrió I, Berná L, Ballester M, et al. Indium-111 antimyosin scintigraphy to assess myocardial damage in patients with suspected myocarditis and cardiac rejection. J Nucl Med 1988; 29:1893–1900.

54. Ballester M, Bordes R, Tazelaar HD, et al. Evaluation of biopsy classification for rejection: relation to detection of myocardial damage by monoclonal antimyosin antibody imaging. J Am Coll Cardiol 1998; 31:1357–1361.

55. Ballester-Rodes M, Carrio-Gasset I, Abadal-Berini L, Obrador-Mayol D, Berna-Roqueta L, Caralps-Riera JM. Patterns of evolution of myocyte damage after human heart transplantation detected by indium-111 monoclonal antimyosin. Am J Cardiol 1988; 62:623–627.

56. Ballester M, Carrió I. Noninvasive detection of acute heart rejection: the quest for the perfect test. J Nucl Cardiol 1997; 4:249–255.

57. Ballester M, Obrador D, Carrió I, Caralps-Riera JM. 111In-monoclonal antimyosin antibody studies in the diagnosis of rejection and management of patients after heart transplantation. In: Khaw BA, Narula J, Strauss WH, eds. Monoclonal antibodies in cardiovascular disease. Philadelphia: Lea & Febiger, 1994:79–98 .

58. Johannson L, Johnsson C, Penno E, Björnerud A, Ahlström H. Acute cardiac transplant rejection: detection and grading with MR imaging with a blood pool contrast agent—experimental study in the rat. Radiology 2002; 225:97–103.

59. Kanno S, Wu YL, Lee PC, et al. Macrophage accumulation associated with rat cardiac allograph rejection detected by magnetic resonance imaging with ultrasmall superparamagnetic iron oxide particles. Circulation 2001; 104:934–938.

60. Wu YL, Foley LM, Hitchens TK, et al. In situ labeling of immune cells with iron oxide particles: an approach to detect organ rejection by cellular MRI. PNAS 2006; 103:1852–1857.

61. Lindner JR, Coggins MP, Kaul S, et al. Microbubble persistence in the microcirculation during ischemia/reperfusion and inflammation is caused by integrin- and complement-mediated adherence to activated leukocytes. Circulation 2000; 101:668–675.

62. Linder JR, Dayton PA, Coggins MP, et al. Noninvasive imaging of inflammation by ultrasound detection of phagocytosed microbubbles. Circulation 2000; 102:531–538.

63. Lindner JR, Song J, Xu F, et al. Noninvasive ultrasound imaging of inflammation using microbubbles targeted to activated leukocytes. Circulation 2000; 102:2745–2750.

64. Weller GER, Lu E, Csikari MM, et al. Ultrasound imaging of acute cardiac transplant rejection with microbubbles targeted to intercellular adhesion molecule-1. Circulation 2003; 108:218–224.

65. Kondo I, Ohmori K, Oshita A, et al. Leukocyte-targeted myocardial contrast echocardiography can assess the degree of acute allograft rejection in a rat cardiac transplantation model. Circulation 2004; 109:1056–1061.

66. Mallory GK, White PD, Salcedo-Salgar J. The speed of healing of myocardial infarction. Am Heart J 1939; 18:647–671.

67. Thakur ML, Gottschalk A, Zaret BL. Imaging experimental myocardial infarction with indium-111-labeled autologous leukocytes: Effects of infarct age and residual regional myocardial blood flow. Circulation 1979; 60:297–305.

68. Caveliers V, Goodbody AE, Tran LL, et al. Evaluation of 99mTc-RP128 as a potential inflammation imaging agent: human dosimetry and first clinical results. J Nucl Med 2001; 42:154–161.

69. Rusckowski M, Qu T, Pullman J, et al. Inflammation and infection imaging with a 99mTc-neutrophil elastase inhibitor in monkeys. J Nucl Med 2000; 41:363–374.

70. Ford-Hutchinson AW. Regulation of leukotriene biosynthesis. Cancer Metastasis Rev 1994; 13:257–267.

71. Dasari VR, Jin J, Kunapuli SP. Distribution of leukotriene B4 receptors in human hematopoietic cells. Immunopharmacology 2000; 48:157–163.

72. Yokomizo T, Izumi T, Shimizu T. Leukotriene B4: metabolism and signal transduction. Arch Biochem Biophys 2001; 385:231–241.

73. Serhan CN, Prescott SM. The scent of a phagocyte: advances on leukotriene B4 receptors. J Exp Med 2000; 192:F5–F8.

74. Brouwers AH, Laveramn P, Boerman OC, et al. A 99mTc-labelled leukotriene B4 receptor antagonist for scintigraphic detection of infection in rabbits. Nucl Med Commun 2000; 21:1043–1050.

75. Riou LM, Ruiz M, Sullivan GW, et al. Assessment of myocardial inflammation produced by coronary occlusion and reperfusion with 99mTc-RP517, a new leukotriene B4 receptor antagonist, that preferentially labels neutrophils in vivo. Circulation 2002; 106: 592–598.

76. van Eerd JEM, Oyen WJG, Harris TD, et al. A bivalent leukotriene B4 antagonist for scintigraphic imaging of infectious foci. J Nucl Med 2003; 44:1087–1091.

77. Christiansen JP, Leong-Poi H, Klibanov AL, et al. Noninvasive imaging of myocardial reperfusion injury using leukocyte-targeted contrast echocardiography. Circulation 2002; 105:1764–1767.

28

Echocardiographic Imaging of Myocardial Inflammation

Flordeliza S. Villanueva

Cardiovascular Institute, University of Pittsburgh, Pittsburgh, Pennsylvania, U.S.A.

INTRODUCTION

Myocardial contrast echocardiography (MCE) is an ultrasound perfusion imaging technique using gas-filled microspheres (microbubbles) as red blood cell tracers. During echocardiographic imaging, ultrasound signals returning from the microbubbles during their microcirculatory transit register as myocardial tissue enhancement, which can be mapped and quantified (1,2). Typically 2 to 5 microns in diameter, these microbubbles move unimpeded through the microcirculation like red blood cells (3), and this intravascular behavior has formed the basis for their use as flow tracers.

Functional ultrasound imaging of tissue using targeted microbubbles differs from the concept that microbubbles passively transit the microcirculation. Targeted ultrasound imaging involves the design and synthesis of microbubbles that adhere to endothelium under disease-specific conditions, such that the molecular process of interest can be ultrasonically detected as tissue contrast enhancement that persists beyond the transient opacification of nontargeted microbubbles. To the extent that the microbubbles are designed to adhere to molecular epitopes on the surface of abnormal endothelium, targeted contrast imaging could provide capabilities for in vivo ultrasonic detection of phenotypic features of endothelium that predate clinical disease and/or are otherwise not detectable using currently available technologies.

One emerging application of this technique is the ultrasonic imaging of myocardial inflammation, or endothelial dysfunction, such as occurs during ischemia-reperfusion, early atherosclerotic disease, and cardiac transplant rejection. Other applications of targeted ultrasound imaging, such as the identification of neovascularization, are described elsewhere in this book. This chapter will provide an overview of the current status of targeted ultrasound contrast imaging of myocardial inflammatory processes. It will first briefly review relevant physiologic aspects of endothelial inflammation pertinent to targeted ultrasound, then introduce the concept of and approach to targeted ultrasound imaging in the setting of inflammation. This will be followed by a summary of the data on ultrasound imaging of endothelial dysfunction/myocardial inflammation by adhesion of mcrobubbles to endothelium through a variety of mechanisms, including ligand-receptor interactions. Potential applications to clinical scenarios will provide a backdrop to the overall discussion.

ENDOTHELIUM IN HEALTH AND DISEASE

A brief review of the changes that occur in endothelium during inflammation is appropriate to frame the discussion on echocardiographic imaging of endothelial function. A comprehensive discussion of endothelial biology is beyond the scope of this text, and the reader is referred elsewhere (see Refs. 4–6 for a review).

The endothelial surface actively plays multiple critical roles in normal vascular physiology (4–6). Normal endothelium regulates tissue perfusion by its control of vascular smooth muscle tone; provides a nonthrombogenic and nonadhesive surface; participates in the regulation and control of coagulation and fibrinolysis; serves as a selective barrier regulating microvascular permeability; and modulates proliferation of underlying vascular smooth muscle.

Unlike its clinical manifestations that typically appear in adult life, atherosclerotic cardiovascular disease is a life-long process that begins at a young age (7). The occurrence of abnormal endothelial function, characterized by a loss or attenuation of the normal functions described above, is a critical early event in atherogenesis that predates ischemic signs of coronary artery disease (8,9). The loss of the normal physiologic activities of endothelium ultimately permits the development of atherosclerotic lesions and may play a role in the occurrence of ischemia by mediating abnormal vasodilatory and vasoconstrictor responses (10–12).

Abnormal leukocyte adherence to the arterial wall is one of the initial cellular events in atherogenesis (8,13). As occurs with inflammation, monocytes cross the endothelium into the subendothelial space and ingest lipids, resulting in the focal accumulation of monocyte-derived foam cells, and leading to formation of early foam cell lesions of atherosclerosis (14). Specific leukocyte adhesion molecules pathologically overexpressed on the surface of endothelial cells regulate different stages of monocyte adhesion (11,12,15–17). These molecules, including intercellular adhesion molecule-1 (ICAM-1), E- and P-selectin, and vascular cell adhesion molecule-1 (VCAM-1) are inducible on endothelial surfaces, colocalize with early fatty streaks in blood vessels, and are upregulated in the presence of coronary risk factors (12,18,19). Because increased surface expression of leukocyte adhesion molecules predates the gross appearance and clinical manifestations of coronary disease, their upregulation on activated cells is a specific indicator of incipient endothelial disease.

Endothelial dysfunction is a hallmark of other cardiovascular disease states. Endothelial disease is not limited to the larger epicardial coronary arteries and there is evidence that functional endothelial derangements in atherosclerotic disease extend distally into the microcirculation (20,21). Endothelial dysfunction occurs in the setting of ischemia-reperfusion (22). Crystalloid hyperkalemic cardioplegia infusions result in endothelial disruption (23), and regenerated endothelium after coronary angioplasty exhibits reduced endothelium-dependent relaxation (24). Hypercholesterolemia results in endothelial dysfunction that may reverse with lipid-lowering interventions (25). Heart transplant rejection is associated with the upregulation of leukocyte adhesion molecules (26,27). Thus, functional aberrations in the endothelial lining of the coronary vascular tree underlie a host of cardiovascular disease states encountered in clinical practice.

CURRENT APPROACHES FOR IMAGING ENDOTHELIAL FUNCTION

Methods for identifying endothelial inflammation in vivo are relatively limited. Because endothelial dysfunction is often a precursor to and/or concomitant with vascular disease, and particularly because it is potentially reversible, such as in the case of hypercholesterolemia (25), early identification of inflammatory endothelial changes could impact on preventative treatment strategies in cardiovascular disease. Furthermore, the identification of endothelial dysfunction could provide an approach to the noninvasive diagnosis of other pathologies in which endothelial dysfunction is a hallmark.

Current clinical methods for studying endothelial function in vivo have limitations. Demonstration of a lack of vasodilator response to intracoronary acetylcholine is a commonly used approach to identifying coronary endothelial dysfunction, but this method is invasive (28). A noninvasive approach employs measurement of endothelium-dependent flow-mediated vasodilatation of the brachial artery following transient occlusion (29,30), which is an indirect measure of coronary vasoreactivity, the precision of which is limited by the spatial resolution of ultrasound.

CONCEPT OF TARGETED ULTRASOUND IMAGING OF INFLAMMATION

The concept that there are conditions in which microbubbles do not transit freely through the microcirculation is based on observations initially made during contrast echocardiographic studies of dogs receiving albumin microbubbles at the time of crystalloid cardioplegia delivery (31). It was noted that there was a prolongation of myocardial microbubble transit at any given cardioplegia flow rate. As cardioplegia delivery is associated with endothelial disruption (23), these observations suggested that red cell and microbubble transit rates were dissimilar during pathophysiologic states associated with endothelial dysfunction. Subsequent MCE studies during cardioplegia delivery in humans undergoing coronary artery bypass grafting confirmed a prolongation of transit rates similar to what was reported in the canine studies (32). These observations gave rise to the hypothesis that myocardial contrast echocardiography could be specifically used to interrogate the inflammatory status of the endothelium.

An ultrasound approach to imaging endothelial inflammation may have advantages over other imaging modalities such as nuclear or magnetic methods. Because microbubbles remain intravascular in location, they are ideally suited for characterizing the molecular and physiologic features of endothelium. Moreover, the portability of ultrasound imaging systems, as compared to gamma cameras or MR magnets, should allow for application that is more widespread across a spectrum of clinical settings. Because microbubbles do not circulate for extended periods of time in blood, the background "noise" of persistent blood activity (which confounds scintigraphic techniques) is of no less concern. The concurrent two-dimensional anatomic information avoids "hot spot" imaging and enables spatial localization of the targeted contrast agent, without difficulties from signal spillover. Another important potential advantage of using targeted microbubbles for imaging is that the "multi-point" targeting (which occurs when high concentrations of ligand are placed on a single microbubble) could result in higher binding of the microbubble agent when compared to other targeted tracers and potentially result in higher signal to noise ratios.

MICROBUBBLE ADHESION TO INFLAMED ENDOTHELIUM WITHOUT THE USE OF TARGETING LIGANDS

Nonspecific Adhesion of Microbubbles to Dysfunctional Endothelium

Microbubble attachment to vascular endothelium was first directly visualized in vitro using a parallel plate chamber lined with cultured human coronary artery endothelial cells perfused with culture medium containing albumin microbubbles (33). Microbubbles adhered to exposed extracellular matrix, and adhesion was increased when the endothelial cells were inflamed prior to denudation (Fig. 1). Microbubble adhesion to denuded endothelium was further confirmed in excised rabbit aorta segments denuded by balloon angioplasty and suspended in a flow circuit perfused with albumin microbubbles. While no microbubbles adhered to intact control vessels, there was adhesion of microbubbles to denuded vessels (Fig. 2) (34). Subsequent in vivo studies suggested that persistent myocardial contrast enhancement seen in cardioplegia-perfused or ischemia-reperfused canine hearts correlated with injury to the glycocalyx (35).

Direct Microbubble Adhesion to Leukocytes

Microbubble adhesion to inflamed endothelium can occur by direct adhesion to activated leukocytes. Albumin and lipid microbubbles were observed during intravital microscopy to stick to activated leukocytes in a murine model of cremaster muscle inflammation (Fig. 3) (36). The mechanism of such adhesion is thought to be integrin- and complement-mediated (36). In vitro studies have shown that activated leukocytes phagocytose the adherent microbubbles, which continue to retain their acoustic activity (37,38).

Lipid microbubbles have been synthesized to augment their complement-mediated avidity to activated leukocytes by the incorporation of phosphatidyl serine into the microbubble shell (39). In a model of murine kidney ischemia-reperfusion, these microbubbles yielded greater persistent videointensity signal compared to standard lipid microbubbles (39). More recently, in a canine model of reperfused myocardial infarction, phosphatidyl serine microbubbles were used to ultrasonically demonstrate the postischemic region of inflamed myocardium (Fig. 4) (40). Similarly, because albumin microbubbles also adhere to activated leukocytes, MCE can be used to identify acute heart transplant rejection, as has been shown in a rat model of heterotopic heart transplantation (Fig. 5).

Target-Specific Adhesion of Microbubbles to Endothelium

An approach using ligand-specific binding interactions could result in greater microbubble binding than can be achieved using nontargeted microbubbles and thus facilitate ultrasound detection of the adhered microbubbles. Furthermore, this approach

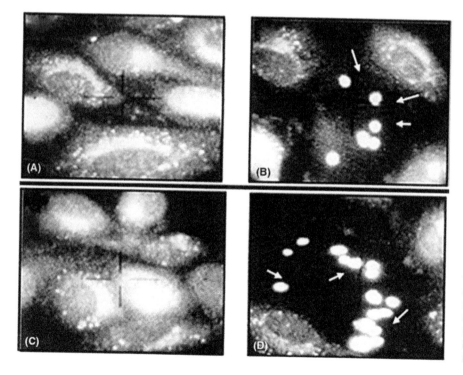

Figure 1 Fluorescent photomicrographs demonstrating fluorescein-labeled microbubble adhesion to denuded endothelium under (**A, C**) basal and (**B, D**) inflammatory conditions. Cultured human coronary artery endothelial cells were labeled with mepacrine. Microbubbles (*arrows*) adhered only to exposed extracellular matrix, and adhesion increased after inflammation was induced. *Source*: From Ref. 33.

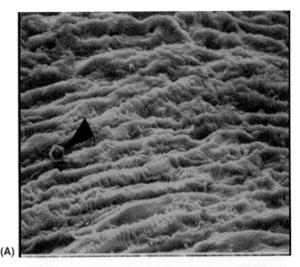

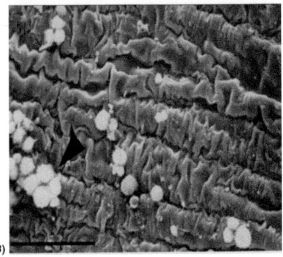

Figure 2 Rabbit aortic segments suspended in an ex vivo flow model (**A**) without and (**B**) with balloon-induced denudation after perfusion with culture medium and albumin microbubbles. There was abundant adhesion of albumin microbubbles to denuded endothelium. *Source*: From Ref. 34.

expands the possibilities for targeting other specific molecular events. The studies described below suggest that targeted ultrasound imaging with a ligand-targeting approach may have promise for clinical application.

Microbubbles bearing targeting ligands on the shell have been synthesized. These lipid-based microbubbles (2–4 μm) bear biotin on the microbubble shell and attachment of biotinylated targeting ligand is achieved using biotin-avidin bridging chemistry (41,42). Lipid microbubbles have also been targeted using direct covalent conjugation of the ligand to the microbubble shell (43,44). Others have developed targeted nongaseous submicron liposomes that have acoustic activity (45–48). Recently, nongaseous perfluorocarbon emulsion nanoparticles have been developed that become ultrasonically detectable when bound in high concentration to the target (49–51).

Ligand-targeting of microbubbles in a model of acute inflammation was first demonstrated using a lipid-based microbubble bearing monoclonal antibody to ICAM-1 on its surface and exposed to cultured human coronary endothelial cells overexpressing ICAM-1 in a parallel plate perfusion chamber (Fig. 6) (52). ICAM-1targeted microbubbles adhered minimally to cultured human coronary artery endothelial cells under basal conditions (Fig. 6, panel C). There was a 40-fold increase in ICAM-1-targeted microbubble adhesion when these cells were stimulated to overexpress ICAM-1 (Fig. 6, panels D, E) while control microbubbles bearing nonspecific IgG on the surface did not adhere to endothelial cells under any condition (Fig. 6, panels A, B).

Subsequent in vivo studies showed that ICAM-1-targeted bubbles can be used to ultrasonically detect inflammation. ICAM-1-targeted lipid microbubbles were intravenously injected into rats undergoing allograft or isograft abdominal heterotopic heart transplantation (53). Using ultraharmonic high mechanical index imaging five days postheart transplantation, it was found that myocardial videointensity after washout of circulating bubbles was higher in rejecting allografts exposed to ICAM-1-targeted microbubbles compared to nonrejecting isografts, and compared to hearts exposed to control bubbles bearing nonspecific IgG on their surface (Fig. 7). All allograft hearts demonstrated grade III–IV rejection and strong immunohistochemical staining for ICAM-1, compared to

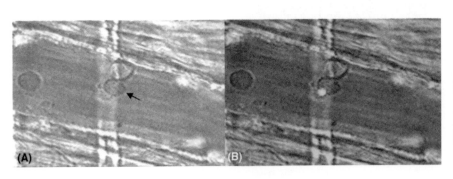

Figure 3 Brightfield (**A**) and fluorescent (**B**) intravital microscopic images of inflamed murine cremaster muscle showing adhesion of microbubbles (green fluorescence) to leukocytes that have adhered to the venular surface. *Source*: From Ref. 38.

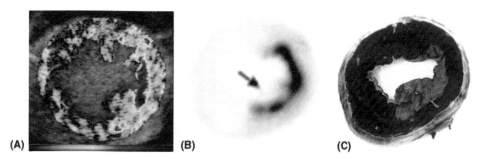

Figure 4   Nonligand mediated adhesion of lipid microbubles to reperfused left circumflex territory (**A**). Short-axis color-coded echocardiogram of canine myocardium after injection of phosphatidylserine-containing lipid microbubbles to enhance microbubble binding to leukocytes. There was persistent contrast enhancement in the circumflex bed, which corresponds to the area of leukocyte accumulation as demonstrated with ^{99m}Tc-RP51, a leukocyte-targeted radionuclide (**B**). (**C**) is the corresponding tetrazolium-stained slice of myocardium showing the region of infarction. *Source*: From Ref. 40.

normal histology and minimal ICAM-1 staining in the control isograft hearts.

Additional in vitro studies using inflamed cultured human coronary artery endothelial cells perfused in a radial flow chamber with ICAM-1-targeted microbubbles suggest that the severity of inflammation may be quantified with targeted microbubbles (Fig. 8) (54). This raises promise for using this approach not only to detect myocardial inflammation during acute cardiac transplant rejection or other inflammatory disease states, but also to monitor the severity of inflammation.

Microbubbles have been developed to target other leukocyte adhesion molecules. In a model of renal ischemia-reperfusion, microbubbles targeted to P-selectin using anti-P-selectin monoclonal antibody were used to ultrasonically identify the postischemic inflamed kidney in mice (Fig. 9) (55). This effect was abolished in P-selectin knock-out mice, indicating specificity of the effect. Selectins have been targeted using the naturally occurring tetrasaccharide ligand, sialyl-Lewis X (sLex). Microbubbles bearing sLex on the shell bound in vitro to inflamed human coronary

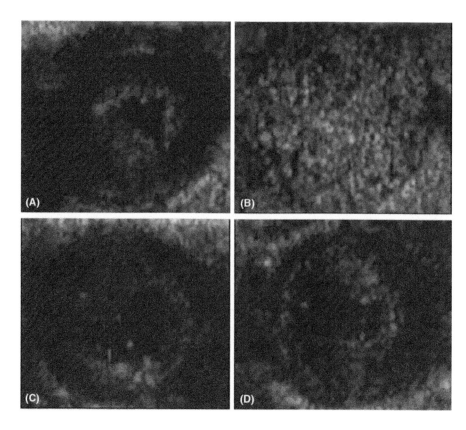

Figure 5   Nonligand mediated adhesion of albumin microbubbles to rejecting transplanted heart. Short-axis pulse inversion echocardiographic images of heterotopically transplanted rat hearts before (**A**, **C**) and 3.5 minutes after injection of nontargeted perfluorocarbon containing albumin microbubbles (**B**, **D**). There was persistent contrast enhancement in the rejecting allograft heart (**B**), which was not seen in the nonrejecting isograft heart (**D**).

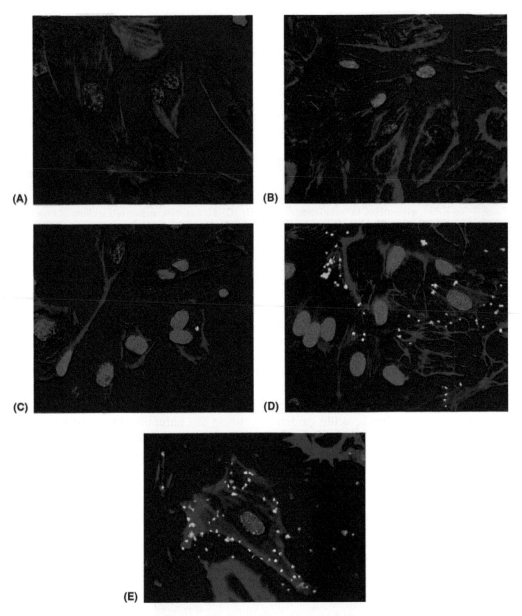

Figure 6   Intercellular adhesion molecule-1 (ICAM-1) targeted fluorescent (*green*) microbubbles binding to activated cultured human coronary artery endothelial cells in a flow chamber. (**A** and **C**) show cultured cells in the basal state; (**B**, **D** and **E**) show activated cells overexpressing ICAM-1 after exposure to interleukin 1-β. Nonspecific IgG-conjugated microbubbles did not adhere to cells under any conditions (**A**, **B**), whereas microbubbles conjugated to monoclonal antibody against ICAM-1 adhered adundantly to activated cells (**D**, **E**) and minimally to nonactivated cells (**C**). Scale bar indicates 10 µm. *Source*: From Ref. 52.

artery endothelial cells cultured in the perfusion chamber described above (56).

Because the selectin family of leukocyte adhesion molecules is acutely upregulated in response to ischemia, microbubble selectin targeting may provide a basis for ischemic memory imaging using ultrasound. Specifically, an imaging approach that can identify acute upregulation of inflammatory markers such as selectins should be helpful in patients presenting with chest pain of uncertain etiology, in whom the detection of upregulated selectins could be used

as a diagnostic marker of a recent ischemic event. In a rat model of myocardial ischemia/reperfusion without infarction, microbubbles conjugated to sLex generated persistent contrast enhancement in the reperfused risk zone after reperfusion, thus identifying that area as having been recently ischemic (Fig. 10) (57). This suggests that ultrasound imaging of acutely expressed inflammatory markers ("ischemic memory imaging") may be helpful in the rapid triage of patients presenting to the emergency room with chest pain of unclear etiology.

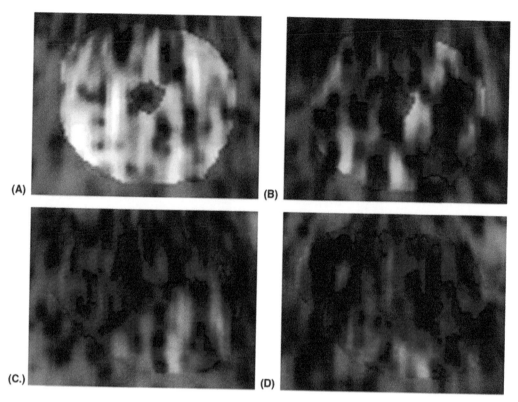

Figure 7 Detection of acute heart transplant rejection using ICAM-1-targeted microbubbles. Color-coded short-axis contrast echocardiographic images of rejecting allografts (**A, B**) and control isografts (**C, D**) after intravenous injection of ICAM-1 targeted (**A, C**) or control nonspecific IgG-conjugated microbubbles (**B, D**). There was greater, persistent opacification of rejecting myocardium after injection of ICAM-1 targeted microbubbles (**A**) compared to the other experimental conditions. *Source*: From Ref. 53.

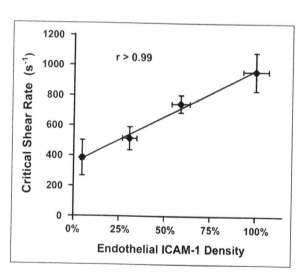

Figure 8 Targeted microbubble binding varies as a function of inflammation severity. Cultured human coronary endothelial cells were exposed for varying periods of time to interleukin 1-β and extent of ICAM1 expression (as percent of maximum) was determined by flow cytometry. Cells were perfused with ICAM-1-targeted bubbles in a radial flow chamber and the critical shear ring of bubble deposition, indicating the maximum wall shear rate at which adhesion still occurs, was measured. Using this as a marker of the magnitude of microbubble binding, it was found that critical shear rate and the degree of inflammation were linearly correlated. *Source*: From Ref. 54.

Figure 9 Targeted ultrasound imaging of postischemic renal inflammation using microbubbles targeted to P-selectin using anti-P selectin monoclonal antibody (**A, B**) or control isotype antibody-conjugated microbubbles (**C**) in wild type (**A, C**) or P-selectin knock out mice (**B**). Greatest persistent contrast enhancement occurred in wild type mice injected with P-selectin targeted microbubbles. *Source*: From Ref. 55.

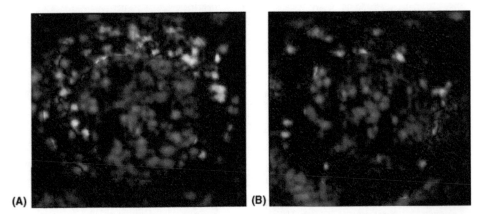

Figure 10   Ischemic memory imaging using P-selectin targeted mirobubbles. Short-axis color-coded echocardiographic images of rat hearts undergoing transient left coronary occlusion and reperfusion after intravenous injection of bubbles bearing sialyl-Lewis-X (**A**) or control sialyl-Lewis C (**B**) on the shell. There is persistent contrast enhancement in the anterior region corresponding to the risk area after injection of sialyl-LewisX-tagged microbubbles (**A**), despite the absence of infarction. There was no persistence in contrast enhancement after injection of sialyl-Lewis C tagged microbubbles (**B**).

MICROBUBBLE DESIGN FEATURES THAT ENHANCE TARGETED BINDING

A requirement to the development and ultimate clinical application of targeted ultrasound imaging is the improvement of the microbubble signal to noise ratio. One approach to augment the signal would be to increase the number of microbubbles adhering to the target. Altering microbubble design to optimize contact between the targeting moiety and the surface target or to increase the number and type of targeting moieties that can be attached to the bubble (58,59) are possible strategies for increasing the extent of microbubble binding. For example, there is a direct relationship between microbubble surface density of targeting ligand and extent of microbubble adhesion (58). Furthermore, an approach using multiple different targeting ligands on a single bubble (e.g., dually targeting selectins and ICAM-1) appears to enhance microbubble adhesion to inflammatory endothelium (54). On the imaging end, systems that can specifically detect adhered microbubble signals distinct from signals generated by freely circulating background microbubbles are needed. Possible approaches to improve signal detection by imaging systems include strategies to pull out nonlinear acoustic responses, which may be unique to adhered bubbles, or to sort returning signals based on motion detection.

SUMMARY

Targeted ultrasound imaging of inflammatory endothelium is possible using microbubbles that adhere directly to activated leukocytes or that bear targeting moieties which bind to adhesion molecules overexpressed by dysfunctional endothelium. This approach has been used to detect acute cardiac transplant rejection and viable postischemic tissue in animal models. Ultrasound imaging of endothelial inflammation as described previously can ultimately have useful application in the surveillance for rejection and follow-up of rejection treatments in patients undergoing orthotropic transplantation of the heart, and possibly of other solid organs such as the kidney. The ability to readily image adhesion molecules that are quickly and transiently elevated in response to acute ischemia, such as the selectins, offers the possibility of detecting recent ischemic events, which may be helpful in the risk stratification of patients presenting to the emergency room with recent chest pain and nondiagnostic electrocardiograms. To the extent that atherosclerotic disease is a diffuse process extending from epicardial coronary artery disease into the microcirculation, echocardiographic imaging of leukocyte adhesion molecules has the potential to identify early endothelial dysfunction at the microvascular level, although this remains to be proven. Steps towards clinical application will require targeting ligands that are safe for human use (e.g., nonimmunogenic), strategies to increase microbubble adhesion to endothelium through innovations in microbubble design, and novel ultrasound imaging systems specifically designed to detect adhered microbubbles.

REFERENCES

1. Kaul S. Myocardial contrast echocardiography: 15 years of research and development. Circulation 1997; 96:745–760.

2. Becher H, Burns PN. Handbook of Contrast Echo-cardiography: LV function and myocardial perfusion. Berlin, Heidelberg, New York: Springer-Verlag, 2000.

3. Jayaweera AR, Edwards N, Glasheen WP, et al. In vivo myocardial kinetics of air-filled albumin microbubbles during myocardial contrast echocardiography: comparison with radiolabeled red blood cells. Circ Res 1994; 74(6):1157–1165.

4. Gimbrone MA. Vascular endothelium: an integrator of pathophysiologic stimuli in atherosclerosis. Am J Cardiol 1995l; 75(6):67B–70B.

5. Vita JA, Keaney JF Jr. Endothelial function: a barometer for cardiovascular risk? Circulation 2002; 106(6):640–642.

6. Widlansky ME, Gokee N, Keaney JF, Vita JA. The clinical implications of endothelial dysfunction. J Am Coll Cardiol 2003; 42(7):149–1160.

7. Scanlon CEO, Berger M, Malcom G, Wissler RW. Evidence for more extensive deposits of epitopes of oxidized low density lipoprotein in aortas of young people with elevated serum thiocyanate levels. Atherosclerosis 1996; 121:23–33.

8. Ross R. The pathogenesis of atherosclerosis: a perspective for the 1990s. Nature 1993; 362:801–809.

9. Nabel EG. Biology of the impaired endothelium. Am J Cardiol 1991; 68(12):6C-8C.

10. Lerman A, Burnett JC. Intact and altered endothelium in regulation of vasomotion. Circulation 1992; 86(suppl III): III-12–III19.

11. Sluiter W, Pietersma A, Lamers JMJ, et al. Leukocyte adhesion molecules on the vascular endothelium: their role in the pathogenesis of cardiovascular disease and the mechanisms underlying their expression. J Cardiovasc Pharm 1993; 22(suppl 4):S37–S44.

12. Nakashima Y, Raines EW, Plump AS, et al. Upregulation of VCAM-1 and ICAM-1 at atherosclerosis-prone sites on the endothelium in the apo-E-deficient mouse. Arterioscler Thromb Vasc Biol 1998; 18:842–851.

13. Libby P, Ridker PM, Maseri A. Inflammation and atherosclerosis. Circulation 2002; 105:1135–1143.

14. Fuster V. Mechanisms leading to myocardial infarction: insights from studies of vascular biology. Circulation 1994; 90(4):2126–2146.

15. Collins T, Palmer HJ, Whitely MZ, et al. A common theme in endothelial activation. Insights from the structural analysis of the genes for E-selecting and VCAM-1. Trends Cardiovasc Med 1993; 3:92–97.

16. Miyasaka M, Kawashima H, Korenaga R, et al. Involvement of selectins in atherogenesis: a primary or secondary event? Annals N Y Acad Sci 1997; 811:25–34.

17. Cybulsky MI, Gimbrone MA. Endothelial expression of a mononuclear leukocyte adhesion molecule during atherogenesis. Science 1991; 251:788–791.

18. Shen Y, Rattan V, Sultana C, et al. Cigarette smoke condensate-induced adhesion molecule expression and transendothelial migration of monocytes. Am J Physiol 1996; 270:H1624–H1633.

19. Li H, Cybulsky MI, Gimbrone MA, et al. An atherogenic diet rapidly induces VCAM-1, a cytokine-regulatable mononuclear leukocyte adhesion molecule, in rabbit aortic endothelium. Arterioscler Thromb 1993; 13: 197–204.

20. Kuo L, Davis MJ, Cannon S, et al. Pathophysiological consequences of atherosclerosis extend into the coronary microcirculation. Restoration of endothelium-dependent responses by L-arginine. Circ Res 1992; 70:465–476.

21. Selke FW, Armstrong ML, Harrison DG. Endothelium-dependent vascular relaxation is abnormal in the coronary microcirculation of atherosclerotic primates. Circulation 1990; 81(5):1586–1593.

22. Sheridan FM, Cole PG, Ramage D. Leukocyte adhesion to the coronary microvasculature during ischemia and reperfusion in an in vivo canine model. Circulation 1996; 93(10):1784–1787.

23. Harjula A, Mattila S, Mattila I, et al. Coronary endothelial damage after crystalloid cardioplegia. Cardiovasc Surg 1984; 25:147–152.

24. Weidinger FF, McLenachan JM, Cybulsky MI, et al. Persistent dysfunction of regenerated endothelium after balloon angioplasty of rabbit iliac artery. Circulation 1990; 81(5):1667–1679.

25. Treasure CB, Klein JL, Weintraub WS, et al. Beneficial effects of cholesterol-lowering therapy on the coronary endothelium in patients with coronary artery disease. N Engl J Med 1995; 332:481–487.

26. Carlos T, Gordon D, Fishbein D, et al. Vascular cell adhesion molecule-1 is induced on endothelium during acute rejection in human cardiac allografts. J Heart Lung Transplant 1992; 11:1103–1109.

27. Ohtani H, Strauss W, Southern JF, et al. Intercellular adhesion molecule-1 induction: a sensitive and quantitative marker for cardiac allograft rejection. J Am Coll Cardiol 1995; 26:793–799.

28. Zeiher AM, Drexler H, Wollschlager H, et al. Modulation of coronary vasomotor tone in humans. Progressive endothelial dysfunction with different early stages of coronary atherosclerosis. Circulation 1991; 83(2):391–401.

29. Celermajer DS, Sorensen KE, Gooch VM, et al. Non-invasive detection of endothelial dysfunction in children and adults at risk of atherosclerosis. Lancet 1992; 340: 1111–1115.

30. Corretti MC, Plotnick GD, Vogel RA. Technical aspects of evaluating brachial artery vasodilatation using high-frequency ultrasound. Am J Physiol 1995; 268: H1397–H1404.

31. Keller MW, Spotnitz WD, Matthew TL, et al. Intraoperative assessment of regional myocardial perfusion using quantitative myocardial contrast echocardiography: an experimental evaluation. J Am Coll Cardiol 1990; 16:1267–1279.

32. Villanueva FS, Spotnitz WD, Jayaweera AR, et al. Myocardial contrast echocardiography in humans: On-line intraoperative quantitation of regional myocardial perfusion during coronary artery bypass graft operations. J Thorac Cardiovasc Surg 1992; 104:1529–1531.

33. Villanueva FS, Jankowski RJ, Manaugh C, et al. Albumin microbubble adherence to human coronary endothelium: implications for assessment of endothelial function using myocardial contrast echocardiography. J Am Coll Cardiol 1997; 30(3):689–693.

34. Basalyga DM, Wagner WR, Beer-Stolz D, et al. Albumin microbubbles adhere to exposed extracellular matrix of perfused whole vessels. Circulation 1998; 98:I-290.

35. Lindner JR, Ismail S, Spotnitz WD, et al. Albumin microbubble persistence during myocardial contrast echocardiography is associated with microvascular endothelial glycocalyx damage. Circulation 1998; 98(20):2187–2194.

36. Lindner JR, Coggins MP, Kaul S, et al. Microbubble persistence in the microcirculation during ischemia/reperfusion and inflammation is caused by integrin- and complement-mediated adherence to activated leukocytes. Circulation 2000; 101(6):668.

37. Dayton PA, Chomas JE, Lum AF, et al. Optical and acoustical dynamics of microbubble contrast agents inside neutrophils. Biophys J 2001; 80 (3):1547–1556.

38. Linder JR, Dayton PA, Coggins MP, et al. Noninvasive imaging of inflammation by ultrasound detection of phagocytosed microbubbles. Circulation 2000; 102 (5): 531–538.

39. Lindner JR, Long J, Xu F, et al. Noninvasive ultrasound imaging of inflammation using microbubbles targeted to activated leukocytes. Circulation 2000; 102:2745–2750.

40. Christiansen JP, Leong-Poi H, Klibanov AL, et al. Noninvasive imaging of myocardial reperfusion injury using leukocyte-targeted contrast echocardiography. Circulation 2002; 105 (15):1764–1767.

41. Klibanov AL, Hughes MS, Marsh JN, et al. Targeting of ultrasound contrast material: an in vitro feasibility study. Acta Radiol Suppl 1997; 38(suppl 412):113–120.

42. Klibanov A, Hughes M, Villanueva FS, et al. Targeting and ultrasound imaging of microbubble-based contrast agents. Magma 1999; 8:177–184.

43. Unger EC, McCreery TP, Sweitzer RH, et al. In vitro studies of a new thrombus-specific ultrasound contrast agent. Am J Cardiol 1998; 81(12A):58G–61G.

44. Schumann PA, Christiansen JP, Quigley RM, et al. Targeted-microbubble binding selectively to GPIIb IIIa receptors of platelet thrombi. Invest Radiol 2002; 37(11):587–593.

45. Alkan-Onyuksel H, Demos SM, Lanza GM, et al. Development of inherently echogenic liposomes as an ultrasonic contrast agent. J Pharmacol Sci 1996; 85(5): 485–490.

46. Demos SM, Dagar S, Klegerman M, et al. In vitro targeting of acoustically reflective immunoliposomes to fibrin under various flow conditions. J Drug Target 1998; 5(6):507–518.

47. Huang S, Hamilton AJ, Nagaraj A, et al. Improving ultrasound reflectivity and stability of echogenic liposomal dispersions for use as targeted ultrasound contrast agents. J Pharmacol Sci 2001; 90(12):1917–1926.

48. Hamilton A, Rabbat M, Jain P, et al. A physiologic flow chamber model to define intravascular ultrasound enhancement of fibrin using echogenic liposomes. Invest Radiol 2002; 37:215–221.

49. Lanza GM, Wickline SA. Targeted ultrasonic contrast agents for molecular imaging and therapy. Prog Cardiovasc Dis 2001; 44(1):13–31.

50. Hall CS, Marsh JN, Scott MJ, et al. Time evolution of enhanced ultrasonic reflection using a fibrin-targeted nanoparticulate contrast agent. Acoustical Soc Am 2000; 108(6):3049–3057.

51. Lanza GM, Trousil RL, Wallace KD, et al. In vitro characterization of a novel, tissue-targeted ultrasonic contrast system with acoustic microscopy. Acoustical Soc Am 1998; 104(6): 3665–3672.

52. Villanueva FS, Jankowski RJ, Klibanov S, et al. Microbubbles targeted to intercellular adhesion molecule-1 bind to activated coronary artery endothelial cells: a novel approach to assessing endothelial function using myocardial contrast echocardiography. Circulation 1998; 98(1):1–5.

53. Weller GE, Lu E, Csikari MM, et al. Ultrasound imaging of acute cardiac transplant rejection with microbubbles targeted to intercellular adhesion molecule-1. Circulation 2003; 108(2): 218–224.

54. Weller GER, Villanueva FS, Tom EM, et al. Targeted ultrasound contrast agents: In vitro assessment of endothelial dysfunction and multi-targeting to ICAM-1 and sialyl Lewis-X. Biotech Bioeng 2005; 92:780–788.

55. Lindner JR, Song J, Christiansen J, et al. Ultrasound assessment of inflammation and renal tissue injury with microbubbles targeted to p-selectin. Circulation 2001; 104:2107–2112.

56. Lu E, Tom EM, Felix MM, et al. In vivo microbubble binding to inflammatory endothelium via selectin targeting by Sialyl Lewis X. J Am Coll Cardiol 2004; 43:8A.

57. Villanueva FS, Lu E, Bowry S, et al. Myocardial ischemic memory imaging using molecular echocardiography. Circulation 2007; 115:345.

58. Weller GER, Villanueva FS, Klibanov AL, et al. Modulating targeted adhesion of an ultrasound contrast agent to dysfunctional endothelium. Ann Biomed Eng 2002; 30:1012–1019.

29

Imaging Matrix Metalloproteinase Expression: Applications for Cardiovascular Imaging

Carolyn J. Anderson
Mallinckrodt Institute of Radiology, Washington University School of Medicine, St. Louis, Missouri, U.S.A.

INTRODUCTION

Matrix Metalloproteinases

Matrix metalloproteinases (MMPs) are zinc-dependent secreted or transmembrane enzymes constituting a family of over 20 proteolytic members that are capable of selectively digesting a wide spectrum of both extracellular matrix (ECM) and nonmatrix proteins (Table 1). MMPs and their tissue inhibitors of matrix metalloproteinases (TIMPs) are central factors in the control of ECM turnover. It has long been known that MMPs play a critical role in tumor growth, angiogenesis, and metastatic processes (1). More recently, the role of MMPs has been explored in remodeling that occurs as a result of cardiac ischemia and infarction (2).

Role of Matrix Metalloproteinases in Myocardial Remodeling

MMP-1 expression is acutely increased in the rat ischemia reperfusion model following one hour of coronary ligation and one h of reperfusion before tissue analysis (3). MMP-1 in this model was cardiomyotoxic, and could be linked to myocardial weakening by favoring collagen scar degradation. Thus, measurement of MMP-1 during the acute phase may be a useful prognostic marker for left ventricular remodeling, and its inhibition may help to prevent myocardial injury and deleterious remodeling (2).

The expression and activities MMP-2, MMP-9 and the TIMPs increase in the heart after myocardial infarction (MI) (4–7). MMP-9 has been implicated in ischemia-reperfusion injury (8). Unlike MMP-1 that is generated by the myocardium, neutrophils are the main source of MMP-9. In mice that were MMP-9 (+/+), MMP-9 (+/−) and MMP-9 (−/−), myocardium was subjected to ischemia-reperfusion (8). Although the three groups of mice displayed a similar-sized ischemic area, in MMP-9 (−/−) mice, there was less neutrophil infiltration in the ischemic region of the heart, and consequently, the myocardial infarct size was reduced

compared to heterozygotes and MMP-9 (+/+) mice. These data suggest that MMP-9 could be a target for prevention or treatment of acute ischemic myocardial injury.

The membrane-type MMPs (MT-MMPs) are a unique class of MMPs that are proteolytically diverse in that protolytic actions serve several biological functions, including degradation of the local ECM (9), activation of other MMPs (10), and processing of other biologically active signaling molecules (10). One of the best characterized MT-MMP is MT1-MMP, well described by Deschamps et al. for its role in ischemia and reperfusion (11). Increased MT1-MMP activity during ischemia is due to a loss of inhibitory control by TIMP-3 and -4, whereas during reperfusion, increased trafficking of newly synthesized MT1-MMP to the membrane sustains the elevation in activity (11).

Increased plasma levels of MMP-9 and TIMP-1 have been demonstrated in acute coronary syndromes in humans compared to normal controls or stable disease (12). Circulating MMP-9 levels appear to rise and fall quickly (by seven days), whereas TIMP-1 rises slowly and remains higher for six months (13). It has been suggested that MMP-9 is an "acute phase MMP," whereas the increased levels of MMP-2 expression is maintained longer. It is unclear whether circulating levels reflect tissue levels, since this will be dependent on whether the markers are metabolized prior to appearing in the circulation (2).

MMP-based imaging agents will potentially address whether tissue levels of the various MMPs correlate with the plasma levels, although it would be important to have imaging agents that are specific for the various MMPs. The following sections of this chapter address the current state of the development of MMP-based imaging agents for myocardial injury. This area of study is relatively new, and there are many obstacles that need to be overcome for the development of specific MMP imaging agents, although progress is being made.

Table 1 Matrix Metalloproteinase Family

Group	Enzyme	Substrate(s)
Collagenases		
Collagenase-1	MMP-1	Collagen I, II, III,VII, X, gelatin, entactin, aggregan, tenascin
Collagenase-2	MMP-8	Collagen I, II, III, gelatin, entactin, aggregan, tenascin
Collagenase-3	MMP-13	Collagen I, II, III,VII, X, gelatin, entactin, aggregan, tenascin
Stromelysins		
Stromelysin-1	MMP-3	Collagen II, IV, V, IX, X, XI, vitronectin, fibrin/fibrinogen, fibronectin, laminin, gelatin, proteoglycan, entactin, tenascin
Stromelysin-2	MMP-10	Collagen III, IV, V, IX, X, XI, fibronectin, laminin, gelatin, proteoglycans, vitronectin, fibrin/fibrinogen, entactin, tenascin
		Fibronectin, laminin, aggregan
Stromolysin-2	MMP-11	Elastin, fibronectin, fibrin/fibrino-
Metalloelastase	MMP-12	gen, laminin, proteoglycan
Gelatinases		
Gelatinase A	MMP-2	Collagen I, IV, V, VII, X, XI, fibronectin, gelatin, elastin, proteoglycan, laminin, aggregan, vitronectin
Gelatinase B	MMP-9	Collagen I, IV, V, VII, X, XI, gelatin, elastin, fibronectin, laminin, aggregan, vitronectin
MT-MMPs		
MT1-MMP	MMP-14	Gelatin, fibronectin, vitronectin, collagen, aggregan
MT2-MMP	MMP-15	Gelatin, fibronectin, vitronectin, collagen, aggregan
MT3-MMP	MMP-16	Gelatin, fibronectin, vitronectin, collagen, aggregan
MT4-MMP	MMP-17	Gelatin
MT5-MMP	MMP-24	Gelatin, fibronectin, vitronectin, collagen, aggregan
MT6-MMP	MMP-25	Gelatin, collagen IV, fibrin, fibronectin, Laminin-1
Matrilysins		
Matrilysin	MMP-7	Collagen III, IV, V, IX, X, XI, fibronectin, laminin, gelatin, procollagenase, proteoglycans, fibrin/fibrinogen, entactin, tenascin, vitronectin
Matrilysin-2/ endometase	MMP-26	Gelatin, collagen IV, fibronectin, fibrinogen

Abbreviation: MMP, matrix metalloproteinase.
Source: From Ref. 2.

MATRIX METALLOPROTEINASES-BASED IMAGING AGENTS

Matrix Metalloproteinases-Based Imaging Agents for Tumor Imaging

There are a number of reports on MMP-based imaging agents for imaging MMP expression in tumors (14). Several synthetic sulfonamide-based MMP inhibitors have been radiolabeled with carbon-11, fluorine-18, or iodine-123; and preliminary evaluations of these radiotracers for tumor imaging have been performed (15–17). However, selective binding of labeled compounds to specific MMPs was not demonstrated, and high nonspecific binding was observed in vivo. A radiolabeled TIMP-2 protein was labeled with ^{111}In for imaging Kaposi sarcoma (KS); however, no KS lesions were clearly identified with this imaging agent (18).

Optical Imaging of Matrix Metalloproteinases Expression in Tumors

In a more successful approach, Bremer et al. reported on a biocompatible near-infrared fluorogenic MMP substrate that is used as an activatable reporter probe to sense MMP activity (19). The rationale for this study is based on the quenching of closely positioned fluorochromes that occurs when the enzyme is cleaved by MMP-2 in vivo. Images of tumors in nude mice were clearly visualized using this near infrared fluorescence (NIRF) imaging agent, while negative control rats that were injected with 150 mg/kg of the MMP-2 inhibitor prinomastat twice a day for two days showed significantly lower tumor uptake.

Radiolabeled CTT for Imaging MMP-2 and MMP-9 Expression in Tumors

Peptide and peptide-like MMP inhibitors have been designed to essentially mimic the collagen substrate of MMPs, and thereby work as potent, competitive inhibitors of enzyme activity. The cyclic decapeptide CTTHWGFTLC (CTT), containing a His-Try-Gly-Phe (HWGF) motif, has been described as a selective MMP-2 and MMP-9 inhibitor that reduced the migration of both human endothelial and tumor cells, and prevented tumor growth and invasion in animal models (20). The peptide CTT was shown to be an inhibitor of MMP-2 and MMP-9 with IC_{50} values of 13 and 15 μM, respectively. Kuhnast et al. recently reported derivatization and radiolabeling CTT with ^{125}I for and in vitro and in vivo studies (21). Unimpaired inhibition of MMP-2 activity was demonstrated in vitro for the derivatized CTT, and the iodinated peptide showed no degradation by activated MMP-2 and MMP-9. However, a partial deiodination of the ^{125}I-D-tyrosine from iodinated peptide was observed in vivo indicating metabolic instability.

Copper-64 is a useful diagnostic and therapeutic radionuclide in nuclear medicine due to its half-life ($t_{1/2}$ = 12.7 h), decay characteristics [$β^+$ (17.4%); $β^-$ (39%)] and the ability for large-scale production with high specific activity on a biomedical cyclotron (22,23). ^{64}Cu-DOTA-CTT (where DOTA is 1,4,7,10-tetraazacyclododecane-1,4,7,10-tetraacetic acid) was evaluated in a tumor-bearing animal model with limited success (14).

Fluorescence Detection and Optical Imaging of Matrix Metalloproteinases Expression

Deschamps et al. probed the expression of MT1-MMP in a pig model of ischemia/reperfusion (I/R) using microdialysis probes inserted within remote and ischemic myocardium and sutured in place (11). The probes were infused with a MTA-MMP fluorogenic substrate [MCA-Pro-Leu-Ala-Cys(p-OmeBz)-Trp-Ala-Arg(Dpa)-NH$_2$], where MCA is the the fluorophore and DPA is the quenching group. When MT1-MMP cleaved the substrate at the Arg-Cys bond, fluorescence was emitted. This study is unique since MT1-MMP was directly measured in the myocardium during I/R. The results demonstrated that MT1-MMP activity increased in a regional and time-dependent manner with acute I/R (90/120 minutes, respectively), which was associated with an increase in total MTA-MMP abundance and a concomitant decrease in the myocardium-specific TIMP-4 as well as TIMP-3 (11). In the prolonged model of I/R (60 minutes/7 days, respectively), MT1-MMP was increased in both the remote and ischemic regions and was associated with LV remodeling (11).

Chen et al. reported the synthesis of a NIRF probe that is activated by proteolytic cleavage by MMP-2 and MMP-9 (24). The sequence, SGKGPRQITA, was identified by screening a phage library of random sequences using MMP-9, and is cleaved between the Gln (Q) and Ile (I) residues. This sequence was used with the linking residues GGPRQITAGK(*FITC*)C to attach the fluorochrome Cy5.5 to a pegylated poly-L-lysine backbone. The NIRF probe was injected i.v. into mice at various times after having a LAD ligation, and the mice were sacrificed and the hearts excised for imaging. Using the NIRF probe in conjunction with zymography and real-time PCR for MMP-2 and MMP-9 expression, it was shown that the NIRF signal at early time-points post-MI was due to MMP-9, whereas the elevated NIRF signal after one-week post-MI was due to MMP-2. With the present optical imaging technology, imaging of the heart in vivo might be challenging due to the difficulties of imaging tissues located deep in the body. Combining NIRF with optical tomography methods could help resolve this problem (25).

Gamma Scintigraphy of Matrix Metalloproteinases Expression in Myocardial Infarction Models

The MMPs are responsible for degradation of the myocardial ECM that is associated with post-MI myocardial LV remodeling and often leads to heart failure, and a relationship between MMPs and LV remodeling has been demonstrated using animal models of heart disease and MMP inhibitors (26). The noninvasive detection and quantification of MMP activity during the process of post-MI remodeling will enable investigators to better understand the role of MMPs in

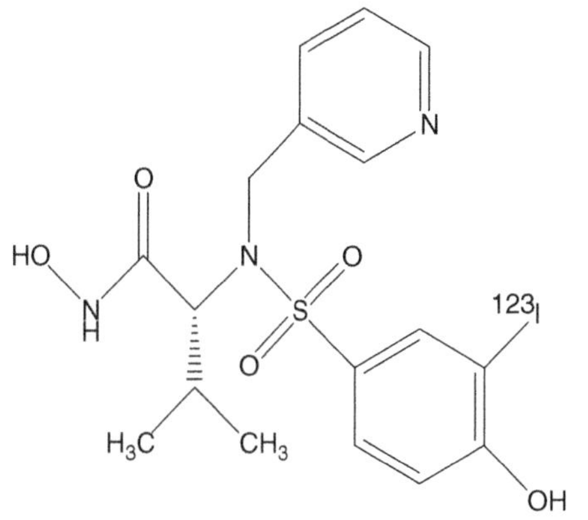

Figure 1   Structure of [^{123}I]I-HO-CGS 27023A.

remodeling. Methods to monitor MMP levels in tissue and plasma samples include in situ zymography, and the use of fluorogenic labeled peptide substrates (11).

In general, there are potential problems with peptide-based constructs, including instability in vivo. Another approach has been the radiolabeling of non-peptide compounds for gamma scintigraphy and single photon emission computed tomography (SPECT) imaging of cardiovascular disease. Schäfers et al. radioiodinated a modified version of the broad-spectrum MMP inhibitor CGS 27023A (methoxy moiety substituted with hydroxy) (Fig. 1) (27). This agent imaged MMP activity in the MMP-rich vascular lesions in an in vivo model of increased MMP activity, which is the arterial lesion developing in cholesterol-fed apolipoprotein-E-deficient mice after ligation of the carotid artery. The uptake of [^{123}I]I-HO-CGS 27023A in the arterial lesions was blocked by injection of unlabeled CGS 27023A 2 h prior to injection of the iodinated tracer, demonstrating specific uptake of the tracer for MMP activity.

Sinusas et al. have investigated a pan-MMP inhibitor (^{111}In-RP782; Fig. 2) developed by Bristol-Myers Squibb labeled with ^{111}In evaluating temporal changes in MMP activation in a murine model of MI using SPECT imaging (26). The imaging data, in addition to microautoradiography and gamma well counting revealed a significant increase in ^{111}In-RP782 retention within the region of the MI.

SUMMARY

There are over 20 members of the MMP family, and several of these are involved in remodeling that occurs as a result of cardiac ischemia and infarction, as well as in atherosclerosis. Higher plasma levels of various MMPs have been linked to acute coronary syndromes; however, it is unclear whether circulating levels reflect

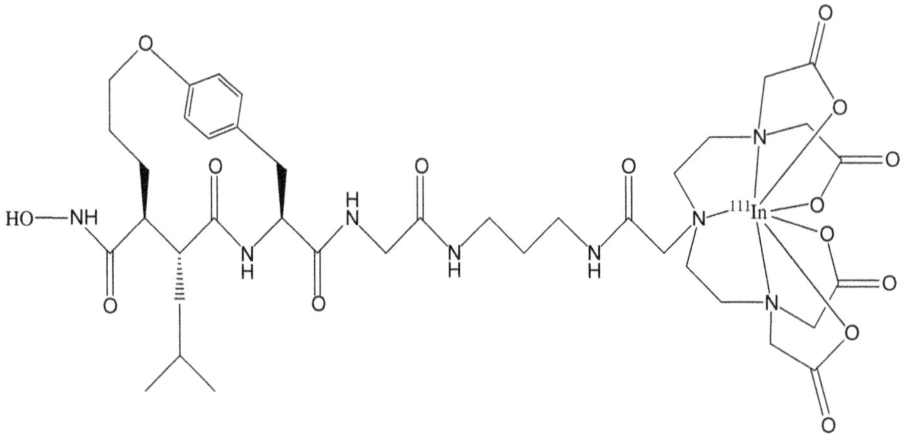

Figure 2 Structure of ^{111}In-RP782.

tissue levels. The development of diagnostic agents that will image levels of the various MMPs after MI or other types of cardiovascular disease is ongoing, and progress has been made in the areas of optical and nuclear medicine imaging modalities. More efforts are needed, however, towards developing a more specific MMP agent that will target only one MMP family member.

REFERENCES

1. Chambers AF, Matrisian LM. Changing views of the role of matrix metalloproteinases in metastasis. J Nat Canc Inst 1997; 89:1260.
2. Tayebjee MH, Lip GYH, MacFadyen RJ. Matrix metalloproteinases in coronary artery disease: clinical and therapeutic implications and pathological significance. Curr Med Chem 2005; 12:917.
3. Chen H, Li D, Saldeen T, Mehta JL. Tgf-b1 attenuates myocardial ischemia-reperfusion injury via inhibition of upregulation of mmp-1. Am J Physiol Heart Circ Physiol 2003; 284:H1612.
4. Spinale FG, Coker FG, Bond BR, Zellner JL. Myocardial matrix degradation and metalloproteinase activiation in the failing heart: a potential therapeutic target. Cardiovasc Res 2000; 46:225.
5. Spinale FG. Matrix metalloproteinases: regulation and dysregulation in the failing heart. Circ Res 2002; 90:520.
6. Lu L, Gunja-Smith Z, Woessner JF, et al. Matrix metalloproteinases and collagen ultrastructure in moderate myocardial ischemia and reperfusion in vivo. Am J Physiol 2000; 279:H601.
7. Danielsen CC, Wiggers H, Andersen HR. Increased amounts of collagenase and gelatinase in porcine myocardium following ischemia and reperfusion. J Mol Cell Cardiol 1998; 30:1431.
8. Romanic AM, Harrison SM, Bao W, et al. Myocardial protection from ischemia/reperfusion injury by targeted deletion of matrix metalloproteinase-9. Cardiovasc Res 2002; 54:549.
9. Woessner JF. MMPs and TIMPs—an historical perspective. Mol Biotechnol 2002; 22:33.
10. Strongin AY, Collier I, Bannikov G, Marmer BL, Grant GA, Goldberg GI. Mechanism of cell surface activation of 72-kDa type iv collagenase: Isolation of the activated form of the membrane metalloprotease. J Biol Chem 1995; 270:5331.
11. Deschamps AM, Yarbrough WM, Squires CE, et al. Trafficking of the membrane type-1 matrix metalloproteinase in ischemia and reperfusion: relation to interstitial membrane type-1 matrix metalloproteinase activity. Circulation 2005; 111:1166.
12. Inokubo Y, Hanada H, Ishizaka H, Fukushi T, Kamada T, Okumaura K. Plasma levels of matrix metalloproteinase-9 and tissue inhibitor of metalloprotease-1 are increased in the coronary circulation in patients with acute coronary syndrome. Am Heart J 2001; 141:211.
13. Kaden JJ, Dempfle C-E, Sueselbeck T, et al. Time-dependent changes in plasma concentration of matrix metalloproteinase-9 after acute myocardial infarction. Cardiology 2003; 99:140.
14. Sprague JE, Li WP, Liang K, Achilefu S, Anderson CJ. In vitro and in vivo investigation of matrix metalloproteinase expression in metastatic tumor models. Nucl Med Biol 2006; 33:227.
15. Zheng Q-H, Fei X, Liu Z, et al. Comparative studies of potential cancer biomarkers carbon-11 labeled MMP inhibitors (s)-2-(4′-[11C]methoxybiphenyl-4-sulfonylamino)-3-methoxyphenyl)sulfonyl]benzylamino]-3-methylbutanamide. Nucl Med Biol 2004; 31:77.
16. Oltenfreiter R, Staelens L, Lejeune A, et al. New radioiodinated carboxylic and hydroxamic matrix metalloproteinase inhibitor tracers as potential tumor imaging agents. Nucl Med Biol 2004; 31:459.
17. Furumoto S, Takashima K, Kubota K, Ido T, Iwata R, Fukuda H. Tumor detection using 18F-labeled matrix metalloproteinase-2 inhibitor. Nucl Med Biol 2003; 30:119.

18. Kulasegaram R, Giersing BK, Page CJ, et al. In vivo evaluation of 111In-DOTA-n-TIMP-2 in kaposi sarcoma associated with HIV infection. Eur J Nucl Med 2001; 28:756.
19. Bremer C, Tung C-H, Weissleder R. In vivo molecular target assessment of matrix metalloproteinase activity. Nature Med 2001; 7:743.
20. Koivunen E, Arap W, Valtanen H, et al. Tumor targeting with a selective gelatinase inhibitor. Nature Biotech 1999; 17:768.
21. Kuhnast B, Bodenstein C, Haubner R, et al. Targeting of gelatinase activity with a radiolabeled cyclic HWGF peptide. Nucl Med Biol 2004; 31:337.
22. McCarthy DW, Shefer RE, Klinkowstein RE, et al. The efficient production of high specific activity cu-64 using a biomedical cyclotron. Nucl Med Biol 1997; 24:35.
23. Obata A, Kasamatsu S, McCarthy DW, et al. Production of therapeutic quantities of 64Cu using a 12 MeV cyclotron. Nucl Med Biol 2003; 30:535.
24. Chen J, Tung C-H, Allport JR, Chen S, Weissleder R, Huang PL. Near-infrared fluorescent imaging of matrix metalloproteinase activity after myocardial infarction. Circulation 2005; 111:1800.
25. Ntziachristos V, Tung C-H, Bremer C, Weissleder R. Fluorescence molecular tomography resolved protease activity in vivo. Nature Med 2002; 8:757.
26. Su H, Spinale FG, Dobrucki LW, et al. Noninvasive targeted imaging of matrix metalloproteinase activation in a murine model of postinfarction remodeling. Circulation 2005; 112:3157.
27. Schafers M, Riemann B, Kopka K, et al. Scintigraphic imaging of matrix metalloproteinase activity in the arterial wall in vivo. Circulation 2004; 109:2554.

30

Multiple Roles of Cardiac Metabolism: New Opportunities for Imaging the Physiology of the Heart

Heinrich Taegtmeyer

Division of Cardiology, Department of Internal Medicine, The University of Texas Houston Medical School, Houston, Texas, U.S.A.

INTRODUCTION

Heart muscle is an efficeint energy converter. It derives the bulk of its energy for contraction from oxidative phosphorylation of ADP. In this system the transfer of energy is, in turn, tied to oxidative metabolsim of energy-producing substrates. In broader terms, and for all practical purposes, there is no life without oxygen. What applies to the body as a whole, applies even more to the heart. No other organ in the mammalian body features a higher rate of oxygen consumption than the heart (1). The tight coupling of myocardial oxygen consumption and the pump action of the heart has by now been recognized for more than a century (2) and includes a tight coupling between myocadial substrate utilization and contractile function (3–6).

The last decade has witnessed a new wave of interest in cardiac metabolism. The purpose of this chapter is to review available evidence that links alterations in metabolism to changes in contractile function of the heart. The first hypothesis is that metabolic changes often antedate functional contractile changes (Fig. 1), and that these changes can be traced by non-invasive imaging methods. A second, well known hypothesis, is that metabolic activity traces viability of stressed or injured myocardial tissue. Lastly, new evidence will be discussed that links metabolism to signaling pathways of cardiac growth and of gene expression in heart. In order to set the stage a brief review of the salient features of cardiac metabolism is appropriate.

SALIENT FEATURES OF CARDIAC METABOLISM

The heart is well designed, both anatomically and biochemically, for uninterrupted, rhythmic aerobic work. In short, the heart makes a living by liberating energy from different oxidizable substrates. The logic of metabolism is firmly grounded in the first law of thermodynamics that states that energy can neither be created nor destroyed (the law of the conservation of energy), and which is based on observations in muscle. In his early experiments on the chemistry of muscle contraction, Helmholtz concluded that "during the action of muscles, a chemical transformation of the compounds contained in them takes place" (7). This work culminated in the famous treatise "On the Conservation of Force." The principle is as valid today as it was at the time it was discovered 150 years ago. In essence, the first law of thermodynamics forms the basis for the stoichiometry of metabolism, for the calculation of the efficiency of cardiac performance, and for the analysis of recorded metabolic activities by non-invasive imaging techniques.

FLEXIBILITY OF CARDIAC METABOLISM

The similarities between the heart and the body as a whole are not limited to the dependency on oxygen. Like the mammalian organism, heart muscle is a metabolic omnivore with the capacity to oxidize fatty acids, carbohydrates and also (in certain circumstances) amino acids either simultaneously or vicariously. Both the complexity and the magnitude of myocardial energy metabolism can be overwhelming. Heart muscle is richly endowed with mitochondrial enzymes, and mitochondria make up more than one-third of the volume of a cardiomyocyte (1). It has been estimated that oxidative metabolism provides the human heart (weight: approximately 300 g) with more than 5 kg of ATP per day.

The intricate network of metabolic pathways that directly or indirectly supports the system of ATP generation seems bewildering (1). It is therefore convenient to group the catabolic reactions of cardiac energy metabolism into three stages (Fig. 2). Such grouping has the advantage that different stages can be probed by different tracers. Stage I includes breaking down of

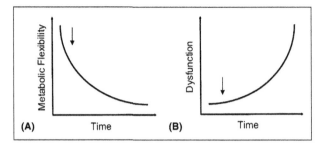

Figure 1 Hypothesis: Metabolic changes antedate functional contractile changes of the heart. See text for discussion.

energy providing substrates from their uptake by the cell to acetyl-CoA (which, in the case of fatty acids, glucose, and lactate, is a highly regulated process), stage II includes the generation of reducing equivalents and CO_2 in the citric acid cycle, and stage III includes the electron transfer, ATP generation and water

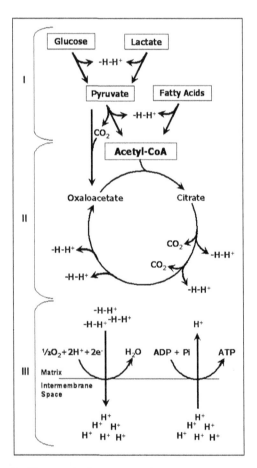

Figure 2 The metabolic reactions of cardiac energy metabolism can be grouped in three stages: Generation of acetyl-CoA (stage I), generation of reducing equivalents in the citric acid cycle (stage II), and electron transfer, ATP generation, and water production in the respiratory chain (stage III). All three stages can be selectively probed with positron labeled tracers (substrates, acetate, and oxygen) or with stable isotopes (^{13}C and ^{31}P).

production in the respiratory chain. The efficiency of the citric acid cycle (8) provides an example for the efficient transfer of energy through a series of moiety-conserved cycles (1). Radiolabeled tracers for imaging stage I include ^{11}C-labeled tracers or ^{18}F-labeled tracer analogs, tracers for imaging stage II include 11C acetate, and the tracer for imaging stage III is $^{15}O_2$.

Although under resting conditions and in the fasted state, the heart prefers fatty acids as its fuel for respiration (3), the heart readily switches to carbohydrates when stressed (6,9,10) or when supplied with glucose, insulin and potassium (11). In short, there is considerable flexibility in the system of energy substrate metabolism of the heart, as the heart responds to changes in the physiological state. Modern metabolism imaging techniques have only recently begun to exploit the opportunities of characterizing the dynamic nature of cardiac metabolism (12). The advantages and disadvantages of the three major imaging modalities for the noninvasive assessment of myocardial metabolism, magnetic resonance spectroscopy (MRS), single photon emission tomography (SPECT), and positron emission tomography (PET) have been discussed elsewhere (12).

Much work has been done in the isolated perfused rat heart to elucidate the mechanisms by which substrates compete for the fuel of respiration. In their celebrated studies in the 1960s, Philip Randle and his group established that, when present in sufficiently high concentrations, fatty acids suppress glucose oxidation to a greater extent than glycolysis, and glycolysis to a greater extent than glucose uptake, these observations gave rise to the concept of a "glucose-fatty acid-cycle" (13). The concept was later modified with the discovery of the suppression of fatty acid oxidation by glucose (14) through inhibition of the enzyme carnitine-palmitoyl transferase I (CPTI) (15). CPTI is, in turn, regulated by its rate of synthesis [by acetyl-CoA carboxylase (ACC)] and its rate of degradation [by malonyl-CoA decarboxylase, (MCD)]. Of the two enzymes, MCD is transcriptionally regulated by the nuclear receptor peroxisome proliferator activated receptor α (PPARα) (16), while ACCβ, the isoform that predominates in cardiac and skeletal muscle, is regulated both allosterically and covalently (17,18). High-fat feeding, fasting, and diabetes all increase MCD mRNA and activity in heart muscle (16). Conversely, cardiac hypertrophy that is associated with decreased PPARα expression (19,20) and a switch from fatty acid to glucose oxidation (21,22) results in decreased MCD expression and activity, an effect that is independent of fatty acids (20). Thus, MCD is regulated both transcriptionally and post-transcriptionally, and, in a feedforward mechanism, fatty acids induce MCD gene expression. The same principle applies to the regulation of other enzymes governing fatty acid metabolism in the heart, including the expression of uncoupling protein 3 (UCP3) (23). Here, fatty acids upregulate UCP3 expression, while UCP3 is downregulated in the hypertrophied heart that has switched to glucose for its main fuel of respiration. Thus, for a given physiologic environment, the heart selects the most efficient substrate for energy production. Fitting examples are the enzymes of glucose and glycogen metabolism, which are highly regulated by either allosteric activation or covalent modification. The regulation of glycogen phosphorylase by AMP and glucose may serve as illustration of how metabolites serve as signals that regulate enzyme fluxes through metabolic pathways. The acute increase in workload of the heart is accompanied by increases in [AMP] and intracellular free [glucose]. While AMP activates phosphorylase and promotes glycogen breakdown, free glucose inhibits phosphorylase and promotes glycogen synthesis (via an increase in glucose 6-phosphate and inhibition of glycogen synthase kinase).

Adaptations to sustained or chronic changes in the environment induce changes of the metabolic machinery at a transcriptional and/or translational level of the enzymes of metabolic pathways. We have therefore proposed elsewhere that metabolic remodeling precedes, triggers, and maintains structural and functional remodeling of the heart (24). Here the nuclear receptor PPARα and its coactivator PGC-1 need to be mentioned again, because they have been identified as master-switches for the metabolic remodeling of the heart (24–27). For example, pressure overload (28) and unloading of the heart (29), hypoxia (30), heart failure (31) and, unexpectedly, also insulin-deficient diabetes (32) all result in the downregulation of genes controlling fatty acid oxidation and in reactivation of the fetal gene program (Fig. 3). These recent observations are in line with earlier work showing increased glucose metabolic activity in the pressure overloaded heart (33), even before the onset of hypertrophy (34). They are also in line with work showing impaired fatty acid oxidation by failing heart muscle in vitro (35), and in vivo (36,37). While we have proposed, largely on theoretical grounds, that metabolic flexibility is lost in diseased heart (38), experimental evidence has recently been provided for the development of myocardial insulin resistance in conscious dogs with advanced dilated cardiomyopathy induced by rapid ventricular pacing (39). Equally important is the observation that metabolic dysregulation in severely obese young women, free of clinical heart disease, is associated with a decrease in cardiac efficiency in the absence of systolic contractile dysfunction (40). Thus, the concept of metabolic flexibility and loss of metabolic flexibility can now be readily tested by dynamic metabolic imaging of the heart.

Coordinated Transcriptional Responses

	Hypertrophy	Atrophy
Protooncogenes		
c-fos	↑	↑
Contractile proteins		
α-MHC	↓	↓
β-MHC	↑	↑
Cardiac α-actin	↓	↓
Skeletal α-actin	↑	↑
Ion pumps		
α₂ Na/K-ATPase	↓	↓
SERCA 2a	↓	↓
Metabolic proteins		
GLUT4	↓	↓
GLUT1	=	=
Muscle CPT-1	↓	↓
Liver CPT-1	=	=
mCK	↓	↓
PPARα	↓	↓
PDK4	↓	↓
MCD	↓	↓
UCP2	↓	↓
UCP3	↓	↓

Figure 3 The transcriptional response of the heart to either pressure overload or unloading is remarkably similar in hypertrophy and atrophy of the heart, showing upregulation of protooncogenes, isoform switches of contractile proteins, and downregulation of ion pumps as well as metabolic proteins. *Source*: From Ref. 24.

NONINVASIVE ASSESSMENT OF CARDIAC METABOLISM

As already mentioned, metabolic activity in the heart can be assessed both by MRS (also termed NMR spectroscopy) by SPECT, and by PET (1,12). Each method offers advantages for addressing specific metabolic questions. The focus of NMR spectroscopy has been the assessment of high-energy phosphate metabolism (using ^{31}P in its natural abundance) and carbon metabolism (using substrates enriched with ^{13}C). Both techniques are highly developed for the isolated perfused heart in vitro and also for the intact heart in vivo (41,42). The focus of PET imaging has been the assessment of intermediary metabolism of glucose and fatty acids in the heart in vivo, most commonly in conjunction with the assessment of myocardial blood flow and myocardial oxygen consumption (12,43,44). Tracers for the assessment of energy substrate metabolism include ^{11}C labeled glucose fatty acids, lactate, and acetate as well as ^{18}F-labeled tracer analogs of glucose and fatty acids. The principal difference between tracer and tracer analogs is depicted in Figure 4. Tracers are taken up by the heart, activated completely, metabolized, and released by the heart as either CO_2 or H_2O. Tracer analogs are taken up by the heart, are activated but only partially metabolized and retained by the heart. The

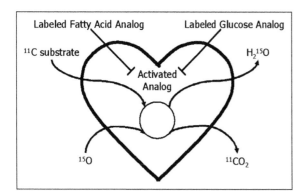

Figure 4 Tracers are taken up by the heart, metabolized, and released as CO_2 and H_2O. Tracer analogs such as the glucose analog [^{18}F] 2-deoxy, 2-fluoroglucose or the fatty acid analog 15[^{18}F] fluoro-3-oxa-pentadecanoate are partially metabolized and retained as "activated analogs."

cumulative radioactivity detected in the heart should therefore reflect substrate uptake and metabolism. A case in point is the linear tracer-time activity curve of [^{18}F] 2-deoxy, 2-fluoroglucose (FDG) accumulation in the isolated working rat heart perfused with glucose (5 mM) under steady state conditions (45) (Fig. 5). As we and others before us (46,47) have shown, prerequisites for the quantification of myocardial glucose uptake are a stable tracer/tracee ratio, and a stable

"lumped constant" (LC), which is a correction factor in the tracer kinetic model for the assessment of glucose uptake and phosphorylation by 2-deoxyglucose that consists of six separate constants and was first developed for brain studies by Louis Sokoloff at the National Institutes of Health (48). Although it is assumed that the LC is also constant in the heart in situ (49,50), the LC changes under nonsteady state conditions, e.g., with the administration of insulin (51) (Fig. 6) and with reperfusion after ischemia (52), and we have proposed a tracer kinetic model that takes into account changes of the LC under nonsteady state conditions (53). Similar considerations involving an LC and its variability also apply to tracer analogs of fatty acids such as ^{18}F-labeled 4-thia palmitate TR (54), and 15[^{18}F] fluoro-3-oxa-pentadecanoate (55), which have been used to assess derangements of myocardial fatty acid metabolism in heart failure (56).

MULTITASKING OF CARDIAC METABOLISM

In recent years it has come to light that the actions of metabolism are more diverse than those found in the network of energy transfer and function of the heart. Indeed, metabolism of substrates (fatty acids, carbohydrates, and amino acids), as well as adenine

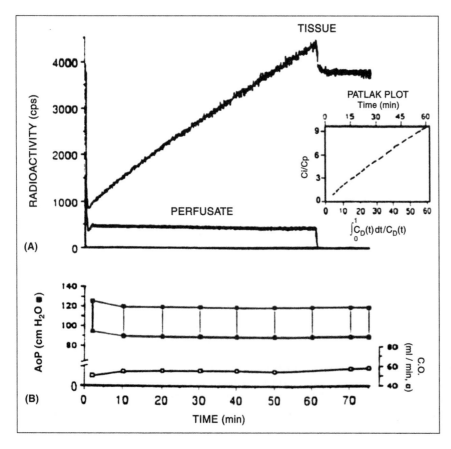

Figure 5 The tracer analog [^{18}F] 2-deoxy, 2-fluoroglucose (FDG) has been validated in the isolated working rat heart. Time-activity curves obtained for a heart perfused at intermediate workload (afterload = 100 cmH$_2$O, preload = 15 cmH$_2$O) with Krebs-Henseleit saline containing glucose (10 mM), plus 2-[^{18}F]fluoro-2-deoxy-D-glucose (2-FDG, 350 µCi/200 mL perfusate) for the first 60 minutes. (**A**, *top tracing*) Myocardial retention of 2-FDG, (*bottom tracing*) 2-FDG activity in recirculating perfusate. A Patlak plot, obtained from graphical analysis of decay-corrected tissue and perfusate curves, is shown in *inset*. (**B**) Physiological performance of heart in terms of aortic pressure (AoP, cmH$_2$O) and cardiac output (CO, mL/min.). Note that at 60 minutes, radioactive perfusate was changed to a nonradioactive medium containing only "cold" glucose as substrate. Radioactivity in the heart remains stable. *Source*: From Ref. 45.

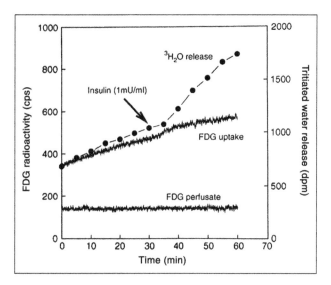

Figure 6 Differential effect of insulin on the uptake of glucose and [¹⁸F]2-glucose-2fluoro-D-glucose (FDG). Time-activity curves of myocardial uptake of FDG are simultaneously compared with time-activity curves of tritiated water released from [2-³H] glucose. The release of ³H₂O from [2-³H]glucose is used to assess glucose transport and phosphorylation. FDG activity in the perfusate (input function) remains constant throughout the experiment. The uptake of [2-³H]glucose increased after a delay of about five minutes when insulin was added at 30 minutes of perfusion (*arrow*). No significant change was observed in the uptake of FDG. *Source*: From Ref. 51.

nucleotides (AMP) provide signals for growth, gene expression, programmed cell death, and program cell survival. The role of metabolic signaling is conceptually long recognized in myocardial stunning and hibernation (57), but less well appreciated in cardiac growth and gene expression. Much has been learned from genetically engineered animal models of altered myocardial growth and metabolism. The ultimate question is, however, to understand whether the mechanism that works in mice also operates in humans. In other words, can the "proof of principle" be applied to human physiology and disease? Metabolic imaging is in a unique position to address this question in the three areas outlined below. As more knowledge is gained, other areas may be added.

Metabolic Signals for Cardiac Growth

A case in point is the mammalian target of rapamycin (mTOR), an evolutionary conserved kinase and regulator of cell growth that serves as a point of convergence for nutrient sensing and growth factor signaling. In recent studies with the isolated working rat heart and in isolated rat ventricular myocytes, we found that both glucose and amino acids are required for the activation of mTOR by insulin (58). We observed an unexpected dissociation between insulin stimulated Akt and mTOR activity, suggesting that Akt is not an upstream regulator

of mTOR. We found that irrespective of the stimulus, nutrients are critical for the activation of mTOR in the heart. Specifically, hexose 6-phosphates and the branched-chain amino acid leucine are required for the activation of mTOR and its downstream targets p70S6K and 4EBP1, both of which are regulators of protein synthesis. These initial observations strongly suggest a link between intermediary metabolism and cardiac function.

Metabolic Signals of Cardiac Gene Expression

A single factor linking myosin heavy chain (MHC) isoform expression in the fetal, hypertrophied, and diabetic heart is intracellular free glucose (59). Compared to fatty acids and their function as ligand for the nuclear receptor PPARα, relatively little is known about the effects of glucose metabolism on cardiac gene expression (59). The precise mechanisms by which glucose availability affects the DNA binding of transcription factors are not known, although it has been shown that glucose availability and/or insulin affect the expression of specific genes in the liver (60). A number of candidate transcription factors have been identified that are believed to be involved in glucose-mediated gene expression, mainly through investigations on the glucose/carbohydrate responsive elements. Carbohydrate responsive element-binding protein (ChREBP) (61), sterol regulatory element binding proteins (SREBPs), stimulatory protein 1 (Sp1), and upstream stimulatory factor 1 (USF1) (62) have all been implicated in glucose sensing by non-muscle tissues (59). It is likely that such glucose-sensing mechanisms also exist in heart muscle.

In preliminary experiments, we observed that altered glucose homeostasis through feeding of an isocaloric low carbohydrate, high fat diet completely abolishes MHC isoform switching in the hypertrophied heart (32). One mechanism by which glucose affects gene expression is through O-linked glycosylation of transcription factors. Glutamine, fructose-6-phosphate amidotransferase (*gfat*) catalyzes the flux-generating step in UDP-N-acetylglucosamine biosynthesis, the rate determining metabolite in protein glycosylation (Fig. 7). In preliminary studies, we observed that overload increases the intracellular levels of UDP-N-acetylglucosamine (unpublished work in collaboration with Dr. Don McClain, University of Utah). Thus, there is early evidence for glucose-regulated gene expression in the heart and, more specifically, for the involvement of glucose metabolites in isoform switching of sarcomeric proteins. More importantly, excess O-GlcNAcylation in the diabetic heart appears to play a significant role in cardiac function because reducing this excess cellular O-GlcNAcylation improves calcium handling and cardiac contraction in diabetic mice (63).

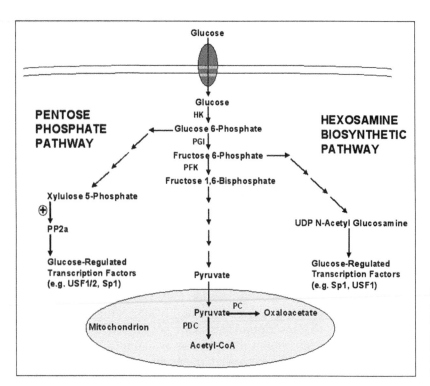

Figure 7 Relative inhibition of glucose oxidation or inhibition of the glycolytic pathway may result in the spill-over of hexose 6-phosphates into the pentosephosphate pathway (*left*) and/or into the hexosamine biosynthetic pathway (*right*). Intermediates of both pathways regulate transcription factors. *Source*: From Ref. 59.

Imaging Viability and Programmed Cell Survival

Perhaps the most dramatic example of chronic metabolic adaptation is the hibernating myocardium. Hibernating myocardium represents a chronically dysfunctional myocardium most likely the result of extensive cellular reprogramming due to repetitive episodes of ischemia or chronic hypoperfusion (57). The adaptation to reduced oxygen delivery results in a downregulation of contractile function and the prevention of irreversible tissue damage. A functional characteristic of hibernating myocardium is the gradual improvement in contractile function with inotropic stimulation or reperfusion. A metabolic characteristic of hibernating myocardium is the switch from fat to glucose metabolism, accompanied by reactivation of the fetal gene program. Because glucose transport and phosphorylation is readily traced by the uptake and retention of FDG, hibernating myocardium is readily detected by enhanced glucose uptake and glycogen accumulation in the same regions (Fig. 8) (64,65). While rates of glucose oxidation are reduced, the glycogen content of hibernating myocardium is dramatically increased (63,66). There is a direct correlation between glycogen content and myocardial levels of ATP (67),

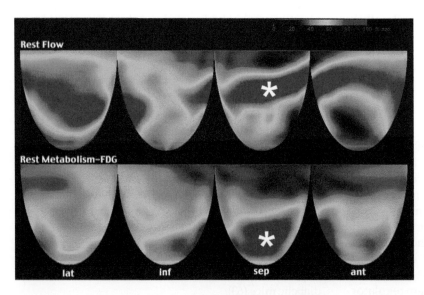

Figure 8 A hallmark of hibernating myocardium is the mismatch of perfusion ($^{13}NH_3$ uptake, *top row* of scans) and metabolism (in this case glucose uptake traced by $[^{18}F]$ 2-deoxy, 2-fluoroglucose, *bottom* scans) in a patient with chronic ischemic heart disease. Positon emission tomography imaging exposes a massive metabolic remodeling and survival program, further discussed in the text. *Source*: Courtesy of Dr. K. Lance Gould, University of Texas Houston Medical School, Houston, Texas, U.S.A.

and one is tempted to speculate that improved "energetics" may be the result of improved glycogen metabolism in hibernating myocardium. However, the true mechanism for "viability remodeling" of ischemic myocardium is likely to be much more complex, and there is good evidence for the activation of a gene expression program of cell survival (63). The similarities between hibernating and fetal myocardia (1) suggest an innate mechanism of myocardial protection.

The vast literature on programmed cell death, or apoptosis (68,69), and recent ideas on programmed cell survival (63,67) support the idea of a direct link between metabolic pathways and the pathways of cell survival and destruction. While the central role of mitochondria in programmed cell death is well defined (68,69), another metabolic mechanism deserves consideration. In cancer cells, there is striking evidence for a link between cell survival and metabolism. Cancer cells not only possess an increased rate of glucose metabolism (70), they are also less likely to "commit suicide" when stressed (71). The same general principle appears to apply to the hibernating myocardium, where the downregulation of function and oxygen consumption is viewed as an adaptive response when coronary flow is impaired (72). In other words, metabolic reprogramming initiates and sustains the functional and structural feature of hibernating myocardium.

The hypothesis finds further support by the observations that insulin promotes tolerance against ischemic cell death via the activation of issue-specific cell-survival pathways in the heart (73). Specifically, activation of PI3 kinase, a downstream target of the insulin receptor substrate (IRS), and activation of protein kinase B/Akt, are mediators of antiapoptotic, cardioprotective signaling through activation of p70s6 kinase and inactivation of proapoptotic peptides. The major mediator is Akt (or protein kinase B). Akt is located at the center of insulin and insulin-like growth factor 1 (IGF1) signaling. As the downstream serine-threonine kinase effector of PI3 kinase, Akt plays a key role in regulating cardiomyocyte growth and survival (74). Overexpression of constitutively active Akt raises myocardial glycogen levels and protects against ischemic damage in vivo and in vitro (75). Not surprisingly, Akt is also a modulator of metabolic substrate utilization (76). Phosphorylation of GLUT4 by Akt promotes its translocation and increases glucose uptake. Although the "insulin hypothesis" is attractive, there is also good evidence showing that the signaling cascade is dependent on the first committed step of glycolysis and translocation of hexokinase to the outer mitochondrial membrane (77,78). These few examples illustrate the fact that any signals detected by metabolic imaging of stressed or failing heart are the product of complex cellular reactions—truly only the tip of an iceberg.

SUMMARY

The last decade has witnessed a new wave of interest in cardiac metabolism. Part of this renaissance is due to the growing realization that energy substrate metabolism and function of the heart are two sides of the same coin, and part of the renaissance is the development of a wide array of molecular and imaging tools that allow the investigator to probe for specific metabolic mechanisms underlying normal and abnormal cardiac function. A new picture of multiple roles for cardiac metabolsim has emerged from these developments: The heart is an adaptable organ with the capacity to change both the nature and magnitude of its metabolism from beat to beat. For a given environment the heart oxidizes the most efficient fuel to support contractile function. For example, the stressed heart readily switches from fatty acids to glucose or lactate as its main fuel for respiration. Substrate switching and metabolic flexibility are therefore features of normal cardiac function. An example for dramatic adaptation of myocardial fuel metabolism is the upregulation of glucose metabolism in the course of myocardial ischemia. This review discusses the hypothesis that loss of metabolic flexibility and metabolic remodeling precede, trigger, and sustain functional and structural remodeling of the heart. The review also highlights the pleiotropic actions of metabolism in energy transfer, cardiac growth, gene expression, and viability. Examples are presented to illustrate that signals detected by metabolic imaging are the result of complex crosstalk between intermediary metabolism and signal transduction pathways. Lastly, it is proposed that the future of metabolic imaging will require some form of reorientation according to the new insights gained from laboratory studies into metabolic adaptation and maladaptation of the heart.

CONCLUSION: THE FUTURE OF METABOLIC IMAGING

The future of metabolic imaging will require some form of reorientation according to a new and more refined understanding of pathways of metabolic adaptation and maladaptation in the heart (Fig. 9). In this paradigm, metabolic signals alter metabolic fluxes and give rise to specific metabolic signals that, in turn, lead to changes in translational and/or transcriptional activities in the cardiac myocyte. In other words, metabolism provides the link between environmental stimuli and signaling pathways of cardiac growth and function. The concept is worth exploring with respect to novel applications for metabolic cardiac imaging. A particularly exciting recent development is the in vivo

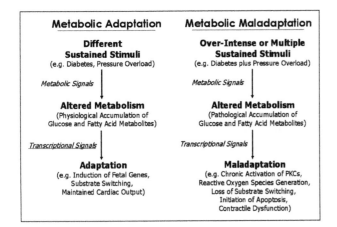

Figure 9 Metabolic signals serve as intermediate steps in the induction of adaptive and maladaptive responses to sustained stimuli. *Source*: From Ref. 59.

imaging of enzymes and their activities using labeled probes acting as substrates for target enzymes (79,80). The future of metabolic imaging could not be brighter (the pun is intended).

ACKNOWLEDGMENTS

The author thanks past and present members of his laboratory for many discussions, and Roxy A. Tate of her editorial assistance, as well as Rebecca L. Salazar for help with the illustrations. Work in the author's laboratory is supported by grants from the National Institutes of Health of the U. S. Public Health Service.

REFERENCES

1. Taegtmeyer H. Energy metabolism of the heart: from basic concepts to clinical applications. Curr Prob Cardiol 1994; 19(9):57–116.
2. Winterstein H. Ueber die Sauerstoffatmung des isolierten Saeugetierherzens. Z Allg Physiol 1904; 4: 339–359.
3. Bing RJ, Siegel A, Ungar I, et al. Metabolism of the human heart. II. Studies on fat, ketone and amino acid metabolism. Am J Med 1954; 16(4):504–515.
4. Neely JR, Liebermeister H, Battersby EJ, et al. Effect of pressure development on oxygen consumption by isolated rat heart. Am J Physiol 1967; 212(4):804–814.
5. Taegtmeyer H, Hems R, Krebs HA. Utilization of energy-providing substrates in the isolated working rat heart. Biochem J 1980; 186(3):701–711.
6. Goodwin GW, Taylor CS, Taegtmeyer H. Regulation of energy metabolism of the heart during acute increase in heart work. J Biol Chem 1998; 273(45):29530–29539.
7. Holmes FL. Between Biology and Medicine: The Formation of Intermediary Metabolism. Berkeley, CA: University of California at Berkeley, 1992.

8. Baldwin JE, Krebs HA. The evolution of metabolic cycles. Nature 1981; 291(5814):381–382.
9. Goodwin GW, Ahmad F, Taegtmeyer H. Preferential oxidation of glycogen in isolated working rat heart. J Clin Invest 1996; 97(6):1409–1416.
10. Goodwin GW, Taegtmeyer H. Improved energy homeostasis of the heart in the metabolic state of exercise. Am J Physiol Heart Circ Physiol 2000; 279(4):H1490–H1501.
11. Korvald C, Elvenes OP, Myrmel T. Myocardial substrate metabolism influences left ventricular energetics in vivo. Am J Physiol Heart Circ Physiol 2000; 278(4): H1345–H1351.
12. Herrero P, Gropler RJ. Imaging of myocardial metabolism. J Nucl Cardiol 2005; 12(3):345–358.
13. Randle PJ, Garland PB, Hales CN, et al. The glucose fatty-acid cycle. Its role in insulin sensitivity and the metabolic disturbances of diabetes mellitus. Lancet 1963; 1:785–789.
14. Taegtmeyer H, Hems R, Krebs HA. Utilization of energy providing substrates in the isolated working rat heart. Biochem J 1980; 186(3):701–711.
15. McGarry JD, Mills SE, Long CS, et al. Observations on the affinity for carnitine and malonyl-CoA sensitivity of carnitine palmitoyl transferase I in animal and human tissues. Demonstration of the presence of malonyl-CoA in non-hepatic tissues of the rat. Biochem J 1983; 214(1):21–28.
16. Young ME, Goodwin GW, Ying J, et al. Regulation of cardiac and skeletal muscle malonyl-CoA decarboxylase by fatty acids. Am J Physiol Endocrinol Metab 2001; 280(3):E471–E479.
17. Hardie DG, Carling D. The AMP-activated protein kinase—fuel gauge of the mammalian cell? Eur J Biochem 1997; 246(2):259–273.
18. Ruderman NB, Saha AK, Vavvas D, et al. Malonyl-CoA, fuel sensing, and insulin resistance. Am J Physiol 1999; 276(1 Pt 1):E1–E18.
19. Barger PM, Brandt JM, Leone TC, et al. Deactivation of peroxisome proliferator-activated receptor-alpha during cardiac hypertrophic growth. J Clin Invest 2000; 105(12): 1723–1730.
20. Taegtmeyer H, Razeghi P, Young ME. Mitochondrial proteins in hypertrophy and atrophy: a transcript analysis in rat heart. Clin Exp Pharmacol Physiol 2002; 29(4):346–350.
21. Allard MF, Schonekess BO, Henning SL, et al. Contribution of oxidative metabolism and glycolysis to ATP production in hypertrophied hearts. Am J Physiol 1994; 267(2 Pt 2):H742–H750.
22. Doenst T, Goodwin GW, Cedars AM, et al. Load-induced changes in vivo alter substrate fluxes and insulin responsiveness of rat heart in vitro. Metabolism 2001; 50(9):1083–1090.
23. Young ME, Patil S, Ying J, et al. Uncoupling protein 3 transcription is regulated by peroxisome proliferator-activated receptor(alpha) in the adult rodent heart. FASEB J 2001; 15(3):833–845.
24. Taegtmeyer H. Genetics of energetics: transcriptional responses in cardiac metabolism. Ann Biomed Eng 2000; 28(8):871–876.

25. Barger PM, Kelly DP. PPAR signaling in the control of cardiac energy metabolism. Trends Cardiovasc Med 2000; 10(6):238–245.

26. Kelly DP. PPARs of the heart: three is a crowd. Circ Res 2003; 92(5):482–484.

27. Huss JM, Kelly DP. Nuclear receptor signaling and cardiac energetics. Circ Res 2004; 95(6):568–578.

28. Lehman JJ, Kelly DP. Gene regulatory mechanisms governing energy metabolism during cardiac hypertrophic growth. Heart Fail Rev 2002; 7(2):175–185.

29. Depre C, Shipley GL, Chen W, et al. Unloaded heart in vivo replicates fetal gene expression of cardiac hypertrophy. Nat Med 1998; 4(11):1269–1275.

30. Razeghi P, Young ME, Abbasi S, et al. Hypoxia in vivo decreases peroxisome proliferator-activated receptor alpha-regulated gene expression in rat heart. Biochem Biophys Res Commun 2001; 287(1):5–10.

31. Huss JM, Kelly DP. Mitochondrial energy metabolism in heart failure: a question of balance. J Clin Invest 2005; 115(3):547–555.

32. Young ME, Guthrie P, Stepkowski S, et al. Glucose regulation of sarcomeric protein gene expression in the rat heart. J Mol Cell Cardiol 2001; 33:A181 (abstract).

33. Bishop S, Altschuld R. Increased glycolytic metabolism in cardiac hypertrophy and congestive heart failure. Am J Physiol 1970; 218(1):153–159.

34. Taegtmeyer H, Overturf ML. Effects of moderate hypertension on cardiac function and metabolism in the rabbit. Hypertension 1988; 11(5):416–426.

35. Wittels B, Spann JF. Defective lipid metabolism in the failing heart. J Clin Invest 1968; 47(8):1787–1794.

36. Sack MN, Rader TA, Park S, et al. Fatty acid oxidation enzyme gene expression is downregulated in the failing heart. Circulation 1996; 94(11):2837–2842.

37. Davila-Roman VG, Vedala G, Herrero P, et al. Altered myocardial fatty acid and glucose metabolism in idiopathic dilated cardiomyopathy. J Am Coll Cardiol 2002; 40(2):271–277.

38. Taegtmeyer H, Golfman L, Sharma S, et al. Linking gene expression to function: metabolic flexibility in normal and diseased heart. Ann N Y Acad Sci 2004; 1015:202–213.

39. Nikolaidis LA, Sturzu A, Stolarski C, et al. The development of myocardial insulin resistance in conscious dogs with advanced dilated cardiomyopathy. Cardiovasc Res 2004; 61(2):297–306.

40. Peterson LR, Herrero P, Schechtman KB, et al. Effect of obesity and insulin resistance on myocardial substrate metabolism and efficiency in young women. Circulation 2004; 109(18):2191–2196.

41. Ingwall JS. ATP and the Heart. Boston: Kluwer Academic Publishers, 2002.

42. Des Rosiers C, Lloyd S, Comte B, et al. A critical perspective of the use of(13)C-isotopomer analysis by GCMS and NMR as applied to cardiac metabolism. Metab Eng 2004; 6(1):44–58.

43. Schelbert HR. PET contributions to understanding normal and abnormal cardiac perfusion and metabolism. Ann Biomed Eng 2000; 28(8):922–929.

44. Dence CS, Herrero P, Schwarz SW, et al. Imaging myocardium enzymatic pathways with carbon-11 radiotracers. Methods Enzymol 2004; 385:286–315.

45. Nguyễn VTB, Mossberg KA, Tewson TJ, et al. Temporal analysis of myocardial glucose metabolism by 18F-2-deoxy-2-fluoro-D-glucose. Am J Physiol 1990; 259(4 Pt 2):H1022–H1031.

46. Phelps M, Hoffman E, Selin C, et al. Investigation of [18F] 2-fluoro-2-deoxyglucose for the measure of myocardial glucose metabolism. J Nucl Med 1978; 19(12):1311–1319.

47. Krivokapich J, Huang S-C, Phelps M, et al. Estimation of rabbit myocardial metabolic rate for glucose using fluoro-deoxyglucose. Am J Physiol 1982; 243(6): H884–H895.

48. Sokoloff L, Reivich M, Kennedy C, et al. The [14C] deoxyglucose method for the measurement of local cerebral glucose utilization: Theory, procedure, and normal values in the conscious and anesthetized albino rat. J Neurochem 1977; 28(5):897–916.

49. Ratib O, Phelps ME, Huang SC, et al. Positron tomography with deoxyglucose for estimating local myocardial glucose metabolism. J Nucl Med 1982; 23(7):577–586.

50. Gambhir SS, Schwaiger M, Huang SC, et al. Simple noninvasive quantification method for measuring myocardial glucose utilization in humans employing positron emission tomography and fluorine-18 deoxyglucose. J Nucl Med 1989; 30(3):359–366.

51. Hariharan R, Bray MS, Ganim R, et al. Fundamental limitations of [18F] 2-deoxy-2-fluoro-D-glucose for assessing myocardial glucose uptake. Circulation 1995; 91(9):2435–2444.

52. Doenst T, Taegtmeyer H. Profound underestimation of glucose uptake by [18F]2-deoxy-2-fluoroglucose in reperfused rat heart muscle. Circulation 1998; 97(24): 2454–2462.

53. Bøtker HE, Goodwin GW, Holden JE, et al. Myocardial glucose uptake measured with fluorodeoxyglucose: a proposed method to account for variable lumped constants. J Nucl Med 1999; 40(7):1186–1196.

54. DeGrado TR, Wang S, Holden JE, et al. Synthesis and preliminary evaluation of(18)F-labeled 4-thia palmitate as a PET tracer of myocardial fatty acid oxidation. Nucl Med Biol 2000; 27(3):221–231.

55. DeGrado TR, Wang S, Rockey DC. Preliminary evaluation of 15-[18F]fluoro-3-oxa-pentadecanoate as a PET tracer of hepatic fatty acid oxidation. J Nucl Med 2000; 41(10):1727–1736.

56. Taylor M, Wallhaus T, DeGrado T, et al. An evaluation of myocardial fatty acid and glucose uptake using PET with [18F]fluoro-6-thia-heptadecanoic acid. J Nucl Med 2001; 42(1):55–62.

57. Depre C, Vatner SF. Mechanisms of cell survival in myocardial hibernation. Trends Cardiovasc Med 2005; 15(3):101–110.

58. Sharma S, Golfman L, Burgmaier M, et al. Nutrient regulation of mTOR in the rat heart (abstract). FASEB J 2005; 19:A693.

59. Young ME, McNulty P, Taegtmeyer H. Adaptation and maladaptation of the heart in diabetes: Part II: potential mechanisms. Circulation 2002; 105(15):1861–1870.

60. Ferre P. Regulation of gene expression by glucose. Proc Nutr Soc 1999; 58(3):621–623.

61. Uyeda K, Yamashita H, Kawaguchi T. Carbohydrate responsive element-binding protein (ChREBP): a key regulator of glucose metabolism and fat storage. Biochem Pharmacol 2002; 63(12):2075–2080.

62. Girard J, Ferre P, Foufelle F. Mechanisms by which carbohydrates regulate expression of genes for glycolytic and lipogenic enzymes. Annu Rev Nutr 1997; 17: 325–352.

63. Hu Y, Belke D, Suarez J, et al. Adenovirus-mediated overexpression of O-GlcNAcase improves contractile function in the diabetic heart. Circ Res 2005; 96(9): 1006–1013.

;64. Mäki M, Luotolahti M, Nuutila P, et al. Glucose uptake in the chronically dysfunctional but viable myocardium. Circulation 1996; 93(9):1658–1666.

65. Depre C, Vanoverschelde JL, Gerber B, et al. Correlation of functional recovery with myocardial blood flow, glucose uptake, and morphologic features in patients with chronic left ventricular ischemic dysfunction undergoing coronary artery bypass grafting. J Thorac Cardiovasc Surg 1997; 113(2):82–87.

66. Depre C, Vanoverschelde JL, Melin JA, et al. Structural and metabolic correlates of the reversibility of chronic left ventricular ischemic dysfunction in humans. Am J Physiol 1995; 268(3 Pt 2):H1265–H1275.

67. Depre C, Taegtmeyer H. Metabolic aspects of programmed cell survival and cell death in the heart. Cardiovasc Res 2000; 45(3):538–548.

68. Gottlieb RA. Mitochondria: ignition chamber for apoptosis. Mol Genet Metab 1999; 68(2):227–231.

69. Downward J. Metabolism meets death. Nature 2003; 424(6951):896–897.

70. Warburg O. On the origin of cancer cells. Science 1956; 123(3191):309–314.

71. Hanahan D, Weinberg RA. The Hallmarks of cancer. Cell 2000; 100(1):57–70.

72. Fallavollita JA, Malm BJ, Canty JMJ. Hibernating myocardiam retains metabolic and contractile reserve despite regional reductions in flow, function, and oxygen consumption at rest. Circ Res 2003; 92(1): 48–55.

73. Sack MN, Yellon DM. Insulin Therapy as an Adjunct to Reperfusion After Acute Coronary Ischemia. A Proposed Direct Myocardial Cell Survival Effect Independent of Metabolic Modulation. J Am Coll Cardiol 2003; 41(8):1404–1407.

74. Matsui T, Nagoshi T, Rosenzweig A. Akt and PI 3-kinase signaling in cardiomyocyte hypertrophy and survival. Cell Cycle 2003; 2(3):220–223.

75. Matsui T, Li L, Wu J, et al. Phenotypic spectrum caused by transgenic overexpression of activated Akt in the heart. J Biol Chem 2002; 277(25):22896–22901.

76. Whiteman E, Cho H, Birnbaum M. Role of Akt/protein kinase B in metabolism. Trends Endocrinol Metab 2002; 13(10):444–451.

77. Gottlob K, Majewski N, Kennedy S, et al. Inhibition of early apoptotic events by Akt/PKB is dependent on the first committed step of glycolysis and mitochondrial hexokinase. Genes Dev 2001; 15(11): 1406–1418.

78. Majewski N, Nogueira V, Robey RB, et al. Akt inhibits apoptosis downstream of BID cleavage via a glucose-dependent mechanism involving mitochondrial hexokinases. Mol Cell Biol 2004; 24(4):730–740.

79. Tung CH. Fluorescent peptide probes for in vivo diagnostic imaging. Biopolymers 2004; 76(5):391–403.

80. Chen J, Tung CH, Allport JR, et al. Near-infrared fluorescent imaging of matrix metalloproteinase activity after myocardial infarction. Circulation 2005; 111(14): 1800–1805.

31

PET Imaging of Heart and Skeletal Muscle: An Overview

Juhani Knuuti

Turku PET Centre, Turku University Hospital, Turku, Finland

INTRODUCTION

Free fatty acids (FFA), glucose and lactate are the main fuels of the heart (1–2). In addition, ketone bodies and amino acids are used in lesser extent. Several factors affect the use of an individual substrate. These include the plasma concentration of the substrate and alternate substrates, myocardial blood flow and oxygen supply, hormone levels and regulatory effects of metabolites arising during degradation of substrates (1). Under normal resting conditions, metabolism is oxidative with FFA being the major source, while carbohydrates contribute only about 30% of substrate to the tricarboxylic acid cycle (2). Dietary and hormonal conditions markedly affect the selection of substrates by the heart. For example, in the fasting state, FFA levels are high and glucose and insulin levels are low. Consequently, the rate of myocardial FFA oxidation is high and inhibits glycolysis and glucose oxidation. After ingestion of carbohydrates however, glucose and insulin plasma concentrations rise. Insulin reduces peripheral lipolysis, thereby lowering plasma concentrations of FFA further and, hence, their availability to the myocardium. Glucose then becomes the dominant substrate for myocardial energy production.

During physiological exercise, when lactate concentrations rise, lactate will become a significant fuel for the heart (3). Like lactate, the uptake of ketone bodies is concentration dependent and their use is increased in uncontrolled diabetes and starvation (2). Amino acid oxidation is enhanced after a protein-rich meal, when their use can account for an increasing amount of total myocardial oxygen consumption (2).

Myocardial blood flow and oxygen consumption are tightly coupled, and changes in the coronary flow rate control the delivery of oxygen. Myocardial oxygen consumption reflects almost totally the overall energy demand of the heart, and is determined by heart rate, systolic wall stress, contractility, myofiber shortening and the oxygen demand necessary to maintain basal myocardial metabolism (1). At rest myocardial oxygen consumption is typically $\sim 10 \, \mathrm{mL} \times 100 \, \mathrm{g}^{-1} \times \mathrm{min}^{-1}$ (1,4).

Myocardial ischemia strikingly alters myocardial substrate metabolism. As blood flow and oxygen supply decline, oxidative metabolism decreases, but still remains the dominant (over 90%) source of residual ATP production (1). Ischemia is associated with increased rates of glycolysis, with glucose transporters translocated to the cell membrane or expressed in higher numbers, serving as possible flux generating steps. However, glycolysis becomes uncoupled from glucose oxidation with excess production of pyruvate and lactate. During states of mild ischemia, lactate continues to be removed from the myocardium by the residual blood flow but accumulates in tissue when blood flow decreases further during more severe states of ischemia. Increased tissue concentrations of lactate and hydrogen ions impair glycolysis, lead to loss of transmembrane ion concentration gradients, disruption of cell membranes and to cell death (5,6).

STANDARDIZATION OF METABOLIC CONDITIONS

The quality of the myocardial image depends on the concentration of tracer in both myocardium and blood. Since several factors affect myocardial substrate metabolism, the metabolic imaging should be performed during standardized metabolic conditions (Table 1).

In the fasting state, FFA uptake and oxidation are high, which provides ideal conditions for FFA tracers, but not for the measurement of glucose utilization. However, also in the fasting state, FFA and glucose concentrations vary significantly leading to variability of substrate utilization, especially in diabetic subjects. Most FFA positron emission tomography (PET) studies have been performed during fasting but also insulin clamping has been successfully used without significant image deterioration, because fractional uptake of FFA tracer is not significantly dependent on substrate availability (7).

Myocardial [^{18}F] 2-deoxy, 2-fluoroglucose (FDG) uptake depends quantitatively on glucose concentrations, the rate of glucose utilization and on the

Table 1 Methodological Aspects that Influence the Results of Metabolic Imaging

Metabolic conditions of the subject
  –Fasting, hot spot imaging
  –Oral glucose loading
  –Insulin clamp
  –Nicotinic acid derivatives
Imaging protocols and data analysis
  –Absolute quantitation and dynamic imaging
  –Semiquantitative analysis, static imaging
  –Visual analysis, static imaging
Interpretation of results
  –Perfusion-FDG mismatch
  –Normalized FDG uptake
  –Absolute glucose utilization
Imaging equipment
  –Dedicated PET
  –511 keV collimated SPECT
  –Coincidence detection gamma camera

relationship between the glucose tracer and tracee as defined by the lumped constant. Glucose uptake in peripheral tissues is the most important clearance mechanism of FDG from the blood for it accounts for most of the whole body glucose disposal. High glucose plasma concentrations lower the fractional utilization of FDG and thus decrease the quality of the myocardial FDG uptake image (8,9). On the other hand, FDG uptake is enhanced by factors that increase regional glucose utilization such as increased myocardial work, catecholamines and oxygen supply and decreased plasma levels of FFA (1–2). Attempts have therefore been made to standardize the metabolic environment for myocardial FDG imaging.

Oral Glucose Loading

Analogous to the glucose tolerance test, oral glucose loading (50–75 g) is commonly used to stimulate insulin secretion and regional glucose utilization and thus myocardial FDG uptake. It enhances the image quality with more homogeneously distributed FDG uptake than observed during the fasting state (10). After a glucose load, plasma insulin concentrations rise, which enhances regional glucose utilization. However, glucose plasma concentrations increase also, so that the fraction of FDG sequestered metabolically into the myocardium and skeletal muscle declines, which in turn may offset the benefits of increased glucose utilization on image quality. Despite the gain in image quality after oral glucose loading, diagnostically unsatisfactory images may still be obtained in 20% to 25% of the patients with coronary artery disease (10). Abnormal glucose handling or even type 2 diabetes that frequently have remained undetected, account for the poor image quality in many of these patients.

Insulin Clamp

Euglycemic hyperinsulinemic clamping as an approach that mimics the postabsorptive steady-state (8,11) has become an alternative approach to oral glucose loading for enhancing glucose utilization (8). Insulin clamping stimulates uptake of both glucose and FDG uptake in the myocardium and in skeletal muscle and yields images of consistently high diagnostic quality, even in patients with diabetes (12). Although insulin clamping apparently provides the best metabolic standardization it can standardize only circulating glucose and insulin concentrations while significant variability in FFA and lactate concentrations still exists. In fact, similar degrees of variability of myocardial glucose uptake in the normal myocardial segments was found both after glucose loading and during insulin clamp (13). In the clinical setting, use of the hyperinsulinemic euglycemic clamp may be too time-consuming and laborious so that oral glucose loading may be more practical, though not optimal in all patients. Supplementation of small doses of regular, short acting insulin may improve image quality in many patients (14).

Nicotinic Acid Derivatives

Oral administration of a nicotinic acid or its derivatives have been shown to provide easy approach to stimulate myocardial glucose utilization and improve image quality (9,15). Nicotinic acid inhibits peripheral lipolysis and, thus, reduces plasma FFA concentrations. Acipimox is a very potent nicotinic acid derivative. The FDG image quality has been reported to be comparable with insulin clamping in most of the patients. Importantly, with the exception of flushing, no side effects of acipimox were observed.

OTHER FACTORS THAT AFFECT MYOCARDIAL SUBSTRATE UTILIZATION

Since myocardial oxygen consumption and substrate utilization reflects tightly energy demand of the heart, it is, on the one hand, important to relate the findings of metabolic imaging to the myocardial work load. Cardiac work is dependent on heart rate, systolic wall stress, contractility, myofiber shortening and the oxygen demand necessary to maintain basal myocardial metabolism (3). Currently, separate measurements of left ventricular (LV) function, in addition to the PET measurements, are required to obtain a more comprehensive understanding of myocardial energetics. The separate measurements of LV function are usually performed with a different imaging method, such as echocardiography or cardiac magnetic resonance imaging. By measuring LV mass, stoke volume and blood pressure, one can obtain a reasonable good estimate of LV work. For the right ventricle, this is much more demanding and MRI may be the method of choice. It has also been proposed that by obtaining electrocardiographically gated PET data, one can simultaneously acquire measurements of myocardial work and metabolism.

In addition to studies at rest, one may be interested in substrate utilization during exercise. Using tracers trapped in the myocardium, physical exercise is also feasible (16), while with other tracers pharmacological stress, e.g., dobutamine is used (17).

OXIDATIVE METABOLISM

The tracers [^{11}C]acetate and [^{15}O]O$_2$ have been used for measuring myocardial oxygen consumption with PET in humans (4,18) (Table 2). The myocardial kinetics of the [^{11}C]acetate correlates closely and directly with myocardial oxygen consumption over a wide range of conditions. The majority of the studies have used simple washout analysis (K$_{mono}$) because of the robustness of the model. The advantage of this simplified analysis is also that it is less sensitive to partial volume effects. Also, right ventricular parameters have been reported (19). A new model for [^{11}C]acetate as a tracer of myocardial oxygen consumption has also been introduced. The greatest drawbacks of this method are its complexity, need for blood sampling and metabolite correction and dependence on partial volume effects. However, this model allows true quantification and has been successfully applied in humans (18).

The model employed with [^{15}O]O$_2$ requires additional measurements of myocardial blood volume and flow, which are needed for corrections of spillover of the cardiac chamber, and corrections of wall motion and wall thickness (4). The model provides absolute values of myocardial regional oxygen consumption and extraction fraction. This model has been applied

successfully in healthy humans (4) and recently in patients with hypertension induced left ventricular hypertrophy (20). However, the method is quite demanding and error sensitive because of many measurements and parameters are needed in the model.

The two methods have been compared in one human study and found to correlate quite nicely (21). Only in areas with significant myocardial scar, some overestimation of oxygen consumption was detected with [^{11}C]acetate as compared to [^{15}O]O$_2$ -model.

MEASUREMENT OF GLUCOSE METABOLISM

Similar to glucose, [^{18}F]FDG is taken up by the myocyte and phosphorylated by hexokinase to FDG-6-phosphate, but is thereafter trapped in the myocyte and provides a strong signal for imaging. With the glucose tracer analog [^{18}F]FDG it is possible to study glucose transport and phosphorylation but not further metabolism. By using simple graphical analysis (22) the glucose uptake rates in the myocardium can be calculated.

The FDG method has been widely used to study myocardial glucose metabolism in various conditions, such as in coronary heart disease, diabetes, hypertension and cardiac failure. In quantitation of myocardial glucose utilization with [^{18}F]FDG is potentially limited by the differences between natural glucose and FDG (Table 3). These differences are accounted by applying a correction factor termed lumped constant (LC). However, LC may not be stable in all physiological and patophysiological conditions (23). There are very limited data about LC in human heart and the single study in humans reported LC of 1.0 in the fasting state and during insulin stimulation (24).

Table 2 The Most Common PET Tracers in Metabolic Studies

Metabolic process	Tracer
Glucose uptake	^{18}F-FDG
	^{11}C-glucose
FFA uptake and oxidation	^{18}F-FTHA
	^{11}C-palmitate
Lactate uptake	^{11}C-lactate
Amino acid uptake	^{11}C-methionine
	^{11}C-MeAIB
Perfusion	^{15}O-H2O
	^{13}N-NH3
	^{82}Rb
Oxygen consumption	^{15}O-O2
	^{11}C-acetate
Hypoxia	^{18}F-fluoromisonidatsole
	^{64}Cu-ATSM

Table 3 Comparison of ^{18}F-FDG and ^{11}C-Glucose as Tracers of Glucose Metabolism

	^{18}F-FDG	^{11}C-glucose
Advantages	Well characterized tracer	Identical to glucose
		No lumped constant needed
	Is retained in the myocardium	Kinetic model published
	Easy acquisition	
	Easy synthesis	Shorter half-life
	Longer half-life	Fate of glucose to glycogen or glycolysis can be measured
Disadvantages	Analog of glucose → Corrections needed (lumped constant)	Limited experience
		Uneasy acquisition with metabolite analysis
	No information on the fate of glucose	Uneasy synthesis
		No clinical data

Although [¹¹C]-Glucose has been introduced already on eighties it has been applied in studies of cardiac metabolism only recently (25). It has been shown to provide accurate quantitation of myocardial glucose uptake (25) and has been successfully applied also in humans (26). The advantage of [¹¹C]-Glucose is that the problems related to LC are avoided. However, the analysis is clearly more demanding and requires blood sampling with metabolite analysis.

FREE FATTY ACID METABOLISM

The natural tracer [¹¹C]palmitic acid has been used traditionally to assess myocardial FFA metabolism by PET. The retention of the tracer serves as an index of FFA uptake and the rapid washout of the tracer is assumed to be associated with the oxidative metabolism of fatty acids and the slower washout with the incorporation to myocardial triglyceride pool (Table 4) (2,27). However, [¹¹C]palmitic acid is distributed between several tissue pools with variable turnover rates, which make it mainly a qualitative tracer in PET studies (27). A model to quantitate [¹¹C]palmitic acid utilization with PET has been introduced (28). This model requires determination of blood metabolites and is still quite complicated. However, it has been recently successfully applied in several human studies of cardiac FFA metabolism (26).

The tracer analog ¹⁸F-labeled 6-thia-heptadecanoicacid ([¹⁸F]FTHA) has also been used to study fatty acid metabolism in human heart (29). [¹⁸F]FTHA is a "false" long-chain fatty acid substrate and inhibitor of fatty acid metabolism (29). After transport into the mitochondria it undergoes initial steps of β-oxidation and is thereafter trapped in the cell, because further β-oxidation is blocked by sulphur heteroatom. Accumulation of [¹⁸F]FTHA has been suggested to be mainly tracing FFA-oxidation in the heart (29–31). Indeed, 80% of tracer is entering and trapped in the mitochondria of myocytes (31). A similar graphical analysis such as used with [¹⁸F]FDG has been successfully applied in quantitation of [¹⁸F]FTHA uptake in heart.

OTHER METABOLIC TARGETS

Lactate has been also labelled with [¹¹C] and human studies have been successfully performed. The label must be attached to 3-carbon of lactate to provide myocardial image since in [1-¹¹C]-lactate label is metabolized rapidly (personal information, Bengt Langstrom, Uppsala PET Centre). Also some studies of cardiac amino acid metabolism have been reported, but uptake of these tracers has been found low.

Imaging of tissue hypoxia is also attractive approach in cardiology. Although hypoxia tracers have been widely used in oncology, the imaging window in cardiology has limited their use in the heart. Recently, Copper-ATSM has been successfully used to image cardiac hypoxia in canine heart (32).

CLINICAL APPLICATIONS OF METABOLIC IMAGING

The assessment of myocardial viability is a clinically important issue for the management of patients with postischemic LV dysfunction and particularly for those with most severe impairment of LV function. The clinical investigations have demonstrated the utility of [¹⁸F]FDG PET for detection of myocardial viability. In addition, [¹⁸F]FDG imaging is able to identify patients at increased risk of having an adverse cardiac event or death (12). The detection of viable myocardium by PET is based on the demonstration of preserved metabolic activity in regions of severely underperfused and dysfunctional myocardium.

SUMMARY

Due to the unique features and large number of existing tracers, PET appears to provide the most complete insight into the myocardial energy metabolism in vivo. In the detection of myocardial viability PET has already regarded as a golden standard.

The future role of these methods in clinical cardiology remains still unclear and further studies are

Table 4 Comparison of ¹⁸F-FTHA and ¹¹C-palmitate as Tracers of Glucose Metabolism

	¹⁸F-FTHA	¹¹C-palmiate
Advantages	Tracer retains in the myocardium Easy acquisition Longer half-life Suggested to depict beta-oxidation	Identical to palmitate Well established tracer Shorter half-life Simplified analysis possible but complete kinetic model also published Fate of FFA to beetaoxidation or triglycerides can be estimated
Disadvantages	Uneasy synthesis Metabolite analysis required Less characterized tracer Analog of FFA No information on the fate of FFA Limited experience	Uneasy synthesis Uneasy acquisition with metabolite analysis if kinetic model is applied

needed. Novel applications depend also on improved instrumentation and development of new tracers.

REFERENCES

1. Opie LH. Fuels: aerobic and anaerobic metabolism. The Heart. Physiology, from Cell to Circulation. Philadelphia: Lippincott Raven Publishers, 1998:295.
2. Taegtmeyer H. Energy metabolism of the heart: from basic concepts to clinical applications. Curr Probl Cardiol 1994; 19:59–113.
3. Kemppainen JT, Fujimoto T, Kalliokoski KK, et al. Myocardial and skeletal muscle glucose uptake during exercise. J Physiol London 2002; 542(2):403–412.
4. Iida H, Rhodes CG, Araujo LI, et al. Noninvasive quantification of regional myocardial metabolic rate for oxygen by use of 15O2 inhalation and positron emission tomography. Theory, error analysis, and application in humans. Circulation 1996; 94:792–807.
5. Camici P, Ferrannini E, Opie L. Myocardial metabolism in ischemic heart disease: basic principles and application to imaging by positron emission tomography. Progr Cardiovasc Dis 1989; 32:217–238.
6. Opie LH. Effects of regional ischemia on metabolism of glucose and fatty acids: relative rates of aerobic and anaerobic energy production during myocardial infarction and comparison with effects of anoxia. Circ Res 1976; 38(suppl I):I-51–I-74
7. Mäki MT, Haaparanta M, Nuutila P, et al. Free fatty acid uptake in the myocardium and skeletal muscle using (18F)Fluoro-6-thia-heptadecanoic acid. J Nucl Med 1998; 39:1320–1327.
8. Knuuti MJ, Nuutila P, Ruotsalainen U, et al. Euglycemic hyperinsulinemic clamp and oral glucose load in stimulating myocardial glucose utilization during positron emission tomography. J Nucl Med 1992; 33: 1255–1262.
9. Knuuti MJ, Yki-Järvinen H, Voipio-Pulkki L-M, et al. Enhancement of myocardial 18-FDG uptake by nicotinic acid derivative. J Nucl Med 1994; 35:989–998.
10. Berry JJ, Baker JA, Pieper KS, et al. The effect of metabolic milieu on Cardiac PET imaging using fluorine-18-deoxyglucose and nitrogen-13-ammonia in normal volunteers. J Nucl Med 1991; 32:1518–1525.
11. DeFronzo RA, Tobin JD, Andres R. Glucose clamp technique: a method for quantifying insulin secretion and resistance. Am J Physiol 1979; 237:E214–E223.
12. Pagano D, Bonser RS, Townend JN, et al. Predictive value of dobutamine echocardiography and positron emission tomography in identifying hibernating myocardium in patients with postischaemic heart failure. Heart 1998; 79:281–288.
13. Knuuti MJ, Nuutila P, Ruotsalainen U, et al. The value of quantitative analysis of glucose utilization in detection of myocardial viability by PET. J Nucl Med 1993; 34: 2068–2075.
14. Lewis P, Nunan T, Dynes A, et al. The use of low-dose intravenous insulin in clinical myocardial F-18 FDG PET scanning. Clin Nucl Med 1996; 21:15–18.
15. Stone CK, Holden JE, Stanley W, et al. Effect of nicotinic acid on exogenous myocardial glucose utilization. J Nucl Med 1995; 36:996–1002.
16. Stolen KQ, Kemppainen J, Kalliokoski KK, et al. Exercise training improves insulin stimulated myocardial glucose uptake in patients with dilated cardiomyopthy. J Nucl Cardiol 2003; 10(5):447–455.
17. Sundell J, Engblom E, Koistinen J, et al. The effects of cardiac resynchronization therapy on left ventricular function, myocardial energetics and metabolic reserve in patients with dilated cardiomyopathy and heart failure. J Am Coll Cardiol 2004; 43(6):1027–1033.
18. Sun KT, Yeatman LA, Buxton DB, et al. Simultaneous measurement of myocardial oxygen consumption and blood flow using [1-carbon-11]acetate. J Nucl Med 1998; 39:272–280.
19. Knuuti J, Sundell J, Naum A, et al. Right ventricular oxidative metabolism assessed by PET in patients with idiopathic dilated cardiomyopathy undergoing cardiac resynchronization therapy. Eur J Nucl Med Mol Imaging 2004; (12):1592–1598.
20. Laine H, Katoh C, Luotolahti M, et al. Myocardial oxygen consumption is unchanged but efficiency is reduced in patients with essential hypertension and left ventricular hypertrophy. Circulation 1999; 100:2425–2430.
21. Ukkonen H, Knuuti J, Katoh C, et al. Use of [11C]acetate and [15O]O2 PET for the assessment of myocardial oxygen utilization in patients with chronic myocardial infarction. Eur J Nucl Med 2001; 28(3):334–9.
22. Gambhir SS, Schwaiger M, Huang SC, et al. Simple non-invasive quantification method for measuring myocardial glucose utilization in humans employing positron emission tomography and fluorine 18 deoxyglucose. J Nucl Med 1989; 30:359–366.
23. Ng CK, Holden JE, DeGrado TR, et al. Sensitivity of myocardial fluorodeoxyglucose lumped constant to glucose and insulin. Am J Physiol 1991; 260: H593–H603.
24. Ng CK, Soufer R, McNulty PH. Effect of hyperinsulinemia on myocardial fluorine-18-FDG uptake. J Nucl Med 1998; 39(3):379–383.
25. Herrero P, Sharp TL, Dence C, et al. Comparison of 1-(11)C-glucose and (18)F-FDG for quantifying myocardial glucose use with PET. J Nucl Med 2002; 43(11): 1530–1541.
26. Davila-Roman VG, Vedala G, Herrero P, et al. Altered myocardial fatty acid and glucose metabolism in idiopathic dilated cardiomyopathy. J Am Coll Cardiol 2002; 40(2):271–277.
27. Schelbert H, Schwaiger M. PET studies of the heart. Positron emission tomography and autoradiography: principles and applications for the brain and heart. Phelps M, Marizziotta J, Schelbert H, eds. New York: Raven Press, 1986:599.
28. Bergmann SR, Weinheimer CJ, Markham J, et al. Quantitation of myocardial fatty acid metabolism using PET. J Nucl Med 1996; 37:1723–1730.
29. DeGrado TR, Coenen HH, Stöcklin G. 14(R,S) [18F]fluoro 6 thia heptadecanoic acid (FTHA): evaluation in mouse of a new probe of myocardial utilization of long chain fatty acids. J Nucl Med 1991; 32:1888–1896.

30. Stone CK, Pooley RA, DeGrado TR, et al. Myocardial uptake of the fatty acid analog 14-fluorine-18-fluoro-6-thia-heptadecanoic acid in comparison to beta-oxidation rates by tritiated palmitate. J Nucl Med 1998; 39:1690–1696.

31. Takala TO, Nuutila P, Pulkki K, et al. 14(R,S)-[18F]Fluoro-6-thia-heptadecanoic acid as a tracer of free fatty acid uptake and oxidation in myocardium and skeletal muscle. Eur J Nucl Med 2002; 29: 1617–1622.

32. Lewis JS, Herrero P, Sharp TL, et al. Delineation of hypoxia in canine myocardium using PET and copper(II)-diacetyl-bis(N(4)-methylthiosemicarbazone). J Nucl Med 2002; 43(11):1557–1569.

32

Single-Photon Metabolic Imaging

Nagara Tamaki and Koichi Morita

Department of Nuclear Medicine, Hokkaido University Graduate School of Medicine, Sapporo, Japan

INTRODUCTION

Alteration of fatty acid oxidation is considered to be a sensitive marker of ischemia and myocardial damage. On the contrary, persistence of glucose utilization is considered as a suitable marker of myocardial viability in the dysfunctional myocardium. While, PET using fluorine-18 labeled fluorodeoxyglucose (FDG) is considered as an accurate means for assessing myocardial viability, FDG-single photon emission computed tomography (SPECT) using ultrahigh-energy collimators can provide similar information as FDG-PET with regard to viability assessment. The role of metabolic imaging for identifying postischemic insult as "ischemic memory imaging" has recently been focused. A number of reports from Japan showed relatively high diagnostic accuracy of iodinated fatty acid analog [^{123}I-labeled beta-methyl iodophenyl pentadecanoic acid (BMIPP)] imaging for detecting coronary patients without prior myocardial infarction. In addition, the recent data indicates BMIPP imaging has a prognostic value when applied in documented or suspected coronary patients. Thus, single-photon metabolic imaging may play a new and important role for assessing myocardial viability, identifying prior ischemia, and assessing the severity in patients with coronary artery disease using a conventional gamma camera.

MYOCARDIAL METABOLISM

Glucose and free fatty acids are major energy source in the myocardium. Each energy source requires enzymatic conversion before its breakdown. While fatty acid oxidation is most efficient for energy production, this process requires a large amount of oxygen. Therefore, under hypoxia or ischemic condition, oxidation of long chain fatty acid is greatly suppressed, whereas glucose metabolism, requiring less oxygen consumption, plays a major role for residual oxidative metabolism (1). No more metabolism is observed in myocardial necrosis. Thus, alteration of fatty acid oxidation is considered to be a sensitive marker of ischemia and myocardial damage. On the contrary, persistence of glucose utilization is considered as a suitable marker of myocardial viability in the dysfunctional myocardium.

Occlusion-reperfusion canine studies with chronic ischemia suggested prolonged metabolic alteration over four weeks after 30 minutes of coronary occlusion which is associated with sustained myocardial dysfunction (2,3). Thus, metabolic imaging may be used to identify prior ischemic insult, as ischemic memory imaging (4). Iodinated fatty acid analog, such as BMIPP, plays an important role to identify the areas of prior ischemia.

FDG as a marker of exogenous glucose utilization has been used for detection of myocardial ischemia. When FDG is administered under fasting state, it accumulates in ischemic regions but not normal or necrotic regions, and therefore, ischemic myocardium is shown as hot spot (5–7). However, there has been a lot of debate for use of glucose metabolic imaging for identifying ischemia (8). In addition, fasting condition may not be reliable to suppress FDG uptake in normal myocardium. Furthermore, this technique remains rather limited availability for applying many patients with coronary artery disease.

Although FDG-PET has become widely used in clinical setting on the oncological patients, there seems to be little slot for cardiac applications. In this sense, SPECT has been expected to probe metabolic alteration in ischemic heart disease and other myocardial disorders in clinical setting. This chapter includes clinical applications of SPECT metabolic imaging using FDG and BMIPP.

FDG-SPECT IMAGING

The accurate assessment of myocardial viability in patients with ischemic heart disease becomes increasingly important in selecting the subpopulation that may require revascularization. Preserved FDG uptake (percentage uptake) in dysfunctional areas indicates a viable myocardium, which has been shown to predict not only the improvement of the regional and global functions after revascularization but also improvement of survival compared with patients treated with medical therapy alone (9–12).

SPECT using ultrahigh-energy collimators permits wide clinical applications of FDG imaging without the use of expensive PET cameras (13,14). Previous studies indicate that FDG-SPECT using ultrahigh-energy collimators yielded similar diagnostic information for identifying the viable myocardium when compared to FDG-PET (15–18). However, the precise evaluation in region has not yet been done.

We evaluated the values and limitations of FDG-SPECT using ultrahigh-energy collimators on a regional basis for assessing myocardial viability in comparison with FDG-PET in 33 patients with ischemic heart disease (Fig. 1) (19). FDG-SPECT and FDG-PET images of the left ventricular myocardium of the patients were divided into nine segments to score regional FDG defect using a four-point scale (0 = normal to 3 = absent). In 297 segments of all the 33 patients, agreement between defect scores based on FDG-SPECT images and those based on FDG-PET images was 70%, and agreement within one rank was 96% (kappa value = 0.52) (Table 1). For semiquantitative analysis of FDG distribution, the circumferential count profiles obtained using FDG-SPECT and FDG-PET short-axis slices were generalized and displayed as polar maps. The FDG uptake was displayed as a peak myocardial activity of 100% to estimate the mean FDG uptake in the nine segments on the polar map display. Regional % uptake based on FDG-SPECT images significantly correlated with that based on FDG-PET images with a high correlation coefficient rate (r = 0.77, p < 0.0001). These data suggest that FDG-SPECT using ultrahigh-energy collimators can be used to assess myocardial viability as accurately as FDG-PET (19).

A more precise evaluation of percentage uptake, there are a minor differences between the FDG-SPECT and FDG-PET data. In the regions with a low tracer uptake, there was a tendency that the percentage uptake based on FDG-SPECT images was higher than that based on FDG-PET images (Fig. 2), mainly due to

Table 1   The FDG Defect Scores on SPECT Studies in Comparison with FDG-PET Studies in 33 Patients with Ischemic Heart Disease

SPECT score	FDG-PET score			
	0	1	2	3
0	141	12	3	0
1	23	24	12	1
2	3	11	27	12
3	3	2	8	15

scatter noise in the SPECT images (19). In addition, the region-to-region study suggested percentage uptake of FDG-SPECT tended to be lower in the inferior regions, as compared to other regions (19). This slight but significant reduction of FDG uptake in inferior region may come from attenuation effect on SPECT images when compared to PET images. Thus, while FDG-SPECT has been confirmed to play an important role for myocardial viability, a quantitative analysis of FDG distribution might be limited by using this system.

Attenuation effect may become quite significant when FDG images are obtained with coincidence imaging system (20–22). This is mainly due to the acquisition of both photons from the long distance through the body, as compared to the single-photon acquisition. Although coincidence FDG imaging has a higher sensitivity than FDG-SPECT imaging using ultrahigh energy collimators, attenuation correction is mandatory for the coincidence imaging. On the contrary, FDG-SPECT has a unique capability for simultaneous acquisition for FDG and perfusion tracers using two difference energy windows. Such combined assessment of perfusion and glucose metabolism may

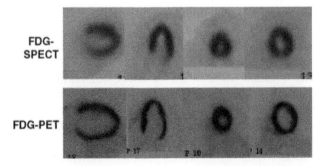

Figure 1   FDG-SPECT and FDG-PET images of a patient with inferior wall myocardial infarction. Preserved FDG uptake is observed by both FDG-SPECT and FDG-PET images (quoted from Ref 19 with permission).

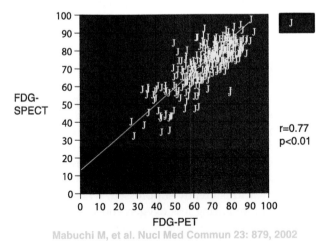

Mabuchi M, et al. Nucl Med Commun 23: 879, 2002

Figure 2   The correlation of regional % uptake on FDG-SPECT and on FDG-PET. A high correlation was observed between the two parameters (r = 0.77, p < 0.0001). *Source:* From Ref. 19.

play an important role for identifying ischemic but viable myocardium and assessing severity of myocardial ischemia.

BMIPP-SPECT IMAGING

Iodine-123 is an appropriate choice for labeling metabolic substrates because of its chemical property for synthesis by halogen exchange reaction in replacing a molecular methyl group and wide clinical application in clinical practice. Thus, iodine-123 labeled fatty acids have received great attention for assessing myocardial metabolism in vivo (23–25).

There are two groups of iodinated fatty acid compounds, including straight chain fatty acids and modified branched fatty acids (Table 2). The straight chain fatty acids are generally metabolized via beta-oxidation and released from the myocardium. Therefore, fatty acid utilization can be directly assessed by the washout kinetics of the tracer, similarly as C-11 palmitate. However, a rapid washout from the myocardium may require a fast dynamic acquisition for imaging following tracer administration. This may become critical problem applied with tomography with a rotating gamma camera. Inadequate image quality with low target-to-background ratio may often be observed. In addition, back diffusion and metabolites should be considered in the kinetic model for quantitative analysis of fatty acid metabolism.

The modified fatty acids are introduced based on the concept of myocardial retention due to metabolic trapping (23). Therefore, an excellent myocardial image is obtained with long acquisition time. On the other hand, their uptake may not directly reflect fatty acid oxidation. Instead, its uptake is based on the fatty acid uptake and turnover rate of lipid pool. Therefore, the combined imaging of the iodinated fatty acid and perfusion is required to demonstrate

perfusion-metabolism mismatch and to characterize fatty acid utilization.

Methyl-branched fatty acid is based on the expected inhibition of β-oxidation by the presence of methyl group in the beta-position. Knapp et al. (26,27) first introduced 15-(p-iodophenyl)-3R,S-methyl pentadecanoic acid (BMIPP) (Fig. 3). The animal experiments showed slow clearance of BMIPP by approximately 25% in two hours. The fractional distribution of these compounds at 30 minutes after tracer injection in rats indicated 65% to 80% of the total activity resided in the triglyceride pool.

A number of experimental studies have been tested to see the tracer kinetics in the myocardium with use of BMIPP. Fujibayashi et al. (28) in the canine study indicated that BMIPP was extracted from the plasma into the myocardium by 74% of the injected dose and was retained about 65% following intracoronary injection of BMIPP with only 8.7% fraction of washout from the myocardium. The slow washout from the myocardium was seen as alpha- and beta-oxidation metabolites (28–30). Hosokawa et al. (31) showed the enhanced rapid washout from the myocardium by long-chain fatty acid transporter inhibitor, etomoxir, which may produce similar condition as myocardial ischemia. Fujibayashi et al. (32,33) indicated that BMIPP uptake correlated with ATP concentration in acutely damaged myocardium treated with dinitrophenol or tetradecyl-glycidic acid, an inhibitor of mitochondrial carnitine acyltransferase I. Nohara et al. (34) also showed the BMIPP uptake correlated with ATP levels in the occlusion and reperfusion canine model and concluded that BMIPP may be useful to differentiate ischemic from infarcted myocardium in their model. These results support the importance of ATP levels for the retention of BMIPP, probably due to cytosolic activation of BMIPP into BMIPP-CoA.

In the occlusion-reperfusion canine model, Hosokawa et al. (35,36) nicely showed this increase in back diffusion of nonmetabolized BMIPP in the coronary sinus sampling after BMIPP administration. Thus, regional uptake of BMIPP may be reduced in the severely ischemic myocardium.

In the clinical studies with BMIPP, a rapid and high myocardial uptake with long retention was observed after BMIPP administration with low background with low uptake in the liver and lung at

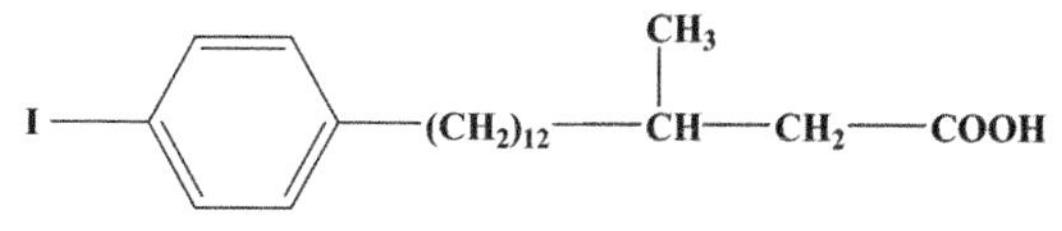

Table 2 Characterstics of the Two Groups of Iodinated Fatty Acid Compounds, Including Straight Chain Fatty Acids and Modified Branched Fatty Acids

	Straight chain FA	Branched FA
PET tracers	^{11}C-Palmitate	^{11}C-βmethyl HDA acid
	^{18}F-FHTA	
SPECT tracers	^{123}I-IPPA	^{123}I-BMIPP
Measurements	Uptake and clearance	Uptake
Advantages	Direct measurement of oxidation	Excellent images Suitable for SPECT
Disadvantages	Dynamic acquisition Back diffusion	Flow dependent Combination with perfusion required

15-(p-Iodo-phenyl)-3-R,S-methylpentadecanoic acid (BMIPP)

Figure 3 Chemical structure of I-123 labeled BMIPP.

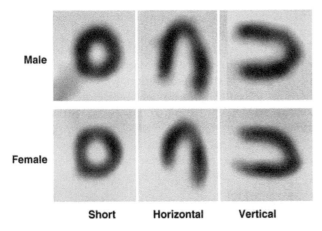

Male

Female

Short   Horizontal   Vertical

Figure 4 SPECT images at 30 minutes after BMIPP administration of two normal subjects. A high and homogeneous uptake of the tracer in the left ventricular myocardium is noted.

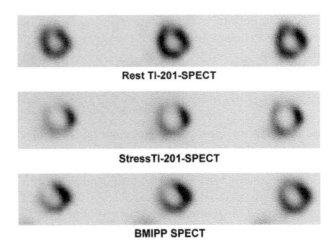

Rest TI-201-SPECT

StressTI-201-SPECT

BMIPP SPECT

Figure 5 Stress and delayed thallium-201 SPECT and resting BMIPP SPECT of a patient with unstable angina. A reduced BMIPP uptake is noted in the corresponding areas with stress-induced ischemia on thallium-201 images.

30 minutes after BMIPP injection (Fig. 4). A high quality of SPECT images can be obtained with collecting myocardial images for approximately 20 minutes. Generally, BMIPP uptake was similar to that of thallium perfusion. Therefore, BMIPP distribution is carefully assessed to identify regional decrease in tracer distribution as an area of altered fatty acid uptake and metabolism. Regional BMIPP uptake is also compared with regional perfusion to detect presence of perfusion-metabolism mismatch. Less BMIPP uptake than perfusion (discordant BMIPP uptake) is often observed in ischemic myocardium.

DIAGNOSTIC VALUE OF BMIPP SPECT

Accordingly, there are a number of reports showing that BMIPP imaging at rest can be used for identifying ischemic myocardium in coronary patients without evidence of myocardial infarction (Fig. 5). Table 3

summarizes the reports showing diagnostic accuracy of BMIPP imaging for identifying coronary abnormalities (38–44). This meta-analysis indicated that BMIPP-SPECT imaging at rest might provide quite acceptable sensitivity (74%) and high specificity (87%) for identifying coronary patients. We would like to emphasize that all of these results were obtained BMIPP imaging at rest, not under stress imaging. The diagnostic accuracy may be similar to those of stress myocardial perfusion imaging. Furthermore, the BMIPP abnormalities seem to be associated with unstable angina, regional wall motion abnormalities, and ECG changes. One important study applied both BMIPP and tetrofosmin perfusion SPECT imaging at rest for patients with acute chest pain with no evidence of acute myocardial infarction (43). BMIPP imaging was more sensitive for detecting organic stenosis and coronary spasm (74%) than tetrofosmin imaging (38%) (p < 0.001) (Fig. 6). Quantitative analysis also suggested that the extent and severity scores of BMIPP abnormalities were

Table 3 Diagnostic Accuracy of BMIPP Imaging at Rest for Identifying Coronary Lesions without Myocardial Infarction

Authors	Ref.	No. of patients	Patients	Diagnostic accuracy sensitivity	Specificity
Nakajima	37	32	Vasospastic angina	25/32 (78%)	NA
Takeishi	38	78	s/o CAD	28/49 (57%)	NA
Fujiwara	39	29	s/o CAD	15/20 (75%)	7/9 (78%)
Tateno	40	31	Angina	27/31 (87%)	NA
Suzuki	41	40	unstable angina	25/28 (89%)	12/12 (100%)
Yamabe	42	104	s/o CAD	38/54 (45%)	16/20 (80%)
Kawai	43	111	s/o CAD (acute chest pain)	64/87 (74%)	22/24 (92%)
Watanabe	44	75	Vasospastic angina	43/50 (86%)	22/25 (88%)
Overall		500		265/351 (75%)	79/90 (88%)

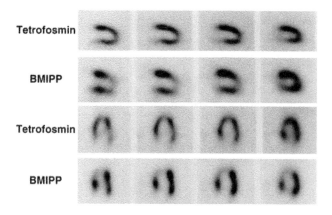

Tetrofosmin

BMIPP

Tetrofosmin

BMIPP

Figure 6 Tetrofosmin and BMIPP SPECT images of a patient with effort angina who come to emergency department due to acute chest pain. Despite minimal hypoperfusion on perfusion SPECT, BMIPP-SPECT obtained on the next day showed definite abnormal uptake in the apical region. His coronary angiogram showed 99% stenosis on proximal LAD. *Source*: From Ref. 43.

greater than tetrofosmin abnormalities in these patients. The specificity of BMIPP was greater than 90% in patients with acute chest pain with normal coronary angiography. Although these data remain to be preliminary, they suggested the diagnostic value for detecting coronary patients with BMIPP imaging at rest. Of particular, this should be of clinical importance in the study of unstable patients who may not be suitable for stress perfusion study, and also vasospastic angina patients where stress perfusion imaging may not identify ischemia. Therefore, BMIPP imaging is considered as a method of choice to identify regional abnormalities as "ischemic memory" noninvasively in these patients.

The overall diagnostic accuracy of BMIPP imaging in the literatures may be comparable to those of stress myocardial perfusion imaging. But we have to remind that the studies in Table 1 include mostly severely ischemic or unstable patients. When another study is designed to focus diagnostic accuracy in rather stable patients with coronary artery disease, the sensitivity may not be as high as those of stress myocardial perfusion imaging. The major advantage of BMIPP imaging is to demonstrate ischemic myocardium as an area of altered metabolism at rest. Therefore, this study is of clinical importance for applying those with elder patients or those not suitable for stress study.

PROGNOSTIC VALUE OF BMIPP SPECT

The areas with less BMIPP than perfusion may represent ischemic and jeopardized myocardium. Accordingly, the combined BMIPP and thallium imaging may have potential values for risk stratification in

coronary patients (45–48). This concept may come from the important prognostic findings reported on the perfusion-metabolism mismatch pattern on FDG-PET studies. Among various clinical, angiographic and radionuclide indices, discordant BMIPP uptake was the best predictor of future cardiac events followed by number of coronary stenosis in the mean follow-up study of 23 months of 50 consecutive patients with myocardial infarction (45). Another study on the multicenter trials showed that BMIPP defect score was the most powerful index for predicting future cardiac events among various clinical and radionuclide parameters (46). These preliminary data suggest that the combined imaging with BMIPP and perfusion imaging may hold a prognostic value for identifying high-risk subgroups among patients with coronary artery disease. On the other hand, many more patients study is warranted to confirm such preliminary findings.

The prognostic study has recently been extended to angina patients without prior myocardial infarction (49). When 167 consecutive patients with angina were followed-up for 48 months, BMIPP defect score at rest, stress perfusion score, diabetes, and LVEF were independent predictors on the multivariate Cox's analysis (48). No hard event was observed with normal BMIPP uptake, whereas two patients with nearly normal stress perfusion with abnormal BMIPP uptake had hard events (49). Although these data remain preliminary, this indicates BMIPP imaging has an important prognostic indicator independent of stress myocardial imaging in patients with angina without prior myocardial infarction.

CONCLUSIONS

Basic studies and preliminary clinical results have raised a number of important clinical roles of metabolic imaging using SPECT. Alteration of fatty acid oxidation is considered to be a sensitive marker of ischemia and myocardial damage. On the contrary, persistence of glucose utilization is considered as a suitable marker of myocardial viability in the dysfunctional myocardium. While, FDG-PET is considered as a accurate means for assessing myocardial viability, FDG-SPECT using ultrahigh-energy collimators can provide similar information as FDG-PET with regard to viability assessment. The role of metabolic imaging for identifying post-ischemic insult as "ischemic memory imaging" has recently been focused. A number of reports from Japan showed quite acceptable diagnostic accuracy of BMIPP imaging for detecting coronary patients without prior myocardial infarction. In addition, the recent data indicates BMIPP imaging has a prognostic value when applied in documented or suspected coronary patients. In conclusion, single-photon metabolic imaging may play a new and important role for identifying prior ischemia and assessing the severity in patients with coronary artery disease using a conventional gamma camera.

REFERENCES

1. Liedke AJ. Alterations of carbohydrate and lipid metabolism in the acutely ischemic heart. Prog Cardiovasc Dis 1981; 23:321–336.

2. Schwaiger M, Schelbert HR, Ellison D, et al. Sustained regional abnormalities in cardiac metabolism after transient ischemia in the chronic dog model. J Am Coll Cardiol 1985; 6:336–347.

3. Schwaiger M, Neese RA, Aroujo L, et al. Sustained non-oxidative glucose utilization and depletion of glycogen in reperfused canine myocardium. J Am Coll Cardiol 1989; 13:745–754.

4. Tamaki N, Morita K, Kuge Y, Tsukamoto E. The role of fatty acids in cardiac imaging. J Nucl Med 2000; 41:1525–1534.

5. McFalls EO, Murad B, Liow JS, et al. Glucose uptake and glycogen levels are increased in pig heart after repetitive ischemia. Am J Physiol Heart Circ Physiol 2002; 282:H205–H211.

6. Aroujo LI, McFalls EO, Lammertsma AA, Jones T, Maseri A. Dypiridamole-induced increased glucose uptake in patients with single-vessel coronary artery disease assessed by PET. J Nucl Cardiol 2001; 8:339–346.

7. He Z, Shi R, Wu Y, et al. Direct imaging of exercise-induced myocardial ischemia with fluorine-18-labed deoxyglucose and Tc-99m-sestamibi in coronary artery disease. Circulation 2003; 108–1208–1213.

8. Gould KL, Taegtmeyer H. Myocardual ischemia, fluorodeoxyglucose, and severity of coronary artery stenosis: the complexities of metabolic remodeling in hibernating myocardium. Circulation 2004; 109:e167–e170 (correspondence).

9. Marshall RC, Tillisch JH, Phelps ME, et al. Identification and differentiation of resting myocardial ischemia and infarction in man with positron computed tomography, 18F-labeled fluorodeoxyglucose and N-13 ammonia. Circulation 1983; 64:766–778.

10. Tamaki N, Yonekura Y, Yamashita K, et al. Positron emission tomography using fluorine-18 deoxyglucose in evaluation of coronary artery bypass grafting. Am J Cardiol 1989; 64:860–865. .

11. Eitzman D, Al-Aouar Z, Kanter HL, et al. Clinical outcome of patients with advanced coronary artery disease after viability studies with positron emission tomography. J Am Coll Cardiol 1992; 20:559–565.

12. Di Carli MF, Davidson M, Little R, et al. Value of metabolic imaging with positron emission tomography for evaluating prognosis in patients with coronary artery disease and left ventricular dysfunction. Am J Cardiol 1994; 73:527–533.

13. Sandler MP, Bax JJ, Patton JA, Visser FC, Martin WH, Wijns W. Fluorine-18- fluorodeoxyglucose cardiac imaging using a modified scintillation camera. J Nucl Med 1998; 39:2035–2043.

14. Sandler MP, Videlefsky S, Delbeke D, et al. Evaluation of myocardial ischemia using a rest metabolism/stress perfusion protocol with fluorine-18 deoxyglucose/technetium-99m MIBI and dual-isotope simultaneous-acquisition single photon emission computed tomography. J Am Coll Cardiol 1995; 26:870–878.

15. Bax JJ, Visser FC, Blankma PK, et al. Comparison of myocardial uptake of fluorine-18-fluorodeoxyglucose imaged with PET and SPECT in dyssynergic myocardium. J Nucl Med 1996; 37:1631–1636.

16. Martin WH, Delbeke D, Patton JA, et al. FDG-SPECT: correlation with FDG-PET. J Nucl Med 1995; 36: 988–995.

17. Bax JJ, Cornel JH, Visser FC, et al. Prediction of recovery of myocardial dysfunction after revascularization: comparison of fluorine-18 fluorodeoxyglucose/thallium-201 SPECT, thallium-201 stress-reinjection SPECT and dobutamine echocardiography. J Am Coll Cardiol 1996; 28:558–564.

18. Srinivasan G, Kitsiou AN, Bacharach SL, Bartlett ML, Miller-Davis C, Dilsizian V. 18F-Fluorodeoxyglucose single emission tomography: Can it replace PET and thallium SPECT for the assessment of myocardial viability? Circulation 1998; 97:843–850.

19. Mabuchi M, Kubo N, Morita K, et al. Value and limitation of myocardial fluorodeoxyglucose single photon emission computed tomography using ultra-high energy collimators for assessing myocardial viability. Nucl Med Commun 2002; 23:879–885.

20. Fukuchi K, Sago M, Nitta K, et al. Attenuation correction for cardiac dual-head gamma camera coincidence imaging using segmented myocardial perfusion SPECT. J Nucl Med 2000; 41:919–925.

21. Nowak B, Zimny M, Schwarz ER, et al. Diagnosis of myocardial viability by dual-head coincidence gamma camera fluorine-18 fluorodeoxyglucose positron emission tomography with and without non-uniform attenuation correction. Eur J Nucl Med 2000; 27:1501–1508.

22. Di Bella EV, Kadrmas DJ, Christian PE. Feasibility of dual-isotope coincidence/single-photon imaging of the myocardium. J Nucl Med 2001; 42:944–950.

23. Knapp FF Jr, Kropp J. Iodine-123-labelled fatty acids for myocardail single-photon emission tomography: current status and future perspectives. Eur J Nucl Med 1995; 22:361–381.

24. Tamaki N, Kawamoto M. The use of iodinated free fatty acids for assessing fatty acid metabolism. J Nucl Cardiol 1994; 1:S72–S78.

25. Tamaki N, Fujibayashi Y, Magata Y, et al. Radionuclide assessment of myocardial fatty acid metabolism by PET and SPECT. J Nucl Cardiol 1995; 2:256–266.

26. Knapp FF. Jr, Goodman MM, Callahan AP, et al. Radioiodinated 15-(p-iodophenyl)-3,3-dimethylpentadecanoic acid: a useful new agent to evaluate myocardial fatty acid uptake. J Nucl Med 1986; 27:521–531.

27. Ambrose KR, Owen BA, Goodman MM, Knapp FF Jr. Evaluation of the metabolism in rat heart of two new radioiodinated 3-methyl-branched fatty acid myocardial imaging agents. Eur J Nucl Med 1987; 12:486–491.

28. Fujibayashi Y, Nohara R, Hosokawa R, et al. Metabolism and kinetics of iodine-123-BMIPP in canine myocardium. J Nucl Med 1996; 37:757–761.

29. Morishita S, Kusuoka H, Yamamichi Y, et al. Kinetics of radioiodinated species in subcellular fractions from rat hearts following administration of iodine-123-labelled 15-(p-iodophenyl)-3-(R,S) methylpentadecanoic acid (123I-BMIPP). Eur J Nucl Med 1996; 23:383–389.

30. Yamamichi Y, Kusuoka H, Morishita K, et al. Metabolism of 123I-labeled 15-p-iodophenyl-3-(R,S)-methyl-penta-decanoic acid (BMIPP) in perfused rat heart. J Nucl Med 1995; 36:1043–1050.

31. Hosokawa R, Nohara R, Fujibayashi Y, et al. Metabolic fate of iodine-123-BMIPP in canine myocardium after administration of etomoxir. J Nucl Med 1996; 37: 1836–1840.

32. Fujibayashi Y, Yonekura Y, Takemura Y, et al. Myocardial accumulation of iodinated beta-methyl-branched fatty acid analogue, iodine-125–15-(p-iodophenyl)-3-(R,S) methylpentadecanoic acid (BMIPP), in relation to ATP concentration. J Nucl Med 1990; 31:1818–1822.

33. Fujibayashi Y, Yonekura Y, Tamaki N, et al. Myocardial accumulation of BMIPP in relation to ATP concentration. Ann Nucl Med 1993; 7:15–18.

34. Nohara R, Okuda K, Ogino M, et al. Evaluation of myocardial viability with iodine-123-BMIPP in a canine model. J Nucl Med 1996; 37:1403–1407.

35. Hosokawa R, Nohara R, Fujibayashi Y, et al. Myocardial kinetics of iodine-123-BMIPP in canine myocardium after regional ischemia and reperfusion: implications for clincial SPECT. J Nucl Med 1997; 38:1857–1863.

36. Hosokawa R, Nohara R, Fujibayashi Y, et al. Myocardial metabolism of 123I-BMIPP in a canine model of ischemia implication of perfusion-metabolism mismatch on SPECT images in patients with ischemic heart disease. J Nucl Med 1999; 40:471–478.

37. Nakajima K, Schimizu K, Taki J, et al. Utility of iodine-123-BMIPP in the diagnosis and follow-up of vasospastic angina. J Nucl Med 1995; 36:1934–1940.

38. Takeishi Y, Fujiwara S, Atsumi H, et al. Iodine-123-BMIPP imaging in unstable angina: a guide for interventional therapy. J Nucl Med 1997; 38:1407–1411.

39. Fujiwara S, Takeishi Y, Atsumi H, et al. Fatty acid metabolic imaging with iodine-123-BMIPP for the diagnosis of coronary artery disease. J Nucl Med 1997; 38: 175–180.

40. Tateno M, Tamaki N, Kudoh T, et al. Assessment of fatty acid uptake in patients with ischmeic heart disease without myocardial infarction. J Nucl Med 1996; 37: 1981–1985.

41. Suzuki A, Takada Y, Nagasaka M, et al. Comparison of resting β-methyl-iodophenyl pentadecanoic acid (BMIPP) and thallium-201 tomography using quantitative polar maps in patients with unstable angina. Jpn Circulation 1997; 61:133–138.

42. Yamabe H, Fujiwara S, Rin K, et al. Resting 123I-BMIPP scintigraphy for detection of organic coronary stenosis and therapeutic outcome in patients with chest pain. Annals Nucl Med 2000; 14:187–192.

43. Kawai Y, Tsukamoto E, Nozaki Y, Morita K, Sakurai M, Tamaki N. Significance of reduced uptake of iodinated fatty acid analogue for the evaluation of patients with acute chest pain. J Am Coll Cardiol 2001; 38: 1888–1894.

44. Watanabe K, Takahashi T, Miyajima S, et al. Myocardial sympathetic denervation, fatty acid metabolism, and left ventricular wall motion in vasospastic angina. J Nucl Med 2002; 43:1476–1481.

45. Tamaki N, Tadamura E, Kudoh T, et al. Prognostic value of iodine-123 labelled BMIPP fatty acid analogue imaging in patients with myocardial infarction. Eur J Nucl Med 1996; 23:272–279.

46. Nakata T, Kobayashi T, Tamaki N, et al. Prognostic value of impaired myocardial fatty acid uptake in patients with acute myocardial infarction. Nucl Med Commun 2000; 21:897–907.

47. Fukuzawa S, Ozawa S, Shimada K, et al. Prognostic values of perfusion-metabolic mismatch in Tl-201 and BMIPP scintigraphic imaging in patients with chronic coronary artery disease and left ventricular dysfunction undergoing revascularization. Annals Nucl Med 2002; 16:109–115.

48. Nanasato M, Hirayama H, Ando A, et al. Incremental predictive value of myocardial scintigraphy with 123I-BMIPP in patients with acute myocardial infarction treated with primary percutaneous coronary intervention. Eur J Nucl Med Mol Imaging 2004; 31:1512–1521.

49. Matsuki T, Tamaki N, Nakata T, et al. Prognostic value of fatty acid imaging in patients with angina pectoris without prior myocardial infarction: comparison with stress thallium imaging. Eur J Nucl Med Mol Imaging 2004; 31:1585–1591.

33

Assessment of Myocardial Metabolism with Magnetic Resonance Spectroscopy

John R. Forder

Advanced Magnetic Resonance Imaging and Spectroscopy Facility, The McKnight Brain Institute, and Department of Radiology, The University of Florida, Gainesville, Florida, U.S.A.

INTRODUCTION

It is becoming increasingly apparent that perturbations in myocardial substrate metabolism are key to the pathogenesis of a variety of cardiac disorders such as coronary artery disease, dilated cardiomyopathy, and diabetic heart disease. Radionuclide approaches such as positron emission tomography can provide quantitative measurements of myocardial oxygen, glucose, and fatty acid metabolism. However, radionuclide methods are limited by relatively low spatial resolution, incomplete characterization at the subsequent metabolic fates of extracted radiolabeled substrate and expensive complex technology requiring highly specialized personnel. NMR spectroscopy can provide highly sensitive and quantitative measurements of multiple metabolic processes nearly simultaneously. Indeed a number of different nuclei are NMR visible, including hydrogen, sodium, fluorine, phosphorus, and carbon, permitting the interrogation of diverse metabolic processes ranging from substrate uptake, turnover, intermediate formation, and storage to energy production. However, at the current time the ability to obtain many of these measurements, particularly non-invasively, is limited. In this chapter, both the current capabilities of NMR spectroscopy and its future potential for quantifying myocardial metabolism will be discussed. Emphasis will be placed on in vivo methods. In addition, for the purposes of this chapter, NMR spectroscopy will be called magnetic resonance spectroscopy or MRS.

BACKGROUND

While magnetic resonance imaging (MRI) routinely uses hydrogen (predominantly from water or lipid) to generate images, there are many other nuclei that can be investigated using MRS. These other nuclei are not routinely used as the basis of images, for two main reasons. One, the concentrations of other nuclei are lower than the concentration of protons. The concentration of water molecules in liquid water is approximately 55 Molar. Two hydrogen atoms on each water molecule means the concentration of protons in water is approximately 110 Molar. Most of the chemical species examined by MRS are 3–6 orders of magnitude lower in concentration. The second reason that other chemical species are not routinely used in imaging is the relatively low sensitivity compared to the sensitivity of protons.

A detailed explanation of the physics of MRS is beyond the scope of this chapter, however, in-depth reviews of this topic can be found elsewhere (1,2). The fundamental principle of MRS is that the magnetic field experienced by an individual nucleus may be modified by the surrounding molecular environment, leading to a change in its resonance frequency. This phenomenon, known as chemical shift, allows the identification of individual components within a NMR spectrum. The measured signal strength is depicted as a function of MR frequency or chemical shift in parts per million (ppm) relative to the resonant frequency of a reference compound at 0 ppm, such as PCr for ^{31}P-MRS or tetramethyl silane for ^{1}H-MRS. The area under the various peaks in a NMR spectrum corresponds to the concentration of the various nuclei.

Biologically ubiquitous nuclei that can be detected with magnetic resonance studies include hydrogen (^{1}H), carbon (^{13}C), fluorine (^{19}F), sodium (^{23}Na), and phosphorus (^{31}P). Studies investigating in vivo metabolism have generally been performed with the use of ^{1}H- and ^{31}P- and, to a lesser extent, ^{13}C-MRS and ^{23}Na-MRS. Because they have high sensitivity and natural abundance, most MRS studies of the heart are performed with ^{31}P and ^{1}H providing measurements of myocardial energetics and metabolism. In vivo cardiac ^{13}C-MRS studies are typically more difficult to perform because ^{13}C is present in very low natural abundance in cells. However, this low natural abundance allows for performance of enrichment studies using ^{13}C-labeled precursors to measure myocardial substrate metabolism. Examples of current applications of MRS in both small animal and human cardiac metabolic imaging are discussed below.

¹H-MRS

As mentioned previously, the hydrogen nucleus (proton) has the highest sensitivity and abundance of the nuclei that are detectable by MRS. The method has been used to assess metabolism in myocardial tissue extracts, in isolated perfused hearts, open-chest animal models, and in vivo in both small and large animals and humans.

Animal Studies

Currently, the genetic manipulation of the mouse is the most commonly used approach for studying mammalian genomics. A key challenge is accurate phenotyping of these mouse models so that their relevance to human disease can be determined. Metabolomics represents the collection of all metabolites in a biological system, which is the end product of gene expression. These metabolic profiles can then be used to provide an integrative phenotypic characterization of a genetic manipulation. High-resolution ¹H-MRS performed in conjunction with multivariable statistical analysis permits such metabolic profiling explanted cardiac tissue with a high degree of sensitivity and specificity (3). These profiles can then be used to accurately metabolically phenotype transgenic mouse models of diverse cardiac disease processes (3). ¹H-MRS of isolated perfused hearts has been used extensively to assess myocardial substrate use. Accelerated myocardial fatty acid uptake, oxidation, and storage (as triacylglycerol or TAG) are the metabolic hallmarks of the diabetic heart and may contribute to the cardiac dysfunction associated with this disease (4,5). Thus, there is great interest in characterizing the mechanisms responsible for the accelerated fatty acid use in these hearts. ¹H-MRS has shown to be useful in this regard. For example, in perfused streptozocin rat heart (a model of type-1 diabetes mellitus) ¹H-MRS quantified myocardial TAG and demonstrated that neutral lipase as opposed to acid lipase was required for lipolysis (6). In vivo ¹H-MRS of myocardial metabolism has also been performed. However, it is technically demanding because the large peak from water needs to be suppressed in order to see important metabolite peaks such as creatine, lipids, and lactic acid. In addition, the myocardium is surrounded by epicardial fat, which produces a substantial peak that can interfere with the ability to visualize proton containing substances of interest such as myocardial lipids and lactic acid. Moreover, ¹H-MRS s is very sensitive to motion and flow and thus excellent gating strategies are required

for good results. Despite these challenges, ¹H-MRS has been performed in mouse heart in vivo (7). In normal C57Bl/6J mice numerous cardiac metabolites, such as creatine, taurine, carnitine, and intramyocardial lipids were successfully detected and quantified relative to the total water content in voxels as small as 2 μL, positioned in the interventricular septum. Moreover, the method demonstrated deceased myocardial creatine levels in a murine model of guanidinoacetate N-methyltransferase (GAMT) deficiency.

¹H-MRS has also been used in large animal imaging such as in the evaluation of the oxygenation status of the myoglobin molecule (8–10). Myoglobin which contains a single heme group (in contrast to hemoglobin's four heme groups), is present in muscle tissue in high concentrations. Although the precise role for myoglobin in the heart remains unknown, the oxygenation state of the tissue can be determined from the frequency shift of the proximal histidyl-NdH of deoxymyoglobin (9,10). This approach has been used to further our understanding of the relationship between tissue oxygen content and oxidative phosphorylation in ischemic and postischemic myocardium (11).

Human Studies

Of note, cardiac applications for MRS become more limited as one moves from rodent to human being, in contrast to other imaging methods such as nuclear imaging, where the reverse occurs. This appears to be a function of both the higher field strength of the small-bore systems and the use of radiofrequency coils that are in closer proximity to the entire heart used in small animal imaging. These advantages overcome the need for markedly improved spatial resolution. Indeed, as opposed to rodent hearts, where measurements of nearly the entire left ventricular myocardium are obtained, measurements in human myocardium are typically limited to the anterior or septal myocardium. Despite these challenges, ¹H-MRS has successfully been performed in humans using conventional 1.5 T systems. For example, ¹H-MRS has been used to measure myocardial creatine levels to provide insights into the impact of diseases such as hypertrophic and dilated cardiomyopathy on myocardial energetics (12,13). One particular area of interest has been in the quantification of myocardial TAG content in a variety of cardiac disease processes such as obesity and diabetes mellitus. In animal models of obesity and type-2 diabetes as well in explanted human hearts from patients with these disorders, increased myocardial TAG has been associated with increased apoptosis and mechanical dysfunction (14,15). Consequently, the availability of a noninvasive approach that quantifies myocardial TAG would have important implications in the management of obese and diabetic patients. Recently, a ¹H-MRS approach has been developed for

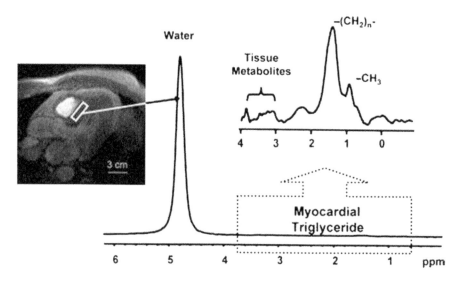

Figure 1 H-1 MRS of the human heart. Selection of volume of interest (equal to 6 mL) in the ventricular septum of the myocardium throughout the experiment is indicated by the *white box* on the cine image (long-axis view, end of systole). The corresponding magnetic resonance spectrum from volume of interest is displayed along with enlarged signal from myocardial triglyceride. *Source:* Courtesy of Wiley InterScience.

this purpose. Consistent with results from experimental models of obesity and diabetes, increased myocardial TAG was associated with left ventricular systolic diastolic function (Fig. 1) (16,17).

31P-MRS

Clinical phosphorus spectra from heart muscle can show up to seven peaks: the three phosphates associated with adenosine triphosphate or ATP (γ, α, and β), phosphocreatine (PCr), inorganic phosphate (Pi), phosphomonoesters (PME), and phosphodiesters (PDE) (Fig. 2). However, it is the ATP and PCr that have been used for most studies to characterize the energy state of the heart with the PCr/ATP ratio directly reflecting the energy status of the tissue.

Animal Studies

31P-MRS has been performed in rodent hearts, both isolated perfusion preparations and in vivo, providing insights into the energetic consequences of manipulations designed to alter key metabolic pathways or control points and in pathologic conditions relevant to human disease. For example, 31P-MRS measurements

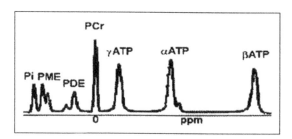

Figure 2 Fitted 31P-NMR spectrum obtained at 4.1 T. Note the clearly defined inorganic phosphate (Pi), phosphomonoester (PME) and phosphodiesters (PDE) peaks.

in isolated perfused mouse hearts deficient in the key glucose transporter, GLUT-4 demonstrated energy depletion and profound mechanical dysfunction when subjected to ischemia (18). The results suggest a key role for GLUT-4 for myocardial protection during ischemia. Results of studies performed in the intact mouse have shown that the PCr/ATP ratio did not change with with dobutamine demonstrating that global levels of myocardial high-energy phosphate metabolites do not regulate myocardial respiration during increased cardiac work (19). In a murine model of pressure-overload left ventricular hypertrophy the decline in PCr/ATP levels were associated with an early increase in LV mass and volume and decline in function. Of note, the early decline in PCr/ATP levels appeared to presage a continued LV remodeling and worsening function suggesting a role for impaired energetics in the morphological and functional abnormalities that occur with pressure-overload left ventricular hypertrophy (20).

In large animals 31P-MRS studies have largely focused on the energetic consequences of myocardial ischemia, infarction, and postinfarction remodeling (11,21–23). During ischemia the PCr/ATP ratio declines in concordance with the decline in blood flow (21–23). However, with the development of left ventricular remodeling and subsequent heart failure, PCr/ATP levels decline independent of the level of blood flow or tissue oxygenation suggesting other mechanisms were responsible for the observed energy deprivation (11). 31P-MRS has been combined with 1H-MRS and water referencing to permit quantification of the concentrations of PCr, ATP, and CR providing a more characterization of energy metabolism than PCr/ATP alone (24). Indeed, reductions in myocardial PCr, ATP, and CR levels measured with MRS correlated with histological measurements in infarcted canine myocardium demonstrating the promise of the method (24).

Human Studies

One of the more unique applications of ³¹P-MRS in humans is the measurement of PCr/ATP ratios at rest and then again during stress induced by either hand-grip exercise or pharmacologically in order to detect myocardial ischemia (25–28). This "³¹P-NMR stress test" has been used to identify physiologically significant coronary artery stenosis (26). Of interest, reduced ratios have identified a subset of women with chest pain but angiographically normal coronary arteries suggesting myocardial ischemia was inducible in these women (Fig. 3) (27,28). Consistent with this premise is that when a reduced PCr/ATP was present it portended a worse prognosis (as defined as hospitalization for unstable angina) when compared with event rates in women where the ratio was maintained (28). Limitations of the method include sampling of only anterior myocardium and the potential for underestimation of the true decline in high energy phosphate stores because of commensurate decreases in both PCr and ATP. The dissemination of high field (≥3T systems) should help increase the sampling of myocardium. As mentioned previously, techniques have been devised to measure the absolute concentration of PCr and ATP.

Initial studies in humans have shown promise in this regard (29).

³¹P-MRS studies have also provided important information on the myocardial energetics in heart failure. For example, humans with heart failure have shown a reduction in phosphocreatine (PCr) concentrations, total creatine, and creatine kinase flux consistent with a reduced energy state (30–32). Moreover, the observed reduction of PCr/ATP correlated with clinical severity and functional parameters (30). ³¹P-MRS studies has also been used to assess the response to therapy. The efficacy of β-blocker therapy in the treatment of the cardiomyopathic patient is well established. Consistent with this observation is a significant increase in PCr/ATP ratios in these patients in response to β-blocker therapy (33,34). This suggests that at least part of the beneficial effect of β-blocker therapy is an improvement in energy state of the myocardium. Studies with ³¹P-MRS have also helped characterize the hearts of diabetic patients. For example, an inverse correlation was observed between PCr/ATP levels and echocardiographic measurements of diastolic function suggesting with abnormalities in energy metabolism may contribute LV dysfunction in these patients (35). Of interest,

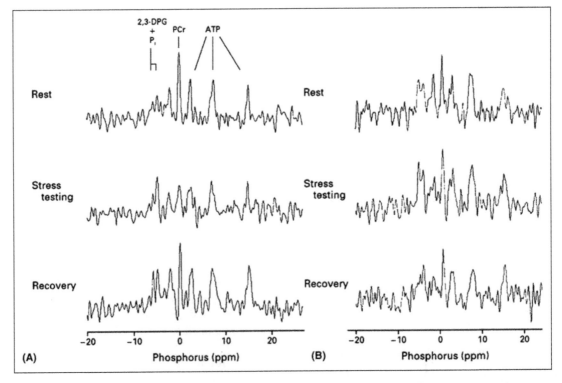

Figure 3 Examples of two different women with chest pain and normal coronary arteries. In (**A**) there is a significant decrease (27%) in the phosphocreatine:ATP ratio during stress testing, whereas in (**B**) there is no decrease (decrease of 1%). The peaks of phosphocreatine (PCr), ATP, and inorganic phosphate (Pᵢ) plus 2,3-diphosphoglycerate (2,3-DPG) are identified in (**A**). In (**B**), there is little change in the phosphocreatine:ATP ratio from period to period and only minor spectral variations in the amount of 2,3-diphosphoglycerate. The presence of this substance reflects the amount of red cells from the ventricular chamber within the area analyzed. *Source*: From Ref. 27.

results of ^{31}P-MRS studies have suggested that the increased plasma fatty acid levels that characterize the diabetic state may have a detrimental effect on myocardial energetics. For example, in diabetic patients there is a negative correlation between the reduction of PCr/ATP ratios and increasing levels of fasting plasma fatty acid levels (36). Thus, there may linkage between the pattern of substrate delivery to the heart, myocardial energetics and LV function in patients with diabetes mellitus.

^{13}C-MRS

A detailed description of the current requirements and potential approaches for using ^{13}C-MRS for assessing myocardial metabolism has been published previously (37). In brief, the sensitivity of the carbon nucleus is much lower than that of phosphorus. However, the relatively low natural abundance (~1.1%) makes this atom ideal for the study of intermediary energy metabolism. Infusion of ^{13}C-enriched substrates facilitates assessing incorporation into metabolite pools with very little background signal. With a few exceptions, this nucleus has not yet been used in clinical MRS studies. With the introduction of higher fields (≥3 T) for clinical magnets, ^{13}C-MRS may become feasible as a clinical tool. In addition, the use of lasers to hyperpolarize the ^{13}C-nucleus results in an increase of signal to noise of approximately 5 orders of magnitude (38). However, whether the ultra-short half-life of these gases will permit assessment of metabolism remains to be determined.

Additional information is available from the J-coupling that is observed between adjacent carbon atoms. When a ^{13}C atom is next to another ^{13}C atom in a molecule, the presence of the adjacent ^{13}C causes the resonance of the first ^{13}C atom to split—two resonances appear, centered on the resonance frequency of the original ^{13}C atom. Although this example is the simplest case, multiplets can be formed from various combinations of labeled atoms within the immediate molecular environment. As a result, each resonance not only provides concentration data for a particular atomic species, but also contains information regarding its neighboring atoms as well. Although currently limited to experimental studies, calculation of the relative contributions of glucose, lipids, and ketone bodies for energy production is possible, and the potential for this tool in the investigation of metabolic abnormalities should not be overlooked.

Overall Oxidative Metabolism

One of the first applications of ^{13}C-MRS was in the assessment of myocardial oxidative metabolism (39,40). Isotopomer analysis of the ^{13}C-^{13}C splitting patterns

(due to J-coupling of adjacent nuclei) of glutamate spectra were used to estimate substrate contribution to the citric acid cycle. By varying the ^{13}C-label distribution in the substrates for energy metabolism, it is possible to distinguish their relative contributions to citric acid cycle turnover. However, direct examination of the citric acid cycle intermediates is not possible with MRS due to the relatively low concentrations for most of these metabolites. However, glutamate is in rapid exchange with α-ketoglutarate, and reflects the label in this intermediate. Different labeling patterns in acetyl-CoA give rise to different labeling patterns in glutamate, permitting the calculation of relative contributions of various substrates.

In addition, the ratio of the steady-state enrichment of either C2 or C3 labeled glutamate to C4-labeled glutamate is routinely used as an index of anaplerosis. Anaplerosis is defined as entry into citric acid cycle at a site other than acetyl-CoA. Since the pool sizes for the citric acid cycle intermediates remain relatively constant, there must be a balance between anaplerosis and exit from the citric acid cycle at a site other than the decarboxylation steps at citrate $\rightarrow$ α-ketoglutarate and succinyl-CoA $\rightarrow$ fumarate. While we know in the isolated rat or rabbit heart that increasing the diversity of substrates increases the amount of anaplerosis (from ~10% with glucose only, to ~20% with glucose, palmitate, and β-hydroxy-butyrate), little is known about the routes of substrate exit from the citric acid cycle. Reduced anaplerosis may lead to reduced citric acid flux and contribute the cardiac manifestations of diseases where fatty acid is used as the predominant energy fuel such as obesity and diabetes mellitus. Conversely, increased anaplerosis may contribute to increased ATP production and cardioprotection during ischemia (41).

Dynamic measurements of intermediary energy metabolism—label incorporation into glutamate pools, for example—are possible. Several groups have used the label incorporation from labeled substrates into the glutamate pool as a way to calculate turnover rates of the tricarboxylic acid cycle. Weiss et al. (42), using a model of low flow ischemia, observed changes in the rate of glutamate enrichment that were evident even in the absence of changes in ^{31}P-MRS spectra demonstrating abnormalities in oxidative metabolism preceded impairment in energy production.

Carbohydrate Metabolism

Results of studies using ^{13}C-MRS have significantly furthered our understanding of myocardial carbohydrate metabolism under both normal and pathological conditions. The strength of the method is its ability to provide detailed information about not only glucose uptake but also the metabolic fate of glucose as it relates to glycogen synthesis, glycolysis, and glucose oxidation (43–47).

High-resolution ¹³C-MRS using higher field systems of myocardial extracts provides exquisite and detailed measurements of the relative contributions of labeled fuels for energy production in the heart. Although less detailed, relative measures of myocardial carbohydrate metabolism can be made under lower-resolution conditions in vivo (47–49). For example, the ¹³C signal from alanine that is produced by the equilibrium reaction with pyruvate produced from glycolysis, provides a readily available signal to use as an internal standard of glycolysis. By using this approach the relative proportion of carbohydrate oxidation to glycolysis can be determined. Furthermore, the appearance of labeled lactate can also be detected with this method, which provides a more comprehensive analysis of glucose metabolism (46,50). Other substrates have also been studied using [3-¹³C]lactate and [3-¹³C]pyruvate in conjunction with MRS. Indeed, using this approach in the isolated hearts perfused with both lactate and glucose, it was demonstrated that simultaneous uptake and efflux of lactate can be modulated by the addition of fatty acids as well as by diabetes. These data support the concept of compartmentalization of lactate metabolism by the heart, where lactated derived from glycolysis is released and simultaneously, extracted exogenous lactate oxidized (51–53).

These approaches have already contributed significantly to our understanding of the perturbations in carbohydrate metabolism that occur in a variety cardiac diseases. Most notable have been the alterations in myocardial substrate use that characterize ischemia and postischemic myocardium (46–48,54,55). Moreover, the importance of the cytosolic redox state and pH on the contractile recovery of the post-ischemic heart has been elucidated (56,57). In addition, the central role of accelerated glucose metabolism in pre-conditioning has been demonstrated based dynamic ¹³C-MRS (44,45). Finally ¹³C-MRS has been used to document the impact of both hypertrophy and diabetes mellitus on myocardial carbohydrate use and the central role of impaired pyruvate dehydrogenase activity in this regard (Fig. 4) (58,59).

Fatty Acid Metabolism

As mentioned previously, abnormalities in myocardial fatty acid metabolism underlie a variety of cardiac disease processes, most notably obesity and diabetic heart disease. Consequently, there is significant interest in developing cardiac MRS approaches to measure myocardial fatty acid use. For example, measurements of myocardial fatty acid transport into the mitochondria and subsequent oxidation have been performed with MRS using ¹³C-palmitate (60). Moreover, with dynamic ¹³C-MRS, it is possible to simultaneously assess fatty acid uptake, oxidation and TAG (61). Indeed, with this approach it has been shown that myocardial oxidation

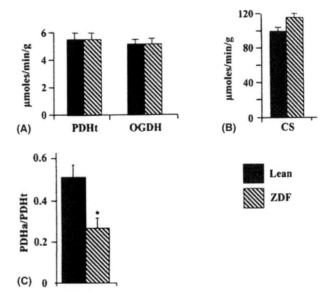

Figure 4 (**A**) Total pyruvate dehydrogenase (PDHt) and oxoglutarate dehydrogenase activity (OGDH). (**B**) Citrate synthase (CS) activity and (**C**) fraction of pyruvate dehydrogenase in the active form (PDHa/PDHt) in hearts from lean and ZDF rats. *Source*: From Ref. 58.

of endogenous fats (from TAG) is increased in a rat model of diabetes mellitus when compared with controls. More recently, similar combined measurements of myocardial fat oxidation and TAG turnover have been performed in isolated mouse heart demonstrating the potential of ¹³C-MRS to accurately phenotype transgenic murine models of cardiac disease (Fig. 5) (62).

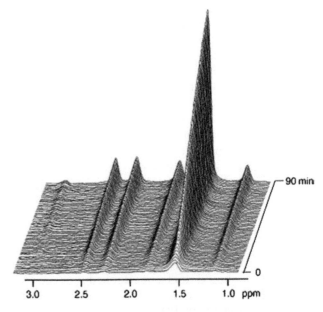

Figure 5 Spectra of an isolated mouse heart supplied with [U-¹³C] fatty acids plus unlabeled lactate, pyruvate, glycerol, glucose, and ketones. Labeled triglycerides are represented by the smaller peaks to the left. *Source*: From Ref. 21.

^{23}Na-MRS

^{23}Na-MRS has been used to examine the changes in intracellular and extracellular sodium content (63,64). By choosing pulse sequences that take advantage of the short relaxation times of ^{23}Na, the increase in signal-to-noise for the ^{23}Na peak can be sufficient for cardiac imaging at fields of 1.5 T and above. This may prove to be extremely useful in the myocardium, because injury (such as ischemia) often can result in a breakdown of ionic homeostasis. As a result, the normal transmembrane gradient of high extracellular sodium and low intracellular sodium may be disrupted, with an increase in the concentration of intracellular sodium. In addition, this may be exacerbated during periods of tissue acidosis due to the activity of the $Na^+/^1H$ exchangers.

A spectrum of ^{23}Na in the heart shows a single peak comprised of two components (intracellular plus extracellular). It is possible to distinguish between the two components. One technique is the use of agents known as "shift reagents," which cause a frequency shift in the extracellular peak (65). These compounds are related to gadolinium, a contrast agent used clinically to increase signal by shortening the apparent T_1 of the tissue. The use of shift reagents is currently limited to experimental studies, due to the toxic effects that most exhibit at effective concentrations. One agent that has not shown toxicity (at least in the rat) is dysprosium triethylenetetraminehexaacetic acid (DyTTHA), but additional work will need to be completed before a clinically sound sodium shift agent is available. The second technique involves the quantum state of ^{23}Na, but the technique is very motion sensitive, which precludes clinical cardiac application (66).

Kim et al. (67) have shown that ECG gated 3D ^{23}Na-MRS of the human heart is feasible with good signal-to-noise ratio and spatial resolution. Because of the relatively short relaxation times associated with this nucleus, fast gradient echo techniques can acquire ^{23}Na images within minutes. Although promising for clinical studies, the worth of ^{23}Na imaging and spectroscopy to cardiac investigations currently awaits validation. The development of a nontoxic shift reagent would substantially improve the application of this technique to more mainstream clinical cardiac practice.

FUTURE DEVELOPMENTS

Discussion of molecular imaging of myocardial metabolism is not complete without consideration of MRS. This technique provides an unprecedented view into metabolic pathways that are critical for maintaining physiological function, and has sufficient sensitivity to reflect alterations on an atomic scale. Recent

developments in carbon labeling include the ability to hyperpolarize carbon, increasing the sensitivity for detection by 5 orders of magnitude. Combined with the shift to higher magnetic fields for routine clinical platforms, these developments hold great promise for the future.

REFERENCES

1. Cox IJ. Development and applications of in vivo clinical magnetic resonance spectroscopy. Prog Biophys Mol Biol 1996; 65:45–81.
2. Dobbins RL, Malloy CR. Measuring in-vivo using nuclear magnetic resonance. Curr Opin Clin Nutr Metab Care 2003; 6:501–509.
3. Jones GL, Sang E, Goddard C, et al. A functional analysis of mouse models of cardiac disease through metabolic profiling. J Biol Chem 2005; 280:7530–7539.
4. Berk PD, Zhou SL, Kiang CL, Stump D, Bradbury M, Isola LM. Uptake of long chain free fatty acids is selectively up-regulated in adipocytes of Zucker rats with genetic obesity and non-insulin-dependent diabetes mellitus. J Biol Chem 1997; 272:8830–8835.
5. Stanley WC, Lopaschuck GD, McCormack JG. Regulation of energy substrate metabolism in the diabetic heart. Cardiovasc Res 1997; 34:25–33.
6. Shen H, Balschi JA, Pohost GM, Wolkowicz P. 1H-MRS detected lipolysis in diabetic rat hearts requires neutral lipase. J Cardiovasc Magn Reson 2001; 35–45.
7. Schneider JE, Tyler DJ, ten Hove M, et al. In vivo cardiac 1H-MRS in the mouse. Magn Reson Med 2004; 52: 1029–1035.
8. Chen W, Zhang J, Eljgelshoven MH, et al. Determination of deoxymyoglobin changes d ing graded myocardial ischemia: an in vivo 1H NMR spectroscopy study. Magn Reson Med 1997; 38:193–197.
9. Kreutzer U, Mekhamer Y, Chung Y, Jue T. Oxygen supply and oxidative phosphorylation limitation in rat myocardium in situ. Am J Physiol Heart Circ Physiol 2001; 280:H2030–H2037.
10. Chung Y, Xu D, Jue T. Nitrite oxidation of myoglobin in perfused myocardium: implications for energy coupling in respiration. Am J Physiol 1996; 271: H1166–H1173.
11. Murakami Y, Zhang Y, Cho YK, et al. Myocardial oxygenation during high work states in hearts with postinfarction remodeling. Circulation 1999; 99: 942–948.
12. Nakae I, Mitsunami K, Matsuo S, et al. Assessment of myocardial creatine concentration in dysfunctional human heart by proton magnetic resonance spectroscopy. Magn Reson Med Sci 2004; 3:19–25.
13. Nakae I, Mitsunami K, Matsuo S, et al. Myocardial creatine concentration in various nonischemic heart diseases assessed by 1H magnetic resonance spectroscopy. Circ J 2005; 69:711–716.
14. Zhou YT, Grayburn P, Karim A, et al. Lipotoxic heart disease in obese rats: implications for human obesity. Proc Natl Acad Sci U S A 2000; 97:1784–1789.

15. Sharma S, Adrogue JV, Golfman L, et al. Intramyocardial lipid accumulation in the failing human heart resembles the lipotoxic rat heart. FASEB J 2004; 18:1692–1700.

16. Szczepaniak LS, Dobbins RL, Metzger GJ, et al. Myocardial triglycerides and systolic function in humans: in vivo evaluation by localized proton spectroscopy and cardiac imaging. Magn Reson Med 2003; 49:417–423.

17. Reingold JS, McGavock JM, Kaka S, Tillery T, Victor RG, Szczepaniak LS. Determination of triglyceride in the human myocardium by magnetic resonance spectroscopy: reproducibility and sensitivity of the method. Am J Physiol Endocrinol Metab 2005; 289:E935–E939.

18. Tain R, Abel ED. Responses of GLUT4-deficient hearts to ischemia underscore the importance of glycolysis Circulation 2001; 103:2961–2966.

19. Naumova AV, Weiss RG, Chacko VP. Regulation of murine myocardial energy metabolism during adrenergic stress studied by in vivo 31P NMR spectroscopy. Am J Physiol Heart Circ Physiol 2003; 285:H1976–H1979.

20. Maslov MY, Chacko VP, Stuber M, et al. Altered high-energy phosphate metabolism predicts contractile dysfunction and subsequent ventricular remodeling in pressure-overload hypertrophy mice. Am J Physiol Heart Circ Physiol 2007; 292:H387–H391.

21. Zhang J, Ugurbil K, From AH, Bache RJ. Use of magnetic resonance spectroscopy for in vivo evaluation of high-energy phosphate metabolism in normal and abnormal myocardium. J Cardiovasc Magn Reson 2000; 2:23–32.

22. Schaefer S, Schwartz GG, Wisneski JA, et al. Response of high-energy phosphates and lactate release during prolonged regional ischemia in vivo. Circulation 1992; 85:342–349.

23. Zhang J, Path G, Chepuri V, et al. Responses of myocardial high energy phosphates and wall thickening to prolonged regional hypoperfusion induced by subtotal coronary stenosis. Magn Reson Med 1993; 30:28–37.

24. Bottomley PA, Weiss RG. Noninvasive localized MR quantification of creatine kinase metabolites in normal and infracted canine myocardium. Radiology 2001; 219:411–418.

25. Butterworth EJ, Evanochko WT, Pohost GM. The 31P-NMR stress test: an approach for detecting myocardial ischemia. Ann Biomed Eng 2000; 28:930–933.

26. Yabe T, Mitsunami K, Okada M, Morikawa S, Inubushi T, Kinoshita M. Detection of myocardial ischemia by 31P magnetic resonance spectroscopy during handgrip exercise. Circulation 1994; 89:1709–1716.

27. Buchthal SD, den Hollander JA, Bairey Merz CNB, et al. Abnormal myocardial phosphorus-31 nuclear magnetic resonance spectroscopy in women with chest pain but normal coronary angiograms. N Engl J Med 2000; 342:829–835.

28. Johnson BD, Shaw LJ, Buchthal SD, et al. National Institutes of Health-National Heart, Lung, and Blood Institute. Prognosis in women with myocardial ischemia in the absence of obstructive coronary disease: results from the National Institutes of Health-National Heart, Lung, and Blood Institute-Sponsored Women's Ischemia Syndrome Evaluation (WISE). Circulation 2004; 109: 2993–2999.

29. Nakae I, Mitsunami K, Omura T, et al. Proton magnetic resonance spectroscopy can detect creatine depletion associated with the progression of heart failure in cardiomyopathy. J Am Coll Cardiol 2003; 42:1587–1593.

30. Neubauer S, Horn M, Cramer M, et al. Myocardial phosphocreatinine-to-ATP ratio is a predictor of mortality in patients with dilated cardiomyopathy. Circulation 1997; 96:2190–2196.

31. Higgins CB, Saeed M, Wendland M, Chew WM. Magnetic resonance spectroscopy of the heart. Overview of studies in animals and man. Invest Radiol 1989; 24:962–968.

32. Neubauer S. High-energy phosphate metabolism in normal, hypertrophied and failing human myocardium. Heart Fail Rev 1999; 4:269–280.

33. Neubauer S, Krahe T, Schindler R, et al. 31P magnetic resonance spectroscopy in dilated cardiomyopathy and coronary artery disease. Altered cardiac high-energy phosphate metabolism in heart failure. Circulation 1992; 86:1810–1818.

34. Neubauer S, Horn M, Pabst T, et al. Contributions of 31P-magnetic resonance spectroscopy to the understanding of dilated heart muscle disease. Eur Heart J 1995; 16(suppl O):115–118.

35. Diamant M, Lamb HJ, Groeneveld Y, et al. Diastolic dysfunction is associated with altered myocardial metabolism in asymptomatic normotensive patients with well-controlled type 2 diabetes mellitus. J Am Coll Cardiol 2003; 42:328–335.

36. Scheuermann-Freestone M, Madsen PL, Manners D, et al. Abnormal cardiac and skeletal muscle energy metabolism in patients with type 2 diabetes. Circulation 2003; 107:3040–3046.

37. Lewandowski ED. Cardiac carbon 13 magnetic resonance spectroscopy: on the horizon or over the rainbow. J Nucl Cardiol 2002; 9:419–428.

38. Johansson E, Olsson LE, Mansson S, et al. Perfusion assessment with bolus differentiation: a technique applicable to hyperpolarized tracers. Magn Reson Med 2004; 52:1043–1051.

39. Chance EM, Seeholzer SH, Kobayashi K, Williamson JR. Mathematical analysis of isotope labeling in the citric acid cycle with applications to 13C NMR studies in perfused rat hearts. J Biol Chem 1983; 258:13785–13794.

40. Malloy CR, Sherry AD, Jeffrey FM. Carbon flux through citric acid cycle pathways in perfused heart by 13C NMR spectroscopy. FEBS Lett 1987; 212:58–62.

41. Lloyd SG, Wang P, Zeng H, Chatham JC. Impact of low-flow ischemia on substrate oxidation and glycolysis in the isolated perfused rat heart. Am J Physiol Heart Circ Physiol 2004; 287:H351–H362.

42. Weiss RG, Chacko VP, Glickson JD, Gerstenblith G. Comparative 13C and 31P NMR assessment of altered metabolism during graded reductions in coronary flow in intact rat hearts. Proc Natl Acad Sci U S A 1989; 86:6426–6430.

43. Laughlin MR, Taylor J, Chesnick AS, DeGroot M, Balaban RS. Pyruvate and lactate metabolism in the in

vivo dog heart. Am J Physiol 1993; 264(6 Pt 2): H2068–H2079.

44. Kalil-Filho R, Gerstenblith G, Hansford RG, Chacko VP, Vandegaer K, Weiss RG. Regulation of myocardial glycogenolysis during post-ischemic reperfusion. J Mol Cell Cardiol 1991; 23:1467–1479.

45. Weiss RG, de Albuquirque CP, Vandegaer K, Chacko VP, Gerstenblath G. Attenuated glycogenolysis reduced catabolite accumulation during ischemia in preconditioned rat hearts. Circ Res 1996; 79:435–446.

46. Damico LA, White LT, Yu X, Lewandowski ED. Chemical versus isotopic equilibrium and the metabolic fate of glycolytic end products in the heart. J Mol Cell Cardiol 1996; 28:989–999.

47. McNulty PH, Jagasia D, Cline CW, et al. Persistent changes in myocardial glucose metabolism in vivo during reperfusion of a limited-duration coronary occlusion. Circulation 2000; 101:917–922.

48. McNulty PH, Cline GW, Whiting JM, Shulman GI. Regulation of myocardial [(13)C] glucose metabolism in conscious rats. Am J Physiol Heart Circ Physiol 2000; 279:H375–H381.

49. Lewandowski ED, White LT. Pyruvate dehydrogenase influences postischemic heart function. Circulation 1995; 91:2071–2079.

50. Rath DP, Zhu H, Tong X, Jiang Z, Hamlin RL, Robitaille PM. Dynamic 13C NRM analysis of pyruvate and lactate oxidation in the in vivo canine myocardium: evidence of reduced utilization with increased work. Magn Reson Med 1997; 38:896–906.

51. Carvalho RA, Zhao P, Wiegers CB, Jeffrey FM, Malloy CR, Sherry AD. TCA cycle kinetics in the rat heart by analysis of (13)C isotopomers using indirect (1)H. Am J Physiol Heart Circ Physiol 2001; 281:H1413–1421.

52. Chatham JC, Des Rosiers C, Forder JR. Evidence of separate pathways for lactate uptake and release by the perfused rat heart. Am J Physiol Endocrinol Metab 2001; 281:E794–E802.

53. Chatham JC, Forder JR. Metabolic compartmentation of lactate in the glucose-perfused rat heart. Am J Physiol 1996; 270: H224–H229.

54. Lewandowski ED, Johnston DL. Reduced substrate oxidation in post-ischemic myocardium: 13C and 31P NMR analyses. Am J Physiol Heart Circ Physiol 1990; 258: H1357–H1365.

55. Johnston DL, Lewandowski ED. Fatty acid metabolism and contractile function in the reperfused myocardium: multinuclear NMR studies of isolated rabbit hearts. Circ Res 1991; 68:714–725.

56. White LT, O'Donnell JM, Griffin J, Lewandowski ED. Cytosolic redox state mediates postischemic response to pyruvate dehydrogenase stimulation. Am J Physiol Heart Circ Physiol 1999; 277:H626–H634.

57. Griffin J, White LT, Lewandowski ED. Proton production determines substrate dependent recovery of stunned hearts during pyruvate dehydrogenase stimulation. Am J Physiol Heart Circ Physiol 2000; 279:H361–H367.

58. Seymour AM, Chatham JC. The effects of hypertrophy and diabetes on cardiac pyruvate dehydrogenase activity. J Mol Cell Cardiol 1997; 29:2771–2778.

59. Chatham JC, Forder JR. Relationship between cardiac function and substrate oxidation in hearts of diabetic rats. Am J Physiol 1997; 273(1 Pt 2):H52–H58

60. O'Donnell JM, Alpert NM, White LT, Lewandowski ED. Coupling of mitochondrial fatty acid uptake to oxidative flux in the intact heart. Biophys J 2002; 82:11–18.

61. O'Donnell JM, Zampino M, Alpert NM, Fasano MJ, Geenen DL, Lewandowski ED. Accelerated triacylglycerol turnover kinetics in hearts of diabetic rats include evidence for compartmented lipid storage. Am J Physiol Endocrinol Metab 2006; 290:E448–E455.

62. Stowe KA, Burgess SC, Merritt M, Sherry AD, Malloy CR. Storage and oxidation of long-chain fatty acids in the C57/BL6 mouse heart as measured by NMR spectroscopy. FEBS Letters 2006; 580:4282–4287.

63. Horn M, Weidensteiner C, Scheffer H, et al. Detection of myocardial viability based on measurement of sodium content: A (23)Na-NMR study. Magn Reson Med 2001; 45:756–764.

64. Kim RJ, Lima JA, Chen EL, et al. Fast 23Na magnetic resonance imaging of acute reperfused myocardial infarction. Potential to assess myocardial viability. Circulation 1997; 95:1877–1885.

65. Pike MM, Frazer JC, Dedrick DF, et al. 23Na and 39K nuclear magnetic resonance studies of perfused rat hearts. Discrimination of intra- and extracellular ions using a shift reagent. Biophys J 1985; 48:159–173.

66. Payne GS, Seymour AM, Styles P, Radda GK. Multiple quantum filtered 23Na NMR spectroscopy in the perfused heart. NMR Biomed 1990; 31:139–146.

67. Kim RJ, Lima JA, Chen EL, et al. Fast 23Na magnetic resonance imaging of acute reperfused myocardial infarction. Potential to assess myocardial viability. Circulation 1997; 95:1877–1885.

Index

Printed and bound by CPI Group (UK) Ltd, Croydon, CR0 4YY

23/10/2024

01778226-0017